P9-AER-828

Mass $\times^3$ (10 kg)	Density (g/cm³)	Escape Velocity (km/s)	Surface	Atmosphere
3.3	5.4	4	silicates	trace Na
48.7	5.3	10	basalt, granite?	90 bar: 97% CO_2
59.8	5.5	11	basalt, granite, water	1 bar: 78% N_2, 21% O_2
0.7	3.3	2	basalt, anorthosites	none
6.4	3.9	5	basalt, clays, ice	0.07 bar: 95% CO_2
18991	1.3	60	none	H_2, He, CH_4, NH_3, etc.
1.1	1.9	2	dirty ice	none
1.5	1.9	3	dirty ice	none
0.5	3.0	2	ice	none
0.9	3.6	3	sulfur, SO_2	trace SO_2
5686	0.7	36	none	H_2, He, CH_4, NH_3, etc.
1.3	1.9	3	ice? hydrocarbons?	1.5 bar: N_2, trace CH_4
866	1.2	21	?	H_2, He, CH_4, NH_3, etc.
1030	1.6	23	?	H_2, He, CH_4, NH_3, etc.
?	?	?	CH_4 ice, liquid N_2 (?)	trace CH_4
0.01	1.7	1	CH_4 ice	trace CH_4

. 8 00000000 l

The Planetary System

David Morrison ◇ Tobias Owen

University of Hawaii   **SUNY · Stony Brook**

Addison-Wesley Publishing Company

Reading, Massachusetts ◆ Menlo Park, California ◆ New York
Don Mills, Ontario ◆ Wokingham, England ◆ Amsterdam ◆ Bonn
Sydney ◆ Singapore ◆ Tokyo ◆ Madrid ◆ Bogotá ◆ Santiago ◆ San Juan

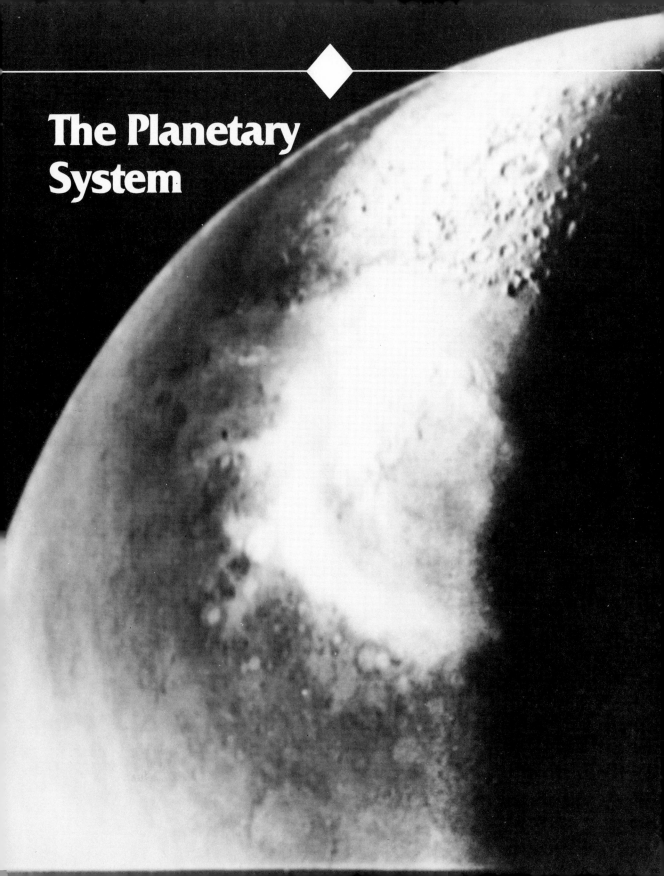

The Planetary System

◆ **Sponsoring Editor**: Stuart Johnson   **Production Supervisor**: Herbert Nolan ◆
Art Consultant: Loretta Bailey   **Editorial and Production Services**: PC&F, Inc.
Text Designer: Melinda Grosser for silk   **Illustrator**: Capricorn Design
Manufacturing Supervisor: Roy Logan   **Cover Designer**: Jean Seal

Library of Congress Cataloging-in-Publication Data

Morrison, David, 1940–
  The planetary system.

  Includes index.
  1. Planets.   I. Owen, Tobias C.   II. Title.
QB601.M76   1988   523.4   87-33265
ISBN 0-201-10487-3

CDEFGHIJ-HA-89

Foreword by Carl Sagan. Copyright © 1987 by Carl Sagan.

Copyright © 1987 by Addison-Wesley Publishing Company, Inc. All rights reserved. No part of this publication may be reproduced, stored in a retrieval system, or transmitted, in any form or by any means, electronic, mechanical, photocopying, recording, or otherwise, without the prior written permission of the publisher. Printed in the United States of America. Published simultaneously in Canada.

FOREWORD

In everyday speech, as in the arcane language of modern physics, we still use "world" to mean "universe." Old habits die hard. There was a time when the Earth was the only place we knew. But by the seventeenth century it had become clear that the planets were not points of light but truly other worlds and that compared to some of them the Earth would be insignificant, a pimple on Jupiter. But such knowledge constitutes an assault on human conceit. It is unwelcome and has only slowly entered into the public consciousness. Our time, though, is an extraordinary age in which vast libraries of data about other worlds are being accumulated by ground-based telescopes, by instrumentation in high-altitude aircraft and Earth satellites, and especially by space vehicles that breach the quarantine of the vast interplanetary distances, flying by, orbiting, and actually setting down on other worlds for a look around.

The pace of human exploration of the solar system has been a little dizzying even for the specialist, and a knowledge of what is out there has not much permeated into the consciousness of nonspecialists. And yet in these investigations of other worlds we are continuing an ancient and deep human tradition of exploration and discovery; we garner vital insights into our own world and what can go wrong with it; and we are setting the stage, if we are not so foolish as to destroy ourselves first, for the next step in the future of our species, ·when we are no longer confined to a single world.

David Morrison and Tobias Owen have been pioneers in the modern exploration of the solar system. This book is marked by a judicious and comprehensive selection of topics, clear qualitative explanations, and instructive and lovely illustrations — all in the context of the great drama of the evolution of our solar system and the worlds it contains.

Laboratory for   Carl Sagan
Planetary Studies
Cornell University

v

PREFACE

The past twenty-five years represent the golden age of planetary exploration, the period in which we have launched our initial exploration missions to the planets and achieved, for the first time, a comparative perspective on the solar system. What has emerged from this exciting period is a new scientific discipline, usually termed "planetary science." This book, *The Planetary System*, is written for a one-semester introductory college course in planetary science. Such courses are taught today in many universities, offered by departments of astronomy, physics, and Earth science or geology. Until now, however, there has been no really satisfactory text at the introductory level that covers the entire field, including both solid surfaces and atmospheres, while bringing together concepts and techniques from astronomy, geology, and atmospheric science. We hope this book will fill the need for a current, comprehensive treatment of planetary science at a level accessible to the non-science undergraduate, yet sufficiently detailed and challenging to be useful to the advanced undergraduate as well.

As an interdisciplinary subject, planetary science defies easy description. Traditionally, of course, the Earth and planets have been studied within the fields of both astronomy and the Earth sciences (geology, geophysics, meteorology, etc.). Only in the past two decades have components of these standard disciplines come together to generate the new perspectives of comparative planetology — the study of the Earth as a planet and of the other planets as worlds. Such a synthesis would not have been possible without the spectacular successes of the planetary spacecraft launched by the U.S. and the USSR. But modern planetary science is based on more than new spacecraft data, for the stimulus of planetary exploration has also created or revitalized such Earth-based activities as planetary astronomy (telescopic study of the planets), meteoritics (now closely allied to the study of returned lunar samples), and theoretical studies in fields as diverse as plasma physics and non-equilibrium chemistry. A new generation of planetary scientists has emerged, with perspectives that extend beyond the traditional disciplinary boundaries.

The subject of this book is the solar system, minus the Sun. The Sun is a star, a complex and fascinating incandescent mass of plasma powered by nuclear reactions, about midway in an evolutionary journey from interstellar cloud to black dwarf. Because it is so different from the rest of the solar system, it could not have been discussed here without increasing the length of the book by 50%. We take the Sun as a given — it has provided the gravitational, radiation, and plasma environment within which the other members of the system have evolved. It is these other darker, colder masses — the planets, satellites, rings, asteroids, comets, and meteorites — whose story fills these pages.

The order of presentation of topics is designed to emphasize the *processes* that formed and influence the members of the planetary system. Part I provides an introduction to the solar system itself, as well as to some of the basic physical and chemical concepts used later. Students who have already had a course in astronomy or the Earth sciences can probably skip directly to Part II, which deals with the simplest, most primitive solid objects in the system: comets, asteroids, and meteorites. We were especially anxious to introduce the meteorites at the beginning of our discussion, since these samples of extraterrestrial material provide an essential "ground truth" for remote studies. In Part III we move one step up the scale of complexity to examine the small cratered worlds, Moon and Mercury, as an introduction to planetary geology. Parts IV and V continue to the terrestrial planets, the Earth-Venus-Mars triad, with their active geological evolution and complex atmospheres. The outer solar system — less well-known than the Earth-like planets — is the subject of Part VI (the outer planets) and Part VII (satellites and rings). Finally, in the Epilogue we review the structure of the entire system in terms of its origin and evolution. Throughout the book we emphasize comparative studies and the common processes acting on the members of the system, but we do so in the context of descriptions of individual objects. Generalizations are important, but it is our feeling that they need to be derived from understanding of the specific, particular objects that we have explored.

Three different kinds of courses could make use of this book. First, we have designed it to be suitable as a first science course for the liberal arts undergraduate — for example, in fulfillment of a general education requirement in science. Used in conjunction with another text on stellar astronomy, it could be part of a one-year sequence in astronomy. Alternatively, with a text on geology, it could form part of an Earth science sequence. Second, we have in mind a more advanced undergraduate course for students who have completed an introduction to astronomy or Earth science. In this case, we suggest that the first two chapters (Part I) be omitted and that the instructor select some supplementary material from the Additional Reading sections. Finally, we recognize the need for a suitable text for advanced science undergraduates, and for this audience we have added a quantitative supplement that presupposes a working knowledge of algebra and an interest in solving some simple quantitative exercises.

Any scientific discipline has a lot of specialized jargon, and planetary science, which draws from several disciplines, may be worse than most. Some of this jargon is useful and even necessary, but much of it serves no purpose in an introductory course. We feel that technical vocabulary should be a tool, not an end in itself. Thus we have tried to practice "jargon control" in this book, introducing only those specialized terms (about 200 in all) that are really necessary for understanding. These terms are printed in boldface where first explained in the text and are defined again in the Glossary.

Another problem for authors concerns the practice of making reference to the work of individuals. We decided that a book at this level should not include specific literature references, although lists of additional reading (both technical and aimed at a general audience) are presented at the end of each part. But we should not neglect the work of individuals entirely, since we feel it would give a false impression of the nature of science to imply that it is an impersonal endeavor. Of necessity, however, we have been able to mention by name only a few of the scientists who have made important contributions to this field, and we apologize to the many others for what may seem to be a rather arbitrary selection of names.

The preparation of this book has stretched over several years, during which we have received assistance and encouragement from many persons. We thank especially Phillip Hagopian and Andrew Crowley of Benjamin/Cummings, who initially expressed interest in publishing a planetary science text and guided us through the early stages (until Benjamin/Cummings merged with Addison-Wesley). We are grateful to our many colleagues who have reviewed chapters and sections of the manuscript in various forms. In particular we thank David Black, Dale Cruikshank, Larry Esposito, Donald Goldsmith, B. Ray Hawke, Sherwood Harrington, Paul Hodge, Richard Sears, Dale Smith, David Tholen, Joseph Veverka, and Laurel Wilkening for their very helpful comments and criticisms. We also appreciate the assistance we have received in assembling the illustrations for this book, especially from Arlene Ammar, Ray Arvidson, Valery Barsukov, Mike Carr, Dale Cruikshank, Mike Gentry, Jim Head, Klaus Keil, Hal Masursky, Al McEwen, Yuri Shukolyukov and Jurrie van der Woude, among many others named individually in the Figure Credits. Morrison thanks Karan Hughes and Dottie Rosinsky for assistance in organizing the manuscript and associated correspondence, and Janet Morrison for comments and critiques of the text throughout the writing process. Owen thanks Diane Meyers and Mildred O'Dowd for their help in manuscript preparation. Finally, we thank our editors at Addison-Wesley and particularly Susan Vicenti and Marc Carver of PC&F for their highly professional efforts to translate a somewhat disorganized manuscript into a proper textbook. We hope that you, the reader, will find the product of all of these efforts to be both useful and enjoyable.

Honolulu, Hawaii   David Morrison
Stony Brook, New York   Tobias Owen

CONTENTS

◆ # The Planetary System ◆

PART ◆ ONE
Introducing the Planetary System

Standing under the stars on a clear, dark night, you are looking out from the surface of our rocky planet into the vast depths of the universe — the grandest assemblage of matter and space that we know. "This majestical roof, fretted with golden fire" has been made familiar to us by the efforts of our ancestors, who gave names to the stars and the patterns they appear to form. Throughout history, humans have continuously struggled to try to understand the place our planet occupies in this starry realm.

The planetary system consists of the nine known planets, the 56 satellites discovered to date, and the innumerable asteroids, comets, and meteoroids that orbit the Sun. We first want to know the shape and size of this system and its internal motions. What moves around what, and why? These puzzles, which have fascinated humans since they first recognized the existence of the planets, will serve as an introduction to the study of planetary science.

The Sun, a rather ordinary star, similar to many of those you see in the night sky, re-

◀ The Great Galaxy in Andromeda. Our familiar Sun is just one of the hundreds of billions of stars that, together with immense clouds of gas and dust, make up the Milky Way Galaxy. The Milky Way itself is simply our insider's view of the galaxy we inhabit, as we look around ourselves in the plane in which most of the stars are located. Beyond the boundaries of our own galaxy, there are billions of other galaxies, such as this one that is seen through the stars in the constellation of Andromeda. It gives us an idea of how the Milky Way Galaxy might look if we could see it from the outside.

quires a separate study that is beyond the scope of this book. Like the other stars, the Sun is composed primarily of hydrogen. It produces energy, including sunlight, by nuclear fusion in its interior where hydrogen is being converted to helium. The Sun, its planets, and the stars, plus great clouds of gas and dust, are part of the enormous system called the Milky Way Galaxy, which is 100,000 light years in diameter and contains more than a hundred billion stars.

One unifying concept between planetary science and the rest of astronomy to bear in mind as we begin to investigate the planetary system concerns the origin of the solar system. We — the planets, their satellites, and ourselves — are all made from the ashes of long-dead stars. Except for hydrogen, the elements that compose our bodies were created in the interiors of stars. All the parts of the solar system, including the Sun, formed from the debris of ancient stellar explosions and the stupendous fireball that gave birth to the universe. Yet the Earth has a very different composition than the stars, consisting mainly of silicon, oxygen, magnesium, and iron instead of hydrogen and helium. We are living on a cosmic cinder that is greatly depleted in the abundant light elements, the hydrogen, helium, carbon and nitrogen that dominate the stars. To put it more attractively, the Earth is a chocolate chip in the great galactic cookie. One of our tasks in this book is to see how this peculiar concentration of heavy elements came about and how likely it is that there may be other chocolate chips out there among the stars.

CHAPTER ONE

Finding Our Place in Space

1.1 Watching the Sky: Sun, Moon, and Stars

Origins of Astronomy

Most of us live in cities these days. We have become unfamiliar with the sky, especially at night when city lights obscure all but the brightest stars. We have thus lost contact with an important window on the universe, one we can look through with our own eyes, unaided by any optical or electronic devices. Even so, some celestial events play an important part in our lives, setting the cycle of the seasons and giving us various measures of time. Occasionally, our attention is focused by the news media on some rare event: a solar eclipse or the appearance of a comet. Otherwise, we are left with random views of the Sun and Moon, paying little enough attention to these.

For many of us, even the time-keeping aspect is rather remote, since we have calendars, watches, radios, television sets, newspapers and magazines to remind us of daily, monthly and yearly time scales and events. But to ancient civilizations without such assistance, the sky was a very important part of the natural environment. It was a source of pleasure as well as fear, and its changing aspects gave them the ability to predict seasonal changes that were vital to their livelihood. In trying to understand the reasons for what they were seeing, these peoples initiated the study of astronomy. We are still struggling with some of the problems they could not solve, such as the origins of the Earth, Moon, Sun and stars, and the question of whether or not there are other worlds like ours.

Many of the phenomena these early observers found mysterious, however, we can now understand. Indeed, we take these solutions so much for granted that it is useful to precede a review of the modern perspective by returning to earlier problems in order to see how they were solved. In the same spirit, we encourage readers to go out and make some of the simple, basic observations themselves — watching how the altitude of the Sun at noon and the point on the horizon where it sets change with the seasons; following the Moon's apparent motion through the sky; learning the constellations and identifying the planets. All this can be done with no optical aid whatsoever, and these simple activities help establish a personal connection with the planetary system we inhabit.

The Constellations

To become acquainted with the night, you need to find a location that is far from any city, with a really dark sky. This is the kind of environment that was always at hand to the ancients. Such a night sky is filled with stars of different brightnesses and colors. With its usual quest for order, the human mind found certain patterns among these apparently random points of light. These patterns are called **constellations** and were given names associated with myths, legends, and historical events in various cultures. Astronomers still use some of these names today. Probably most of you at one time in your lives have identified the Big Dipper, part of the Roman *Great Bear* or the Old English *Hay Wain*, and

Orion, the brilliant constellation of cold winter nights (Fig. 1.1). (All these descriptions are given for observers living at latitudes from 20° to 50° north of the equator. The sky is very different for people in the southern hemisphere.)

In summer, Orion is replaced by other constellations: Lyra and Cygnus nearly overhead and Scorpius and Sagittarius toward the south. Each time these seasons return, we can find

FIGURE 1.1 The constellation Orion showing the brightest stars and the location of a giant gas cloud, the Orion Nebula, where new stars are forming at the present time.

these same constellations in the same places in our skies. While this grand seasonal cycle is taking place, the positions of the stars within each constellation remain fixed. The Big Dipper looked the same to Charlemagne (742–814 AD) as it does to you. One of the earliest mysteries, then, was the reason for this repetition. The obvious solution was to suggest that the Earth was in the center of a set of giant spheres that slowly rotated about it, giving rise to day and night and the seasonal change in the visible constellations.

Apparent Motion of the Stars

Like the Sun, the stars appear to rise in the east and set in the west. This diurnal, or daily, motion can be noticed in an hour or less, and by the end of the night it has brought many new constellations into view to replace the stars present at dusk. A slower seasonal cycle is also apparent, however, if observations are extended over several weeks.

If you go out each evening at the same time for several weeks, you will notice that the eastern constellations are farther up in the sky and the western ones farther down than when you first began looking. This seasonal shift ultimately brings back the same constellations to the same place at the same time each year (Fig. 1.2). It is the manifestation in the night sky of the seasonal motion of the Sun through the background of stars.

The Sun's Seasonal Motion

The Sun's daily motion through the sky is also subject to seasonal changes. In winter the days are shorter than in summer. The Sun is lower in the sky at noon and casts long shadows. Thus a careful observer can detect a seasonal motion of the Sun by plotting its height above the horizon at noon. This motion comes to a stop and re-

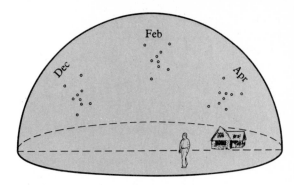

FIGURE 1.2 The appearance of Orion in the sky at 9 p.m. on the first of December, February, and April. The seasonal change in the appearance of the night sky is caused by the revolution of the Earth about the Sun.

verses at the **summer solstice** (Sun-stationary) when the Sun is highest in the sky, and again six months later at the **winter solstice,** when the Sun is at its lowest point (Fig. 1.3). Halfway in between we have the spring (or vernal) and autumn **equinoxes,** when days and nights are equal in length.

In repeating this regular cycle, the Sun appears to move slowly through a certain set of constellations. The apparent path of the Sun through the constellations in the course of its seasonal round is an imaginary line in the sky known as the **ecliptic.** It may seem strange that the constellations through which the Sun moves can be determined, since it is impossible to see the stars during the day, but it is simply a matter of watching which constellations are near the Sun in the twilight skies just after it sets and before it rises. This is something you could do yourself, if you have access to clear horizons. When you know the configuration of constellations over the entire sky, you can tell where the Sun is by figuring out the position halfway between the stars seen in morning and evening twilight.

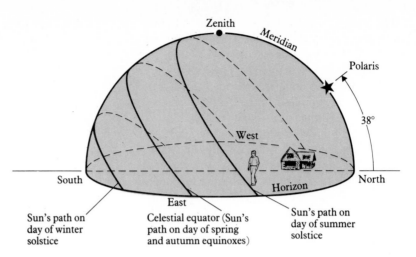

FIGURE 1.3 A person living at the north temperate latitudes of the United States will see the Sun rise directly in the east at the spring or fall equinoxes and set directly in the west. At the summer solstice, the Sun rises north of east and sets north of west, reaching its highest noontime elevation in the sky. At the winter solstice, when days are shortest, the Sun rises south of east and sets south of west, reaching its lowest noontime altitude.

The Zodiac

It is easier to see which constellations form the backdrop for the motions of the Moon. It turns out that the Moon moves through the same constellations as the Sun. The Moon nearly follows the ecliptic, only deviating from it by a few degrees, sometimes appearing above it, sometimes below. The set of constellations through which the Sun and Moon appear to move is called the **zodiac,** from the same Greek word that gives us zoo and zoology. That relationship tells you right away that most of these constellations are named for various real and imaginary animals (Fig. 1.4).

The Sun requires one year or 365.26 days to make its slow, eastward journey through the constellations of the zodiac, returning to its starting point. It is this journey, in fact, that defines our year. The cycle of the Moon through the zodiac is much faster, giving us the period of time we call the month.

Phases of the Moon

During the course of its journey through the zodiac, the Moon goes through a complete set of phases, from new to full and back to new (Fig. 1.5). **New moon** occurs when the Moon is between us and the Sun. In this configuration, the Moon rises and sets with the Sun and is only above the horizon in the daytime sky. The side of the Moon that faces us receives no sunlight and we cannot see it. The night after new moon, we see a thin crescent in the western sky in twilight. The next night the Moon has moved farther east and the crescent has grown. When the crescent is far enough from the Sun to be seen against a dark sky, we can sometimes see the rest of the Moon faintly illuminated by sunlight reflected from the Earth (Plate 1), a phenomenon sometimes called "the old Moon in the new Moon's arms."

The **first quarter moon** is half-full and rises at noon. At nightfall it is halfway between the eastern and western horizons, and at midnight it sets. The **full moon** rises at sunset and sets at sunrise. The Moon has now moved to a position opposite the Sun, so it is fully illuminated. But we are still looking at the same hemisphere that faced us when the Moon was new and we could not see it. Only the illumination has changed. Since the full moon is opposite the Sun, in winter full moons occur with the Moon high in the sky, the position the Sun occupies in summer. Similarly, summer full moons are low, in the winter position of the Sun. This opposition of Sun and Moon creates a striking winter effect in places where it snows, since the snow that resists melting because it lies in shadows during the day is brightly illuminated by the full moon at night.

Moving still farther east in the sky, the waning Moon again reaches half-full phase, rising now at midnight with its curved edge facing the eastern horizon instead of the western one. This is called **last quarter.** And finally we have a thin crescent in the morning sky, rising just before the Sun.

This complete cycle happens every 29.5 days, giving us the basic period of our month. Nature was not quite kind enough to make the year exactly divisible by the month, but twelve is clearly the closest multiple. Thus we end up with our present patchwork of twelve stretched and diminished months in one year. And we have twelve constellations in the zodiac, each one corresponding approximately to one of these months (Fig. 1.4). By knowing where the Sun was with respect to these constellations, the ancients had a good monthly calendar available.

Within the months, the phases of the Moon provided a cycle of days.

It should be evident from this discussion that it is quite easy to account for the apparent motions of both the Moon and the Sun by having each of them revolve around the Earth, just like the great sphere holding the stars. This was the popular ancient belief, later codified by Ptolemy of Alexandria and enforced by the medieval Catholic church. But we know today that in fact the Sun is at the center of our system, and the Earth travels around it in one year, turning daily on its axis while the Moon makes its monthly circuit of our planet.

Eclipses

The circuit of the Moon around the Earth sometimes produces the spectacular events called eclipses. There are two quite different types of eclipses. The most dramatic, but also most rarely seen, is an **eclipse of the Sun.** If at new moon our satellite is exactly between the Earth and the Sun, its shadow falls on the Earth, blotting the Sun from the view of those in its path

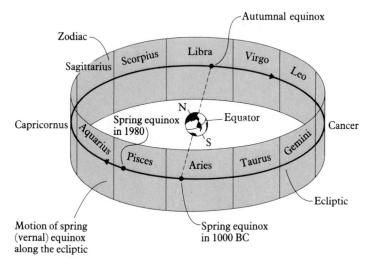

FIGURE 1.4 The planets, the Sun, and the Moon all appear to move through a narrow band of 12 constellations called the zodiac.

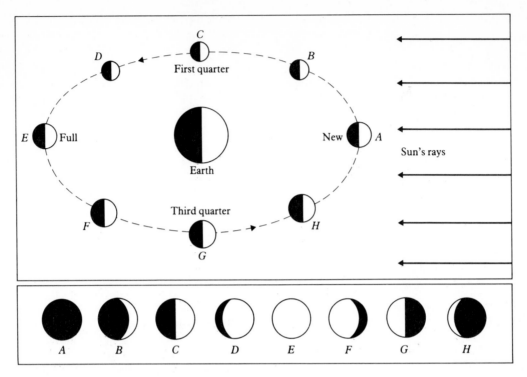

FIGURE 1.5 The phases of the Moon as seen from space (top) and the corresponding appearance of the Moon seen from Earth (bottom). (Figure not drawn to scale.)

(Fig. 1.6). It is a remarkable coincidence of nature that at the present time, the Moon's distance from the Earth is such that its apparent size is approximately equal to that of the Sun. The match is so close that the slight deviation of the lunar orbit from an exact circle can carry the Moon far enough from us to appear a little smaller than the Sun. If a solar eclipse occurs when the Moon is at this point in its orbit, a ring of sunlight appears around the Moon and the eclipse is called annular.

The complement to an eclipse of the Sun is an **eclipse of the Moon,** which occurs when the Moon passes through the Earth's shadow. This can only occur when full moon is exactly lined up with the Earth and Sun (Fig. 1.6). But the Earth's shadow is not as dark as the Moon's, since our planet's atmosphere scatters sunlight

into the shadow, illuminating the eclipsed Moon with a dull, coppery glow. Unlike an eclipse of the Sun, which is only visible along the narrow path of the Moon's shadow, a lunar eclipse can be seen from the entire night side of the Earth.

Eclipses also occur on other planets. In Fig. 1.7 we see the shadow of the martian satellite Phobos on Mars. This event, which would be an eclipse of the Sun for any Martians who were present, was actually detected by the U.S. Viking lander that happened to be in the shadow's path. On the giant planet Jupiter, the dark shadow of one of its satellites would produce an eclipse of the Sun for Jovians floating above that planet's colorful clouds. Similarly, these satellites go into eclipse themselves when they pass into their planets' shadows. They can even eclipse each other.

◆

1.2 Watching the Sky: The Planets
Stars That Move

While most of the points of light we see in the night sky are fixed stars that form the unchanging patterns of the constellations, there are five bright ones that move. *Wild sheep*, the Babylonians called them, while the Greeks used the term *planēt* or wanderer, from which we get our word planet. These wild sheep do not roam over the entire heavenly meadow. Like the Sun and the Moon, they confine their wandering to the twelve constellations of the zodiac, and their apparent motions differ greatly from one another. Our night sky also contains airplanes, whose flashing lights reveal their motion in a few seconds, and artificial satellites, whose motion is also quickly discernible. So it is important to stress that while planets most definitely move, it usually takes several nights to detect their motion with respect to the fixed stars.

FIGURE 1.6 Looking down at the Moon's orbit from space (top). An eclipse of the Sun *can* occur at new moon when the Moon is directly between the Earth and the Sun (*A*). An eclipse of the Moon occurs at full moon *if* the Moon passes into the Earth's shadow (*E*). (This would be experienced as an eclipse of the Sun by an astronaut standing on the Moon.) Side view (bottom). Because the Moon's orbit is slightly inclined relative to the orbit of the Earth about the Sun, eclipses *do not occur* at each new and full moon. In the configuration shown here, both shadows miss their targets.

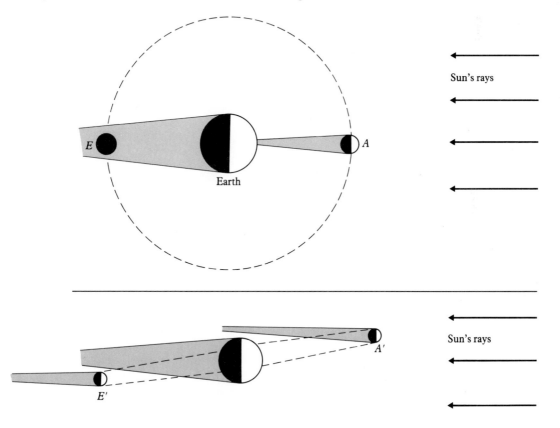

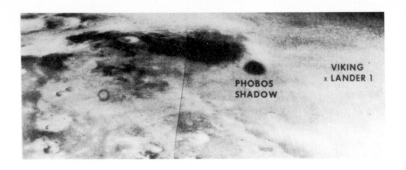

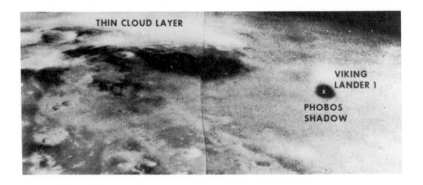

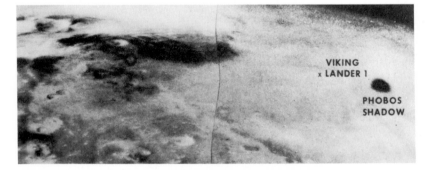

FIGURE 1.7 An eclipse of the Sun for Martians. The shadow of Phobos moving across the surface of Mars as it passes directly over the Viking 1 Lander.

Mercury and Venus

The fastest moving planet was appropriately named Mercury (Greek Hermes) after the swift messenger of the gods. This planet is only visible close to the sunlit horizon in morning and evening twilight. For this reason, it is not easy to see, and some cultures thought it was actually two objects seen at different times of year: one visible only in the morning and one in the evening.

This same duplicity was sometimes assumed for Venus, which also stays close to the Sun. Venus is the brightest object in the sky after the Sun and Moon, so bright that it is sometimes visible during the day, if you know exactly where to look. Sailing serenely in the silent sky, Venus shines like a brilliant jewel, marking the transition between day and night. The ancient Greeks and Romans must have been very impressed, for they named this planet after their goddess of love and beauty (Greek Aphrodite). Venus appears farther from the Sun than Mercury, but never so far that it can be seen in the middle of the night.

Mars

The other three planets easily visible to the naked eye do not exhibit this close connection with the Sun, and they appear to move through the constellations of the zodiac at a more leisurely pace. There are two unusual characteristics of Mars (Greek Ares) that must have led to its association with the god of war. First, its distinctly reddish hue singles it out from all of the other planets, conjuring up images of blood and war. Secondly, Mars exhibits apparently erratic, unpredictable motions that must have been puzzling and unsettling to ancient observers, another appropriate characteristic for the god of war.

Retrograde motion is simply a reversal of the normal direction of movement. Like the Sun and the Moon, the planets all move slowly eastward through the constellations of the zodiac. Each day they rise and set with the other stars, but from one night to the next, you can notice that they have shifted their position eastward, except that sometimes they reverse themselves and move toward the west! Thus Mercury and Venus will first move eastward until they reach their maximum separation from the setting Sun in the west, then they head west, disappearing in the twilight glow and reappearing in the eastern sky before dawn, where they proceed to achieve their maximum western separation from the Sun before reversing and repeating the cycle.

Mars, Jupiter, and Saturn behave differently. We might first see Mars in the late night sky, moving slowly eastward through the constellations of the zodiac. As the changing seasons cause the stars to rise earlier and earlier, Mars rises earlier with them, becoming visible in the middle of the night. We now say it is at opposition, since it is on the opposite side of the Earth from the Sun. A few months before Mars reaches opposition, its apparent motion gradually slows down and then reverses. The planet speeds up along a retrograde path, then slows down and returns to its normal direction (Fig. 1.8). Its path appears as a large loop in the sky, and in traversing this loop, near opposition, Mars becomes very much brighter than usual. Although this pattern must have been apparent to ancient observers, its causes remained obscure.

Jupiter and Saturn

The brightest of the planets after Venus, Jupiter (Greek Zeus) was named for the king of the gods. Jupiter takes twelve years to complete one circuit of the zodiac, and exhibits only a small retrograde loop each year (Fig. 1.8). Saturn (Greek Chronos) is even more dignified. This planet takes a full thirty years to travel through the zodiac. Saturn was Jupiter's father and the ancient god of time. Since thirty years was (and still is) a large fraction of a human lifetime, the name is again appropriate.

Ancient Interpretations of Planetary Motion

The various apparent motions of the stars and planets provided the first observations that peo-

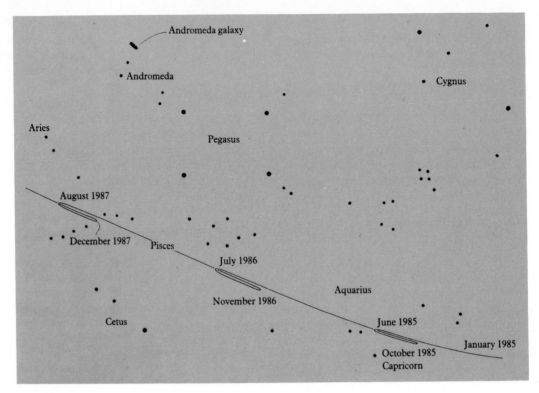

FIGURE 1.8 The retrograde loops made by Jupiter as it wanders among the stars. The apparent motion of the planet slows down, reverses, slows again and continues in the original direction.

ple could use to understand what the Earth was and to establish its place in the universe. With the Sun, Moon and stars apparently circling the Earth in some kind of spherical system, it was natural to try to add the planets to this group. Another constraint was provided by the desirability of having everything move along circular paths. The circle was regarded by the ancient Greeks as the most perfect geometric figure, and casual observations of the sky indeed suggested that circular paths were entirely appropriate for the bright objects it contained. The Sun and Moon were clearly circular themselves, giving added support to this approach.

With our space-age perspective, we may regard these ideas as hopelessly naive, but a moment's reflection will indicate that all of your daily needs can be met by this simple model of the universe — a flat Earth around which everything else revolves. It is only if you take a long trip or stand at the seashore watching boats disappear over the horizon (or see a picture of Earth taken from space) that you need to question the flat Earth. History shows us that it was not until the fifteenth and sixteenth centuries that these ideas were abandoned, as a part of the European artistic and intellectual rebirth known as the Renaissance.

1.3 The Slow Progress of Reason
Copernicus and the Heliocentric System

Among the basic questions that have been asked in almost all human civilizations are those that concern the size and shape of the Earth, the motions of the stars and planets, and the distances to the Sun, Moon, and planets. These basic elements in the structure of our solar system are now taught in our elementary schools. Nevertheless, these "facts" were only slowly established, and even today few of us could actually cite the evidence that proves this modern perspective is correct.

The slow growth of understanding that led to our present view is a fascinating story, but we can only mention a few highlights here. It is useful to review these because they offer some lessons about how we should approach new, unsolved problems today, and they show how the perspective we now take for granted was achieved. Here we shall consider the work of just four people: Copernicus, Tycho, Kepler, and Galileo, moving on to Newton in the next section (Fig. 1.9).

Nicholas Copernicus was a Polish astronomer born in 1473. He was a contemporary of some of the great Italian painters of the Renaissance, such as Botticelli and Michelangelo, and was nineteen when Columbus made his famous voyage to the New World. Copernicus spent some time as a student in Italy before returning to Poland in 1503, where he was given a position in the church. He had become very interested in the **geocentric** (or Earth-centered) astronomical system (Fig. 1.10). This geocentric system had been thoroughly developed by the Alexandrian Greek, Claudius Ptolemy, in 200 AD, and it was still the accepted view 1300 years later. In this system, the motionless Earth stood in the center of the universe and everything else circled around it.

Copernicus didn't like the geocentric system because it seemed too complex and aesthetically unappealing. To him, it made more sense for the radiant, life-giving Sun to be in the center, with the planets revolving around it (Fig. 1.11). This bold hypothesis allowed him to explain the apparent motions of the planets much more easily than the geocentric system could. It also allowed the stars to be motionless. This pleased Copernicus since he felt it was more reasonable for the small Earth to move about the Sun than for the great vault of heaven to turn about the Earth. This new perspective was called **heliocentric,** or Sun-centered.

It is hard for us to recapture the challenge to human thought that the Copernican system required. We are so accustomed to the modern view that we find the geocentric system absurd. The geocentric system, however, had been supported by most of the philosophers of ancient Greece and Rome as part of a broader framework that argued for a stationary Earth and unchangeable perfection in the heavens. To worsen the situation, these ideas had become entrenched in the medieval Catholic church, which was then the dominant intellectual and political force in Europe.

Copernicus had no proof that his heliocentric system was correct. To him it was both clearer and more pleasing than the geocentric system. Others felt differently. As we said earlier, our senses tell us that the Earth is flat and everything revolves around it, not an easy point of view to abandon. Nor, in that period of history, were people accustomed to the idea of using experiments or observations of nature to distinguish between different philosophical systems. But consider just one simplifying stroke that Copernicus achieved. Previously it had been thought that the motions of the planets as seen in the sky were real. Thus when a planet looked as if it were moving backwards (retrograde motion) it really was. The retrograde motion was

(a)

(b)

(c)

(d)

FIGURE 1.9 a) Copernicus, b) Tycho, c) Kepler, and d) Galileo — a gallery of the four men who changed our perception from an Earth-centered, totally mysterious universe to a Sun-centered solar system surrounded by distant stars.

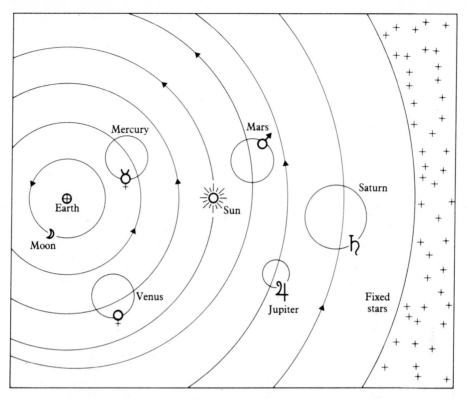

FIGURE 1.10 The Earth-centered or Ptolemaic system to explain the apparent motions of the Sun and planets. Each object except the Earth must move in small epicycles in addition to moving about the Earth.

thought to be caused by the planets' motion on small circles called epicycles that were centered on their circular orbits. Thus when a planet needed to go backwards, it just traversed the small circle in a direction opposite to that in which the big circle carried it. This complex arrangement may be compared with what happens when the Sun is at the center. Now the retrograde motion of the planet is not real, it is simply the effect seen by an observer on Earth as we overtake the planet in our journey around the Sun (Fig. 1.12).

Note that both the geocentric and the heliocentric models for the solar system explain retrograde motion. Copernicus simply does it in a less complex, more elegant way. This is a fre-

FIGURE 1.11 A copy of the original sketch of his plan for the solar system published by Copernicus; note that only the Moon orbits the Earth, while all the planets move around the Sun.

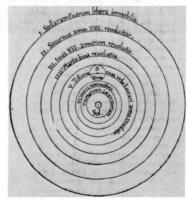

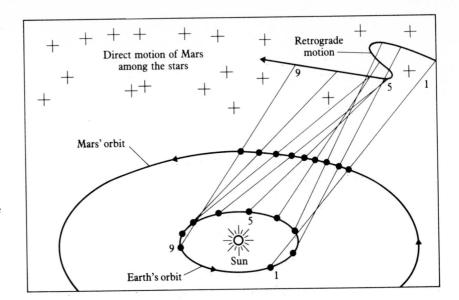

FIGURE 1.12 Retrograde motions as explained by the Copernican system: outer planets appear to move backwards as the faster Earth overtakes them.

quent situation in science, where more than one theory can sometimes explain the same set of observations. However, when the theories are as different as these, they will make different predictions about other phenomena that have not yet been observed. These predictions can then be tested by new observations which will provide a way of choosing between the two theories.

Tycho and Kepler

In the absence of observations to distinguish the heliocentric from the geocentric systems, the ideas of Copernicus remained highly controversial. One of the astronomers who found them interesting but not entirely satisfactory was Tycho Brahe (1546–1601). Born in Denmark just three years after the death of Copernicus, Tycho's early education was in jurisprudence. An eclipse of the Sun he had seen as a boy of fourteen made a profound impression on him, however, and he soon abandoned law to become the most accomplished observer of the heavens in the era before the invention of the telescope.

Using instruments of a large scale and high angular precision but with no optical parts, Tycho surveyed the heavens, measuring the positions of the stars and planets with a precision far greater than anyone before him. To indicate the need for this work, it suffices to point out that the apparent close approach to one another by Jupiter and Saturn in 1563 was actually several days later than predicted by the best tables of positions then available. Tycho was determined to improve this situation. He was also a colorful character, rather vain, who wore a silver nose as a result of a dueling injury. Among his own discoveries, his observational proof that a "new star" that appeared in 1572 in the constellation of Cassiopeia was farther away than the Moon was an important step away from ancient ideas. It showed that the heavens were not immutable, that the stars themselves could change.

It was actually Tycho's successor, Johannes Kepler (1571–1630), who extracted the hidden science from these remarkably precise observations of planetary positions. Kepler was taught the Copernican system at the University of Tübingen, where he was studying to become a Lutheran minister. But the faculty there, recognizing his intellectual gifts, persuaded him to take a post as a mathematician at the University in Graz, Austria, in 1594. (To place the period, note that in England Shakespeare was writing his plays and sonnets at this time.) In addition to exhibiting a real gift for mathematics, Kepler was strongly influenced by a mystical sense of the intrinsic harmony and order in the universe. It was this sense that provided his main motivation for studying the motions of the planets: he felt that there must be an underlying set of simple principles that governed the structure of the solar system.

Kepler was very fortunate in that his early work came to the attention of Tycho Brahe. Tycho recognized the younger man's mathematical abilities and invited Kepler to join him in Prague in 1600. Tycho died a year later, and Kepler took over his post. Kepler was fascinated by the apparent harmony and order in the Copernican system. Because of his own mystical beliefs, he was certain that there must be a way of expressing this harmony in simple mathematical relationships. This proved very difficult to do, but nine years after his arrival in Prague, Kepler published a book containing the first two principles of planetary motion that we now refer to as Kepler's laws.

Kepler's Laws of Planetary Motion

Kepler had been trying to analyze Tycho's observations of Mars, fitting them to the system of orbits described by Copernicus. It simply wouldn't work. The problem lay in the circular orbits that the Copernican system still used. Mars obviously didn't move about the Sun in a circle, not even a circle in which the Sun was not at the center. So Kepler tried a variety of other geometrical shapes, until he finally found the ellipse. This worked, not just for Mars, but for all of the planets.

At first, Kepler was unhappy with the ellipse because it doesn't have the perfect symmetry of the circle. The Sun is at one focus, but there is nothing at the center of the ellipse and nothing at the other focus (Fig. 1.13). The ellipse is still elegantly simple, however, and it worked. This then is Kepler's first law: *The planets move in elliptical orbits about the Sun, with the Sun at one focus of the ellipse.*

Another problem that had vexed astronomers was the fact that at some times the planets moved faster in their orbits than at others. This seemed random and unexpected, until Kepler found another principle that made the behavior intellectually acceptable. With the planets moving in elliptical orbits, it turns out that the fastest motion occurs when they are nearest to the Sun. Consequently, an imaginary line drawn between the Sun and the planet will sweep out the same area within the ellipse during the same interval of time, regardless of where the planet is in its orbit (Fig. 1.14). This discovery restored a cer-

FIGURE 1.13 The circle, ellipse, and parabola represent an increase in eccentricity from 0 (circle) to 1 (parabola). Some comets have elliptical orbits with eccentricities of 0.999, while Earth's orbit has an eccentricity of only 0.02, and that of the orbit of Venus, the most circular of the planets, is only 0.007.

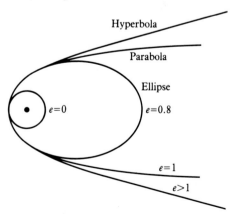

FIGURE 1.14 Kepler's second law. A planet moves fastest when it is closest to the Sun (perihelion). Yet, the area created by a line connecting the planet to the Sun is always the same for a given period of time. Thus the areas of the light and dark triangles are all equal.

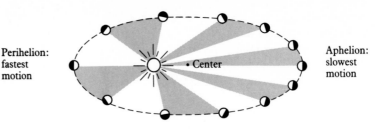

Perihelion: fastest motion

• Center

Aphelion: slowest motion

tain order to nature and became another simple mathematical description of what was observed. It is Kepler's second law: *Imaginary lines from the Sun to the planets sweep out equal areas in equal times as the planets move about the Sun.*

The best was yet to come. Kepler was still searching for some principle that would relate the motions of the planets to one another. He now had an accurate description of how they moved, but he was certain there must be a harmony in those motions that would explain why the planets moved the way they did. Ten years after publishing his first two laws, he found that harmony. What he had discovered was a simple mathematical relationship between the period of a planet's revolution about the Sun and its distance from the Sun. This is Kepler's third law: *The cube of the distance from the Sun divided by the square of the time required to traverse the orbit is a constant, the same for every planet* (Fig. 1.15).

This third law represented a remarkable achievement. The underlying harmony of the solar system had been found, and Kepler was overjoyed. Obviously his work had greatly strengthened the Copernican system, providing the predictive power it had lacked in its original form when the planets were thought to move in circles. These laws, however, are still descriptive. Kepler's work could be viewed as simply providing a much better way to calculate the positions of planets. The proof that would clearly distinguish between the geocentric and heliocentric systems was still lacking.

Galileo and the Foundation of Modern Science

Kepler was a contemporary of Galileo Galilei (1564–1642), who was at this time laying the foundations of modern physical science. Galileo's experiments and observations, just as much as Copernicus' and Kepler's theories, were responsible for overturning the geocentric model of the solar system.

Galileo, the son of a musician, began his studies as a candidate for a degree in medicine but was soon attracted to mathematics. In 1609,

FIGURE 1.15 A logarithmic graph illustrating Kepler's third law: the square of a planet's period (P^2) is proportional to the cube of its distance from the Sun (a^3). Periods are plotted in years and distances in units of the Earth's distance from the Sun.

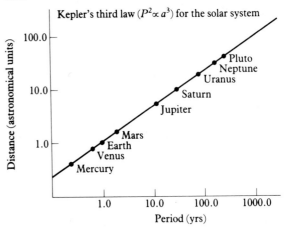

Kepler's third law ($P^2 \propto a^3$) for the solar system

he was teaching mathematics in Padua when he became aware of the invention of the telescope. Unlike many of his contemporary academics, he recognized that the combination of lenses enabled one to see better, not just differently (as with a distorting mirror). He quickly made some telescopes of his own and began using them to study the night sky. What he found provided the first experimental demonstration that the geocentric universe and its philosophical underpinnings were not simply complex and inelegant; they were wrong!

One of Galileo's earliest discoveries was the presence of some small "stars" very close to Jupiter. Observing night after night, he found that these stars changed their positions with respect to the planet in an orderly, repetitive way. Evidently, they were satellites moving in four separate orbits around Jupiter. This was a major blow to the geocentric solar system, since that system required that everything must orbit the Earth. Galileo saw that what he had found resembled a miniature Copernican solar system itself, in which Jupiter played the role of the Sun and the four satellites that he had found resembled the planets. We shall see in Chapter 13 that this similarity was even greater than Galileo thought.

The evidence for a heliocentric system was further strengthened when Galileo discovered that the planet Venus exhibited phases like those of our Moon. If this planet orbits the Sun, it also should become larger in apparent size as it nears the Earth, at the same time shrinking to a thinner and thinner crescent. Conversely, as Venus moves around to the other side of the Sun we can see more of it because of the favorable viewing angle, but it appears much smaller than when it is a crescent since it is now much farther away from us (Fig. 1.16). This behavior, exactly what Galileo observed, is impossible in the geocentric system, which would not permit an observer on Earth to see Venus fully illuminated.

So the geocentric system was clearly untenable, at least to those who had looked through Galileo's telescope or who believed what he wrote. But the Church did not accept these proofs and forced Galileo to deny their validity in 1633, fourteen years after Kepler (who did not live in a Catholic country) had published his third law. It is also worth noting that Kepler had to interrupt his work at one point to save his mother from being burned as a witch. And only nine years before Galileo began experimenting with telescopes, Giordano Bruno was burned at the stake for his heretical views of the universe. Galileo was treated much less severely, being sentenced simply to house arrest for the last eight years of his life. One wonders if he knew that it made no difference how the Church courts ruled, that others would prove him to be correct. In any event, that is exactly what happened. The stage was now set for the appearance of Isaac Newton.

◆

1.4 Newton and the Law of Gravitation

The Importance of Newton

Isaac Newton was born in 1643, a few months after the death of Galileo (Fig. 1.17). He therefore grew up at a time when the basic discoveries we have just described should have been part of his college curriculum. As an undergraduate at Cambridge from 1661 to 1665, though, Newton found that the University was still teaching the geocentric view of the universe. Nevertheless, Newton managed to learn about the new ideas, to which he added his own insights. Today undergraduates interested in science and mathematics study calculus. Newton began *inventing* calculus as an undergraduate, while also formulating ideas about optics and mechanics. Throughout a long career (he died at 84), New-

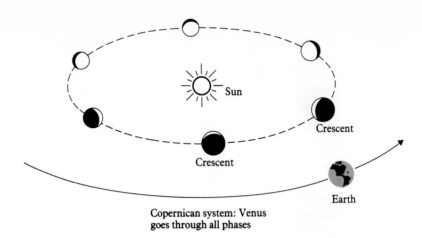

Copernican system: Venus
goes through all phases

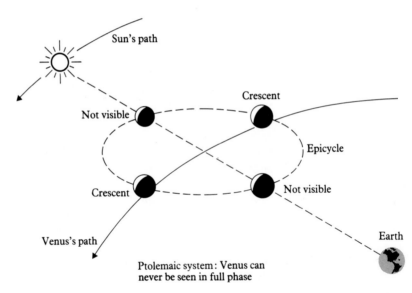

FIGURE 1.16 The Ptolemaic (Earth-centered) view of the solar system could explain crescent phases for Venus (bottom) but would not predict that a full phase would occur (Venus would have to appear in the midnight sky, which is not observed). The Copernican (Sun-centered) view (top) correctly predicts the phases and also the observed fact that the planet at full phase appears much smaller than it does as a crescent — it is simply much farther away.

Ptolemaic system: Venus can
never be seen in full phase

ton made many fundamental contributions to several areas of physics and astronomy. He was one of the first scientists to use **quantitative,** or mathematical, arguments to test scientific ideas. Because he was so successful, he altered human perceptions of the universe and its workings in many basic ways. The impact of his accomplishments was summed up in the following couplet by a contemporary poet, Alexander Pope:

"Nature and Nature's laws lay hid in night:
God said, let Newton be! and all was light."

FIGURE 1.17 Isaac Newton, the man who discovered the physical laws governing planetary motions, thereby bringing a sense of order to an apparently random universe.

The Laws of Motion

For our purposes, Newton's most important contributions were his three laws of motion and the law of gravitation. The first law of motion may be stated as follows:

1. *Every body continues in its state of rest or of uniform motion in a straight line unless it is compelled to change that state by the action of some outside force.*

The first part of this law seems pretty obvious. Things don't move unless they are forced to do so, but the second may not be as evident. There are no familiar examples of an object moving uniformly in a straight line. Everything we start moving ultimately stops. Newton realized that this was because the force of friction would slow things down. As friction is reduced, motion continues for a longer time, and one can imagine a situation with no friction at all in which a moving object coasts on forever. How can we express the relationship between the forces acting on bodies and the motions of these bodies? Newton's second law provided the answer:

2. *The change of motion (acceleration) is proportional to the force acting on the body and inversely proportional to the mass of the body.*

When expressed in mathematical form, this law provides a way to calculate the motion of a body from the force exerted on it. Or, conversely, by tracking the motion it allows us to calculate the forces. These calculations, carried out today with large computers, are still the way we navigate our spacecraft on their interplanetary journeys. The larger the force, the greater the acceleration. Also, a given force will have more effect on a body of small mass than on a large one. All of this is well known from everyday experience, where we deal frequently with objects (automobiles for example) that change speeds. What may be less obvious is that an object moving at constant speed in a circular or elliptical path is also accelerating. Changing direction is a form of acceleration, which means that a force is acting on the body to produce the curved path.

Newton's third law takes us back to the issue of balancing forces:

3. *To every action there is an equal and opposite reaction.*

This says that when an apple falls toward the Earth, the Earth must also move toward the apple. But since the Earth has about 10^{26} times as much mass as the apple, if the force acting on the two of them is the same, the Earth accelerates 10^{26} times less. That is a small number indeed! To see this law in action, you might find a partner and try pushing each other on roller skates or while you are standing on skate boards. This law is also what sometimes makes getting in or out of a canoe rather wet work, while getting in and out of a big boat is easy.

Newton's brilliant solution to these dilemmas was the law of gravity. If an apple falls on Earth, it is being attracted to the Earth by a force. Perhaps that same force is attracting the Moon. The Moon doesn't fall to the Earth because it is moving sideways, approximately at right angles to the force of gravity. The attractive force changes the motion of the Moon from a straight line to a closed curve, an orbit around the Earth. In effect, the Moon is "falling" around the Earth. Newton's law of gravitation may be written as follows:

> *The gravitational force between two objects is proportional to the product of their masses and inversely proportional to the square of the distance between them.*

This is an extremely powerful law. With it, we can determine the mass of a planet or the mass of the entire galaxy (Fig. 1.18). It governs the flights of both spacecraft and pole-vaulters, and it explains why the planets move the way they do. The planets are "falling" around the Sun in the same way the Moon moves around the Earth. Left to their own devices, they would move off into space in straight lines. The acceleration toward the Sun that they experience from gravity is made manifest by the acceleration produced by their orbital motion. Hence they are condemned to move forever in the orbits in which we find them.

FIGURE 1.18 A couple holding each other at the intimate distance of 1 cm are exerting a gravitational attraction on one another that is about equal to the attraction of the Moon on either one of them. Fortunately, (1) the gravitational force on both of them exerted by the Earth is about 100 thousand times greater and (2) there are other forces operating between people that are stronger than gravity.

The Law of Gravitation

The framework provided by these three laws of motion seems very close to providing an understanding of why the planets move around the Sun in the way that Kepler described. Obviously, a force is acting on them, or they would be moving in straight lines. But what is that force? And what is the balancing force that allows them to continue this motion instead of slowing to a stop or falling into the Sun?

Science Since Newton

"If I saw a little farther than others," Newton wrote in a letter, "it was because I was standing on the shoulders of giants." This modest appraisal of his extraordinary achievements is nevertheless an accurate description of how science proceeds, each step forward building on the work of those who came before. This was particularly the case after Newton, for his carefully reasoned, mathematical approach has set the tone for most subsequent scientific enquiry.

Galileo and Newton ushered in a new world of experimental and quantitative thinking. During the following decades, the pace of science quickened, and one discovery followed another. Ever more elegant and powerful mathematical tools were developed, just as scientific instruments became more complex and precise. Even though space travel was still more than two centuries into the future at Newton's death in 1727, the way toward the modern world was clearly indicated.

A further major change in our conception of the physical universe came in the early part of this century with Einstein's theory of relativity, atomic and nuclear physics, and quantum mechanics. We shall discuss the aspects of these developments that we need to know in the next chapter. First we will describe the picture of the solar system that emerges from the post-Newton perspective.

◆

1.5 The System Revealed

The Shape of the Solar System

It should now be clear why the planets appear to move the way they do. The fact that they confine their motions to a narrow band along the ecliptic is a natural consequence of having their orbits nearly in the same plane. Imagine yourself as a letter in a word on the middle of a single page, taken from this book. Stretching away from you on all sides are the letters that make up the other words you are now reading, so you would be surrounded by words, but only in the plane of the paper. Looking up or down, you would see no writing.

It is the same for us in the solar system. If we stood on the surface of Mars and studied the night sky, we would see the same constellations we see from Earth, and we would find that the planets were still moving through the zodiac. When we look out toward the planets moving in front of these constellations, we are seeing an edge-on view of the solar system, just as the Milky Way represents an edge-on view of our galaxy. From Mars, the Earth would be a gorgeous blue-white planet in the twilight sky, exhibiting the same kind of apparent motions with respect to the Sun that Venus does from our terrestrial perspective.

The two planets that stray farthest above and below the plane defined by the Earth's orbit are Mercury and Pluto. Their orbits have **inclinations** of 7° and 17°, respectively (Fig. 1.19). All of the other planets have orbital inclinations within 4° of Earth's orbit. The orbits of Mercury and Pluto also have the greatest **eccentricity.** That is, the ellipses along which these planets move are more elongated than the paths of the other planets.

We can also gain an approximate idea of the relative distances and sizes of the planets from simple visual observations, once we have adopted the Copernican point of view. Jupiter appears to move more slowly through the zodiac than Mars, so it must be farther away. The fact that it is nevertheless the brightest object in the sky after Venus suggests that it must be unusually large. Saturn, still slower, is even farther away and must also be big to be as bright as it is. Mars is much closer than these two giants and must therefore be relatively small.

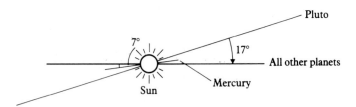

FIGURE 1.19 All of the orbits of the planets except for those of Mercury (closest to the Sun) and Pluto (most distant) lie essentially in the same plane.

Directions of Motions

Which way is up? The convention that has come down to us from experience on the Earth is to say that north is up. North in the sky is defined as the direction in space toward which the Earth's north pole points. If we imagine ourselves located in space somewhere *above* the solar system in this sense, we would find when we looked down at the planets that they were all moving about the Sun in a counterclockwise direction. Looking down at the Earth from above its north pole, we would see that our planet is also rotating on its axis in a counterclockwise direction, in the same direction that the Earth moves around the Sun. The same is true for most of the other planets.

Looking down on Venus from our lofty perch north of the system, we would find that the planet was rotating in the opposite direction, clockwise instead of counterclockwise, even though its revolution about the Sun is in the same direction as all the other planets. Uranus and Pluto would appear tipped over on their sides, with their north poles (defined by rotation) actually a little below their orbital planes. They too are technically rotating backwards compared with Earth and the other planets.

The various satellite systems show a different kind of symmetry. Except for our Moon — whose motion is influenced by the proximity of the Sun — most of the larger satellites stay close to the equatorial planes of their parent planets. The same is true for systems of rings. In the case of Saturn, for example, the ring system and the satellite orbits all share the planet's inclination of 27° to the plane of Saturn's orbit. Most satellites revolve around their planets in the same counterclockwise direction (as seen from the north) that the planets move around the Sun, but four of Jupiter's moons and one each in the Saturn and Neptune systems revolve in the opposite sense. (Clockwise rotation or revolution is called **retrograde,** the same word used in Section 1.2 for the apparent backward motion of a planet seen in the sky.)

The asteroids are a diverse lot of small bodies with diameters less than 1000 km that orbit the Sun primarily between Mars and Jupiter. There are tens of thousands of them, but their combined mass is less than 1/10 the mass of the Moon. Some asteroids have orbits inside the orbit of the Earth, while at least one, Chiron, circles the Sun beyond the orbit of Saturn. The orbits show a variety of eccentricities and inclinations, but on average they are inclined about 8° and are more eccentric than most planetary orbits.

The orbits of the comets are even more diverse (Fig. 1.20). In their pristine state, these bodies orbit the Sun at immense distances, roughly 1,000 times the distance of Pluto. Unlike the planets, their orbits show a completely random distribution of inclinations. When a comet appears in our night skies, it is therefore rarely in a constellation of the zodiac. Its orbit may be perpendicular to the plane of the planets, or it might move around the Sun in a retrograde direction. For us to see it, the comet must have sailed in from its distant domain, perhaps perturbed by the gravity of a passing star. If the comet passes close to one of the major planets on this journey, it can be perturbed again, this time into a relatively small elliptical orbit. We then say it has been captured, and its period of revolution will now be just a few years instead of a few million years, the period corresponding to its initial orbit.

A Scale Model

To gain an appreciation of the size of the solar system and the distances between the various planets, we can make use of a scale model. Let's reduce every dimension in the solar system by a factor of 200 million. On this scale, the Earth is the size of an orange, and our Moon would be a grape, orbiting the orange at a distance of six feet. The Sun would be a little less than half a mile away, and it would have a diameter of

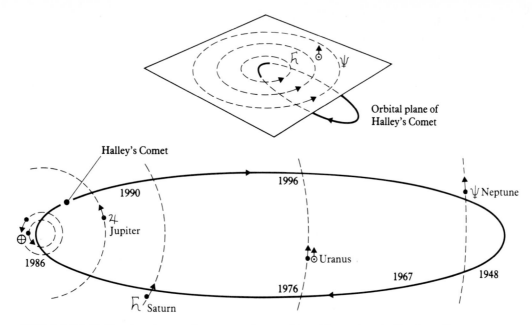

FIGURE 1.20 Unlike the planets, the comets have orbits with random inclinations. Some, such as that of Comet Halley, are inclined by more than 90°; such comets move around the Sun in the opposite direction to that of the planets.

twenty-three feet, the height of a two-story building. It is the great distance of the Sun from the Earth that makes it appear to be about the same size as the Moon when we see both of them in our skies.

Although the Earth and other inner planets in this model are only the size of pieces of fruit, we do have some bigger planets. At this scale, Jupiter would be a large pumpkin, still small compared to the Sun but eleven times larger than Earth. Considering the Saturn system, we find that if we measure it from one edge of the rings to the other, the planet and its rings would just fit between the Earth and the Moon. Saturn's own large satellites would lie far beyond the Moon. Obviously scales in the outer solar system are considerably grander than those we find for the planets close to the Sun.

Moving farther out, we pass Uranus and Neptune, finally encountering Pluto, another

grape like our Moon at an average distance of 18.5 miles from the twenty-three-foot Sun. This is still not the edge of the solar system, however. That lies at a point where the gravitational field of the Sun is challenged by the fields of passing stars. Here is where the great spherical cloud of comets lies, at a distance of some 20,000 miles from our scale model Sun. This means that our scale model solar system is about four times the size of the real Earth.

This exercise illustrates how large the Sun is, how much space exists between the planets, even in the inner solar system, and to what an enormous distance the Sun's gravitational influence extends. The latter simply emphasizes once again how huge the distances are between the stars. The nearest star, Alpha Centauri, is five times as far from the Sun as the comets; to include it, our model would have to extend to a dimension of 100,000 miles.

The real solar system is described in Tables 1.1 and 1.2. It is useful to try to keep a few dimensions in mind. The Earth is 1.496×10^8 kilometers (93×10^6 miles) from the Sun. For convenience, this distance is commonly referred to as the **astronomical unit,** abbreviated AU. Jupiter, the closest of the giant outer planets, is about five times as far from the Sun as we are. The next giant planet, Saturn, is about as far from Jupiter as Jupiter is from us. The next step out, to Uranus, again nearly doubles the size of the system. Neptune and Pluto are closer together, but the average distance between the orbits of these two planets is about equal to the distance of Saturn from the Sun.

The Titius-Bode Rule

Given Kepler's success in deriving laws governing planetary motion, it was natural for others to attempt to find some law that determined why the planets have the spacing we observe. The most famous attempt to unravel this mystery was that of Johann Daniel Titius in 1766. Titius found a relatively simple numerical relationship among the distances of the known planets, as described in Table 1.1. This relationship is sometimes called "Bode's law" or the "Titius-Bode law," since a more famous astronomer, Johann Bode, played a major role in publicizing it.

In fact, it is not a law at all, since there is no physical reason for this particular numerical progression. It would probably be better to call it the **Titius-Bode rule.** Despite this shortcoming, it played an important part in the history of astronomy. At the time Bode popularized it, there were no known asteroids, and the solar system ended at Saturn. William Herschel (1738–1822) discovered Uranus in 1781, adding a planet to the solar system just where the "law" said its orbit should be. This made the gap between Mars and Jupiter all the more evident, and sure enough the first asteroids were discovered at this position at the beginning of the nineteenth century. But Neptune doesn't occur where it should, nor does Pluto. So Bode's "law" seems to be nothing more than one of nature's surprising coincidences, very different from the laws of Kepler and Newton.

TABLE 1.1 Dimensions of the planetary system

Planet	Distance from Sun (million km)	Distance from Sun (AU)	Predictions[a] from Titius-Bode Rule (AU)
Mercury	58	0.39	0.4
Venus	108	0.72	0.7
Earth	150	1.00	1.0
Mars	228	1.52	1.6
Asteroid belt	330–500	2.20–3.30	2.8
Jupiter	778	5.20	5.2
Saturn	1429	9.55	10
Uranus	2875	19.22	20
Neptune	4504	30.12	} 39
Pluto	5900	39.44	

[a]Distance (AU) $= 0.4 + 0.3\ (2^n)$, where $n = -\infty$ (Mercury), 0(Venus), 1(Earth), 2(Mars), 3(Jupiter), and so on for the other planets.

1.6 Basic Properties: Mass, Size, and Density

What are these whirling worlds really like? Before we begin the detailed answer to this question, let's take a quick look from a very basic point of view. We shall describe the sizes and masses of these bodies, and from these two characteristics, we can then derive the densities. Even before we study the planets with spectrographs and spacecraft, a knowledge of the densities gives us an important clue as to their composition.

Density As a Guide to Composition

Density is simply a measure of the amount of mass contained in a given volume. In the metric system we express the units of density as grams per cubic centimeter or, equivalently, as tons per cubic meter. In these units, water has a density of 1.0. Ice, since it floats in water, must have a lower density. In fact, it is only a little lower: 0.92 g/cm³. That's why only the tip of an iceberg shows above the surface of the ocean in which it floats. On the other hand, a piece of pine wood has a density of 0.5 g/cm³, and a piece of the porous volcanic rock called pumice may have a density of 0.7 g/cm³. Both the wood and this unusual rock float better than ice. So would the planet Saturn, if we could build a big enough bath tub, since its density is only 0.7. Metals, in contrast, have high densities; lead has a density of 11, indicating 11 times as much mass in the same volume as a gram of water (Fig. 1.21).

Intuition tells us that we could expect a planet like the Earth to be at least as dense as the rocks we find on its surface. They typically have densities on the order of 2.5–3.5 g/cm³. In fact, the Earth's density is nearly twice as great. The reason is twofold: the central core of our planet is composed of iron and nickel, obviously more dense than rock, and even these metals are compressed to a greater than normal density by the

TABLE 1.2 The planets

	Radius[a]	Mass[a]	Density
Mercury	0.38	0.06	5.4 g/cm³
Venus	0.95	0.82	5.3
Earth	1.00	1.00	5.5
Mars	0.53	0.11	3.9
Jupiter	11.2	318	1.3
Saturn	9.5	95	0.7
Uranus	4.1	15	1.2
Neptune	3.9	17	1.7
Pluto	0.19	0.002	1.7

[a]Both relative to the Earth = 1:
Earth equatorial radius = 6378 km
Earth mass = 5.98×10^{24} kg

weight of the overlying material. The result is a density for the Earth as a whole of 5.5 g/cm³. For more massive planets, we expect higher densities, since the central compression will be greater; the converse is true for the smaller planets. Any deviations from this expectation must indicate differences in composition.

The Densities of Planets

The density of a planet can easily be calculated once we know the mass, which can be estimated from its gravitational influence on its satellites or other bodies in the solar system, and its size, derived from telescopic observations. The technique requires Kepler's laws, Newton's laws, some simple trigonometry, and a telescope. We need no fancy astrophysical equipment, and the picture of the solar system we can now unfold was already available in the eighteenth century, except, of course, for the new planets that were discovered later. Let's see what this picture looks like.

A summary of the masses, sizes, and densities of the planets is given in Table 1.2. It is evident that the planets can be divided into three categories according to these characteristics,

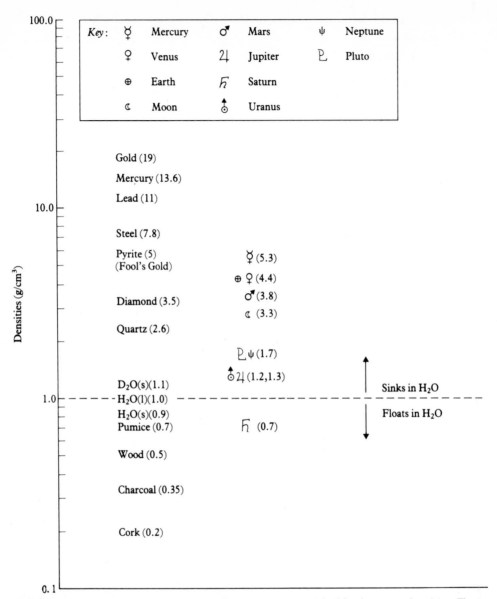

FIGURE 1.21 The densities of various substances compared with planetary densities. The densities of the inner planets are corrected for the effect of compression. Note that just as pumice and Saturn could float in water, steel and Mercury could float in mercury (pumice and Saturn could also)!

which turn out to be correlated with distance from the Sun: inner planets, outer planets, and Pluto. Our Moon may also be thought of as an inner planet in terms of its composition, but it is distinctly different from Earth, as we'll see in Chapter 5.

Considering only the densities, the inner planets seem very similar. A closer look, however, reveals that even with this very simple approach it is possible to find some distinguishing characteristics. Thus we expected bodies smaller than Earth to have smaller average densities, since their centers will be less compressed. This is certainly true for Mars, but not for Mercury, whose density of 5.4 g/cm^3 is nearly equal to that of our planet. As we shall see in Chapter 6, the solution to this apparent paradox is to assume that Mercury has a disproportionately large core of iron and nickel.

We confront the opposite problem when we consider the planets in the outer solar system. Here we are dealing with bodies far larger than the Earth, and we would expect them to have much greater densities. Instead, we find that these objects are less dense than any of the inner planets. The conclusion is inescapable: they cannot be made of rock and metal. The only way to construct such gigantic planets with such low densities is to make them predominantly of the two lightest elements: hydrogen and helium. Thus the outer planets are fundamentally different from the inner ones. In terms of their composition, they are more closely related to the stars than to the rocky bodies in our immediate neighborhood.

Finally, there is Pluto. This planet is about the size of our Moon, but its much lower density (1.7 vs. 3.3 g/cm^3) suggests that about half of Pluto's mass is probably water ice. Therefore, it resembles one of the large icy satellites of the giant planets more closely than it resembles any of the inner or outer planets. It may also be thought of as a giant comet nucleus, since they are also predominantly icy objects.

Satellites, Asteroids, and Comets

Among the satellites of the outer planets, we find an interesting parallel to the solar system itself when we look at the densities of the four moons of Jupiter discovered by Galileo. In order of increasing distance from the planet, the names and densities of these objects are Io (3.3), Europa (3.0), Ganymede (1.9), and Callisto (1.9). Just as in the case of the planets, the satellites exhibit decreasing densities with increasing distance from the center of their system. Evidently, proximity to Jupiter has a comparable effect to proximity to the Sun. You might also ask yourself what such low densities for Ganymede and Callisto imply about the composition of these objects. They are far too small to be made of hydrogen and helium.

The satellites of Saturn and Uranus also have densities below 2.0 g/cm^3. In these systems there is no apparent variation of density with distance, indicating that Saturn and Uranus had very different histories than Jupiter. Could this difference also account for Saturn's magnificent system of rings? We shall return to this question in Chapters 11 and 15.

Our information about the densities of asteroids and comets is much less secure. Nevertheless, it is clear from the evidence we do have that the asteroids are predominantly composed of rock, while the comet nuclei are mainly made of ice.

———————————◇———————————

Summary

Before studying the detailed nature of the individual planets as other worlds, it is useful to look up into the sky and try to reconstruct the evolution of ideas about the solar system from the period before spaceflight, or even before the invention of the telescope. The most basic properties of the sky are its apparent daily and seasonal motions. The stars remain in fixed patterns as they rise and set, but the more complicated motions of the Sun, Moon, and planets challenged the ingenuity of the ancients to find an adequate interpretation. This was not just an intellectual challenge, but a practical one as well. A working knowledge of the seasons, in particular, was a requirement for the successful development of agricultural society. Two thousand years ago, the Greco-Roman world had developed a sophisticated geocentric view of the heavens and their relationships to human activity. They understood time-keeping and seasons, the phases of the Moon and the causes of eclipses. Their view prevailed in the western world until about 300 years ago.

The modern view of the planetary system was developed between the Renaissance and the eighteenth century, primarily by five extraordinary scientists. Copernicus (1473–1543) devised a heliocentric theory of the solar system, which he advocated on grounds of simplicity and aesthetic appeal. The heliocentric theory was verified by the first telescopic observations of the planets, carried out in 1610 by Galileo (1564–1642), the founder of modern experimental science. Meanwhile, Kepler (1571–1630), using a remarkable body of pre-telescopic measurements of planetary positions made by Tycho (1546–1601), placed the heliocentric theory on a sound mathematical basis by developing his three laws of planetary motion, which still form the foundation for the description of the orbits of planets, satellites, comets, and other solar system bodies. Kepler's laws are purely descriptive; they tell how the planets move, but not why. The unifying theories of celestial mechanics were the product of Newton (1642–1727), who established the physical laws that govern the motion of all bodies, developed the theory of gravitation (a force that acted equally on falling apples and celestial objects), and invented the mathematics required to calculate trajectories and orbits.

Modern observations leave no doubt that the heliocentric model for the solar system is correct. The true distances, motions, and sizes of the planets have been determined, and these properties easily account for the observations so puzzling to our ancestors. Furthermore, we can use this information to begin studying the planets as worlds. The simple concept of density by itself reveals three classes of bodies of very different composition: the inner planets, the outer planets, and Pluto. In subsequent chapters we will focus on the physical and chemical nature of these bodies, rather than on their motions.

Key Terms

astronomical unit	inclination
constellation	last quarter moon
eccentricity	new moon
eclipse of the Moon	quantitative
eclipse of the Sun	retrograde motion
ecliptic	retrograde rotation
epicycle	summer solstice
equinox	Titius-Bode rule
first quarter moon	vernal equinox
full moon	winter solstice
geocentric	zodiac
heliocentric	

mass,
density
Kepler's law
Newton's laws

CHAPTER TWO

Getting to Know Our Neighbors

2.1 Building Blocks: Atoms, Isotopes, Compounds, Minerals

Planets are places, made of rock and metal, ice and various gases. Some are similar to the Earth; others are very different. We begin with a description of the basic components of matter itself and then proceed to a discussion of the materials that form the planets. We want to know how the composition varies from one object to another, and how closely individual planets represent the mixture from which the solar system was originally made.

Atoms and Isotopes: The Origin and Structure of Matter

The matter that makes up the Sun, Moon, and planets consists of pure substances called elements and of chemical compounds of these elements. The creation of the elements is a natural process in stars, resulting from the nuclear reactions that also generate stellar energy. For ordinary stars like the Sun, the dominant nuclear reaction is the fusion of hydrogen into helium. Every second, 600 million tons of hydrogen are destroyed in the core of the Sun, and a nearly equal quantity of helium is created. In some older stars, other more complex reactions take place, forming elements heavier than helium. In

the catastrophic explosions that terminate the lives of massive stars, these reactions can create all of the ninety-two elements. Thus stars do what the medieval alchemist could not: they create new matter from old, converting one element into another.

The age of our galaxy is estimated to be about 12 billion years. In contrast, the Sun and planets are relatively recent arrivals, having formed only 4.5 billion years ago. Therefore, several generations of element building took place before our system formed. As a result, about one percent of the matter from which the planetary system formed was made of elements heavier than hydrogen and helium. Table 2.1 lists the proportions of elements that we believe were available to form the building blocks of the Sun and planets.

The details of element formation are the subject for another book. In order to understand the composition of the planetary system, however, we need to know something about these fundamental constituents of matter. The ninety-two naturally occurring elements are composed of atoms that consist of nuclei surrounded by orbiting electrons. The nucleus of an atom contains almost all of the mass. It is made up of particles of nearly equal mass called protons, which have a positive charge, and neutrons, which have no charge at all. An electron carries negative charge and has 1/1836 the mass of a proton. Each atom has as many electrons as it has protons, so

TABLE 2.1 Cosmic abundances of the major elements

Element	Symbol	Atomic Number	Number of Atoms per Million Hydrogen Atoms
Hydrogen	H	1	1,000,000
Helium	He	2	68,000
Carbon	C	6	420
Nitrogen	N	7	87
Oxygen	O	8	690
Neon	Ne	10	98
Sodium	Na	11	2
Magnesium	Mg	12	40
Aluminum	Al	13	3
Silicon	Si	14	38
Sulfur	S	16	19
Argon	Ar	18	4
Calcium	Ca	20	2
Iron	Fe	26	34
Nickel	Ni	28	2

the atom itself exhibits no net charge in its neutral state. If an electron is removed, however, the atom is said to be ionized and now exhibits a positive charge. The simplest atom is hydrogen with a single proton as its nucleus and one electron in orbit about it.

Imagine adding a neutron to the nucleus of a hydrogen atom (Fig. 2.1). The mass of the atom is changed by a factor of two, but since the charge of the nucleus is the same, the same single electron is all this new atom needs to remain electrically neutral. This new atom is an **isotope** of hydrogen, that is, a form of an element that differs from others only in the number of neutrons in the nucleus. The isotope of hydrogen with one neutron in the nucleus is called **deuterium,** abbreviated D. Since deuterium still has just one electron, its chemical properties are virtually identical with those of ordinary hydrogen (H), despite its greater mass. Thus deuterium can combine with oxygen to form water, but we now give it the symbol D_2O instead of the familiar H_2O. The surprise comes when an ice cube of D_2O is placed in ordinary H_2O. It sinks! The greater mass of the nucleus of deuterium atoms leads to an increased density for D_2O. More mass is packed into the same volume compared with ordinary H_2O. It is for this reason that D_2O is often called "heavy water."

This is not just an idle thought experiment, since deuterium is found abundantly in nature. Other elements also exhibit more than one isotope. Ordinary carbon, for example, that exists in pencil lead, diamonds, coal, and humans, also has two stable isotopes, one of which, with six neutrons and six protons, is eighty-nine times as abundant as the other, which has seven neutrons and six protons. The different isotopes are identified by the masses of their nuclei on a scale in which the proton and neutron each have a mass of 1.0. Thus the two stable carbon atoms are designated C-12 and C-13. Oxygen has three stable isotopes, O-16, O-17, and O-18.

Compounds

Most of the matter we encounter is not in the form of pure substances. Water, carbon dioxide, alcohol, and quartz are all examples of **compounds,** substances that are composed of more than one element. Just as the smallest unit of an element that still preserves the element's chemical identity is called an atom, the smallest unit of a compound is called a **molecule.** A molecule is formed from the atoms of the various elements making up the compound. Collisions between atoms can form molecules, and ultraviolet light can break molecules apart. Like atoms, molecules can be ionized by losing one or more electrons. The tails of some comets, for example, contain ionized molecules of water, written H_2O^+. The gases around the comet's head contain H and OH, fragments of H_2O molecules that have been broken apart by ultraviolet light.

Both compounds and elements can change their state, becoming gases, liquids, or solids depending on temperature and pressure. Of the three possibilities, the liquid state is the most rare, since it usually requires a rather restricted range of temperatures. Water is unusual in that it remains a liquid over a range of 100 Celsius degrees. Ammonia, for example, is liquid only from -78 C to -33 C,† or less than half the range of water.

Rocks and Minerals

Some of the most important solids we will encounter in our exploration of the solar system are rocks. Rocks are compounds formed with the element silicon. Living on the Earth, we tend to

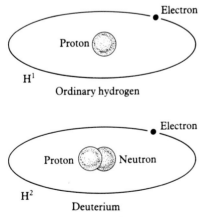

FIGURE 2.1 Ordinary hydrogen atoms consist of one proton as the nucleus and one electron in orbit around it. Adding a neutron to the nucleus changes the mass of the atom but not its electrical charge. The new isotope is called deuterium (D).

Electron

Proton

H^1
Ordinary hydrogen

Electron

Proton   Neutron

H^2
Deuterium

†Throughout this text, we will use the convention of C to signify degrees Celsius and K to signify degrees Kelvin.

take rocks for granted. As we have seen, however, hydrogen and helium are far more abundant than silicon in the universe at large, so rocks are actually rather rare. Even if we restrict ourselves to solid compounds, we can see that water ice must be far more common than rocks, since oxygen is eighteen times more abundant than silicon in the universe (Table 2.1). This means that oxygen could combine with every silicon atom in the universe to make quartz, SiO_2, and there would still be plenty of oxygen left over to make an even larger quantity of water ice. When we explore the outer solar system where ice is stable, we will find that ice is indeed the dominant solid. Meanwhile, let us consider the rocks.

The rocks we see around us in the familiar landscapes of our planet are composed of assemblages of compounds or elements called **minerals.** The principal difference between a rock and a mineral is in their homogeneity. A mineral is composed of a single substance, whereas a rock may be made up of several different minerals. Only a few elements occur in nature in a pure state, so that most minerals are compounds. Gold and silver are probably the most famous examples of single element minerals in the United States. Free sulfur, copper, and carbon in the form of graphite and diamonds are also rare, and as such are economically important. Minerals composed of a single compound are much more common. Quartz (SiO_2), hematite (Fe_2O_3), iron pyrite ("fool's gold," FeS_2), and calcite ($CaCO_3$) are all relatively familiar. These four minerals are examples of **silicates,** oxides, sulfides, and carbonates, respectively, the four most common types of rock-forming minerals. Rocks can be classified in terms of the minerals they contain. However, it is convenient instead to divide them into three large categories on the basis of their origins: igneous, sedimentary, and metamorphic.

The **igneous rocks** are those that have formed directly by cooling from a molten state.

A rock picked up from the slope of a volcano in Hawaii is an obvious example (Fig. 2.2). This is a representative of the family of igneous rocks called basalts, which we shall encounter repeatedly during our studies of the inner planets (Plate 2). Igneous rocks make up roughly two thirds of the Earth's crust.

Sedimentary rocks are made up of fragments of other rocks that are cemented together. On Earth, the fragments are produced by various weathering (erosional) processes that break up the parent rocks. The most effective weathering processes on our planet involve liquid water, with sandstone being a well known example of the end result. Fragmentation can also be produced when one rock bangs into another one, creating a different type of sedimentary rock on the lunar surface known as a breccia.

Finally, the **metamorphic rocks** are produced from either igneous or sedimentary rocks that have been buried far below the Earth's surface, modified by the high pressures and temperatures they encountered there, and then returned to the surface. This process of burial and return is part of the great cyclical movement of the Earth's continental plates. The marble used to make statues (Fig. 1.18) is a metamorphic rock. At the present time, we are not certain that any other planet in our solar system exhibits this phenomenon, so Earth may be the only place where metamorphic rocks are present on the surface.

Primitive Rocks and the Origin of the Earth

In this review of rocks and minerals, we have actually been talking about reprocessed material, since all the rocks on Earth were melted after the planet formed. Are there more ancient rocks in the solar system? There certainly are. We can establish a fourth category called **primitive rocks** which have never melted and which have

FIGURE 2.2 a) This ropy pahoehoe lava on the floor of Hawaii's Kilauea volcano was molten rock just a few weeks ago. The crack in the left foreground is about 2 cm wide. b) Astronomer Dale Cruikshank holding a 2-year-old rock (Hawaiian lava) in his left hand, and a 4.5-billion-year-old rock (a primitive meteorite) in his right hand.

(a)

(b)

been affected only moderately by chemical and physical processes since the solar system began. We expect such rocks to have the same abundances of the non-volatile elements (elements that are solid rather than gaseous at low temperatures) that exist in the Sun or the giant planets. Examples of such rocks are found among the meteorites. There are also primitive rocks lodged in the ices of the comets, and some asteroids and many of the smaller planetary satellites must also be composed of this material.

The reason we don't find primitive rocks on Earth has to do with the way in which a planet forms and evolves with time. Current theories for the formation of the inner planets start with solid material. This material is indeed primitive in the sense described above, and the meteorites are remnants of it. However, as this dust and rocky debris come together to form a planet, the solid body that develops begins to heat up. This heating results in part from the energy released by the impacting material that is bombarding the forming planet and causing it to grow, and partly from radioactivity deep inside the body. A large planet can retain heat better than a small one. The generation of heat usually depends on the volume of the planet, which is proportional to the cube of the radius (remember Volume = $4/3\pi$ (radius)^3). But the planet can only lose heat through its surface, and the surface area is proportional to the square of the radius (Area = 4π (radius)^2). Thus since the generation of heat increases faster than its loss, a large planet will tend to grow warmer than a small one.

If the internal temperatures rise enough, the rocks melt and the central part of the planet becomes a liquid. At this stage, the denser materials are free to migrate to the center, and the lighter materials rise to the top. This process is known as **differentiation,** and it leads to the development of differences in internal composition. All of the inner planets have undergone differentiation, as we would expect. An interesting puzzle, however, is posed by the presence of as-

teroids that have also undergone differentiation. What processes have heated these tiny bodies? We shall return to this problem in Chapters 3 and 4. Meanwhile, we can see why our own planet does not have any primitive rocks. In fact, we even have trouble finding *old* rocks on Earth, since an active geology and erosion have effectively eliminated the first billion years of our planet's geological record. To probe this ancient history, we must go to other places, such as the Moon, Mars, and the asteroids, where less alteration of primordial conditions has taken place.

◆

2.2 Four Types of Matter: Plasma, Gas, Ice, and Rock

A Simple View of Planetary Composition

There was a school of ancient Greek philosophers which thought matter could be divided into just four categories: air, water, earth, and fire. As we saw in Section 2.1, matter is actually composed of atoms of pure substances called elements, of which there are many more than four. Yet this simple Greek view is a useful tool for organizing the objects in the solar system. For our purposes, it is helpful to change the categories somewhat. We shall consider the following four forms: plasma, gas, ice, and rock. Using these four components, we can then classify the various members of the solar system according to the relative proportions of these forms of matter that they contain.

Plasma: Electrically Charged Matter

Of the four, the least familiar to us on Earth is **plasma.** In the sense we are using it here, a plasma is simply a gas that has been at least partially ionized. That means that electrons carrying negative charge have been stripped from some of the atoms and molecules in the gas, leaving them with a positive charge. So the plasma consists of some neutral gas plus ions and electrons. Unlike a purely neutral gas, a plasma interacts with electric and magnetic fields and with other plasmas, all of which can control its motions.

For example, the gas in the tail of a comet is a plasma, and by watching the behavior of knots of material in comet tails, Ludwig Biermann was able to deduce the existence of a **solar wind** in the early 1950s, long before spacecraft were

available to detect it (Fig. 2.3). The solar wind itself is a completely ionized plasma that continuously streams out from the Sun. The outermost layer of the solar atmosphere is so hot that the energies of ions and electrons there allow them to escape from the solar gravitational field. Because stars are mainly plasma, this may well be the dominant state of matter in the universe. Aside from the Sun itself, however, most of the mass in the solar system is in a neutral (un-ion-

FIGURE 2.3 a) A photograph of Comet Halley, when it visited the inner solar system in 1986, shows a straight plasma tail with considerable structure. The tail is glowing with light from ionized carbon monoxide gas molecules that are being blown away by the solar wind. b) A painting of Donati's Comet as it appeared to the naked eye on 4 October 1858, over La Cité in Paris. Note the two straight plasma tails and the curved dust tail. The bright star near the comet's head is Arcturus in the constellation Bootes. Evidently the Parisian night sky was much darker 130 years ago!

(a)

(b)

ized) state. We will be considering plasmas again when we discuss the comets and the planetary magnetospheres — those domains around the planets where the behavior of charged particles is determined by planetary magnetic fields. But the overwhelming majority of the mass that we discuss in this book is in the other three states we have defined: gas, ice, or rock.

Gas: The Most Primitive Matter

We all know what gas is, and we know that planetary atmospheres are composed of gas. We have also noted in Section 1.6 that the giant planets Jupiter and Saturn are predominantly composed of the gases hydrogen and helium. These planets preserve the most **primitive** material in the planetary system, that is, the material that has been least modified chemically since the formation of the solar system.

Most scientists believe that the solar system began as a cloud of neutral gas and dust, commonly called the primordial **solar nebula.** The abundances of the elements we find in the atmosphere of our central star are the same as those that existed in the solar nebula (Table 2.1). This follows directly from the fact that when the solar nebula collapsed to form the solar system, most of the material gathered to form the Sun. Deep in the solar interior, nuclear reactions are obviously changing this primordial composition, but in the outer atmosphere that we observe, the Sun can serve as a standard for the composition of the matter from which the entire solar system formed. The one notable exception to using the Sun as a standard is provided by deuterium, the heavy isotope of hydrogen. Deuterium is converted to helium by nuclear fusion at lower temperatures than hydrogen and is therefore depleted in middle-aged stars like the Sun.

Are there any planets whose composition resembles the Sun's? Jupiter and Saturn are the best candidates, but as we shall see in Chapter 11, even these planets seem to be slightly enriched in heavy elements as compared to the solar mixture.

Ice: Frozen Volatiles

No other planets come as close as Jupiter and Saturn to having a composition like that of the Sun. The next closest by a wide margin are Uranus and Neptune. These two giants have atmospheres that are still dominated by hydrogen and helium, but include a higher proportion of compounds containing heavy elements. Their interiors consist of dense, ice-enriched mantles and cores that account for much of the total mass of each planet. However, the atmospheric hydrogen and helium are in solar proportion, at least on Uranus.

Water ice and other **volatiles** condensed in the cooler parts of the solar nebula. If it was cold enough for ice, then rock and some metal should also be present, since we are considering a solar mixture from which we have simply stripped away some of the lightest gases, the hydrogen and helium. Since there are more molecules of water than of any other compound, the total mass of water or ice is about equal to the mass of the (much denser) rocky material. We think that the dense cores of the outer planets consist of a mixture of rock and ice, although at their very high pressures and temperatures they do not resemble the kinds of rock and ice we normally encounter. Probably these cores are actually liquid.

The next step away from the primordial composition is to eliminate the hydrogen and helium entirely. This would leave us with ice and rock. If such bodies are large enough, they have the potential to retain atmospheres of heavier gases. In our solar system, the large icy satellites of Jupiter, Saturn, and Neptune satisfy this description, although only Triton and Titan have significant atmospheres.

Rock: Silicates and Other Solid Material

If we began with objects composed of mixtures of ice and rock and subsequently heated them, we can imagine the ice evaporating to leave behind rocky objects. (Condensing matter from the solar nebula in a warm environment will have the same effect.) If they are only heated moderately, they will still contain large amounts of condensed substances such as various compounds of carbon. The satellites of Mars and some of the asteroids correspond reasonably well to this description. Planetary scientists often call these primitive bodies, even though they are less primitive than the ice/rock objects, and much less primitive than Jupiter or Saturn.

Heating the residue further, we will drive off the carbon compounds and other substances that evaporate at low temperatures, until we eventually produce a dry, bare rock. At still higher temperatures, the rock itself undergoes chemical and structural transformations. Condensation at *high* temperatures will produce the same result, and Mercury is probably an example of this process.

Density, Composition, and Evolution

We have described a progression from gas to rock, which corresponds to an increase in the amount of heating or other processing that original, primitive material has undergone. This progression is also correlated with the composition of the gases we find around those bodies capable of holding atmospheres. In the cold, outer reaches of the solar system, atmospheres are rich in hydrogen and/or hydrogen compounds. In the warm, inner solar system, hydrogen is no longer present, and we find oxygen compounds in planetary atmospheres. The reasons for this difference are described in the next section.

If you recall the discussion of density in Section 1.6, you will expect a general progression from low to high density as we move from primitive to more evolved objects. This is a natural consequence of losing the abundant light elements, hydrogen and helium, and the most abundant compound, water ice. Thus a good determination of an object's density can tell us a great deal, before we begin the much more difficult work required to make a detailed analysis of chemical composition. These various trends are illustrated in Table 2.2, which provides a useful one-page introduction to solar system chemistry and planetary evolution.

2.3 Planetary Atmospheres

The Origins of Atmospheres

An important characteristic that distinguishes one body from another in Table 2.2 is the presence or absence of an atmosphere. Earth is eleven times smaller than Jupiter and over four times as dense, with a totally different composition. Yet both bodies have atmospheres, although the composition of these atmospheres is very different. On the other hand, Jupiter's satellite Ganymede has no atmosphere while Saturn's slightly smaller moon Titan has an atmosphere denser than Earth's. What accounts for these differences?

There are two basic ways in which a planet can obtain an atmosphere: it can form with one (a primordial or **captured atmosphere**), or it can produce one from the material of which it is made (secondary or **outgassed atmosphere**). Often when there are two explanations, a third can be created by combining the first two. In fact, capture plus outgassing appears to be the most likely process for the formation of the atmospheres of the giant planets. They are a blend of captured solar nebula gases and gases produced from the planetary cores. The smaller inner planets and the satellites appear to have produced the gases we find around them today entirely by outgassing.

TABLE 2.2 Planets and satellites: overview of composition

Object	Distance from Sun (AU)	Density (g/cm³)	Composition
Mercury	0.4	5.4	iron, nickel, silicates
Venus	0.7	5.3	silicates, iron, nickel
Earth	1.0	5.5	silicates, iron, nickel
Moon	1.0	3.3	silicates
Mars	1.4	3.9	silicates, iron, sulfur
Jupiter	5.2	1.3	hydrogen, helium
Callisto	5.2	1.9	water ice, silicates
Ganymede	5.2	1.9	water ice, silicates
Europa	5.2	3.0	silicates, water ice
Io	5.2	3.4	silicates
Saturn	9.6	0.7	hydrogen, helium
Titan	9.6	1.8	water ice, silicates
Uranus	19.2	1.2	ices, hydrogen, helium
Neptune	30.1	1.7	ices, hydrogen, helium
Triton	30.1	?	water ice, silicates
Pluto	39.4	1.7	water ice, silicates

Maintaining an Atmosphere

In order for a planet to hold on to an atmosphere over the 4.5-billion-year life span of the solar system, the molecules in that atmosphere must not move fast enough to escape from the planet's gravitational field. In other words, their speed must be less than what we call the **escape velocity,** a concept we return to in Section 2.6. Otherwise they will soar off into space from the "top" of the atmosphere, eventually leaving a denuded planet behind.

The top of the atmosphere is defined for this purpose as the layer that is so tenuous that a molecule moving in an upward direction will not encounter any other molecules. It is free to leave the planet if it is moving fast enough. Scientists have called this layer of the atmosphere the exosphere, since the gases here can exit from the planet if they have escape velocity. Physics tells us that the velocities of molecules in a gas are determined by the temperature of the gas and its composition. If the temperature is high enough and/or the gas is sufficiently light, individual atoms and molecules in the exosphere may reach escape velocity and be lost. For a planet to keep an atmosphere, it should have a large mass and thus a high escape velocity. It should also be cold, so that velocities of the gas molecules will be low (Fig. 2.4). These two conditions are well met in the outer solar system where we find the giant planets. These bodies have such high escape velocities (typically tens of kilometers per second) that they can retain thick atmospheres of the light gases hydrogen and helium.

Saturn's satellite Titan is not massive enough to retain hydrogen, but it is both sufficiently massive and cold to keep an atmosphere of heavier gases, such as the nitrogen and methane that we find there. Similar conditions apply

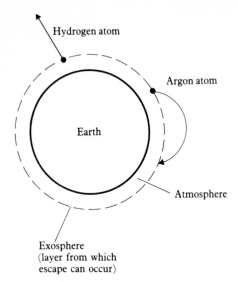

FIGURE 2.4 The light hydrogen atoms (atomic mass = 1) move much faster than the heavier argon atoms (atomic mass = 40) at the same temperature. Thus hydrogen can escape from the Earth, while argon cannot.

to Ganymede, so the reason this massive satellite of Jupiter does not have an atmosphere must be found elsewhere. Our Moon and the planet Mercury are both too warm and insufficiently massive to maintain atmospheres.

Why Atmospheres Are Different

There are, therefore, several reasons for the diversity of atmospheres that now exist around various bodies in the solar system, when they exist at all. We will come back to this discussion as we consider each planet individually, but let's briefly consider the implications of what we have learned.

Large planets form with sufficient mass to capture hydrogen and helium (and everything else) from the solar nebula, which is why they exhibit hydrogen-rich atmospheres today. Small planets must produce their own atmospheres, so the composition of these atmospheres will de-

pend on the materials that make up the planet and on the planet's geological and chemical evolution. Thus we find atmospheres on Mars and Venus that are dominated by carbon dioxide. Earth should exhibit the same thing. The fact that it doesn't is a result of the presence of liquid water and life on our planet. Why is methane much more abundant in Titan's atmosphere than carbon dioxide? We will return to this question when we discuss this fascinating satellite in Chapter 13.

2.4 Radiation and Spectroscopy: Studying Matter from a Distance

We have reviewed several basic characteristics of the planets and some of the properties of the matter that composes them. We now want to tie these two themes together by asking what we can learn about these distant worlds by studying the light, heat, and radio emissions that they send us across the vast emptiness of space.

Remote Sensing

The process of investigating distant objects by the analysis of their radiation is often referred to as **remote sensing,** to distinguish it from in situ (or "on site") studies. Astronomy is almost entirely a remote sensing science, although astronomers rarely use the term. Today, we study the members of the planetary system by a combination of astronomical studies from Earth, remote sensing carried out from flyby or orbiting spacecraft, and direct studies from entry probes and landers. Most planetary missions do not include landers or atmospheric probes, however. These sophisticated devices are very expensive and require a rather detailed knowledge of the target planet's environment before they can be successfully deployed. At the time of this writing, the only objects on which our probes have landed are the Moon, Mars, and Venus; thus most of the material in this book is based on remote sensing studies of some type.

Remote sensing can be considered as the analysis of **electromagnetic radiation.** The light illuminating this page as you read it is an example of electromagnetic radiation; radio and television broadcasts are others. This radiation may be generally defined as the propagation of energy through space by varying electric and magnetic fields. It can be produced and absorbed by interactions of these fields with atoms and molecules. Electromagnetic radiation can be understood in terms of a wave-like motion of electric and magnetic fields, with each kind of radiation

having its own wavelength. But electromagnetic radiation also behaves as if it were made up of particles called **photons.** These particles have no mass, but they do carry energy. The energy of an individual photon is inversely proportional to its wavelength, so long waves correspond to low energies and short waves to high energies.

The Spectrum and Spectroscopy

The **electromagnetic spectrum** (Fig. 2.5) is a way of describing the energy range of electromagnetic radiation. It extends from radiation

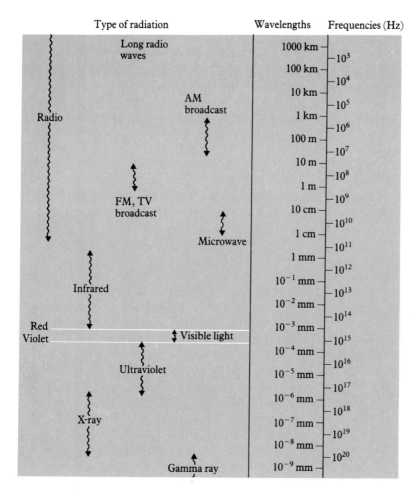

FIGURE 2.5 The electromagnetic spectrum extends from gamma rays with wavelengths shorter than one billionth of a centimeter to radio waves with wavelengths measured in millions of centimeters. All of this radiation travels through space at the speed of light. Only certain wavelengths are able to pass through a planetary atmosphere (see Fig. 2.7).

with very short wavelengths (gamma rays at 10^{-10} cm, X-rays at 10^{-8} cm) through the so-called visible region of the spectrum to which our eyes are sensitive (.35 $\times$ 10^{-4} cm to .76 $\times$ 10^{-4} cm) and on through the infrared to the radio region, where wavelengths are measured in meters and kilometers. This can also be thought of as a progression from very high energy (gamma rays) to very low energy (radio waves). All of this radiation travels at the velocity of light, $c = 3 \times 10^8$ m/s.

The primary tool for the analysis of electromagnetic radiation is **spectroscopy.** Everyone is familiar with rainbows, the beautiful array of colors created when sunlight passes through a mist of water droplets. The yellow-white light of the Sun is suddenly split apart into its component colors, which are spread out for us to admire. This is a very simple spectrum, in which radiation is actually sorted by energy, from violet to red. We can achieve the same effect in our laboratories by using a prism or a diffraction grating instead of raindrops. Adding a narrow opening to limit the overlapping of the radiation, some lenses to make images, and a detector to record them, we have constructed a **spectrograph** (Fig. 2.6a). Now we can spread out sunlight and examine it in great detail.

Formation of Spectral Lines

What we find is that amidst the beautiful display of colors, which faithfully reproduces the range from violet to red exhibited by rainbows, there are large numbers of dark lines (Fig. 2.6b). These lines represent wavelengths where energy is being removed from sunlight by atoms or molecules in the Sun's atmosphere. We can think of sunlight as basically being radiant energy produced deep in the Sun's interior by nuclear reactions. This radiation gradually works its way to the Sun's outer layers, from which it escapes into space. During this last stage, it encounters the cooler outer envelope of the Sun, and it is the effects of absorption by atoms and molecules in this solar atmosphere that we see as lines in the spectrum.

Why are there discrete spectral lines instead of a general decrease in the intensity of the light, as we experience on a cloudy day on Earth? In other words, why do atoms and molecules of gas absorb only specific energies or wavelengths, unlike cloud droplets, which absorb all colors? The reason is found in the quantum theory of matter. This powerful theory, which is at the foundation of most modern physics, explains mathematically the discrete nature of the absorption and emission of energy by atoms and molecules.

A violin string produces an approximate analogy to the behavior of an atom of gas. If you pluck the string, you get only one note since the string vibrates at only one frequency. An atom is a bit more complex, but it also operates at only a few specific frequencies or wavelengths, which can be calculated from quantum theory. You can, however, change the violin note (frequency) by changing the length of the string. This would correspond to changing the internal configuration of an atom or molecule, for instance by shifting an electron into a different orbit. In such a changed state, it interacts with a different set of wavelengths or frequencies of electromagnetic radiation. Atoms and molecules in the solid or liquid state also interact with electromagnetic radiation to produce a characteristic spectral signature. The atomic configuration is generally different for a solid or liquid, however, so the spectrum is distinct from that of the gaseous form. Because the solid or liquid is less finely tuned, the spectral features are often broad and fuzzy. To pursue our musical analogy, these forms respond like a bass drum rather than a violin. Therefore, spectroscopy is a less precise tool for analyzing solids and liquids than it is for gases, and consequently remote sensing has told us less about planetary surfaces than about planetary atmospheres.

Since the spectrograph that recorded the spectrum in Figure 2.6b was located on the

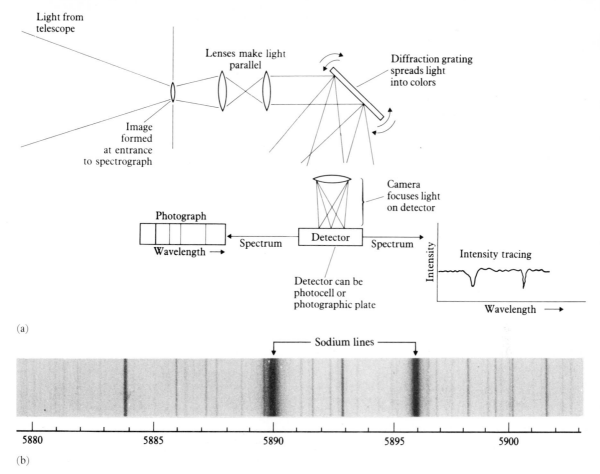

FIGURE 2.6 a) A spectrograph spreads out the light from a source according to its wavelengths, which we perceive as colors in visible light. b) Bright or dark lines in the spectrum reveal the presence of gases in the source (or between the source and the observer) that absorb or emit specific wavelengths of light. This portion of the visible region of the Sun's spectrum shows dark lines caused by atoms in the outer atmosphere of the Sun. The two prominent absorptions at 5890 Å and 5895 Å are caused by atoms of sodium. (One angstrom (Å) = 10^{-8} cm.)

Earth's surface, we might expect to find absorption lines produced by gases in our planet's atmosphere also. After all, the sunlight has to pass through the Earth's atmosphere to reach the spectrograph, and along the way the photons are going to encounter a lot of atoms and molecules. In fact, the abundant nitrogen in our atmosphere does not absorb visible light, but oxygen and water vapor do.

Long-wave and Short-wave Spectra

Leaving the visible spectrum and considering shorter wavelengths, we find that we cannot detect sunlight at the Earth's surface in the ultraviolet region. This is not because the Sun does not radiate here, but because of a layer of **ozone** in the Earth's atmosphere. Ozone has the chemical formula O_3. It is formed from oxygen molecules in the Earth's upper atmosphere and shields the Earth's surface from high energy ultraviolet photons. Life on Earth benefits from the ozone, since this high-energy radiation is capable of destroying many of the molecules that make up living organisms.

Moving past the red end of the visible region of the spectrum, we encounter the infrared, first detected by Herschel, best known for his discovery of Uranus. Here again we encounter absorption lines. It turns out that water vapor and carbon dioxide have particularly strong absorptions in this region of the spectrum. Thus despite their relatively low abundances in the Earth's atmosphere, these two gases play a dominant role in absorbing infrared radiation. Unlike the short wavelength end of the spectrum, however, some infrared light can penetrate our planet's atmosphere and reach the Earth's surface. This occurs in regions of the spectrum that lie between the strong absorption bands of water vapor and carbon dioxide. Such regions of transparency are called **atmospheric windows.** The atmosphere acts as a kind of color filter, letting some wavelengths (colors) through and absorbing others (Fig. 2.7).

FIGURE 2.7 The Earth's atmosphere absorbs radiation at most wavelengths at various altitudes above our planet's surface. Only visible light, and some infrared and radio waves, can penetrate the atmosphere and reach the surface. These regions of the spectrum in which radiation can pass through the atmosphere are called atmospheric windows. The same situation (with variations depending on atmospheric mass and composition) will exist on all other planets with atmospheres.

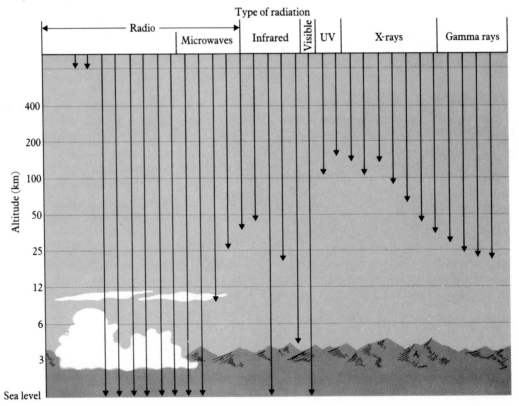

This pattern of windows and absorptions continues out to longer and longer wavelengths. The atmospheric absorption grows progressively stronger until it suddenly stops at a wavelength of about one millimeter. The atmosphere is transparent to radiation of longer wavelengths, providing another clear window on the universe. The long-wave limit on this window is at about 30 cm. This is the region of microwaves, radar, television, and FM broadcasts, which are often lumped together as "radio." Not only can we detect emissions from planets and stars in this region, but they can detect our signals also. It is arresting to realize that a Martian (if there were any!) could be puzzling over the significance of our military radar defense system, or trying to interpret rock music or professional football.

Spectra of Other Planets

Now that we have some knowledge about the electromagnetic spectrum and spectroscopy, we can attempt to use this information to study the planets. Orbiting the Sun, the planets shine by reflected light. Just as sunlight is absorbed at certain wavelengths by the gases in our planet's atmosphere, we can expect the same thing to occur in the atmospheres of the other planets. By examining the reflected sunlight, we can then hope to discover new absorption lines that will tell us what gases are present on that planet (Fig. 2.8).

Not all the sunlight that strikes the planets is reflected, however. Our Moon, for example, reflects only 11% of the sunshine that illuminates

FIGURE 2.8 Sunlight, whose spectrum (*A*) contains absorptions (*S*) from gases in the Sun's atmosphere, penetrates the atmosphere of another planet, where some of it is absorbed by planetary gases, producing new lines (*P*) in the spectrum of the planet (*B*). Reflected by clouds or the planet's surface, the light continues its journey to Earth, where some of it may be absorbed by gases in our planet's atmosphere, producing new lines (*E*) before it reaches a telescope and spectrograph to form the observed spectrum (*C*).

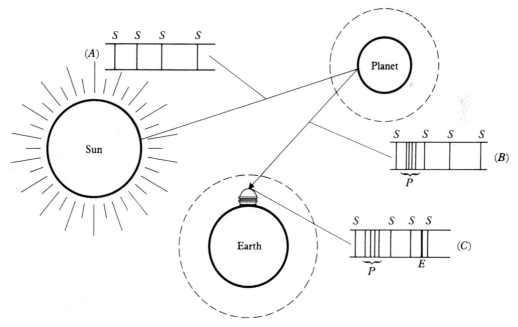

it, despite the fact that it seems so bright in our night skies. This low reflectivity is characteristic of an exposed, rocky surface. In contrast, cloud-covered Venus sends back 75% of the incident sunlight, while icy Enceladus reflects nearly 100%. What happens to the radiation that is not reflected? It is absorbed by the planet, causing it to warm up. Thus each planet, satellite, asteroid, etc., assumes a temperature that will depend on its distance from the Sun, its reflectivity, and what kind of an atmosphere it has, if any. If there is a large internal source of heat, that will also be important, as we'll see in the case of some of the giant planets.

When the surface of a planet absorbs sunlight and becomes warm it begins to radiate energy at infrared and radio wavelengths. Now imagine you were studying the spectrum of such a planet. After looking at the visible region, where you would be analyzing reflected sunlight, you might move on to the infrared. Here you detect thermal radiation from the planet itself. Since there are no solar lines in the spectrum to confuse you, the absorptions you find in the planet's infrared spectrum are caused by gases in its atmosphere or in the atmosphere of the Earth.

Spectroscopy provides a powerful tool for studying planets from a distance. By analyzing the light they reflect as well as the radiation they emit, we can learn about the composition of their atmospheres and the nature of their surfaces. In fact, we can do even more than this. The strengths of absorption lines can tell us the amount of the absorbing gas. By observing many lines of the same gas, we can often derive an average atmospheric pressure and temperature. Comparing the intensities of thermal emission at several different wavelengths, it is possible to determine physical and chemical properties of the emitting surface. In short, there is a huge bag of tricks that can be played with planetary radiation, once we have enough of it to work with. To achieve that, we need telescopes.

2.5 Telescopes and Observatories

The Need for Large Instruments

The previous section describes some of the characteristics of a planet that astronomers can deduce from a study of its electromagnetic spectrum. To perform such studies with maximum effect, however, it is necessary to collect as much radiation from the planet as possible, to view the planet from all angles, and to be able to study small regions, not just the entire globe. There are two ways to do this: a) use a large collecting area that focuses all the radiation falling on it into an image that can then be examined (but this doesn't broaden the angle of view), or b) go closer to the object. The first approach employs a telescope; for the second a spacecraft is essential. In practice, the optical instruments on spacecraft also use small telescopes, thereby further improving their performance.

The reasons scientists need to make this effort to study the planets are quite simple. First, even the brightest planets don't send us very much light. Second, when a spectrograph spreads out the planetary radiation to form a spectrum, it is essentially sorting photons by wavelength or energy. Instead of simply detecting the planet by adding up all the photons in the visible spectrum, which is what our eyes do, the detector in the spectrograph is only sampling a few discrete energies. Unless a lot of radiation is collected before it is spread out in this way, there simply won't be enough photons to stimulate the detector and allow a spectrum to be recorded.

Telescopes

An optical telescope performs two critical functions for the astronomer: it collects light, and it forms a magnified image of the object being studied. The collection of light is accomplished

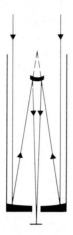

FIGURE 2.9 The Cassegrain focus of a modern reflecting telescope. Light from a planet (or star, etc.) is reflected by the large primary mirror to a smaller secondary that sends it back through a hole in the primary. Instruments to examine the light may then be hung on the back of the telescope (see Fig. 2.10).

posed of different lenses, different magnifications can be achieved. A good pair of binoculars is a small but sophisticated pair of telescopes. Typically each has a main lens 35 mm in diameter, collecting several hundred times more light than the naked eye. The magnification might be 6, in which case we would call this a pair of 6 × 35 binoculars. An amateur astronomer's telescope, such as are sold widely in camera stores, might typically have a main mirror diameter of 10 to 20 cm and magnifying powers from 20 to 100. Binoculars can just be held in the hand, but large astronomical telescopes require permanent mountings (Fig. 2.10). The motion of the telescope is computer controlled to permit precise

FIGURE 2.10 The 4-meter (157-inch) Mayall telescope of the Kitt Peak National Observatory. The telescope tube is mounted in a yoke and is in a vertical position in this picture. Note the figure on the platform at lower right for scale.

by the mirror or lens that forms the principal optical element of the telescope. This lens or mirror is the heart of the telescope, and its diameter or **aperture** is the measure of telescope power. Thus, for example, we speak of the 200-inch telescope on Palomar Mountain, or the 10-meter telescope being constructed on Mauna Kea.

All the light from a single source such as a planet that falls on the entire area of the lens or mirror is brought to a focus at a single point. When you look at such a planet without a telescope, your retina is detecting only the amount of energy that passes through the few square millimeters of area corresponding to the size of the lens in your eye. But gazing through the eyepiece of a 2-meter reflecting telescope, you have access to all of the light falling on the three million square millimeters of the primary mirror (Fig. 2.9). The eyepiece, in turn, is constructed to magnify further the image of the planet formed by the telescope. Using eyepieces com-

pointing and tracking. The telescope, its computer drive system, and all the auxiliary instrumentation are housed in an observatory. This is a building with a dome that has a slot in it through which the telescope can be pointed to view the heavens. The dome rotates to give the telescope complete access to the sky.

With appropriate modifications, these same considerations apply to radio telescopes. Once again a large collecting area is used to gather the incident radiation. In this case the collectors may be an array of antennas or a parabolic dish, shaped to bring the radiation to a focus. The long wavelengths and low energies of radio photons lead to a requirement for very large aperture antennas or even collections of antennas linked by computer to allow studies of just a small area in the sky (Fig. 2.11).

Seeking Better Observatory Sites

To take full advantage of a telescope's radiation-collecting ability, it is necessary to observe from a favorable location. If you have ever looked at a distant scene near the surface of the Earth on a hot day, you will have noticed that motions in the air through which you looked distorted the appearance of what you were trying to see. In extreme cases, layers of air at different temperatures can produce mirages. A good observatory site should minimize these distortions, and have as many cloudless nights as possible.

When you look at a star or a planet in the night sky, you are looking through the Earth's entire atmosphere, so the light reaching your eyes has passed through many different layers of air, at various temperatures and moving at vari-

FIGURE 2.11 The Very Large Array (VLA) of radio telescopes of the National Radio Astronomy Observatory near Socorro, New Mexico. Each of the 27 telescope antennas has a diameter of 25 meters (82 ft). They can be moved on railroad tracks laid out in the shape of "Y" to produce an effective aperture of nearly 35 km. Even though this aperture is not filled by the antennas, they are linked by a computer so the whole array functions as a single instrument.

ous speeds from various directions. The effect of all this is to make the stars appear to twinkle, when actually they are shining with a very steady light.

A telescope magnifies these effects, since it is collecting light over a much greater area than the eye, thus allowing even more random properties of the atmosphere to affect the image. That is why even the best pictures of planets taken from telescopes on Earth appear blurry. This situation can be improved by placing the telescope at a high-altitude location, e.g., on a mountain top, where the air is thinner and steadier. Such a location has several additional advantages. Mountain tops are usually far from cities so the sky is darker and clearer, free from dust, smog, and the light from street lamps, cars, and buildings. Furthermore, a sufficiently high altitude also takes the observer above most of the water vapor in the Earth's atmosphere, thereby opening the atmospheric windows in the infrared spectrum still wider. Some mountains are better than others because the local topography and wind patterns reduce turbulence and yield sharper images.

The best sites for astronomical observations in the world today are both high and far from civilization. Examples are Mauna Kea in Hawaii, the Pic du Midi in France, and Cerro Tololo in Chile. Mauna Kea, with an altitude of 14,000 feet (4.2 km) above sea level, is now generally acknowledged to be the premier site on Earth for astronomical observations, and most of the very large telescopes under construction or planned for the near future are to be located there (Plate 1).

Even a mountain top is not ideal. The next step is to use a telescope in an airplane in order to reach still higher altitudes. This has proved particularly effective for infrared studies, since the absorptions by Earth's water vapor are reduced dramatically. To do still better, one must use balloons, rockets, and ultimately send a tele-

scope into orbit about the Earth, totally outside our planet's atmosphere. We will discuss how this is done in Section 2.6.

Orbiting Observatories

In the discussion of good sites for observatories, we have concentrated on observations made in the visible and infrared regions of the spectrum. What about the ultraviolet? The ozone layer is so high that even an airplane doesn't get above it, and only instruments in rockets or satellites will do. Without the distorting effects of the atmosphere, a space telescope can see fainter sources and distinguish finer detail than its counterpart on Earth, as well as having access to the entire electromagnetic spectrum.

While this approach has obvious advantages, putting it into practice is not easy. Launching a large optical instrument into orbit with the ability to find and track faint sources, to detect and analyze their radiation, and then to send the information to Earth, are all major challenges. Nevertheless, these challenges have been met, although the resulting space observatories are many times more expensive than those built on the ground.

As we write this, the International Ultraviolet Explorer (IUE) is completing its sixth year of service. This satellite contains a 70-cm telescope and two spectrographs for recording ultraviolet radiation. The Infrared Astronomy Satellite (IRAS) carried out a survey of the entire sky in 1983. Among many other discoveries, it found a fast-moving comet that passed only five million km from the Earth.

The best is yet to come. In late 1988 or early 1989, the United States will put a 2.4-meter telescope into orbit. Called the Hubble Space Telescope after the astronomer Edwin Hubble who discovered the expansion of the universe, this instrument will bring astronomy across a new frontier (Fig. 2.12). It will be equipped with four instruments initially, as indicated in Table 2.3 which lists their capabilities and illustrates what this telescope will be able to do. While the main emphasis of the observing program will be deep space observations of distant galaxies, the Hubble Telescope will certainly reveal much new information about the solar system as well. For example, this instrument will be able to record images that directly reveal the disk of Pluto for the first time.

2.6 Escaping from Earth
Rockets

Whether we are launching a space telescope into low Earth orbit or sending a probe to the outer planets, we need to overcome the force of gravity exerted by the Earth. To do that, we must lift a spacecraft above the atmosphere (so friction will not slow it down) and supply it with an amount

FIGURE 2.12 This is a diagram of the Hubble Space Telescope deployed in orbit. The solar panels on either side of the main telescope tube provide the necessary electrical power to operate this sophisticated instrument.

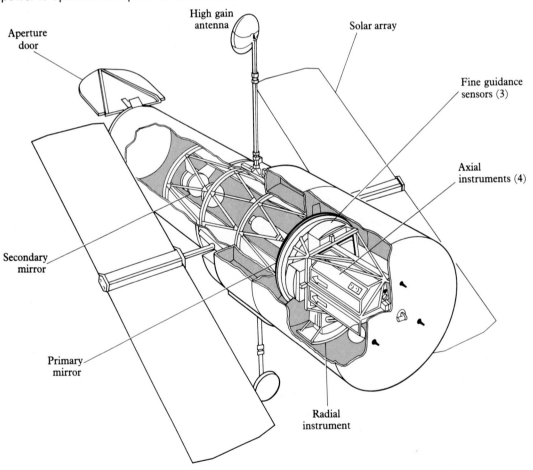

TABLE 2.3 Space telescope instruments

Instrument	Function
Wide-Field Planetary Camera (WF/PC)	Wide field photography of stellar and planetary objects. Many filters available to permit spectrophotometry.
Faint Object Camera (FOC)	Highest spatial resolution, 48 selectable filters.
Faint Object Spectrograph (FOS)	Low and moderate spectral resolution of objects fainter than those accessible from Earth.
High Resolution Spectrograph (HRS)	Highest spectral resolution of all instruments, primarily in ultraviolet region of spectrum.
High Speed Photometer (HSP)	Time resolution of 16 μs with photometric precision of 0.2%.
Fine Guidance System (FGS)	Keeps telescope guiding on targets (including moving targets); can also be used for astrometry.

of energy sufficient to keep it from falling back. The best machine to impart this energy to a spacecraft is a rocket, still a primitive and unreliable toy just fifty years ago.

A rocket accelerates a spacecraft by expelling gases backwards at high speed. This is an excellent example of Newton's third law; the exhaust goes in one direction and the reaction drives the spacecraft in the opposite direction (Fig. 2.13). A jet engine works on the same principle, but it draws oxygen from the atmosphere to burn with its petroleum fuel. A rocket carries its own oxidizer as well as its fuel, and therefore it can operate in the near-vacuum of space.

Today there are many kinds of rockets that have been developed for spaceflight. For large, brute-force applications there are solid fuel rockets, which look like giant firecrackers and are nearly as explosive. The external boosters of the Space Shuttle are solid fuel rockets. More easily controlled are liquid fuel rockets, such as those of the three Shuttle main engines. Many liquid fuel rockets are designed to be restarted in space, a requirement if we are to make in-flight adjustments on interplanetary trajectories. The most efficient liquid fuel rockets burn hydrogen and oxygen, but these two gases must be kept ex-

tremely cold to be used in liquid form and as such are not suitable for engines that must be restarted. Less efficient fuels that remain liquid at less extreme temperatures are needed for flights to deep space; a nitrogen compound called hydrazine is one of these.

Chemical rockets burning either solid or liquid fuels are the workhorses of space travel, but with the development of liquid hydrogen en-

FIGURE 2.13 Newton's third law applies both to tennis and rockets. When the girl puts the ball in play with a forehand shot to the left, she and the cart will move to the right.

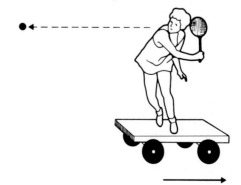

gines they are reaching the limits of their effi-
ciency and usefulness. To achieve greater effi-
ciency, we must look to non-chemical systems.
One possibility is a solar-powered concept called
the "ion drive." An ion drive rocket works by
using sunlight to generate electric energy, which
in turn accelerates ions to much higher speeds
than can be achieved with chemical rockets.
Small ion drive motors have been tested on the
Earth and in space, and a larger system is being
studied in Germany for possible use on interplanetary flights during the 1990s. If nuclear energy were to be used instead of sunlight, such a
system could provide continuous power for missions to the outer solar system, far from the Sun.

Going into Orbit

Suppose that a rocket has lifted a spacecraft
above most of the Earth's atmosphere and is
ready to inject it into orbit. Enough horizontal
speed must be imparted to keep the new artificial satellite from falling back to the Earth. To
understand how the trajectory depends on the
speed, imagine a mountain that extends above
the atmosphere, and ask yourself how fast you
must throw a ball in the proper direction to keep
it in orbit (Fig. 2.14). The motion of the ball
results from the pull of gravity in combination
with its forward speed. If the ball is dropped instead of thrown, it simply falls toward the center
of the Earth. A moderate forward speed, say a
few thousand kilometers per hour, results in a
curving fall that takes the rocket part way
around the world, on a path similar to that of an
intercontinental missile. But when the speed
reaches a critical value called the **orbital velocity,** the surface of the Earth curves away as fast
as the ball falls, resulting in a low-altitude circular orbit. For Earth, the orbital speed is about
8 km/s, or about 28,000 km/h.

Calculations using Newton's laws show that
the square of the orbital velocity is proportional

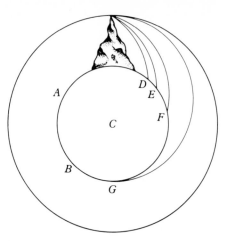

FIGURE 2.14 Throwing a ball horizontally from a
mountain top with greater and greater velocity
produces trajectories D, E, F, G that will ultimately
become circular (or elliptical!) putting the ball into
orbit. We assume here that the mountain is on the
small martian moon called Phobos, whose
gravitational field is so weak that someone with a
strong arm could in fact put a baseball into orbit.

to the mass of the planet and inversely proportional to the distance above the center of the
planet. Thus a smaller planet with its weaker
gravity requires a lower speed to achieve orbit,
just as common sense would suggest. Similarly,
the higher the satellite orbit, the slower its
speed. At the distance of the Moon, the orbital
speed is only about 1.3 km/s.

It is easy to calculate the orbital period of an
Earth satellite. Since the circumference of the
Earth is about 40,000 km, a satellite orbiting just
above the atmosphere (at an altitude of about 200
km) will take

circumference/orbital-speed = 40,000/28,000
                            = 1.4 hours

or about 90 minutes to circle the planet. We can
sometimes see these satellites moving against the
pattern of fixed stars in our night sky. As it happens, this period for a low orbit is nearly the

same for any planet. From a tiny asteroid to giant Jupiter, the minimum orbital period is between one and two hours.

We know the Moon takes about 29 days to orbit the Earth at a distance of 384,000 km, so there must be a distance between 200 km and 384,000 km where the period of an orbit is just equal to 24 hours, the length of our day. Of course there is, and it is 35,680 km. A spacecraft put into an orbit at this altitude will appear stationary to an observer on Earth. In fact only observers who are on the appropriate side of the Earth could see such a satellite; it will never become visible to inhabitants of the other hemisphere. This kind of orbit is called geostationary. It is very useful for communications, reconnaissance, and meteorological satellites on Earth. When we study Pluto, we shall find that its natural satellite occupies just such a stationary orbit.

Escape Velocity

Return to Fig. 2.14 and consider what happens when we throw a ball into orbit with greater speed. As the speed increases, the orbit becomes more and more elongated. The part of the ellipse that is closest to the Earth is called the **perigee,** and the part at the greatest distance (on the opposite side of the Earth) is the **apogee.** (If the orbit were around the Sun, these points would be called **perihelion** and **aphelion** respectively.) As the initial speed increases, the perigee moves farther and farther out, until we reach the point where the ball escapes completely.

The escape velocity described above is exactly the speed required for a ball projected straight up from the surface to escape entirely from the planet. This is also the same as the speed introduced in Section 2.3 to explain the loss of gases from planetary atmospheres. At this speed, the energy is so large that gravity, pulling backward, can slow but never quite stop the outward motion. Mathematically, the escape velocity is just equal to the orbital velocity multiplied by the square root of two, or about 1.4. Thus for the Earth, the escape velocity is 40,000 km/h, or just over 11 km/s. On giant Jupiter, it is 60 km/s, which explains how Jupiter can retain an atmosphere of hydrogen while Earth cannot. Escape from a planet becomes easier if the mass of the planet is small. For example, the escape velocity from Phobos, one of the small satellites of Mars, is only about 60 km/h. This means that you could literally throw a ball into space from Phobos if you have a strong arm, or you could ride into space on a motorcycle, if you found a smooth highway to serve as your launch pad.

Interplanetary Flight

All of the spacecraft that completely leave our planet are moving like bullets shot from a gun. They are given a terrific boost to escape from the Earth's gravitational field, but then they coast most of the way to their targets. Small thruster rockets are fired occasionally to provide midcourse corrections, nudging the spaceships into more accurate trajectories. If one of these robot vehicles is to go into orbit around the Moon or another planet, it must reduce its speed and change its direction as it approaches. Otherwise, it will fly past in a trajectory that becomes curved in response to the pull of gravity, or it will simply crash into the target. This is one reason that it is difficult for a planet to capture a satellite. Such captures require a free roaming body (the satellite) to lose energy — either by frictional drag in a planet's atmosphere or by encountering a third body — in order to move into a stable orbit about the planet.

An important technique that is used to change the velocity of a spacecraft without using up precious fuel is to perform a close flyby of one planet in order to reach another one. This kind of cosmic billiard shot (without the impact!)

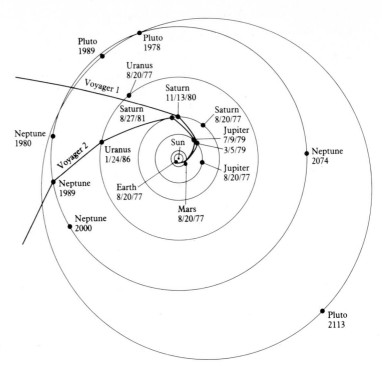

FIGURE 2.15 The two Voyager spacecraft were both launched in 1977 but have taken divergent paths in their journeys past the planets (see Chapters 11–14).

helped Mariner 10 travel from Venus to Mercury, sent Pioneer 11 completely across the solar system from Jupiter to Saturn, and is minimizing the travel time to the outer planets for the Voyager spacecraft (Fig. 2.15). These **gravity-assisted trajectories** work by using the target planet's moving gravitational field (as the planet itself moves in its orbit about the Sun) to accelerate the spacecraft. This acceleration can be both a change in direction and in speed. Gravity-assist has proven to be a very effective technique.

Both of the Voyager spacecraft as well as Pioneer 10 and 11 have achieved escape velocity with respect to the Sun, not just with respect to Earth. Thus they are on trajectories that will take them out of the solar system, into the vast realms of space among the stars. But while we may feel proud to be starting on this adventure,

we should understand just how inadequate our current technology is for the task. The nearest star, Alpha Centauri, is 4.3 light years away, about 25 trillion miles. If these speedy spacecraft were directed toward that star (and none of them is), it would take them about 100,000 years to get there. In fact, we will lose communication with them early in the next century.

Our technology has obviously not yet reached the level imagined by authors and moviemakers who have their space crews happily zipping from star to star in the course of an evening's entertainment. We should recognize, however, that human travel to the stars is not intrinsically impossible; it simply requires some formidable new propulsion technology and/or some new ways of packaging and caring for the crew.

2.7 Exploring the Planetary System
Spacecraft and What They Can Do

Even the Hubble Space Telescope is limited by the immense scale of the solar system. To see the craters on the moons of Jupiter, to explore the rings of Saturn, to search for seas on Triton, to sift the sands of Mars for signs of life — all these things are beyond the reach of this marvelous instrument. To accomplish objectives like these, we have no choice: we have to go there. Direct planetary exploration, begun in 1959, has by now resulted in spacecraft visits to every planet known to the ancients. Men have walked on the Moon and brought nearly half a ton of lunar rocks and soil back to Earth. Instrumented probes have landed on the surfaces of Mars and Venus, and in a few years a probe will be sent into Jupiter's atmosphere. The Voyager 2 space-craft extended our reach to Uranus in January 1986 and will cruise past Neptune in August 1989. By the end of the current decade, we will know every planet in our solar system except Pluto as a familiar, three-dimensional world.

Direct exploration provides us with many advantages. Getting closer allows a much more detailed inspection of the target object. Equipped with telephoto lenses, the cameras on the space-craft can record details that even the Hubble Space Telescope will not discern. Spectrographs can study small areas of a planet, to see whether the composition of one lava flow or type of cloud resembles that of another. Spacecraft can also view planets from angles we can never achieve from Earth. The farside of the Moon was the first such realm to be so explored, in 1959 by the Soviet spacecraft Luna 3. A more recent example is shown in Fig. 2.16, where we see a view of Saturn we can never achieve from Earth in a

FIGURE 2.16 Looking back at Saturn from Voyager 2; a view from behind and below the planet showing the unlit side of the rings, as the spacecraft set forth on its journey to Uranus.

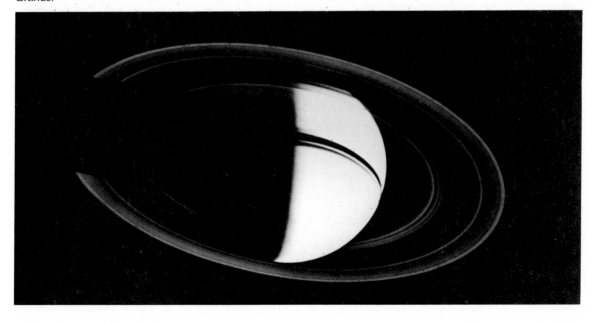

farewell image recorded by Voyager 2 in 1981 as it left the Saturn system on its way to Uranus and Neptune.

How a Spacecraft Works

In the previous section, we discussed the way in which a spacecraft is launched on a trajectory to a planet. The trip can span a few months, if the target is Venus for example, or last twelve years as in Voyager 2's journey to distant Neptune. These spacecraft come in all different shapes and sizes, depending on the functions they are to perform, how far they are traveling, the capability of the launch vehicle, and such practical things as the funding available to build them. The Viking landers (Fig. 2.17) that are resting

FIGURE 2.17 The science test model of the Viking landers that reached the surface of Mars in 1976. Viking Project Manager Jim Martin provides a useful scale. This lander is currently on display in the Smithsonian Air and Space Museum in Washington, D.C.

on the surface of Mars are about the size of a subcompact car and weigh about a ton. Comparatively, the Venus Pioneer spacecraft was smaller, and the Voyagers are somewhat larger (Fig. 2.18).

These spacecraft are filled to capacity with sophisticated instrumentation, the electronics and computers required to make them run, the antennas for communication with Earth, and the panels of solar cells or tiny nuclear generators that provide the power to make everything work. The amount of power required is amazingly small, since everything has been carefully miniaturized. A 100-watt light bulb consumes more energy than some of these spacecraft.

What happens when one of these remarkable devices reaches its intended target depends on its instrumentation. Usually it measures the intensity of the magnetic field and the density of charged electrons and protons in the interplanetary medium all along its trajectory. Once at its destination, it continues measuring these quantities, and the magnetosphere of the planet is mapped out in detail. These kinds of measurements require no particular pointing. Indeed, it is helpful for some of them to be mounted on a spacecraft that spins slowly around one axis so they can sample the full 360° of angle around the trajectory.

Recording images or making spectroscopic observations, however, requires a movable platform that is attached to a spacecraft that does not spin. Such a spacecraft uses star sensors and small rockets (or thrusters) to maintain a fixed orientation as it proceeds along its track. The large radio antenna that it carries to communicate with Earth remains pointed toward our planet, while the scan platform carrying the cameras and spectrographs is turned toward the target planets and satellites in accord with a sequence of commands stored in the on-board computer.

FIGURE 2.18 Linda Morabito with a picture of Io in front of the science test model of the Voyager spacecraft. It was with this picture that she discovered the plumes indicating the intense volcanism that wracks this satellite of Jupiter.

Orbiters, Probes, and Landers

If orbit around another planet is desired, an additional engine must be carried to slow down the spacecraft just enough to permit the planet's gravitational field to capture it. In this case, the spacecraft is called an orbiter. Examples include the U.S. Pioneer Venus and the Soviet Veneras 15 and 16. Among other instruments, these three spacecraft were all equipped with radar antennas and transmitters, in order to penetrate the cloudy veil of Venus and gradually produce topographic maps of the planet's surface after many orbits. The U.S. Mariner 9 and Viking orbiters and the Soviet orbiter Mars 3 performed the same function for Mars, using cameras instead of radars.

An atmospheric probe or a lander might be carried on the orbiter, from which they are released to coast down to the planet itself. The USSR has mounted a very successful series of Venus probe missions over the last fifteen years, bringing us much new knowledge about that planet's atmosphere and surface. The more recent probes could properly be called landers, since they survived on the surface, sending back pictures and compositional analyses. The U.S. has also sent probes to Venus and two highly successful Viking landers to Mars. A Soviet mission to Mars planned for 1988 will feature an orbiter that comes within fifty meters of the moon Phobos, passing by it at the speed of a fast walk while zapping it with lasers and ion beams to determine its composition, and deploying one or more landers. The U.S. Galileo mission to Jupiter will include an orbiter plus a probe into the giant planet's atmosphere. Both countries are planning future missions that could bring samples back to Earth from comets and the surface of Mars. If all of these plans mature, the last decade of this century will see a flood of new results that will exceed the riches we know now.

---◇---

Summary

As a first attempt at studying what planets are composed of, we can look at four types of matter: plasma, gas, ice, and rock. Plasma can exist in low-density regions such as the solar wind. For the planets, a simple measurement of density already gives important clues concerning composition, allowing us to distinguish a rocky from an icy world or to deduce the overall composition of the giant planets.

The inner planets are relatively small, dense objects composed predominantly of rock and metal. Any primitive material they once contained has been strongly modified, producing the familiar igneous, sedimentary, and metamorphic rocks. In contrast the outer giants have low densities and are made primarily of hydrogen and helium, the two lightest and most abundant elements in the universe. These giant planets captured most of their hydrogen and helium from the primordial solar nebula. These planets are surrounded by satellites that form systems reminiscent of the solar system itself.

The distinction in composition between inner and outer solar system reflects a difference in the degree to which the matter in the planets has been processed. The primitive outer planets and their satellites are rich in gases and ices, reflecting more closely the original composition of the nebula. The rocky inner planets are differentiated bodies and are much more evolved. Some have outgassed thin atmospheres that are severely depleted in hydrogen and helium.

The ability of a planet to retain an atmosphere depends on its mass and temperature. Large, cold planets do better than small, warm ones. The gases in planetary atmospheres were captured from the original nebula or outgassed from the material making up the planets themselves.

To learn about the rocks and gases of another planet, we must study the radiation that planet sends us, either reflected sunlight or thermal emission. Using a spectrograph to analyze the radiation, we can determine the nature of the reflecting or emitting surface and the composition, density, and temperature of any surrounding atmosphere.

We are helped enormously in such studies by powerful telescopes that collect the radiation we wish to investigate. These telescopes must be located in observatories on remote mountain tops if they are to realize their full potential. To do still better, we must leave the Earth and its obscuring atmosphere, and launch our telescopes into space.

Most of what we are learning about the planets today is the result of visits by instrumented spacecraft. With spacecraft cameras we can see details we could never hope to glimpse from Earth, even using a telescope in orbit. Spacecraft also offer the opportunity to send probes and landers to a planet to sample its environment directly. Ultimately, such missions can even bring samples back from some distant world, allowing detailed studies in our laboratories.

Key Terms

aperture

aphelion

apogee

atmospheric window

captured atmosphere

compound

deuterium

differentiation

electromagnetic radiation

electromagnetic spectrum

escape velocity

gravity-assisted trajectories

igneous rock

isotope

metamorphic rock

mineral

molecule

orbital velocity

outgassed atmosphere

ozone

perigee

perihelion

photon

plasma

primitive

primitive rock

remote sensing

sedimentary rock

silicates

solar nebula

solar wind

spectrograph

spectroscopy

volatiles

atom
ion
element
electron
proton
neutron

PART ONE
REVIEW QUESTIONS

1. Take a good look at the night sky and also visit a planetarium if there is one near you. Make sure you understand the apparent motions of the Sun and stars, both daily and seasonally.

2. What is retrograde motion of planets? Explain how this phenomenon was interpreted in both the geocentric and heliocentric systems.

3. What were the main contributions of each of the five scientists discussed in Chapter 1: Copernicus, Tycho, Kepler, Galileo, and Newton? To what extent was each aware of the accomplishments of the others, and how did they use previous discoveries as a basis for their own contributions?

4. Why is density such an important quantity to measure for a planet? List the densities of some common materials, including gold and lead. Determining whether an item is made of pure gold has been of interest for many centuries. Look up how the ancient Greek scientist Archimedes solved this problem based on a determination of density and relate this to the problem of determining the composition of distant planets.

5. What are the differences between atoms, compounds, and minerals? List some common examples of each kind of substance.

6. Compare the cosmic abundances of the elements given in Table 2.1 with their abundances on the surface of the Earth. Can you see a pattern that will help explain how the Earth arrived at such a different composition from the rest of the universe?

7. What is an isotope? How can you distinguish one isotope of the same element from another?

8. Distinguish between rocks and minerals. What are some common examples of each? On a cold planet (such as the moons of the outer planets) ices of various kinds are present; are these rocks or minerals?

9. Why are there no primitive rocks on the Earth? Where in the solar system are primitive rocks likely to be found? Why is the search for them important?

10. What is differentiation? Explain why a large planet is more likely to differentiate than a small one. What other properties, in addition to size, are likely to be important in determining whether a planet differentiates?

11. Distinguish among plasma, gas, ice, and rock, giving examples of each. Which of these dominate in the various members of the planetary system?

12. Explain the origin and loss of planetary atmospheres. Is it clear why some objects have atmospheres while others do not?

13. What are the differences between direct measurements and remote sensing? Explain how each might be used to study a planetary surface, a planetary atmosphere, or the plasma that makes up a planetary magnetosphere.

14. What are the various advantages and disadvantages of studying the planets by ground-based telescopes, by space telescopes, by flyby spacecraft, by orbiters, or by probes/landers?

15. How is an interplanetary spacecraft launched toward its target? Describe each step: achieving Earth orbit, escaping the Earth, using gravity-assist techniques, and finally orbiting the target.

16. What evidence could you offer to a friend to support the idea that the Earth revolves around the Sun?

ADDITIONAL READING

Abell, G.O., D. Morrison, and S.C. Wolff. 1987. *Exploration of the Universe*, 5th ed. Philadelphia: Saunders.

Beatty, J.K., B. O'Leary, and A. Chaikin, eds. 1982. *The New Solar System*, 2nd ed. Cambridge, MA: Sky Publishing Corp.

Boorstein, D.J. 1983. *The Discoverers*. New York: Random House.

Frazier, K. 1985. *Solar System*. Alexandria, VA: Time-Life Books.

Goldsmith, D. and T. Owen. 1980. *The Search for Life in the Universe*. Menlo Park, CA: Benjamin-Cummings.

Koestler, A. 1959. *The Sleepwalkers*. New York: Macmillan.

Krupp, E.C. 1983. *Echoes of the Ancient Skies*. New York: Harper and Row.

Sagan, C. 1973. *The Cosmic Connection*. New York: Doubleday.

Sagan, C. 1982. *Cosmos*. New York: Random House.

PART ◆ TWO
Remnants of Creation

One of the most fundamental questions about the planets concerns their origin. When was the solar system created, and how? Are the processes that formed the planets unique, or might they be a common feature of the birth and evolution of stars, implying that multitudes of planetary systems might exist throughout the galaxy? Unfortunately, the planets themselves tend to be mute on such questions of origin because they and their larger satellites are geologically and chemically active bodies that have evolved considerably since their formation. Much more informative is the surviving matter in smaller, unmodified objects — planetary building blocks — that have changed little since the formation of the solar system. These primitive materials are found among the comets, asteroids, and meteorites, the subjects of Chapters 3 and 4.

A piece of debris in space, before it encounters the Earth, is called a **meteoroid.** Small meteoroids striking the Earth's atmosphere are consumed by atmospheric friction,

burning up in a flash of glowing gas called a meteor or shooting star. If a meteoroid survives its brief passage through the air, we call it a meteorite. With the exception of the lunar rocks brought back by the Apollo and Luna missions, meteorites are the only extraterrestrial materials that we can presently study directly in the laboratory.

In order to interpret the scientific message of the meteors and meteorites, we need to have some idea of their origin. The two classes of objects in space that are the likely source of most meteoroids are the comets and the asteroids. Both are thought to be remnants of the population of small bodies that date back to the formative stages of the planetary system, differing from each other primarily in composition and place of origin. The asteroids are rocky objects, composed of the same sorts of materials as the inner planets. Comets, in contrast, contain a substantial quantity of water ice and other frozen volatile materials, which evaporate when heated by the Sun to produce the comet's head and tail.

Often the comets, asteroids, meteors, and meteorites are referred to as "debris." However, we prefer to think of them as artifacts from which events of the distant past may be deduced. We study them in much the same way an archaeologist sifts through the ruins of past civilizations, in the hope that we can find a Rosetta Stone that will unlock some of the secrets of the birth of the planetary system.

◀ The nucleus of Comet Halley is about twice as big as the island of Manhattan. It is composed primarily of water ice, but frozen CO_2 and other gases are also present, all covered by a layer of dark, carbon-rich material like a giant, celestial Eskimo pie. This view was produced from sixty separate images obtained by the Giotto spacecraft in March 1986. The smallest details visible are only 60 m across. Jets of dust and gas that produce the comet's spectacular tails (Figs. 2.3a and 4.23) emanate from at least three bright regions along the left (sunward) side of the nucleus. (See Fig. 4.20 for an explanatory diagram.)

CHAPTER THREE

The Meteorites: Samples of Cosmic Material

3.1 The Solar Nebula

Fundamental Properties of the Solar System

As already noted, the planetary system began with the primordial solar nebula, the cloud of gas and dust out of which the solar system formed approximately 4.5 billion years ago. Astronomers can identify many similar condensing clouds of interstellar material in our galaxy today, at locations where other stars and perhaps other planetary systems are now being born. Before we begin to study the individual objects in our planetary system in greater detail, we need to sketch some current ideas on the origin of our solar system.

It is easy to give a name to the solar nebula. Characterizing it is harder. In order to deduce the properties of the solar nebula, we must work our way backward from conditions in the solar system today. More than 99% of the material in the solar system is in the Sun, which is composed almost entirely of the two lightest gases, hydrogen and helium (Section 2.2). All of the other elements make up only about 1% of the mass of the Sun. In this respect, the Sun is like the other stars and is indeed like the interstellar material between the stars. Thus we can conclude that the solar nebula also had approximately this same cosmic composition.

The largest planets, Jupiter and Saturn, have almost the same composition as the Sun. The smaller bodies, however, are very much depleted in hydrogen, helium, and other light gases. In addition, the inner planets have been formed without many ices or other volatile materials. Thus it appears that some processes involved in planetary formation sometimes concentrated the less volatile matter, permitting lighter materials to escape. In other words, while the Sun may have the same composition as the original solar nebula, the processes that gave rise to smaller solid bodies were highly selective as to the building materials they employed.

Another basic property of the solar system is its rotation. All of the planets have elliptical orbits of small inclination and eccentricity; that is, the orbits are approximately circular and lie roughly in the same plane. All the planets revolve in the same direction around the Sun. Further, the Sun itself rotates in this same direction, and its equator lies essentially in the same plane as the orbits of the planets. Evidently, all of these bodies formed from a solar nebula that was rotating. Further, the material that formed the planets must have been confined to a disk in order for their orbits to have settled so closely to the same plane.

These are, therefore, the most fundamental properties of the solar system that must be related to the solar nebula: its rotation, the fact that the orbits of the planets define a disk, and its chemical composition, with the Sun composed of unmodified cosmic material while the solid planets and smaller members of the system are made up primarily of rarer rocky, icy, and metallic substances.

Formation and Condensation of the Nebula

In order for the Sun to form at all, the thin clouds of interstellar gas and dust must have become concentrated enough to collapse under their own weight. Astronomers see many similar gravitationally contracting clouds at sites of active star formation today, but we still do not understand just what triggers the collapse (Fig. 3.1). In any event, it does occur and so we start our story of the solar system at the stage of a contracting solar nebula with a mass perhaps twice that of the Sun today and a composition representative of our corner of the galaxy.

As the solar nebula collapsed, its central parts were heated by the infalling material. Such a contraction always generates heat, as the energy of falling material is transformed into heat

FIGURE 3.1 The Great Nebula in Orion (Fig. 1.1) is a giant cloud of gas and dust 1,500 light years away. Deep within this huge complex — partially illuminated by light from embedded and nearby stars — new stars, planets, comets, asteroids, and meteoroids are forming.

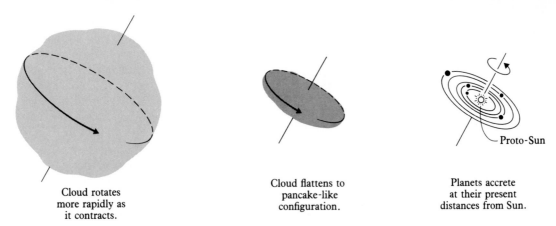

FIGURE 3.2 As the cloud of gas and dust that formed the solar system contracted, it rotated more rapidly leading to the formation of a flattened disk, which in turn broke up into discrete condensations that became the planets and their satellites. The Sun, accumulating most of the mass of the system, formed at the center.

energy. At the same time, the shrinking nebula began to spin faster, and its outer parts flattened into a disk.

The formation of a rapidly spinning disk of material is an example of the principle called conservation of **angular momentum.** This familiar physical law is based on the fact that the angular momentum of a body depends on three quantities: mass; rotation rate; and size. In order for angular momentum to be conserved or held constant, if one of these quantities decreases another must increase proportionately. In our example the size of the nebula decreases, but its mass remains constant, so the law tells us that the rotation rate must increase, and it gives us a means to calculate the rate of change. Just as a turning ballet dancer on point increases her spin by drawing in her arms, the nebula speeds up as it contracts. Thus the contracting solar nebula develops a hot, rapidly growing central core surrounded by a rapidly spinning disk. The core will become the Sun when it gets hot enough and dense enough to sustain thermonuclear reactions, while the disk must give rise to the planets. Let us focus our attention on the disk (Fig. 3.2).

Not all parts of the disk will have the same temperature. The inner parts are heated by the proto-Sun, just as the temperatures of the inner planets are higher than those in the outer solar system today. This temperature difference had a profound effect on the physical and chemical processes taking place in the nebula. At each point, solid grains began to condense like hailstones from a storm-cloud, but they were composed only of those chemical substances that had freezing temperatures below the local temperature of the nebula. Thus, for instance, water ice could condense at a distance of 5 AU from the proto-Sun, but not at 2 AU. Similarly, many common silicate rocks could form at 1 AU but not at the higher temperatures closer to the center. Inside of about 0.2 AU it never got cool enough for any solids to form, and the solar nebula remained gaseous. In this way, a sequence of chemically distinct grains formed at different distances from the center of the nebula. This temperature-related chemical condensation sequence provides the explanation for most of the differences we see today in the compositions of the planets and other members of the planetary system.

Accretion and Fragmentation

We now have a solar nebula in which solid grains have formed, chemically sorted according to distance from the proto-Sun. The next step in the formation of the planetary system required that these grains came together to form larger aggregates. Initially, much of this aggregation simply resulted from bonding as the grains bumped into each other. Soon the largest bodies began to attract their neighbors gravitationally. Thus began the process of **accretion,** in which the particles of the solar nebula grew by their mutual gravitational attraction. Not all theorists who have studied the problems of accretion agree on the details of the process, but ultimately the innumerable tiny grains are swept up into bodies that reached hundreds and then thousands of kilometers in diameter — the proto-planets.

The two largest proto-planets eventually became so big that they were able to attract and hold the uncondensed gas in their part of the nebula. When this point was reached, they quickly grew to giant proportions, becoming the Jupiter and Saturn we see today. Some of the loose gas was also attracted to Uranus and Neptune. Elsewhere in the nebula, however, only the condensed solids were available as building materials for accretion.

As the number of planetary objects became smaller and their individual sizes larger, they exerted stronger gravitational forces upon each other. They particularly affected their smaller neighbors, which were jostled about like floating ping-pong balls stirred with a stick. As a result, the speeds at which one body impacted another grew higher, and the violence of the impacts increased sharply. When a small body struck a large one, it gouged out a crater or, if the body were not solid, splashed liquid fragments in all directions. When two bodies of similar size crashed, the results were even more catastrophic, ending in mutual disruption.

These processes, by which impacts break down objects rather than building them up, are called **fragmentation.** As the system evolved, fragmentation became more important, ultimately dominating over accretion for all but the largest bodies. Thus the big got bigger, but most of the smaller proto-planets were broken back down into fragments. Only a relatively few survived in the form of the asteroids and comets we see today.

Final Stages

In the end, three processes eliminated the gas and dust of the solar nebula. One was condensation and accretion, which led to the formation of proto-planets. A second process, which accounted for a much greater loss of the original material, was the increasing activity of the young Sun at the center. Early in their lifetimes, stars go through a stage of mass loss in which they expel material at high speed from their surfaces. These streams of outflowing solar material, a kind of super solar wind, effectively swept away most of the gas in the solar nebula, as well as any grains that had not yet accreted into the larger bodies. Finally, gravitational interactions with the newly formed planets eliminated most of the remaining solid fragments. Either these fragments impacted the solid surfaces of the planets or they came sufficiently close to be expelled gravitationally from the planetary system.

We have provided this sketch of solar system formation as an introduction to the meteorites. Much of what we know about the processes of condensation, accretion, and fragmentation in the solar nebula has been learned from these fragments from comets and asteroids, the rare survivors of the vast numbers of objects that formed from the solar nebula and became the building blocks of the planets. When you contemplate a meteorite, don't think of it as just a piece of ugly dark rock that fell from the sky; think of it as a survivor from the time when the planets were forming, which provides us a unique perspective on our origins.

3.2 Classification of Meteorites

Early History

Meteorites are defined as those extraterrestrial fragments that collide with the Earth and survive to reach the surface. Until the last century, however, the idea that extraterrestrial materials were reaching the surface of the Earth was scoffed at by educated persons, who placed stories of falling stones in the same category with tales of fairies and dragons. U.S. President Thomas Jefferson, himself a distinguished amateur scientist, is reported to have reacted to information of an 1807 meteorite fall in Connecticut by commenting that he could more easily believe that Yankee professors would lie than that stones would fall from the sky. Such events were so infrequent and unpredictable, and so rarely observed by "reliable" witnesses, that it was easy to dismiss them (Fig. 3.3).

By the end of the eighteenth century, however, the special and very un-Earth-like compositions of some meteorites were becoming recognized, and a case was made that they came from elsewhere. For most scientists of the time, the proof of extraterrestrial origin came in April 1803, when a fall of stones was reported in the village of l'Aigle, France. The French Academy of Science sent a team of reputable scientists to investigate, interview the witnesses, and collect the fallen stones; further investigation confirmed that these stones were unlike any ordinary rocks. Thus was the authenticity of meteorites established.

Meteoritic materials are constantly reaching the Earth, and several falls are observed and recovered every year. Most of these are stones or metallic masses of only a kilogram or two, small enough to be held comfortably in your hand. The rarer, larger falls are produced when a mass of hundreds or thousands of kilograms strikes

FIGURE 3.3 An old woodcut showing the fall of a meteorite near Ensisheim, France, in 1492. The German caption reads: "Of the thunder-stone (that) fell in xcii (92) year outside of Ensisheim."

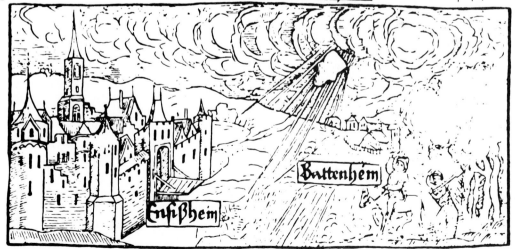

the atmosphere, often breaking up to scatter fragments over many miles. (The 1803 meteorite fall in France was of this type.) Even larger projectiles can also strike at intervals of thousands of years, producing impact craters when they crash into the surface. We will return to this process of impact cratering in Chapter 5, when we consider the heavily scarred surface of the Moon.

Irons, Stones, and Stony-Irons

A surprising variety of rocks are in our meteorite collections, suggesting the existence of many different parent bodies. The traditional descriptions of these meteorites are based on their appearance and bulk composition. Three classes are generally used: iron, stony, and the rarer stony-iron types.

The **irons** are nearly pure metallic nickel-iron and are readily recognized from their high density of more than 7 g/cm^3. Their extraterrestrial origin is obvious when we recall that iron and most other metals normally occur on Earth in the form of oxides rather than in the pure metallic state (Section 2.5). In fact, iron meteorites were one of the first sources of this metal, and their use helped to stimulate the transition from the bronze age to the iron age cultures. The second group, the **stones,** more closely resemble terrestrial rocks, and they are not generally recognized as being of extraterrestrial origin unless their fall was witnessed. The third group, the **stony-irons,** contain a mixture of stone and metallic iron, as their name implies.

Primitive and Differentiated Meteorites

A more useful categorization of the meteorites is based on the history of their **parent bodies,** the asteroids and/or comets of which the meteorites are fragments. If the chemistry of a meteorite indicates that it is representative of the original materials out of which the solar system was made, little altered by the subsequent chemical evolution of its parent body, we refer to it as a **primitive meteorite.** All primitive meteorites are stones, although not all stones are primitive. Since the majority of the meteorites that reach the Earth are primitive, we can be confident that material still exists in the solar system that has remained relatively unchanged since before the planets were formed.

Although the surface of a meteorite is heated to incandescence during its brief plunge through the atmosphere, the heat pulse penetrates no more than a few centimeters into the interior, most of which remains cool and undisturbed. Even the outer layers have normally cooled by the time the meteorite strikes the ground, as shown by objects that have fallen on ice or snow without melting it. Thus primitive meteoritic material remains largely unaltered by its violent arrival at Earth. Those meteorites that have experienced major chemical or physical change since their formation are called **differentiated meteorites.** Like the igneous terrestrial rocks, they solidified out of a molten state. Differentiated meteorites appear to be fragments of differentiated parent bodies that experienced major episodes of heating, along with loss of volatile materials. All of the iron and stony-iron meteorites, and many of the stones as well, are examples of differentiated meteorites.

While many meteorites are lumps of material of fairly uniform composition, others of both the primitive and differentiated types show evidence of having been once broken, mixed, and welded by impact processes on their parent bodies. These fragmented and recemented rocks are called **breccias** (Fig. 3.4). Brecciated rocks, whether meteorites or lunar samples, have experienced impact cratering in the crustal soil of their parent bodies. Some meteorite breccias are especially interesting, because each contains fragments from a variety of regions on the crust of the parent body, thus providing a sampling of a wide area rather than a single rock type.

FIGURE 3.4 A meteorite sawed in half to show the appearance of its interior. It is a breccia, consisting of light and dark rocky components. The dark material contains chondrules.

Falls and Finds

Meteorites can also be distinguished in terms of the way they are identified on Earth. The most obvious meteorites are ones that are seen falling. A flaming plunge through the atmosphere may terminate in an aerial explosion that scatters fragments over many square kilometers. Meteorites located in this way are termed **falls.**

A second group of meteorites are termed **finds.** These are objects whose falls are not witnessed but which are later recognized to be of extraterrestrial origin. Because stony meteorites look superficially like ordinary rocks, they are rarely recognized unless their fall is seen or they land in an unusual location, such as a snowfield. Most finds therefore consist of the much more obvious irons and stony-irons.

Most of the meteorites exhibited in museums are irons, in part because they are more easily identified, and in part because they are more spectacular in appearance. Table 3.1 illustrates the frequency of the different types of meteorites in three population groups: finds, falls, and the Antarctic meteorites discussed in the next section.

The Antarctic Meteorites

Recent discoveries in the Antarctic have established an important new source of meteorites that does not really conform to the traditional distinction between falls and finds. In some parts of the Antarctic continent, the slow movement and subsequent evaporation of ice transports and concentrates any meteorites that fall over areas of tens of thousands of square kilometers (Plate 5a). In these "blue ice" regions thousands of meteorites have been identified and collected, representing the accumulation of hundreds of thousands of years.

Although they are technically finds, not falls, the Antarctic meteorites do not suffer from the selection effects that cause stony meteorites, including nearly all of the most primitive samples, to be overlooked among meteorite finds. The fact that essentially every meteorite that has fallen within the area of ice flow can be spotted and collected is one reason why the Antarctic meteorites comprise such a valuable addition to our inventory of extraterrestrial materials.

The first Antarctic meteorites were discovered in 1969 by a team of Japanese scientists. Since then, expeditions have been sent out nearly every southern summer to search by helicopter and on land for the stones, many of which are only slightly larger than pebbles. Because they have remained frozen in the ice since their fall, these meteorites have suffered relatively little weathering and atmospheric con-

TABLE 3.1 Abundances of major meteorite types (as percent of total)

	Falls	Finds	Antarctic
Primitive stones	87%	52%	85%
Differentiated stones	9%	1%	12%
Irons	3%	42%	2%
Stony-irons	1%	5%	1%

tamination. In the U.S., many of the Antarctic meteorites are being stored under carefully controlled conditions at the NASA Johnson Spacecraft Center in Houston, along with the lunar samples brought back by the Apollo and Luna missions. From there, these samples are distributed to scientists for analysis in their laboratories. As of the end of the 1986/87 season, a total of 4000 Antarctic meteorites had been collected, about half by American scientists and about half by Japanese scientists. The total number of meteorites that had been collected prior to the Antarctic discoveries was under a thousand (where fragments from the same fall are not counted separately).

Nomenclature

Each individual meteorite is given a name, usually for a town or other geographic feature near its point of recovery. For instance, we have the Allende meteorite for a large collection of stones that fell in 1969 near Pueblito de Allende in northern Mexico. In the case of the Antarctic meteorites, where thousands have been found in just a few locations such as the Yamato Mountains and the Allan Hills, a number is also used

FIGURE 3.5 The fifth meteorite found at Allan Hills (Antarctica) in 1981 (ALHA 81005). A sample of the lunar surface delivered to Earth without human intervention.

for identification, such as ALHA 81005, a unique meteorite believed to be a fragment of lunar material (Fig. 3.5). The first two digits give the year of the find, in this case 1981, while the last three represent a running index. Decisions on meteorite nomenclature are made by a committee of the Meteoritical Society, an international organization of meteorite researchers. Table 3.2 summarizes the types of meteorites, proceeding from primitive to modified.

TABLE 3.2 Characteristics of the main meteorite types

	Composition	Age (billion years)
Primitive meteorites		
Carbonaceous meteorites	silicates, carbon compounds, water	4.5
Other primitive stones	silicates, iron	4.5
Differentiated meteorites		
Differentiated stones	igneous silicates	4.4 – 4.5
Stony-irons	igneous silicates, iron, nickel	4.4 – 4.5
Irons	iron, nickel	4.4 – 4.5

Orbits and Origins

In order to determine where the meteorites are coming from, we would very much like to know their orbits before they strike the Earth. Of course, once a meteorite has fallen, it is too late to go back and reconstruct its path before its orbit intersected that of the Earth, and the meteorites we deal with are all far too small to have been sighted telescopically before impact. The only solution is to make sufficiently accurate measurements of the path of the meteorite through the atmosphere during its fall to reconstruct its pre-impact orbit.

Four major efforts have been made to acquire such data, in the United States, Canada, England, and Czechoslovakia. Of course, we cannot know when a meteorite is coming, so we must set up our cameras, photograph the sky continuously every clear night, and hope for the best. During about twenty years of such searches, three meteorites have been photographed in flight and subsequently recovered: Pribram in Czechoslovakia (1959), Lost City in the United States (1970), and Innisfree in Canada (1977). These meteorites proved to be on eccentric orbits that carried them in from the main asteroid belt between the orbits of Mars and Jupiter to cross the orbit of the Earth (Fig. 3.6). While not identifying any specific parent bodies, these three observations do suggest that the asteroids might have been the source of these meteorites.

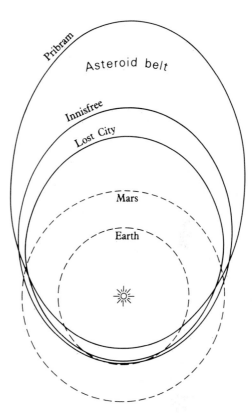

FIGURE 3.6 The reconstructed orbits of three meteorites, Lost City, Innisfree, and Pribram, whose tracks through our atmosphere were photographed. All three orbits extend into the asteroid belt.

3.3 Ages of Meteorites and Other Rocks

Natural Radioactivity

We have already asserted several times that the solar system, including the Earth, was formed 4.5 billion years ago. The ability to date the formation of rocks and planets is one of the triumphs of modern science, with implications for many fields ranging from astrophysics to paleontology. Rock dating, for example, provides the absolute time scale for biological evolution on Earth, permitting us to determine the time intervals between different eras originally assigned only relative ages on the basis of geological context and fossil remains.

The dating of rocks depends on the properties of radioactive decay, discovered early in the twentieth century. **Radioactivity** is the natural process whereby some isotopes of certain elements spontaneously change into other isotopes (Section 2.5). This is a one-way process, accompanied by the release of energy in the form of gamma rays, electrons, or alpha particles (the nuclei of helium atoms). It is these emitted radiations that make high levels of radioactivity damaging to living tissue.

Today much of our concern about atomic radiation involves radioactivity from nuclear reactors or nuclear bombs. Such artificially created radioactive materials can pollute the environment and, in the aftermath of a nuclear war, could result in the extinction of much of the life on our planet. The radioactivity that we will be discussing here though is naturally occurring, the product of minute quantities of uranium, thorium, potassium, and other radioactive elements that are distributed throughout most rocks and minerals.

As a radioactive atom decays to produce a new, nonradioactive isotope, the concentration of the original, or **parent,** material decreases and that of the new, or **daughter,** material increases. The relative concentrations of parent and daughter are the basic information from which a **radioactive age** can be derived.

The Rate of Radioactive Decay

In order to interpret measurements of the parent/daughter ratios, we must know the rate at which radioactive decay is taking place. Fortunately, this rate is a fixed property of the parent material and is independent of external conditions, such as temperature, pressure, or even the chemistry of the minerals in which the radioactive material is located. Although there is no way to predict when any particular individual atom will undergo transmutation, we know that in a fixed interval of time some specific fraction of a collection of similar atoms will decay.

Usually the radioactive decay rate is expressed in terms of a **half-life,** defined as the time for one half of the atoms to decay into daughter products. If a rock began, for instance, containing 64 micrograms of a radioactive isotope with a half-life of 1 million years, we would find that only 32 micrograms remained at the end of the first million years, 16 at the end of 2 million, 8 at the end of 3 million, 1 at the end of 6 million years, etc. The daughter product would, of course, increase as the parent declined.

To determine the ages of meteorites and other ancient rocks we require parent elements with half-lives in the range of hundreds of millions to tens of billions of years, commensurate with the ages of the rocks we are studying. The most useful reactions, together with their half-lives, are given in Table 3.3.

Different Ages

Suppose that you make a careful measurement of the concentrations of one of the parent/daughter pairs in Table 3.3 and calculate an age from the given half-life. Just what does such an age

TABLE 3.3 Radioactive decay reactions used to date meteorites and planets

Parent	Daughter	Half-life
Samarium (Sm-147)	Neodymium (Nd-143)	106 by
Rubidium (Rb-87)	Strontium (Sr-87)	48.8 by
Thorium (Th-232)	Lead (Pb-208)	14.0 by
Uranium (U-238)	Lead (Pb-206)	4.47 by
Potassium (K-40)	Argon (Ar-40)	1.31 by

mean? This is an important question, especially since different parent/daughter pairs may not yield the same calculated age. Essentially, the age measured is the time interval over which the daughter product has been able to accumulate undisturbed from the decaying parent. Since in a liquid the parent and daughter tend to become mixed with other materials, the measured age is the time since the material became solid, and it is referred to as a **solidification age.**

In the case of the potassium-argon pair, the daughter product is a gas, and the age measured is a **gas retention age.** Thus, if we calculated a meteorite age as 18 million years by the K/Ar method and as 4.3 billion years by the Rb/Sr method, we would probably conclude that the rock solidified from the liquid 4.3 billion years ago but that a subsequent shock (perhaps the breakup of the parent body) allowed the argon gas to escape 18 million years ago, in effect resetting the clock that measures the gas retention age.

Measured Ages

In order to realize the full potential of radioactive age measurement techniques, we must have a way to determine the original ratio of parent to daughter isotopes in the rock at the time of solidification. This ratio is calculated using other nonradioactive isotopes of the same elements. These sister isotopes behave in exactly the same way as the radioactive isotopes in any chemical processes that alter the composition of the rock melt, thereby removing this source of uncertainty. Additional checks can be applied by making independent age measurements for different mineral grains from a single sample.

One of the main centers for measuring the ages and the detailed chemical properties of meteorites and lunar samples is at the California Institute of Technology. In a series of sterile laboratories jokingly called "the Lunatic Asylum,"

scientists and technicians in white coveralls and surgical masks analyze grains of material almost too small to be seen without a microscope. The leader of this team, Gerald Wasserburg, is a perfectionist when it comes to laboratory measurements (Fig. 3.7).

Our confidence in the measured ages of meteorites and lunar rocks is the result not only of the high precision achieved in the measurement of the concentrations of parent and daughter products, but also of the number of checks that

FIGURE 3.7 Professor Gerald Wasserburg of the California Institute of Technology, in his laboratory that became known as "the Lunatic Asylum" during the heyday of the Apollo missions to the Moon. One of the strongest advocates of returning samples from other planets, he is said to have claimed that given a one-gram sample of any object, the measurements made in the Lunatic Asylum could establish the chemical and geological history of the parent object.

can be carried out to establish the consistency of the results. Usually it is possible to apply three or more separate techniques, based on different radioactive elements with very different half-lives, and also to apply the measurements to a variety of mineral grains of differing chemistry. Generally, these results agree to within a few percent in the derived age. Where differences do arise, they can usually be attributed to subsequent events, such as the example cited above where the gas retention age was shorter than the solidification age as a consequence of impact resetting of the gas retention clock.

Table 3.4 illustrates the solidification ages derived for a variety of meteorites. Almost all of them cluster near 4.5 billion years. These values provide the best measure of the age of the solar system, defined as the time interval since its most primitive constituents condensed out of the cooling gas and dust of the solar nebula.

TABLE 3.4 Measured solidification ages of primitive meteorite groups[a]

H chondrites	4.50 ± 0.04 by
L chondrites	4.43 ± 0.05 by
LL chondrites	4.51 ± 0.03 by
E chondrites	4.45 ± 0.03 by

[a]Rubidium-strontium ages, adapted from data given in Robert T. Dodd *The Meteorites: A Petrologic-Chemical Synthesis* (Cambridge University Press, 1981) 368 pp.

3.4 Primitive Meteorites

The Oldest Meteorites

Primitive meteorites have chemical compositions that are relatively unchanged since they formed in the cooling solar nebula about 4.5 billion years ago. Except for a shortage of gaseous and other volatile constituents such as hydrogen, helium, argon, carbon, and oxygen, the composition of the primitive meteorites is thought to be the same as that of the Sun. In fact, these meteorites are used to define the presumed solar abundances for some rare elements that cannot be observed directly in the Sun.

The primitive meteorites are also called chondrites, named for the small round **chondrules** that they contain. These chondrules, typically about one millimeter in diameter, appear to be frozen droplets that condensed from molten nebular material (Fig. 3.8). Because not all chondrites contain chondrules, we prefer the more general term primitive meteorite. In appearance, the primitive meteorites are light to dark grey rocks, often with a darker crust produced during their fiery descent through the atmosphere. Many are breccias. Their densities are about 3 g/cm^3, similar to that of many crustal rocks on the Earth.

Many of the primitive meteorites contain grains of metallic iron as a major constituent, 10% to 30% by weight. This leads to a useful classification scheme in terms of metallic iron content. Thus scientists speak of H chondrites for those with high iron content, L chondrites for those with low iron, and so on. The most abundant elements after iron are silicon, oxygen, magnesium, and sulfur.

It is interesting to note the absence of any young meteorites in our collections, as well as of any older than 4.5 billion years (Table 3.4). Clearly, these extraterrestrial samples are all associated with our own solar system. None has intruded from elsewhere. This fact indicates just how isolated the planetary system is within the Milky Way Galaxy, even though our system has completed some twenty revolutions around the center of the galaxy in the 4.5 billion years of its existence.

Carbonaceous Meteorites

The most primitive meteorites are a special group, called the **carbonaceous meteorites,** some of which have no chondrules at all. These carbonaceous meteorites are relatively rich in carbon (a few percent by weight), as their name implies, and also in volatile compounds such as water, which combines chemically with other minerals to form clays. They contain very little metallic iron. Carbonaceous meteorites (Fig. 3.8a; Plate 5b) are dark grey to black in color and are physically very weak; a fragment can generally be crushed between the fingers. Because of their relatively high proportion of water, they are less dense than other meteorites, typically about 2.5 g/cm³. From their composition we conclude that they were formed in a cooler region of the solar nebula from the other primitive meteorites.

One of the special properties of the carbonaceous meteorites is the evidence that some of them have been altered by liquid water, which has dissolved certain minerals and redeposited them in veins. Apparently there existed parent bodies within which water could assume a liquid form, at least for some stage of early solar system history. Many also contain up to 3% of complex carbon compounds, as discussed next.

Organic Matter in Meteorites

Carbon compounds in which carbon is joined with hydrogen and other elements are called **organic compounds,** since scientists once thought that only living organisms could produce them. On Earth, natural carbon minerals are rare except for calcium carbonate, and these complex

FIGURE 3.8 One of the meteorites from the Allende shower. a) The part of the stone facing us has been broken off and shows unaltered meteoritic material. b) An enlargement of a thin section of Allende showing chondrules (one to a few millimeters across) embedded in a dark matrix.

(a)

(b)

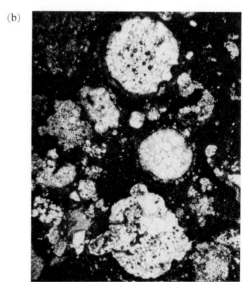

compounds, which make up coal and petroleum, among other things, are indeed produced as a part of the life process. The organic carbon compounds in meteorites, however, are not an indication of life in the primordial solar nebula. Many chemical reactions could have taken place in the hydrogen-rich environment of the solar nebula to produce organic compounds. This kind of matter is actually rather common in the solar system, among the asteroids and comets and in the satellite families of the outer planets. Let us look at the organic materials in carbonaceous meteorites and see what they can tell us about the origin of the planetary system.

Most of the carbon compounds in the primitive meteorites are complex, tar-like materials that defy exact characterization. They also include substances that we recognize as being of fundamental importance for life, components of proteins and nucleic acids. These were first identified with certainty in the Murchison meteorite, a carbonaceous meteorite that fell in Australia in 1969, a few months after Allende was recovered in Mexico. Murchison and Allende are probably the two best-studied meteorites ever to fall to Earth. Murchison yielded sixteen separate amino acids, including eleven that are rare on Earth. The most remarkable thing about these and other amino acids in meteorites is that they include equal numbers of left-handed and right-handed forms. Amino acids can have either of these kinds of symmetry, but life on Earth has evolved using only the left-handed versions of these compounds to make its proteins. The fact that both symmetries are present in the meteorites demonstrates that no contamination has taken place since they arrived on Earth, and suggests that these organic compounds formed in space without the intervention of living things.

Peculiar Inclusions

A few carbonaceous meteorites contain remarkable evidence on the very earliest periods of planetary history, stretching back perhaps to even before the solar nebula formed. Much of this evidence is to be found in small grains of foreign matter called inclusions. These inclusions are light colored irregular grains, not to be confused with the round chondrules. The most revealing results on the light colored inclusions have been derived from studies of the Allende meteorite, but a number of other carbonaceous meteorites are also contributing to this rapidly growing field of research.

The most important way in which these meteorites differ from terrestrial and lunar rocks is in their isotopic compositions. In all previous measurements of materials from the Earth, the Moon, or other meteorites, the ratios of the three common oxygen isotopes, O-16, O-17, and O-18, have followed the same pattern, independent of the chemical context or the place of origin. Inclusions in some of the carbonaceous meteorites, however, show different isotopic ratios, suggesting that a mysterious source of nearly pure O-16 has been mixed with the "normal" type of oxygen. Apparently this anomalous oxygen was injected into the solar nebula at the time these meteorites were forming, a topic we will return to in Chapter 15.

Magnesium is another anomalous element. Some of the Allende inclusions show an enhancement of the isotope Mg-26, identified as the daughter product of the decay of Al-26, a radioactive isotope with a half-life of only 720,000 years. This was the first direct evidence for the existence of this short-lived isotope in planetary material. The decay of Al-26 is commonly invoked as a major source of the energy required for heating (and differentiating) solid bodies during the formation of the planetary system.

Because the carbonaceous meteorites are the oldest samples of solar system material available for laboratory study, they reveal the most about these earliest stages of the planetary system. In the search for a key to unlock the secrets of the past, Allende and Murchison and other carbonaceous meteorites are proving to be our most valuable artifacts.

3.5 Differentiated Meteorites

Indications of Differentiated Parent Bodies

A wide variety of meteorites, including stones, irons, and stony-irons, show evidence of substantial chemical modification since their formation. Although they may be very old, they are not primitive in a compositional sense. Since these changes suggest that their parent bodies underwent some degree of differentiation (Section 2.3), we call them collectively the differentiated meteorites. Often they are also referred to as the achondrites, a term that signifies the absence of the chondrules that are characteristic of most primitive meteorites.

The chemical changes seen in the differentiated meteorites apparently took place when their parent bodies were heated. Under such circumstances, the more volatile elements can be lost, while the remaining material may melt and rearrange itself through the process of planetary differentiation. If several differentiated parent bodies were subsequently broken up, the resulting fragments could explain most of the differentiated meteorites that strike the Earth today.

Irons and Stony-Irons

The most obviously evolved meteorites are the irons, which consist of almost pure metallic nickel-iron, with trace quantities of sulfur, carbon, and metals such as platinum (Fig. 3.9a). The nickel content is usually about 10% by weight. Iron meteorites make up only 4% of the falls and of the Antarctic meteorites, which probably represent an unbiased sample of meteoritic material. These meteorites can be quite beautiful when cut and polished, and etching further reveals a unique crystalline pattern, created by slow cooling of the nickel-iron melt over millions of years (Fig. 3.9b). Presumably the iron meteorites are fragments of the metal cores

FIGURE 3.9 a) The Henbury iron meteorite, showing a perfectly preserved pattern of "thumb prints." This pattern is caused when the thin skin of melted iron produced by the friction of the meteorite's passage through the atmosphere is pushed back away from the direction of motion. b) A slice of an iron meteorite that has been polished and then etched with dilute nitric acid to show the criss-cross pattern produced by intersecting plates of the different alloys composing the meteorite. This is the so-called "Widmanstätten" pattern.

(a)

(b)

of their parent bodies. The cores of the Earth and other intact planets will remain forever beyond our direct investigation, but these meteorites provide a unique glimpse into the very heart of a former differentiated planetary body. Detailed chemical analysis indicates that there were a number of different parent bodies, at least several dozen, for the known iron meteorites.

Related to the iron meteorites are the stony-irons, composed of a mixture of nickel-iron and silicate minerals. The most spectacular of these meteorites, called pallasites, consist of large crystals of olivine, a green semi-precious stone, in a setting of shiny meteoritic iron. The pallasites are thought to be fragments from the interface between the core and mantle of their parent bodies. Fewer than 1% of the meteorites reaching the Earth are stony-irons.

Age measurements cannot be made for the iron but can be carried out for the silicate fraction in stony-irons or for small silicate inclusions found in some iron meteorites. These ages average about 4.5 billion years, suggesting that the processes of heating, differentiation, and cooling of their parent bodies all took place very early in solar system history.

Basaltic Meteorites

The primary group of differentiated stony meteorites appear to be derived from the crusts of their parent bodies. These rocks were formed by crystallization from a cooling body of basaltic lava, which in turn represents the lower-density material that rose to the surface during differentiation. **Basalt** is the most common form of lava, and it makes up more than half the crust of the Earth. As we will see in Chapter 5, basalt is also the material that makes up the dark lunar "seas." Most of the basaltic meteorites are breccias, further indicating their long residence near the surface of their parent bodies where they were fragmented and mixed together by impact cratering.

The best studied group of basaltic meteorites are the **eucrites** (Fig. 3.10), of which more than thirty examples are known having identical oxygen isotopic ratios and closely related compositions. Their solidification ages are all 4.5 billion years, with evidence that they were located in the crust of their parent body for an additional 1.5 billion years before being ejected through breakup or major cratering. It is generally believed that the eucrites are derived from a single parent body, and that their compositions are representative of a series of different lava flows, much as lunar and terrestrial lavas show some chemical diversity.

Another group of about a dozen unusual basaltic meteorites are called the **SNC meteorites,** short for the names of the prototype subgroups: shergotites, nakhlites, and chassignites. Like the eucrites, these are chemically re-

FIGURE 3.10 An example of a eucrite showing the glossy black crust caused by melting during atmospheric entry and the underlying grey rocky material, which is free of chondrules. The eucrites are composed of basalt.

EETA79001

1 cm

FIGURE 3.11 An example of one of the SNC meteorites. You may be looking at a rock from Mars, altered by the impact that sent it to us and by the heating it experienced upon entry to the Earth's atmosphere.

lated lavas, apparently derived in this case from a source region in the mantle of their parent body (Fig. 3.11). All show evidence of violent shocks, perhaps associated with the impact that liberated them. Most remarkably, however, these meteorites show unusually recent solidification ages of about 1.4 billion years. If these ages are correct, they suggest that the parent body remained volcanically active until relatively recently. In fact, the favored parent body of the SNC meteorites is the planet Mars, although it is difficult to understand how meteorites could be ejected from a body this large. If this identification can be verified, it will be interesting to see what we can conclude about the history of Mars from analyzing the SNC meteorites.

3.6 Meteorite Parent Bodies

Aside from the SNCs, both comets and asteroids have been suggested as meteorite parent bodies. An exploded planet has also been hypothesized. Before pursuing these ideas further, it would be useful to summarize what we know about the parent bodies from our study of the meteorites.

Sources of Primitive Meteorites

Primitive meteorites must have originated in or on relatively small bodies that formed directly from dust condensing out of the cooling solar nebula. Large parent bodies are not possible because they would have retained too much internal heat generated by natural radioactivity; the thicker layers of material act as a blanket to keep the radioactive heat confined instead of allowing it to escape to space. Calculations show that in order to have escaped unacceptable heating levels, the parent bodies must have been no more than a few hundred kilometers in diameter. Some of these must have included organic chemicals and liquid water. The chemical and isotopic variety among the primitive meteorites indicate that many different parent bodies are represented in our meteorite collections.

Sources of Differentiated Meteorites

The differentiated meteorites are fragments of differentiated parent bodies. From detailed chemical analysis, it is clear that several dozen distinct parent bodies were involved. We can set some limits on their sizes from the iron meteorites, which retain in their crystal patterns indications of the rates at which they cooled (Fig. 3.9b). These cooling rates indicate objects no more than about a hundred kilometers in diameter. Thus the differentiated parent bodies were at least as small as the parent bodies of the primitive meteorites, and a characteristic other than size was responsible for the fact that one group of objects differentiated while the other remained in a primitive state. No one knows why one set of parent bodies differentiated and the other did not.

From the existence of basaltic meteorites we conclude that some parent bodies experienced surface volcanism, and in the case of the eucrite parent body the chemistry of these lava flows is well defined. On the SNC parent, volcanism persisted at least 3 billion years after formation, suggesting a planetary-size object, perhaps Mars. And several Antarctic meteorites have been identified as having originated on the Moon.

Asteroids or an Exploded Planet?

As we will see in Chapter 4, most of the meteoritic evidence points toward the asteroids as parent bodies, although comets are possible as parents for some of the primitive meteorites. What is clearly excluded, however, is the old idea of an exploded planet as the source of both the asteroids and the meteorites. It is easy to see why the exploded planet theory fails.

More than 90% of the meteorites show evidence of parent bodies no more than a couple of hundred kilometers in size, and several lines of evidence suggest dozens of distinct parent objects. In addition, the prevalence of breccias points to the impact-stirred surfaces of airless precursors as a common environment for many meteorites. None of these observations would be true if the meteorites were derived from a single large planet as has sometimes been proposed. In addition, the breakup or explosion of such a planet seems to be an intrinsically unlikely, and probably impossible, event.

Summary

The meteorites are rocky and metallic fragments that have reached the Earth and survived their plunge through the atmosphere. They are the only extraterrestrial materials we have other than the lunar samples, and they are providing most of the information available on conditions at the birth of the solar system, when solid material condensed out of a contracting and cooling solar nebula of interstellar gas and dust. Generally they are fragments of parent bodies that formed very early in the history of the planetary system.

Most of the meteorites that fall on Earth are primitive in composition, meaning that they and their parent bodies have not suffered significant changes since their birth. In this respect, they differ from the rocks of Earth and Moon, both of which are the product of a history of planetary heating and geological activity. The most important of the primitive meteorites for understanding the solar nebula are the carbonaceous meteorites, which apparently formed at temperatures corresponding to the present location of the asteroid belt. Some of these even contain small inclusions that appear to have survived from the period before the solar nebula.

Differentiated meteorites are samples from differentiated parent bodies which, like the Earth and other planets, have experienced chemical changes due to melting. The evidence indicates, however, that the parent bodies of these meteorites were not full-fledged planets, but small bodies that were heated and differentiated very early in solar system history and subsequently broken up by impacts. The iron meteorites appear to be fragments from the metallic cores of such bodies, and the stony-irons are from the boundary between core and silicate mantle.

A few differentiated meteorites have come to us from the crusts of objects that experienced surface volcanic activity. These basaltic meteorites include samples from the Moon, the eucrites which are widely thought to originate on the asteroid Vesta, and the SNC meteorites, which some scientists believe to be samples of martian material.

Today there are thousands of individual meteorites in scientific collections around the world, and hundreds of scientists are working on problems of their analysis. In some laboratories, instruments of tremendous sophistication, many developed for study of the Apollo lunar samples, can determine the detailed chemical and isotopic composition of even a single crystal or grain almost too small to be seen with the naked eye. In the filtered air of these super-clean laboratories, some of the most important results of modern planetary science are produced.

Key Terms

accretion

angular momentum

basalt

breccia

carbonaceous meteorite

chondrule

differentiated meteorite

eucrite

fall

find

fragmentation

gas retention age

half-life

iron meteorite

meteorite

meteoroid

organic compound

parent/daughter isotopes

parent body

primitive meteorite

radioactive age dating

radioactivity

SNC meteorite

solidification age

stony meteorite

stony-iron meteorite

CHAPTER FOUR

Small Bodies: The Asteroids and Comets

4.1 Discovery of the Asteroids

The First Four Asteroids

There is a pattern to the orbits of the planets. As we discussed in Section 1.5, the distances of most of the planets from the Sun can be represented by a numerical progression called the Titius-Bode rule. However, this rule creates a problem: it predicts a planet at 2.8 AU from the Sun where none exists.

In earlier times, before scientists had developed a sense of the workings of physical laws, such empirical numerical schemes were taken very seriously. Thus late in the eighteenth century many astronomers became disturbed by this problem, and a group of six German observers set out on a systematic search for this "missing" planet. Just as their effort was beginning, however, they were scooped by an Italian, Giuseppe Piazzi, who was measuring the positions of stars from his observatory in Palermo, Sicily.

On New Years Day, 1801, Piazzi discovered the first asteroid, which he named Ceres for the Roman patron goddess of Sicily. This faint object, invisible without a telescope, had just the expected orbital distance of 2.8 AU, and it was at first hailed as the missing planet. The German astronomers did not give up, however, and soon they discovered three more minor planets — Pallas, Juno, and Vesta — also orbiting between Mars and Jupiter. These were even smaller than Ceres, although Vesta is slightly brighter due to its more reflective surface. Even when combined, however, the masses of these four objects came nowhere near that of the Moon, let alone adding up to a real planet.

Modern Asteroid Searches

The next asteroid discovery did not occur until 1845, but from that date on they were sighted regularly. By 1890 the total number had risen to 300. At that same time visual searches were replaced by photographic patrols, and the number of known objects rapidly increased, reaching the 1000th (named Piazzia) in 1923, 2000th in 1973, and 3000th in 1984. While not entirely confined to the space between Mars and Jupiter, the great majority of asteroids do occupy this part of the solar system, with a **main asteroid belt** existing between 2.2 and 3.3 AU.

Today asteroids are relatively easy to find, and accidental photographic discoveries often annoy stellar astronomers by cluttering their long-exposure images of stars and galaxies (Fig. 4.1). To be entered on the official list of asteroids, an object must be observed well enough to establish its orbit and permit its motion to be accurately calculated many years into the future. Thus fewer than 4000 objects have been cataloged so far, although many thousands more could be added to this list if we wished.

The responsibility for cataloging asteroids and approving new discoveries is assigned to the

International Astronomical Union Minor Planet Centers in Cambridge, MA, and in Leningrad. As soon as a discovery is made and a preliminary orbit calculated, a temporary designation is assigned and predictions of future positions are circulated to interested astronomers. Only after observations extend to a second observing season, usually two years after discovery, is a permanent number assigned.

In addition to their numbers, which are assigned in order of discovery, most asteroids have names, usually selected by the discoverer. Initially these were the names of Greek and Roman goddesses, such as Ceres and Vesta, and later expanded to include female names of any kind. When masculine names were applied, they were given the feminine Latin ending, as in the example of 1000 Piazzia mentioned above. More recently the requirement of a feminine name has been relaxed, and asteroids today are named for a bewildering variety of persons and places, famous or obscure. (For instance, asteroid 2410 is named Morrison, after one of the authors of this book.)

Basic Asteroid Statistics

While many thousands of small asteroids remain undiscovered, we can draw some general conclusions from present data. Ceres is the largest asteroid with a diameter of just under 1000 km. The next largest objects are about half this size (Fig. 4.2). The total mass of the asteroids amounts to only $\frac{1}{2000}$ of the mass of the Earth, indicating that the gap in the planetary system between Mars and Jupiter is real, a topic we return to in Chapter 15.

Our census of the larger asteroids is by now fairly complete. It is estimated that about 99% of the objects 100 km or more in diameter are known, and discovery should be at least 50% complete for diameters down to 10 km. Our knowledge is much more complete for the closer asteroids in the inner part of the asteroid belt, and most of the larger undiscovered bodies are probably beyond 3 AU from the Sun.

FIGURE 4.1 A photograph of the night sky taken with a telescopic camera that was slowly moved to compensate for the Earth's rotation. The stars appear as dots in this time exposure, while the rapidly moving asteroid 2062 Aten appears as a streak.

A few attempts have been made to sample the fainter asteroids. In the early 1950s sky photographs were taken for this purpose at McDonald Observatory in Texas, and another photographic survey was made a decade later in a joint project between Palomar Observatory in California and the University of Leiden in the Netherlands. The most comprehensive effort, however, did not involve ground-based telescopes at all. In 1983, the first infrared astronomical satellite (IRAS, a joint U.S.-Netherlands-U.K. project) detected thousands of new asteroids from their heat radiation. While none of these surveys was designed to yield the precise orbital information required to add to the catalog

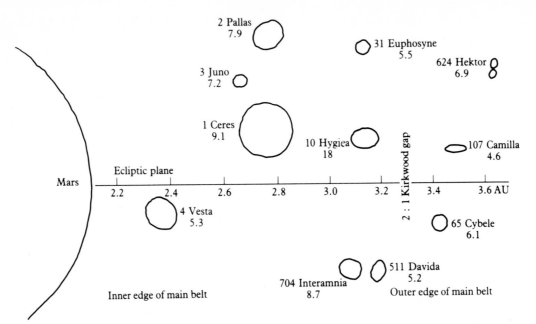

FIGURE 4.2 This scale drawing shows the relative sizes of some of the larger asteroids compared with the planet Mars. The numbers next to the names were assigned in order of discovery; the numbers below the names give the rotation periods in hours. The horizontal scale gives the mean distance from the Sun in astronomical units (AU). (Adapted from A. Chaikin.)

of numbered asteroids, they have produced valuable statistical data on the number and distribution of the smaller objects.

Size-Frequency Distribution

Many more small asteroids exist than do large ones. An estimate of the relative numbers of objects of each size is interesting as a characterization of the asteroid population, and it will also prove important when we look at the size distribution of lunar craters produced by asteroidal and cometary impacts. As a rule, many processes in nature, including those of fragmentation, result in approximately equal amounts of material in each size range. Consider asteroids with diameters of 100 km, 10 km, and 1 km respectively. Each 100-km asteroid has 1000 (10 × 10 × 10) times the mass of a 10-km asteroid,

which in turn has 1000 times the mass of a 1-km object. Therefore, if the total mass is to be the same in each size range, there must be 1000 times more 10-km objects than there are 100-km ones, and a million (1000 × 1000) more at 1 km than 100 km.

Actual measurements of the asteroids (Fig. 4.3) indicate that the numbers do not rise quite as fast with declining size, resulting in a distribution that has most of the mass in the larger objects. (This is why we are relatively certain of the total mass of the asteroids, even without having counted all of the small ones.) However, this simple rule does give us an idea of the way numbers increase very rapidly toward smaller sizes in a naturally occurring population of fragments. As we shall see later, this type of distribution applies to the much smaller particles in planetary rings as well as to asteroids.

Physical Studies: Size and Reflectivity

As seen through a telescope, an individual aster-oid is an unresolved star-like point. Before about 1970 almost nothing was known about the phys-ical nature of asteroids, and research was con-fined to discovery, charting of orbits, and deter-mining rates of rotation and shapes. More recently, however, the application of new ob-serving techniques and larger telescopes has re-vealed a great deal about the physical and chem-ical nature of the asteroids.

One of the first problems was that of mea-suring the size and reflectivity of an object too small to appear as a disk in the telescope. The most powerful means of achieving this involves timing the passage of an asteroid in front of a star. For an asteroid with a well-determined or-bit, we know exactly how fast it is moving against the stellar background, and measuring how long the star is obscured yields an accurate size. If timings of the same event made from dif-ferent locations on Earth are combined, the pro-file of the asteroid can also be derived. An ex-ample of the results for one asteroid, 3 Juno, are

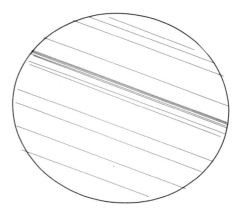

FIGURE 4.4 The approximate shape of the asteroid 3 Juno as determined from its passage in front of a star. This *occultation* was observed from several different locations on Earth, hence from several different viewing angles. Each straight line in the figure represents the apparent path of the star as viewed from each observing location on Earth as the asteroid passed in front of it.

shown in Fig. 4.4. Unfortunately, however, such events are rare, and only half a dozen asteroids have been measured successfully in this way.

More generally, we would like to have a way to determine if an asteroid of a given brightness is large and dark or small but highly reflective. The most useful technique that can be applied to large numbers of asteroids involves the measure-ment of the heat radiated as well as the solar light reflected (Section 2.4). A dark object absorbs most of the incident sunlight and re-emits it in the infrared as heat; it will appear relatively bright in the infrared and faint in the visible. A highly reflective object will be bright in the vis-ible but a much weaker source of heat (Fig. 4.5).

Application of techniques based on this principle has resulted in the determination of di-ameters and reflectivities for more than a thou-sand asteroids. In addition, the rotation period of an asteroid and its departure from a spherical shape can be established if an extended series of measurements of its brightness is made. The variation with rotation of the brightness of an elongated object is readily apparent, and the

FIGURE 4.3 The number of asteroids increases with decreasing size.

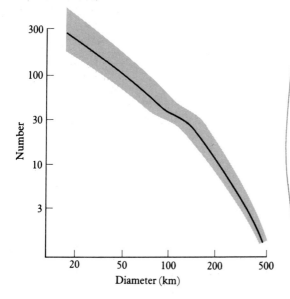

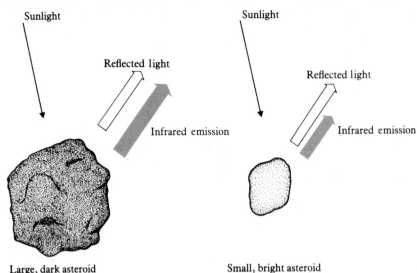

FIGURE 4.5 Comparison of a large, dark asteroid with a small, bright (highly reflective) asteroid. In this example, both reflect the same amount of sunlight, so both appear equally bright in visible light. The large, dark asteroid emits much more infrared thermal radiation, owing to its larger size and its hotter surface. Thus measurement of reflected and emitted radiation allows us to determine both size and reflectivity.

amplitude of the variation directly measures the changing cross-section of the rotating body (Fig. 4.6).

Physical Studies: Spectroscopy

A second important type of information about asteroids is obtained from observing the spectrum of reflected sunlight. Variations with wavelength in the reflectance of the surface material can indicate the composition of the asteroid. While not as rigorously diagnostic as the sharp spectral lines produced when light passes through a gas, the broad absorption features in the spectrum of an asteroid are often sufficient to identify the major minerals present. This identification is aided by the spectral similarity between the reflectance of asteroids and of many meteorites, permitting a laboratory comparison between individual meteorites and asteroids (Fig. 4.7).

Much of the pioneering work in remote sensing of asteroids was carried out by Thomas B. McCord, while at the Massachusetts Institute of Technology, and his colleagues and students. This work began in 1970, when McCord's group obtained visible and near-infrared spectra of asteroid 4 Vesta and found features diagnostic of basaltic lavas (Section 4.3). During the next few

FIGURE 4.6 The change in brightness with time of asteroid 3 Juno as the asteroid rotates on its axis. The fact that there are two maxima and two minima means that this asteroid has an irregular shape and/or dark patches on its surface (see Fig. 4.4). (Adapted from Groenveld and Kuiper.)

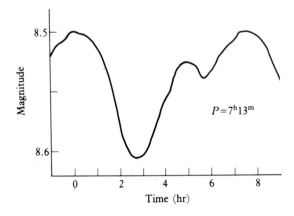

years this team, led by McCord's former student, Clark R. Chapman (Planetary Science Institute, Tucson), measured about 200 asteroids, finding a surprising variety of spectral types. By 1975, a data base sufficient to begin the classification and interpretation of asteroid mineralogy existed.

The development of asteroid studies during the 1970s illustrates how rapidly a new field can blossom even without spacecraft missions. McCord, Chapman, and their coworkers (Fig. 4.8) required only five years from first measurements to the statistical data base and classification system discussed in the following pages. By 1979, a 1181 page book on asteroids could be published with 69 scientists as chapter authors. Contrast this history with the slow pace of scientific progress in previous centuries (Sections 1.3 and 1.4).

Statistical Studies

The use of spectral and reflectance data to characterize asteroids has yielded preliminary determinations of composition for nearly a thousand objects. In most cases the minerals inferred for the asteroids are similar to those in the stony meteorites. Exact identifications are difficult, however. The tools we have for studying the asteroids do not provide a unique fingerprint. Our current state of knowledge is more nearly equivalent to the verbal description of a criminal suspect. We can categorize in terms analogous to height, weight, age, sex, and hair color, but we cannot specify the unique properties that identify an individual. Contemporary asteroid research, therefore, tends toward broad statistical studies rather than detailed investigation of particular objects.

FIGURE 4.7 A comparison of reflection spectra from four asteroids (circles) with laboratory-determined spectra of meteorites (solid lines).

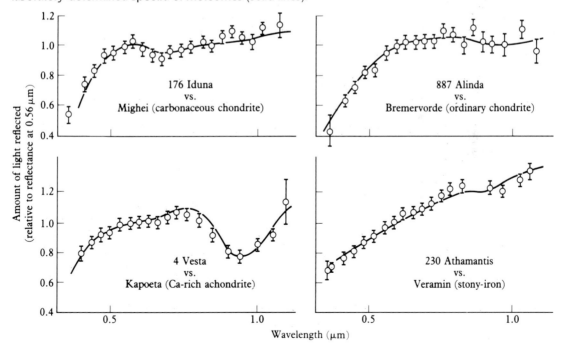

FIGURE 4.8 Three pioneers in the study cf asteroid sizes and compositions, from right to left: Clark Chapman, Benjamin Zellner, and David Morrison.

---◆---

4.2 Main Belt Asteroids

Orbits

As previously noted, the great majority of the nearly 4000 numbered and cataloged asteroids are located in the asteroid main belt, at average distances from the Sun between 2.2 and 3.3 AU. From the observed distribution of sizes we can estimate that there are approximately 100,000 down to a diameter of a kilometer or so.

Although one hundred thousand sounds like a lot of objects, space in the asteroid belt is still empty. The belt asteroids occupy a very large volume, roughly doughnut shaped, about 100 million km thick and nearly 200 million km across. Typically they are separated from each other by millions of kilometers. They pose no danger to spacecraft passing through the belt en route to the outer planets; in fact, it has proved difficult to locate even one asteroid near enough to a randomly chosen spacecraft trajectory to allow a picture or other measurements to be made.

The orbits of the belt asteroids are for the most part stable, with eccentricities less than 0.3 and inclinations below 20° (Fig. 4.9). In the past, when there were presumably more asteroids in this region of space, collisions may have been common, but by now the population has diminished to the point where each individual asteroid can expect to survive for billions of years between collisions. Still, with 100,000 objects, a major collision somewhere in the belt is expected every 100,000 years or so. Such collisions, as well as lesser cratering events, presumably yield some of the fragments that eventually impact the Earth as meteorites.

FIGURE 4.9 a) The number of asteroid orbits having various inclinations. b) The number of asteroid orbits having various eccentricities.

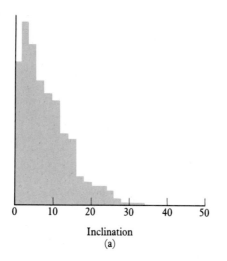

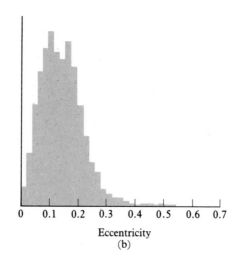

Inclination
(a)

Eccentricity
(b)

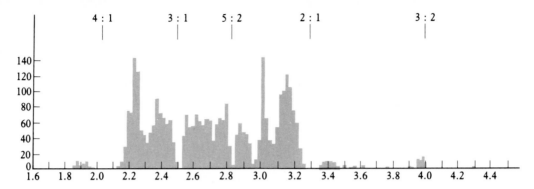

FIGURE 4.10 The number of asteroid orbits with various average distances from the Sun. The positions of orbital resonances with Jupiter are shown (2 : 1 = ½ Jupiter's orbital period, etc.).

Resonance Gaps in the Belt

The orbits of asteroids within the main belt are not evenly distributed. As shown in Fig. 4.10, some orbital periods seem to be preferred, while others are nearly unpopulated. These unpopular sections of the belt are **resonance gaps,** also known as the Kirkwood gaps for the nineteenth century American astronomer who discovered them.

The gaps in the asteroid belt are at orbital periods that correspond to **resonances** between the asteroid and Jupiter. When an orbital resonance occurs, repeated small gravitational effects can add to alter the orbit, in much the same way that repeated small pushes on a child in a swing, applied at the proper times, lead to a large oscillation. In gravitational theory, such a resonance takes place when the orbital period of one body is an exact fraction of the period of another. In this case, the underpopulated asteroid orbits correspond to periods that are ½, ⅓, ¼, etc., that of the period of Jupiter.

Consider an asteroid that orbits the Sun exactly twice for each orbit of Jupiter. Its distance from the Sun would be 3.3 AU. It is closest to the giant planet once each 12 years, at exactly the same place in its orbit. (In 12 years Jupiter circles the Sun once, the asteroid twice.) The cumulative effects of Jupiter's gravity, always applied at the same place, will alter the orbit of the asteroid; in contrast, an object on a nearby orbit that is not resonant with Jupiter experiences gravitational nudges all around its orbit with no net effect. In this way Jupiter will eliminate asteroids in resonant orbits. Indeed, the period corresponding to 3.3 AU is one of the observed gaps, the one that marks the outer edge of the main asteroid belt. Similar processes involving the satellites of Saturn play an important role in creating gaps in Saturn's rings (Chapter 14).

It is important to recognize that the resonant gaps represent orbital *periods* that are missing, rather than physical gaps in the belt. Since most asteroids have orbits of significant eccentricity, their in-and-out motions as they orbit the Sun effectively fill in the physical (spatial) gaps. In contrast, the resonance gaps in the rings of Saturn show up as *physical* gaps also, since the Saturn ring particles are in essentially circular orbits.

Asteroid Families

Asteroid orbits display other patterns in addition to the resonance gaps. An **asteroid family** is defined as a group of objects with similar orbits, suggesting a common origin. About half of the known belt asteroids are members of families, with nearly 10% belonging to just three: the Koronos, Eos, and Themis families. Although not clustered together in space at the present, the members of an asteroid family were all at the same place at some undetermined time in the past.

Members of the same family tend to have similar reflectivities and spectra. Apparently the family members are fragments of broken asteroids, shattered in some ancient collision and still following similar orbital paths. According to some estimates, almost all of the asteroids smaller than about 200 km in diameter were probably disrupted in earlier times, when the population of asteroids was larger. The families we see today may be remnants of the most recent of these inter-asteroid collisions.

The Largest Asteroids

The largest asteroid in the main belt is 1 Ceres, the first to be discovered. Its diameter is just under 1000 km, and its mass of $\frac{1}{5000}$ that of the Earth amounts to nearly half of the total mass of all the asteroids. Two other asteroids are about 500 km in diameter: 2 Pallas and 4 Vesta. The asteroids known to be larger than 225 km in diameter are listed in Table 4.1. In this list large catalog numbers reveal that many of the larger asteroids, such as 704 Interamnia, 511 Davida, and 451 Patientia, were not among the early discoveries. These objects are in the outer part of the belt and have low reflectivities, making them faint in spite of their relatively large sizes.

Compositional Classes

In order to characterize the asteroids further, we utilize the classifications based on measurements of reflectivity and spectrum discussed in the previous section. As soon as the data on a substantial sample of asteroids were examined, it became apparent that most of them fell into one of two classes based on their reflectivity: they were either very dark (reflecting only about 3% to 6% of incident sunlight) or moderately bright (15% to 25% reflectivity). A similar distinction exists in their spectra; the dark asteroids are fairly neutral reflectors with no major absorption bands to reveal their composition, while most of the lighter asteroids are reddish and show the spectral signatures of common silicate minerals. Since the dark grey asteroids had spectra similar

TABLE 4.1 The largest asteroids

Name	Diameter (km)	Distance from Sun (AU)	Class
1 Ceres	940	2.77	C
2 Pallas	540	2.77	*a*
4 Vesta	510	3.36	*a*
10 Hygeia	410	3.14	C
704 Interamnia	310	3.06	C
511 Davida	310	3.18	C
65 Cybele	280	3.43	C
52 Europa	280	3.10	C
87 Sylvia	275	3.48	C
3 Juno	265	2.67	S
16 Psyche	265	2.92	M
451 Patientia	260	3.06	C
31 Euphrosyne	250	3.15	C
15 Eunomia	240	2.64	S
324 Bamberga	235	2.68	C
107 Camilla	230	3.49	C
532 Herculina	230	2.77	S

*a*Two of the largest asteroids do not fit into the common C or S classes: Pallas has a dark surface but shows some evidence of modification by heating; the surface of Vesta is covered with basalt from a time when this asteroid was volcanically active.

to the *carbonaceous* meteorites, they were called **C-type** asteroids. The lighter class was named the **S-type,** indicating *silicate* or *stony* composition.

Table 4.1 includes the assigned classes for the asteroids larger than 225 km. 1 Ceres is in the C class, but the other two largest asteroids, 2 Pallas and 4 Vesta, fit into neither class. The special case of Vesta is discussed in Section 4.3. Given the uncertainties inherent in deducing composition from such limited data, few scientists are willing to state positively that C-type asteroids are made of exactly the same thing as carbonaceous meteorites, but the connection seems probable. Less certain is the identification of the S-type asteroids with the primitive stony meteorites, but that is the best hypothesis available today. Alternatively, they may be igneous stony-iron objects. In spite of careful comparisons between the spectra of the asteroids and of various types of meteorites, this problem remains unresolved at this writing.

Relationships to the Meteorites

Given the current classification of nearly a thousand asteroids, we can look at the distribution in space of the C and S types. At the inner edge of the belt, near 2.2 AU, the S asteroids predominate by a small margin. Moving outward, the fraction of C-type objects increases steadily, and in the belt as a whole, these dark, possibly carbonaceous objects make up 75% of the population, compared to 15% S and 10% other types. Clearly, most asteroids are composed of dark, chemically primitive material. This distribution presumably is also telling us something important about the solar nebula. Since carbonaceous meteorites formed at lower temperatures than the other primitive stones, the concentration of C-type asteroids in the outer belt is consistent with their formation farther from the Sun, where the nebular temperatures were lower.

If the C and S asteroids both represent the *primitive* meteorites, where did the *differentiated* meteorites come from? Are there enough differentiated parent bodies among the asteroids to account for these meteorites? We still aren't sure. Another class of asteroids, called the M-type asteroids, has been proposed to consist of metallic objects and to represent the parents of the iron meteorites. Several other minor classes of asteroids might also be evolved objects, but the best case for a differentiated asteroid is 4 Vesta, which is discussed in detail in the next section.

4.3 Vesta: A Volcanic Asteroid

Uniqueness of Vesta

If planetary scientists were asked to pick a favorite asteroid, most would surely name 4 Vesta, the brightest, third largest, and the fourth discovered of the minor planets. Vesta is unique, and largely for that reason we know the most about it. It is the only asteroid confidently identified as the parent body of a group of meteorites, and it is the one differentiated asteroid known to have experienced volcanic activity on its surface.

The uniqueness of Vesta derives not from its position, which is in the midst of the main belt, but from its surface composition. It is one of the most reflective (30%) of the asteroids, and its spectrum (Fig. 4.7) exhibits two deep and well-defined absorptions at wavelengths near 1 and 2 μm. These absorption bands indicate the presence of minerals associated with igneous rocks, basalt in particular. A careful search of the spectra of hundreds of other asteroids has failed to identify another object with this same basaltic surface composition.

Similarities to the Eucrites

Recall that in Section 3.5 on the differentiated meteorites we introduced the *eucrites* as a related group of about thirty basaltic meteorites. When the spectra of the eucrites were measured in the laboratory, it was found that they matched the spectrum of Vesta closely, although not exactly. Subsequent, more precise spectral measurements of Vesta then showed that the shapes and wavelengths of its prominent spectral features changed slightly with rotation, indicating that the surface composition is not uniform.

The eucrites are not all exactly alike either. A more careful comparison now makes it appear that there are regions on the surface of Vesta that have the compositions of the main subgroups of the eucrites. As the asteroid rotates, these regions — essentially large lava flows of slightly different composition — are viewed successively. The new data therefore provide an even more convincing association between the eucrites and Vesta.

The Eucrite Parent Body

But perhaps this spectral similarity is fortuitous, and the eucrites are fragments of another asteroid similar to Vesta rather than samples of Vesta itself. This seems like a reasonable hypothesis, and to eliminate it we must apply a careful chain of logic. The lava rock of the eucrites represents the crust of their parent body. From the chemistry of the eucrites it is possible to predict the composition of the much thicker mantle of the parent. Yet an examination of our meteorite collections reveals no meteorites with this predicted mantle composition. Since there would be more mantle debris than crust debris following the total breakup of this hypothetical parent, it follows that no such breakup took place. Therefore, the "eucrite parent body" is still intact, and the eucrites themselves were chipped from its crust by impact cratering.

If we accept this argument for an intact eucrite parent body and combine it with the fact that no other large asteroid exists with the same type of lava surface as Vesta, we are drawn to the conclusion that Vesta really is the parent body of the eucrites. This conclusion is reinforced when we consider that only a rather large asteroid could have been heated sufficiently to generate a lava surface, so we need not worry about some very small, undiscovered asteroid as the parent body.

Establishing a connection between Vesta and the eucrites permits the meteoritic evidence to be applied to this asteroid. Thus we can date the Vesta lava flows at 4.5 billion years ago, the solidification age of the meteorites, and fix the impact that released the meteorites at about 3 billion years ago from the gas retention age. Whatever triggered the volcanic activity of this asteroid was short-lived, ceasing within a hundred million years of its formation.

Although a great deal has been learned about Vesta, this knowledge raises more questions than it answers. Why did Vesta become volcanically active while its larger neighbors, Ceres and Pallas, did not? Why is Vesta unique? Were there once other similar differentiated asteroids long since destroyed by collisions? What were the parent bodies of the iron and stony-iron meteorites? These are questions that trouble scientists trying to understand the processes of heating and differentiation of planets and asteroids.

---◆---

4.4 Asteroids Far and Near
The Outer Asteroid Belt

Beyond 3.3 AU from the Sun, the population of asteroids quickly thins out. The distribution of the asteroids also changes character, with most of the objects concentrated at just a few orbital periods with wide gaps in between. The compositions of these more distant asteroids also appear to be different from those of main belt objects.

The asteroids between the main belt at 3.3 AU and the orbit of Jupiter at 5.2 AU are very dark (reflectivity less than 4%), but they differ in spectra from the C-type asteroids that predominate in the main belt. Their colors are redder, and they do not look the same as any known carbonaceous meteorite. Because these objects do not seem to be represented in our meteorite collections, scientists hesitate to commit themselves concerning the composition of these asteroids. It is generally believed, however, that they are primitive objects, and that a fragment from one of them would be classed as a carbonaceous meteorite, although of a different kind from those already encountered.

Lagrangian Asteroids

A particularly interesting group of dark, distant asteroids is orbitally associated with Jupiter. Although the gravitational attraction of this giant planet generally has the effect of making nearby asteroid orbits unstable, exceptions exist for objects with the same orbital period as Jupiter, but leading or trailing it by 60° (Fig. 4.11).

These two stable regions are called the leading and trailing Lagrangian points, named for the French mathematician Joseph Louis, Comte Lagrange, who demonstrated their existence in 1772. An object in one of the Lagrangian points occupies one corner of an equilateral triangle, with the Sun and Jupiter at the other two points.

The regions of stability around the two Lagrangian points are quite large, and each contains several dozen known asteroids. The first of these Lagrangian asteroids was 624 Hektor, discovered in 1907. All of them are named for the heroes of the *Iliad* who fought in the Trojan War, and collectively they are known as the Trojan asteroids. Their spectra are distinctive, suggesting that they represent a pool of special, primitive objects that have been trapped in this region of space since the birth of Jupiter.

Since there are no stable orbits between Jupiter and Saturn, the Trojans are the most distant group of asteroids known. Searches have been made for similar Lagrangian asteroids associated with Saturn, but none has been found. Beyond Saturn, our telescopes are not powerful enough to see any but the largest asteroidal bodies.

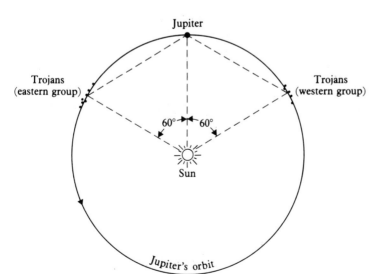

FIGURE 4.11 The Trojan asteroids are located 60° ahead of, and behind, Jupiter in that planet's orbit. We will find small satellites in the Saturn system occupying similar positions with respect to Saturn (in the role of the Sun) and a larger satellite (in the role of Jupiter) (see Chapter 14).

Hidalgo and Chiron

There are two known asteroids that venture beyond Jupiter, and both deserve description. The first is 944 Hidalgo, a dark object occupying a comet-like orbit of high inclination and eccentricity that carries it from the inner edge of the main belt out almost as far as Saturn. Although orbitally similar to a comet, it does not develop a cometary atmosphere and is therefore classed as an asteroid. If it is an extinct comet nucleus, its diameter of 50 km makes it surprisingly large.

Still more mysterious is 2060 Chiron, discovered in 1977. Its orbit carries it from 8.5 AU, near Saturn, out to 19 AU, near Uranus. Its composition and reflectivity are unknown, but if its surface is dark, its diameter could be as large as 300 km. Since the search for objects beyond Jupiter is very incomplete, we do not know how many other, smaller objects of this type might be present. Chiron is in an unstable orbit, however, and ultimately it will either impact a planet or be gravitationally ejected from the solar system.

Earth-Approaching Asteroids

While only a small fraction of the asteroids stray out of the asteroid belt and approach the Earth, we have a special interest in those that do. The Earth-approaching asteroids are potentially useful to humanity as sources of raw materials in space, and they also pose a possible threat to us if they should collide with the Earth. The evidence for past impacts from asteroids is discussed in Chapters 5 and 6.

The first asteroid found on an orbit that crossed the Earth's orbit (coming within less than 1 AU of the Sun) was 1862 Apollo, discovered in 1948, and thus the Earth-crossing objects are often called the Apollo asteroids. Here we use the somewhat broader term **Earth-approaching asteroid** to include potential Earth-crossers, objects whose orbits do not now intersect that of the Earth but which can do so under the shifting gravitational nudges of the other planets.

Continuing searches for Earth-approachers carried out primarily at Palomar Observatory and at the University of Arizona yield from one to three new objects of this class each year. Today more than fifty are known. The largest is 433 Eros, a highly elongated asteroid about 30 km in length, but most of the known Earth-approachers are no more than a few kilometers across. The asteroid that has the smallest perihelion distance is 1566 Icarus, which comes inside the orbit of Mercury. It is estimated that something over a thousand Earth-approaching asteroids exist down to one kilometer in diameter.

The roughly thirty Earth-approaching asteroids that have measured reflectivities and spectra comprise a varied lot. Most are S-type objects, but there are some of C-type and others with spectra that are uncommon among main belt asteroids. Since only about 3% of existing Earth-approachers have been measured, it is probably premature to try to compare them in detail with the meteorites.

Origin and Fate of the Earth-Approachers

The orbits of Earth-approaching asteroids are unstable, as a result of the constantly varying gravitational influence of Earth, Mars, and Venus. Any object that finds itself caught in such an orbit will experience one of two fates, each about equally probable. It may be gravitationally ejected as the result of a near miss of a planet, or it may not miss, dramatically terminating its existence in a crater-forming impact. Most of the craters on the Moon and the terrestrial planets are probably caused by impacts with Earth-approaching asteroids.

Calculation shows that an Earth-approaching asteroid will meet one or the other of these fates after an average interval of about a hundred million years. A hundred million years may seem like a long time to us, but it is short in comparison with the 4.5-billion-year history of the planetary system. Any asteroids that began as Earth-approachers at the time the solar system was formed would long since have impacted a planet or have been ejected. Therefore the Earth-approachers present today must have come from somewhere else, and there must be a continuing source for these objects.

One probable source for the Earth-approachers is collisions among main belt asteroids, several of which take place every million years. Another possible source is the comets. If a rocky core remains after a comet has exhausted its volatiles, this core might be indistinguishable from an asteroid. This latter possibility was strengthened by the discovery beginning in 1983 of several Earth-approaching asteroids with orbits similar to those of the Jupiter family comets, which we will discuss below. The IRAS orbiting observatory also discovered an Earth-approaching asteroid, 3200 Phaethon, in association with a meteor stream (the Geminids), another property normally associated with comets.

4.5 The Moons of Mars

Why discuss the satellites of Mars, Phobos and Deimos, in a chapter devoted to comets and asteroids? Because these two small moons are probably captured asteroids, acquired by Mars shortly after the formation of the planet. Although they probably differ in some important ways from the main belt asteroids, Phobos and Deimos may be able to tell us something about their asteroid cousins.

Discovery of Phobos and Deimos

After Galileo's discovery of the four large satellites of Jupiter, scientists began to speculate about moons orbiting other planets. Since Venus had no satellite, Earth had one, and Jupiter had four, some concluded that Mars, the planet between Earth and Jupiter, should have two (another example of numerology). Kepler was one of those who suggested the existence of two martian satellites, and these hypothetical objects were widely publicized when Jonathan Swift included them in his popular satire *Gulliver's Travels*, published in 1726. Swift wrote of the astronomers of his mythical Laputia that they had discovered

> two satellites, which revolve about Mars, whereof the innermost is distant from the center of the primary planet exactly three of the diameters, and the outermost five; the former revolves in the space of ten hours, and the latter in twenty-one and a half; so that the squares of their periodical times are very near in the same proportion with the cubes of their distance from the center of Mars . . .

Note that Swift is explaining Kepler's third law to his readers! It was to be another 150 years before these fictional moons were actually discovered, however.

The finding of the two satellites of Mars was among the last important planetary discoveries made by the human eye, looking through a telescope. After many nights of careful searching, the American, Asaph Hall, spotted both of them during the favorable opposition of 1877, the same year that Schiaparelli first reported canals on Mars (Chapter 9). The satellites were named Phobos (fear) and Deimos (panic), for the horses that pulled the chariot of Ares (Mars) in Greek mythology.

Because Phobos and Deimos are small, dark, and close to the planet, they are difficult to study telescopically; more difficult, certainly, than the larger main belt asteroids. It was not until the Mariner and Viking spacecraft acquired close-up photographs during the period 1971–1978 that scientists realized how interesting these two remarkable objects are.

Orbits and Origins

Phobos, the larger of the two satellites (about 23 km diameter), is closest to Mars; its distance is 9378 km from the center of the planet, and its orbital period is 7 h 39 min. Because Phobos revolves faster than Mars rotates, the satellite is seen, and was photographed by the Viking landers, to rise in the west and set in the east. Deimos, which is about half the size of Phobos, is 23,459 km from Mars and has an orbital period of 30 h 18 min. Both satellites always keep the same side turned toward Mars, as our Moon does toward the Earth.

Phobos and Deimos both orbit in the equatorial plane of Mars, but this does not mean that they formed at the same time as the planet. Being so close to Mars, they are strongly influenced by its gravity, including the tug of Mars' equatorial bulge. Thus they would be expected to end up in equatorial orbits, even if they had begun with high inclination.

Because Phobos and Deimos seem to have very little in common with Mars, and because scientists would not expect the processes that formed Mars to generate two small satellites as well, these objects are assumed to have been captured from elsewhere. Certainly there were plenty of asteroids whizzing past Mars early in its history, while the inner solar system was filled with remaining debris. The trick is to find a way for the planet to have captured some of these asteroidal objects into stable orbits.

To be captured, a passing asteroid must be slowed near the planet, losing energy and falling into orbit — the energetic equivalent of firing a retrorocket on a spacecraft (Section 2.6). Probably the capture of Phobos and Deimos took place while Mars still had a large but tenuous early atmosphere, and friction with this envelope of gas provided the capture mechanism. We can imagine many objects approaching Mars while the atmosphere was denser and being slowed to the point where they crashed into its surface. Later, the atmosphere was too thin to affect passing bodies. Thus Phobos and Deimos represent just two of many potential captured bodies, the two which happened to come along at the right moment in the evolution and loss of the early atmosphere of Mars.

Composition and Appearance

Both martian satellites are dark, cratered, irregular objects that have been compared to potatoes (Figs. 4.12 and 4.13). With reflectivities of about 6%, Phobos and Deimos are about as dark as the C-type asteroids. Their colors are similar to those of these asteroids and of carbonaceous meteorites. The Viking orbiters were able to measure the densities of both satellites, which came out to be slightly under 2.0 g/cm^3. This density is below that of most silicates, but can be matched by some of the most volatile-rich carbonaceous meteorites. From these three lines of evidence, it seems clear that Phobos and Deimos are primitive objects. The two satellites seem similar enough that they might be fragments of the same parent, perhaps broken apart in the capture process.

Phobos has a longest diameter of 28 km, an intermediate diameter of 23 km, and a shortest diameter of 20 km. Its silhouette is dominated by three craters, one of which, named Stickney, is 10 km across, nearly half the average diameter of the satellite itself. The other two are each 5 km across. Hundreds of smaller craters are also visible. Deimos is smaller and less irregular in shape. Its longest diameter is 16 km and its

FIGURE 4.12 A view of Phobos from one of the Viking orbiters. The largest crater is named Stickney, after the wife of Asaph Hall, the man who discovered Phobos.

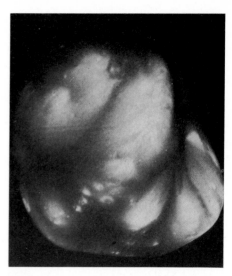

FIGURE 4.13 A full-moon view of Deimos, showing its irregular shape and its whale-like appearance at this (full) phase.

shortest 10 km. Deimos is less cratered, with a largest recognizable crater just 2.3 km in diameter. The craters on Deimos seem subdued or softened, in contrast to the sharp craters on Phobos.

Geological Features

Late in the Viking mission to Mars, special efforts were made to obtain very high resolution photographs of the satellites. The spacecraft orbits were adjusted to be resonant with the satellite periods (3/1 for Phobos and 5/4 for Deimos), and with careful navigation a flyby of Phobos at just 88 km distance was achieved in February 1977, followed in May by a 28-km flyby of Deimos. This Deimos encounter, just 23 km above the surface, is the closest any planetary spacecraft has approached an object without actually hitting it.

The highest-resolution picture of Deimos is shown in Fig. 4.14, showing an area about 2 km

FIGURE 4.14 The highest-resolution picture of Deimos obtained to date. The smallest details are only 5 meters across.

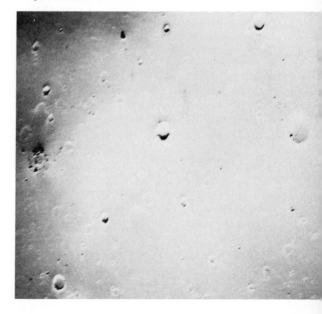

on a side. The many craters are subdued, as if the surface were blanketed with dust. Individual boulders the size of small houses can be seen. The blanketing of the surface of Deimos, and to a smaller extent also of Phobos, is a result of these objects' orbits close to Mars. Because both satellites are small, dust and debris ejected from their surfaces by impact will tend to escape. The martian gravity, however, continues to hold this ejected material, forming tenuous dust rings around the planet in approximately the satellite orbits. Eventually the ejecta returns to the satellite, where it is deposited as a porous dust layer perhaps many tens of meters thick. In this important respect the martian satellites are very different from similar-sized asteroids, which will not accumulate a thick layer of surface dust.

Close-up views of Phobos (Fig. 4.15) are even more spectacular than those of Deimos. In addition to its heavy cratering, Phobos displays a unique set of parallel grooves or valleys, typically several hundred meters across and several tens of meters deep. These grooves form a strong pattern, radiating away from the 10-km crater Stickney. Apparently the grooves are fractures, related to the impact that formed Stickney. Indeed, this crater is so large relative to the size of the satellite that its formation must very nearly have broken Phobos apart. The extensive fracture system bears witness to this nearly catastrophic event early in the history of the satellite. It would perhaps not be surprising to find similar grooves on a few asteroids that have suffered similar huge impacts, although the details of such groove formation remain poorly understood.

Future Missions

The martian satellites are ideal staging areas for the eventual human exploration of Mars. Rather like natural space stations, they provide a platform for observation of the planet, a docking point for descent of small craft to the surface, and even a potential source of rocket fuel, which could be generated by converting the water be-

FIGURE 4.15 Seen in this close-up view, the surface of Phobos is marked with a series of linear grooves that appear to be surface fractures caused by the impact that formed the crater Stickney (see Fig. 4.12).

lieved to be present in their surface rock to liquid oxygen and liquid hydrogen.

Space scientists in the USSR have shown particular interest in these two satellites, as part of their long-term effort toward manned Mars missions. The first Soviet mission to Phobos, involving a pair of spacecraft, is planned for 1988. Whereas the Viking orbiter only achieved a 3/1 resonance with the satellite, the Soviet spacecraft will try to match orbits with Phobos, permitting a very slow controlled flyby. The speed of the craft relative to the surface will be less than 1 m/s, about human walking speed, and the distance will be well within one hundred meters. If the first flight is successful, the second spacecraft will probably concentrate on Deimos.

From their hovering spacecraft, the Soviet experimenters can carry out a variety of studies of the martian satellites. These include several active compositional measurements, in which the satellite surface will be zapped with lasers or charged particle beams to release a cloud of plasma that can be analyzed directly on the spacecraft. If successful, this will be the first mission in which such active methods, as opposed to remote sensing, will be used in a flyby. In addition, there will be one or more instrument packages that will be dropped directly on the surface.

The Soviet scientists are talking about extending these techniques to missions directed at the asteroids. An international mission called "Vesta," anticipated for 1994, will visit one or more main belt asteroids, including Vesta itself.

4.6 Comets Through History
Signs of Evil

Approximately one comet visible to the naked eye appears each year, and a really bright comet comes along an average of one per decade. Writ-

FIGURE 4.16 The representation of Halley's Comet on the Bayeux Tapestry commemorating the conquest of England in 1066 AD. For King Harold of England (shown seated here) it was a bad omen, but presumably William the Conqueror felt differently about it. The Latin words say, "They marvel at the star."

ten records of these long-tailed wanderers go back to at least 1140 BC in the Middle East, and nearly as far in China. Perhaps because the times of their appearances were unpredictable and their forms were constantly variable, comets have always been regarded with apprehension (Fig. 4.16). Often the apparition of a comet was believed to herald some remarkable event on Earth, usually unfavorable. Thus Shakespeare wrote in *Julius Caesar*: "When beggars die, there are no comets seen / The heavens themselves blaze forth the death of princes"; while Milton characterized Satan as a comet that "from its horrid hair / Shakes pestilence and war." No one knows how comets acquired such a bad reputation, but particularly during the medieval period in Europe it seems that no good could be associated with them. In Asia, however, comets were treated more like other celestial phenomena, capable of association with either good or evil happenings.

The scientific study of comets may be traced back two thousand years to the Greeks and Romans. Because of their belief that the celestial realm was unchanging, most of the classical philosophers thought that comets were located in the atmosphere, and efforts were made to use them to predict the weather. As late as the sixteenth century, the scientists of Renaissance Europe were arguing over whether comets were astronomical or meteorological phenomena.

Tycho and Halley

A turning point in cometary studies came with the Great Comet of 1577, which was extensively investigated by Tycho Brahe, greatest of the pretelescopic observers (Section 1.3). He carefully measured the position of the comet against the background of stars, finding that it did not shift back and forth with the Earth's rotation as would a nearby object. Thus Tycho concluded that this comet was not in the atmosphere of the Earth, but well beyond even the Moon.

At about the same time that Tycho was studying the comet of 1577, others demonstrated that comet tails always point away from the Sun, rather than back along the direction of motion. After Johannes Kepler's confirmation of the heliocentric motion of the planets, it was quickly realized that comets also were bound to the Sun, although their orbits were so eccentric they were essentially parabolic rather than nearly circular ellipses.

One astronomer whose name has been immortalized for his work on comets was Edmund Halley, a contemporary of Newton in late seventeenth century England. It was Halley who first realized that cometary orbits could be closed ellipses, with a given comet reappearing at regular intervals. He reached this conclusion from a study of the Great Comets of 1531, 1607, and 1682, all of which had similar orbits. Halley concluded that these were successive appearances of the same comet, which he predicted would return in 1758 (Fig. 1.20).

Although Halley did not live to see his calculations verified, the comet returned just as he had expected, and it was quickly hailed as Halley's Comet, a name it retains today. Comet Halley, with its seventy-six-year orbit, is the brightest of the short-period comets. It appeared twice during the twentieth century, in 1910 and in 1986. Not until 1820 was a second periodic comet identified: the much fainter Comet Encke, with a period of only 3.3 years.

Understanding the physical nature of comets has proved even more difficult than pinning down their elusive motions. Tycho's determination that the comet of 1577 was beyond the Moon also implied that it was very large, comparable in size to the Earth. Soon astronomers were measuring cometary tails millions of kilometers in length, and public fear of plagues from comets was replaced by fear of collision with them.

Comets in the Nineteenth and Twentieth Centuries

From the fact that stars could easily be seen through the cometary head and tail, it soon became clear that in spite of their large size, comets were basically rather tenuous objects. Isaac Newton was among the first to suggest that most of what was seen was thin gas expelled from a solid nucleus as it was heated by the Sun. Nineteenth century astronomers also correctly attributed the streaming of the tail away from the Sun to the force of solar radiation on the insubstantial atmosphere of the comet.

During the second half of the nineteenth century, great public interest in comets throughout the United States and parts of Europe provided a substantial boost to astronomy, leading to the endowment of a number of observatories. In 1910, Comet Halley was particularly well placed for viewing, and the Earth even passed through the tail of the comet on May 20 of that year (Fig. 4.17). A few months earlier, an even brighter new comet passed through the inner solar system, making 1910 a memorable year for comet watchers.

Since the early part of this century, the gradual but steady increase in smog and light pollution has blotted the night sky from the view of city dwellers, and several recent fine comets (Bennett in 1969 and West in 1976 — Fig. 4.29) have gone practically unnoticed by the public. The 1986 return of Halley was also a disappoint-

ment in comparison to its appearance in 1910. The comet was very poorly placed, appearing on the opposite side of the Sun from Earth when at its brightest. Although there was a great deal of public interest and media attention, many persons who tried to see the comet, especially from Canada, Europe, and the northern parts of the U.S., were frustrated in their efforts. In the southern hemisphere, the comet was easy to see during March and early April, although not as bright as it had been 76 years earlier.

FIGURE 4.17 A series of photographs of Halley's Comet in 1910 showing the changes in the comet's appearance as it approached and then receded from the Sun.

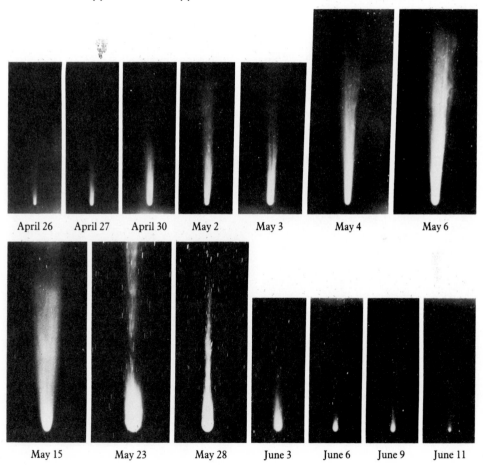

The primary interest in the most recent appearance of Comet Halley did not depend, however, on ground-based observations. The pending arrival of this famous comet stimulated the USSR, the European Space Agency, and Japan all to send instrumented spacecraft to fly into the comet a few weeks after perihelion. Of the major space-faring nations, only the U.S. failed to contribute to this Halley "armada." Later in this chapter we will discuss the results from these missions (Figs. 4.18 and 4.19).

In spite of the difficulty of seeing comets in our modern world, a devoted group of comet hunters continues to discover several new ones each year, as well as to greet the return of short-period comets. This interest is encouraged by the International Astronomical Union policy of naming comets for their discoverers, a custom that began about a century ago. Up to three independent discoverers can be recognized, leading to some awkward names, such as Comet Honda-Mrkos-Pajdusakova or Comet IRAS-Araki-Alcock, named for the IRAS satellite as well as two astronomers. Discovering a comet is the only sure way of getting your name assigned to an astronomical object.

FIGURE 4.18 Two Halley missions: a) The VEGA spacecraft launched by the USSR dropped a probe at Venus (including the release of balloons into that planet's atmosphere) en route to Halley's Comet. b) An artist's conception of the Giotto spacecraft encountering Comet Halley. Solar cells are wrapped around the exterior of this cylindrical spacecraft, which spun slowly around its long axis to maintain stability. Relative sizes of comet and spacecraft are not to scale.

(a)

(b)

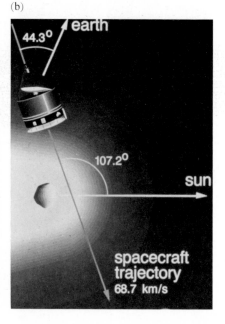

(a)

(b)

FIGURE 4.19 a) Fred Whipple (left), the originator of the icy-snowball model for comet nuclei, listening to Roald Sagdeyev (right), the director of the VEGA mission and of the USSR Institute for Space Research, as the data from the spacecraft were arriving (March 1986). The VEGA pictures were the first visual evidence that Whipple's model — conceived 36 years earlier — was indeed correct (see page 60 and Fig. 4.20). b) Part of the European team that developed the Giotto mission, celebrating a successful launch in French Guiana: in back, F. D'Allest; middle row, left to right, J. Lions, G. Delouzy, A. Ammar, and R. Lüst (Director of ESA); front row, R. Bonnet, A. Brahic, and G. Vedrenne.

Orbits of Comets

Comets are classified on the basis of their orbits as long-period comets or short-period comets, with the division placed rather arbitrarily at 200 years. Most **long-period comets** are on extremely eccentric orbits, falling in toward the Sun from a great distance and returning again to the depths of space. Their ellipses are so elongated that they can be approximated very well by a parabola with the Sun at one focus. Of the 658 comets discovered through 1979, 545 are long-period comets.

Comets with periods shorter than 200 years are called **short-period comets.** Thus both Halley and Encke, with their periods of 76 and 3.3 years, respectively, are short-period comets. As of 1979, 113 short-period comets were known. The great majority of these have periods of 5 to 8 years, indicating that their aphelia are near the orbit of Jupiter; they are called Jupiter family comets. Among the short-period comets, Halley, with its aphelion near Neptune's orbit, is unusual (Fig. 1.20).

Because the paths of comets cut across those of the planets, their orbits are inherently unstable. Like the Earth-approaching asteroids, they must be coming from elsewhere. As we will discuss below under the heading of cometary evolution, scientists believe there is a large, diffuse cloud of comets at the outer fringe of the solar system, containing perhaps a trillion (10^{12}) objects. The comets we see represent the tiny fraction that leak from this immense reservoir.

Unlike any other group of planetary objects, the comets do not share the normal, direct sense of rotation of the system. On their first trip through the inner solar system, they are just as likely to have high inclination as low, or to orbit in the retrograde direction as directly. From this aspect of their behavior, we come to one of three conclusions: (1) comets either come from outside the solar system; (2) they formed before the solar nebula collapsed and spun itself up; or (3) their history since formation has been able to shuffle or randomize their orbits so that they retain no memory of the original spin of the solar nebula.

The Parts of a Comet

As it nears the Sun, a comet develops the extended atmosphere that is its trademark. The small solid body from which this atmosphere is released is called the nucleus. This is the real heart of the comet, even though it is generally too small to be seen. The atmosphere that surrounds the nucleus is called the head or coma of the comet, and the long streamers of less substantial gas and dust sweeping away from the Sun are called the tail. By definition, all comets have a nucleus and a coma, and all except the faintest have tails.

4.7 The Cometary Nucleus

The Dirty-Snowball Model

The most important part of a comet is its **nucleus**; the rest is insubstantial show. We know a nucleus must exist, since the transient gases of the atmosphere come from somewhere. Telescopic observers cannot see it directly, however, since it is lost in the glare of the head of an active comet. Only recently has radar penetrated the atmospheres of several comets to provide unambiguous detection of their nuclei, while the spacecraft that flew through Comet Halley in 1986 were able to photograph the nucleus of that comet at close range.

Before we recognized that comets contain a great deal of water ice, which evaporates under solar heating to generate the atmosphere, some scientists thought that the nucleus consisted of a loose agglomeration of grains and boulders similar in composition to the primitive meteorites. Because this meteoritic material contains limited quantities of trapped gas, it was necessary to hypothesize a great number of particles to account for the observed gas production near the Sun.

In 1950, however, Fred L. Whipple of Harvard University made a fundamental step forward when he developed a model for the nucleus as a single, relatively small "dirty snowball," made up of roughly equal quantities of silicates and ice. Whipple suggested that the rock and ice were intimately mixed, as might be expected for a primitive body that formed directly from the solar nebula or even from dust and gas clouds in interstellar space (Fig. 4.19a).

Composition of the Nucleus

At first very little was known about the composition of the nucleus, although Whipple concluded that water (H_2O) ice was likely to be the major constituent. This hypothesis was supported by observations that most cometary activity turns on as the approaching comet reaches about 3 AU from the Sun. At this distance we can calculate that the surface temperature, warmed by the approaching Sun, should reach the point (about 210 K) at which water ice begins rapid evaporation. Other ices, with different evaporation rates, would begin activity at a different temperature and hence a different distance from the Sun.

The central role of water ice has been confirmed by observations of the spectra of various comets. When water vapor is released from the nucleus, it is quickly ionized to become H_2O^+, and it is then broken down by sunlight into its constituent parts: hydrogen (H) and hydroxyl

(OH). Emission by OH molecules is a common feature in spectra of bright comets. In recent years both H and OH have been detected from observatories in space, while emission from H_2O^+ ions has been detected optically and measured by radio astronomers. Moreover, in 1986 neutral water itself was detected in the atmosphere of Comet Halley using the telescope of the NASA Kuiper Airborne Observatory.

In addition to water, emissions have been seen from carbon gases: C, CO, and CO_2. These indicate the presence of additional ices, either carbon dioxide (CO_2) or carbon monoxide (CO) or both. Other smaller quantities of nitrogen compounds indicate the presence of ammonia (NH_3). These constituents are summarized in Table 4.2.

In addition to water ice and probably also frozen carbon dioxide, carbon monoxide, and ammonia, the cometary nucleus also contains substantial quantities of dark carbonaceous and silicate dust. As the ices evaporate, they release the dust particles, which stream into space carried along by the flow of gas. This dust contributes to the cometary tail. As we will see in Section 4.9, it may also be the origin of the meteors. Larger rocky masses may also be present in the comet, but whether these are represented in our meteorite collections remains an open question.

Physical Nature

How large is the nucleus of a comet? For most comets only a rough estimate can be made. Once the atmosphere of a comet forms, the tiny nucleus is lost in the midst of the bright dust and gas cloud that forms the comet's head, and its size cannot be measured.

One way to penetrate through the obscuring gas and dust is to use radar. Radar is a powerful tool for the astronomer because the exact properties of the transmitted radio pulse can be controlled by the experimenter. (We will return to this subject in more detail in Chapter 6.) Its primary disadvantage is that it requires that a small target must be relatively close to the transmitter, closer than most comets come to the Earth. The first radar signal was bounced from the nucleus of Comet Encke in 1980, and in 1983 two faint Earth-approaching comets were picked up in quick succession. In each case, the derived diameter of the nucleus was between 1 and 4 km.

The only cometary nucleus that has been photographed in detail is that of Halley. During the week beginning 6 March 1986, three spacecraft flew into the heart of this comet. First were the Soviet spacecraft VEGA 1 and VEGA 2 (Fig. 4.18a), which each came within approximately 8000 km and photographed a thick envelope of dust surrounding the nucleus with several bright

TABLE 4.2 Atoms and molecules observed in cometary spectra

	Neutrals	Ions
Atoms	H, O	
	C, S	C^+
	Na, K, Ca,	Ca^+
	V, Cr, Mn,	
	Fe, Co, Ni, Cu	
Diatomic Molecules	OH	OH^+
	CH, NH, CN	CH^+
		CN^+ ?
	C_2	
	CO, CS, S_2	CO^+
		N_2^+
Polyatomic Molecules	NH_2, C_3, HCO ?	CO_2^+
	H_2O, CO_2	H_2O^+
	HCN, H_2CO ?	
	NH_3 ?, CH_3CN ?	
	Silicates	

The presence of a species shown with a question mark needs to be confirmed (adapted from C. Arpigny).

jets of gas and dust erupting from it. Using navigation data from the VEGAs, the European Giotto craft targeted toward only 500 km from the nucleus for its 14 March encounter (Fig. 4.18b). Its cameras pierced the dust fog and imaged the nucleus itself, an irregular dark mass about $16 \times 8 \times 8$ km in size. The reflectivity of this nucleus was only about 3%, similar to the darker asteroids (page 60; Fig. 4.20).

The very low reflectivity of Halley, indicative of primitive, carbonaceous material, apparently confirms the suggestions made by Hawaii astronomer Dale P. Cruikshank and a few other observers that most comets are covered by dark material similar to that found on some distant asteroids. The nucleus is a very dirty snowball, indeed. Presumably the loss of ice and other volatiles from the upper layers of the nucleus allows a crust of dark dust to accumulate over the entire surface. A similar effect darkens the toe of a receding glacier on Earth by concentrating nonvolatile material as the ice evaporates.

Halley's diameter of somewhat more than 10 km may be typical of bright, active comets, while the dimmer Jupiter family comets seem to be smaller. A few spectacular comets from past centuries are estimated to have been much larger than Halley — perhaps 50 km or more in diameter. The corresponding range in masses is from about 10^{10} to 10^{12} tons.

Cometary Activity

As a typical comet nucleus approaches the Sun, the rapid evaporation of its ices begins at a surface temperature of a little above 200 K, out in the main asteroid belt. By the time it crosses the orbit of Mars, the comet begins to develop a full-scale atmosphere and tail. Nearer the Sun, the solar energy evaporates more and more ice as well as heating the surface. Eventually, most of the energy goes into evaporation, although the darker parts of the cometary crust can become quite warm.

Measurements of the total brightness of comets have yielded estimates of the mass of gas and dust lost during one trip through the inner solar system. Typically this is about ten million tons for an active comet, corresponding to something like 0.1 percent of the total cometary mass. Clearly, at this rate of loss the comet will not last forever, but it will exhaust its store of ices after only a thousand passes through the inner solar

FIGURE 4.20 A schematic diagram illustrating the main features visible on the Halley nucleus as photographed by the Giotto spacecraft (see page 60).

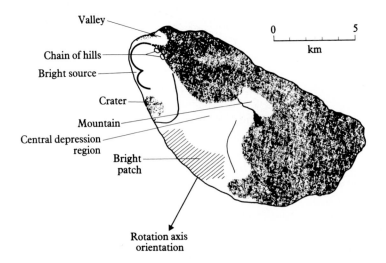

Valley

Chain of hills

Bright source

Crater

Mountain

Central depression region

Bright patch

0   5
km

Rotation axis orientation

system. If the solid or dirty component is carried away with the evaporating ices, the comet will simply shrink down to nothing. On the other hand, if a residual core of solid material remains after the ices are gone, this core would be indistinguishable from a dark Earth-approaching asteroid. We do not know which of these scenarios for the death of a comet is correct. Nor are they mutually exclusive; some comets may end one way, some the other.

Non-Gravitational Forces

While a comet remains active, the escaping gases can have an effect on its orbit. Recall our discussion in Section 2.6 of the way a rocket engine operates, generating thrust from the ejection of mass. The same thing occurs with a comet. The evaporation of ice takes place primarily in jets on the warmest part of the comet, which corresponds to the "afternoon" side facing the Sun. As dust and gas stream away from the nucleus at a speed of hundreds of meters per second, the reaction creates a force in the opposite direction, away from the Sun (Fig. 4.21). Acting continuously over a period of weeks, such a "non-gravitational" force can have an easily measurable effect on the orbit, and the influence of such forces must be taken into account when predicting the future paths of comets.

One of the motivations for Whipple's icy-snowball model for the nucleus was its ability to explain non-gravitational forces by the rocket effect. A larger nucleus, consisting of a loose collection of gravel and boulders, would not behave in this way. In his original 1950 paper on the cometary nucleus, Whipple showed how the rocket effect could explain the observed changes in the orbit of Comet Encke, which had been tracked by astronomers for more than a century. Subsequently, analysis of the non-gravitational forces on comets by Whipple and others has also led to determinations of the spin rate of the nucleus of Encke and other short-period comets,

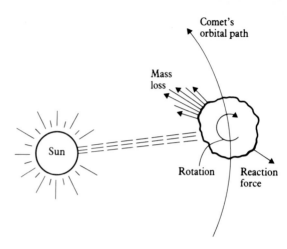

FIGURE 4.21 A schematic illustration of the rocket effect exerted by gases escaping in jets from a comet nucleus.

demonstrating that comets, like asteroids, typically spin in periods that range from a few hours to a few days. Comet Halley, as measured by the Japanese spacecraft in 1986, has a rotation period of 2.2 days, although other periods have been reported by other investigators.

While the icy-snowball model provides a satisfactory understanding of many cometary phenomena, it also leaves important questions unanswered concerning the details of the activity. What would it really be like to ride along on a comet as it approaches the Sun? The Halley probes seemed to indicate a highly dynamic environment, with huge jets of material like geysers dominating the activity near perihelion. Since the spacecraft gave only brief snapshots, we do not know the stability of these centers of activity. Perhaps eruptions burst out from different spots on the nucleus, as the dark surface absorbs sunlight and heats the ices below. Even if we did understand Halley, other comets probably behave differently. We need close encounters with several comets, including some "fresh" ones from the outer reservoir, before we can be sure of any generalizations.

4.8 The Comet's Atmosphere

The part of a comet that is visible from Earth is its atmosphere, a term we use to include both the coma and the tail. It is the atmosphere that struck fear into the minds of watchers on hilltops in the ancient world, and it is also the atmosphere that generated the poison-gas panic that afflicted parts of Europe and the U.S. in 1910 when the Earth passed through the tail of Comet Halley. It is also the atmosphere that gave rise to the word comet, from the Greek for long-haired.

The Comet Head

The brightest part of a comet is the inner **head** or **coma,** consisting of gas and dust recently ejected from the nucleus (Fig. 4.22). Sometimes there appears to be a starlike condensation of bright material at the center, fooling observers into thinking they are seeing the nucleus itself, when actually it is only the brightest inner part of the atmosphere, a few hundred to a few thousand kilometers in diameter. Sometimes this inner atmosphere is symmetric about the nucleus, but more often it is brightest in the direction of

FIGURE 4.22 The head of Halley's Comet as seen on 8 May 1910

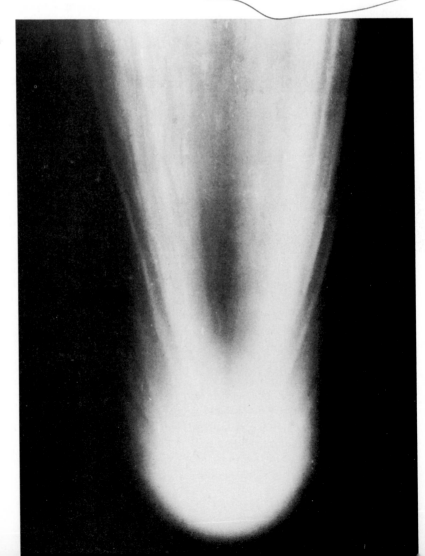

the Sun, and frequently it displays structure in the form of fans of denser material apparently streaming away from the hottest part of the nuclear surface.

Although astronomers presume that the primary constituents of the innermost atmosphere are the gases produced by the evaporation of the cometary ices — water vapor (H_2O) with perhaps carbon dioxide (CO_2), carbon monoxide (CO), ammonia (NH_3), and others — these gases have only been detected recently, many of them by the spacecraft that penetrated deep into the atmosphere of Comet Halley. Instead, the spectroscopist detects the gases listed in Table 4.2, from which we must infer the nature of the original material.

Apparently the gases released from the nucleus are quickly broken down by ultraviolet sunlight to create the molecular fragments also listed in Table 4.2. Most of these reactions are straightforward and easy to calculate, but others remain mysterious. There is a great deal of complex chemistry that takes place in the inner atmosphere of a comet, within minutes of the release of gas from the surface, and not all of it is understood. Thus the identities of some of the so-called "parent" molecules, from which the fragments in Table 4.2 are produced, are still unknown.

It is possible that some of the silicate material in comets predates the origin of the solar system, having originated in the interstellar material before the solar nebula formed. Let us examine Table 4.2 again. Most of the molecules listed are unstable fragments of some larger compound. CN, for example, is not a stable molecule. Evidently the parent compounds are quickly broken down by solar ultraviolet light. Among the known or suspected parent molecules, HCN and CH_3CN both seem reasonable sources of CN. H_2O will produce H, O, and OH. C_2 and C_3 are harder; are the parents hydrocarbons like C_2H_2, or are they complex organic molecules?

It is worth noting that many of the cometary molecular fragments have also been identified by radio astronomers in interstellar "molecular clouds" throughout our galaxy. It is possible that some of the cometary volatiles are materials frozen out directly from such interstellar sources. If this idea is correct, it implies that comets may be vehicles for carrying molecules created out among the stars to the surfaces of satellites and planets throughout the solar system, and includes the possibility that they brought organic material to the early Earth.

The Plasma Tail

Near the surface of an active comet, the gas density is only about a millionth as great as the density of the Earth's atmosphere, and this density falls off rapidly with distance as the gas streams away from its source. At distances as far as 10,000 km the gas flows smoothly; beyond this range, the individual molecules cease interacting with each other and with the embedded dust. Surrounding the rest of the atmosphere is a cloud of glowing hydrogen atoms that can extend more than a million kilometers from the nucleus. Beyond a hundred thousand kilometers, however, most of the gas molecules become ionized, and this plasma is swept away by the blast of the solar wind.

Now controlled by the solar wind, the charged molecular debris rapidly assumes the characteristic wispy form of the cometary tail. This tail is called a **plasma tail.** We readily identify plasma tails by the blue glow of ionized carbon monoxide (CO^+) stimulated to fluorescence, although molecular emissions due to water, carbon dioxide, and other ionized gases have also been detected. Plasma tails are straight, point away from the Sun, and usually are made up of individual streamers or rays only

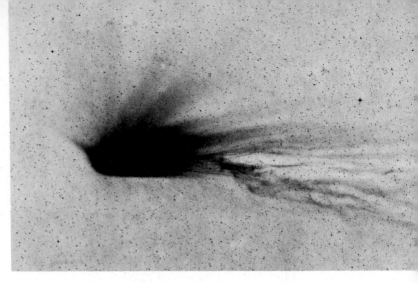

FIGURE 4.23 Comet Halley as photographed from Australia with the U.K. 1.2-m (47-inch) Schmidt telescope on 22 February 1986, just 13 days after perihelion. (The picture is printed as a negative — black is actually white — in order to bring out faint details.) The "spike" pointing to the upper left from the head of the comet is dust in its orbit seen in projection. Several dust tails and multiple streamers in the plasma tail are visible.

a few thousand kilometers across. An active comet such as Halley can develop a plasma tail a hundred million kilometers in length, practically long enough to stretch from the Sun to the Earth (Fig. 4.23).

Within a plasma tail the actual density of material is incredibly low, typically only a few hundred molecules per cubic centimeter. Thus a giant tail tens of millions of kilometers long contains no more molecules than a modern supertanker — perhaps half a million tons of mass. For most comets, the mass of material in the tail is far smaller.

Since the plasma tail is driven by the solar wind, it provides a means of recording the speed and direction of flow. By tracking individual kinks and twists as they move downstream, astronomers could measure properties of the solar wind long before direct observation of this outflow of gases from the Sun was possible by spacecraft. Typical solar wind velocities determined from comets are about 400 km/s, sufficient to traverse the entire length of the comet tail in less than a day. It is not surprising, therefore, that plasma tails can alter in form from hour to hour and that the overall appearance of a comet even as seen by the naked eye can change dramatically from one night to the next.

The Dust Tail

Most comets have two tails: the plasma tail, which we have been discussing, consisting of charged molecules caught in the solar wind, and a second tail consisting of dust grains, appropriately called the **dust tail.** Although generally shorter than plasma tails (usually less than 10 million km long), dust tails can be as bright or even brighter. They are readily distinguished by their color (yellow-white, from reflected sunlight) and from the fact that they are curved rather than straight (Fig. 4.24).

Dust tails consist primarily of small grains only a few micrometers in size, indicating that much of the solid material in the nucleus must be of the same consistency, similar to the finest cake flour. Once decoupled from the expanding and thinning gas, the dust grains follow their own orbits, moving under the joint influence of solar gravity and of pressure generated by solar radiation. Because they move relatively slowly, the dust grains tend to mark the location of the comet at the time of their release. Thus the dust tail traces the path of the comet across the sky. Measurements of the material in dust tails suggest a mass that is roughly comparable to the gas in the plasma tail, indicating that similar quantities of ice and dust are present in the nucleus.

4.9 The Comet-Meteor Connection

Meteors

The dirty-snowball model for the comet nucleus predicts that comets will release large quantities of dust or larger solid material into the inner solar system. The production of solid material is further supported by the curving dust tails of comets. What is the ultimate fate of this dust? A tiny fraction of it strikes the Earth, burning up in the atmosphere to produce the **meteors.**

Meteors or shooting stars can be seen on any clear night, usually at a rate of several per hour. A typical meteor that can be spotted by the naked eye is no larger than a pea, but this is sufficient to generate a much larger cloud of glowing gas high in our planet's atmosphere that can be seen as far away as 200 km. Perhaps as many as 25 million meteors bright enough to be seen strike the Earth every day, amounting to hundreds of tons of cosmic material added to our atmosphere every twenty-four hours.

Many meteors are produced by bits of material in random orbits. These can come from any direction at any time, and are called sporadic meteors. Sometimes, however, the Earth encounters a stream of particles moving together along similar orbits around the Sun which we recognize as a **meteor shower.** It is the shower meteors that are most directly linked to the comets.

Meteor Showers

To illustrate the connection between cometary dust and meteors more specifically, we can look to one of the famous comets of the last century called Comet Biela, discovered in 1826 on a 6.75 year orbit. In 1846 this comet split in two, and upon its next return in 1852 both components

FIGURE 4.24 Comet Mrkos photographed during six nights in 1957. Note the large changes in the plasma tail that occur from night to night, while the curved dust tail remains relatively unchanged.

| August 22 | August 24 | August 26 | August 27 |

were again present, separated by 2 million kilometers. Neither part of Comet Biela was ever seen again; between 1852 and 1866, it simply ceased to exist. Nevertheless, astronomers watched with interest when the Earth passed through the orbit of Comet Biela in 1872, and they were not disappointed. Instead of the comet they saw a wonderful meteor shower, with thousands of meteors visible from any spot on Earth during the night the Earth crossed the comet's orbit. The comet had transformed itself into a stream of meteoric particles (Fig. 4.25).

Many other extant comets have been identified with meteor showers, although a number of other showers are without known cometary associations. Comet Halley, for example, seems to be the source of two such showers annually, produced when the Earth crosses Halley's orbit each May and October. The showers themselves vary in intensity from just a few meteors per hour (hardly deserving the name shower), to the spectacular, once-in-a-lifetime events in which the sky is filled with silent, flashing meteors as numerous as falling snowflakes. One of the most vivid displays occurred in 1899, when the Earth crossed the orbit of Comet Tempel-Tuttle, in which up to 100,000 meteors per hour were observed.

The most dependable meteor shower, although certainly not the most spectacular, can be seen each year between 9 and 13 August. Called the Perseid shower, this stream of debris is associated with a bright comet of a century ago named Comet 1862 III. Since the particles seem to be evenly distributed along the comet's orbit, we encounter about the same number each year. In contrast, the really spectacular but unpredictable showers take place when the Earth moves through a grouping of particles along the path of the meteor stream. Table 4.3 lists the most dependable annual meteor showers and their associated comets, when known.

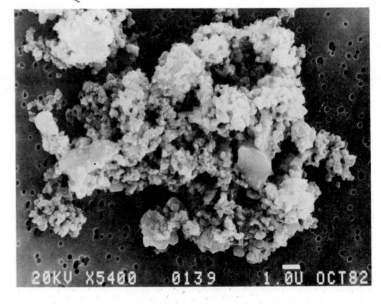

FIGURE 4.25 These meteoric particles (often called "Brownlee particles" after their discoverer, Donald Brownlee) were captured in the Earth's stratosphere by a U-2 aircraft equipped with a special dust sampling device. A steady rain of particles like these (here magnified × 10,000) is constantly falling on the Earth. We are thus continuously breathing in "comet dust" along with everything else our air contains.

20KV X5400   0139   1.0U OCT82

TABLE 4.3 Meteor showers

Shower Name	Date of Maximum	Associated Comet	Comet's Period
Quadrantid	January 3	—	—
Lyrid	April 21	1861 I	415 yr
Eta Aquarid	May 4	Halley	76
Delta Aquarid	July 30	—	—
Perseid	August 11	1862 III	105
Draconid	October 9	Giacobini-Zinner	7
Orionid	October 20	Halley	76
Taurid	October 31	Encke	3
Andromedid	November 14	Biela	7
Leonid	November 16	1866 I	33
Geminid	December 13	3200 Phaeton[a]	1.4

[a]Phaeton is actually an asteroid, i.e., a small body showing no cometary activity at present.

Most of the meteors that strike our atmosphere are associated with showers, and therefore with meteor streams in space. Perhaps most of the non-shower, or sporadic, meteors are the remnants of dispersed meteor streams, and most known meteor streams in turn are associated with comets.

If most meteors are really cometary dust, we naturally ask if meteorites might not also be from comets. However, meteorites are not associated with meteor showers, and even on the rare occasions when the sky is filled with "falling stars" no actual meteorites fall to Earth. The shower meteors are not simply small meteorites that do not reach Earth's surface. They are fundamentally different material.

Cometary Meteors

A clue as to the nature of the cometary meteors has been inferred from observations made during their flights through the atmosphere. Their densities, calculated from the rate at which they slow due to atmospheric friction, are less than 1 g/cm^3, in contrast to the densities of meteorites, which are between 3 and 7 g/cm^3.

Efforts have also been made, with some success, to collect tiny meteoric particles from the upper atmosphere. These fragments of cosmic dust are fluffy bits of chemically primitive matter (Fig. 4.25); probably, they are our only true samples of comets. Keep in mind, however, that simply because most meteors are of different composition than the meteorites, we cannot be sure that some meteorites might not also have a cometary origin.

The Zodiacal Dust Cloud

Cometary dust also contributes to another phenomenon of the planetary system called the **zodiacal dust cloud.** In reflected sunlight, this cloud appears as a faint, diffuse band of light stretching around the ecliptic, called the zodiacal light. You can see the zodiacal light just after twilight, if you have a clear, dark horizon well away from city lights. In the thermal infrared part of the spectrum, the glow of the zodiacal cloud is one of the brightest features of the entire sky. Both the reflected sunlight and the thermal glow originate in small dust particles that fill the inner solar system, including the asteroid belt, near the plane in which the planets orbit the Sun.

Maintenance of the zodiacal dust cloud requires the release of about 10 tons of dust per second into the inner solar system. Much of this dust comes from comets, but a part of it may also originate in the asteroid belt. This connection is further strengthened by the 1983 discovery by the IRAS infrared satellite observatory of several distinct dust bands within the asteroid belt (Fig. 4.26). As with the meteorites, we find that we cannot make a definitive choice between comets and asteroids as the source of the dust in the solar system.

FIGURE 4.26 The IRAS image in which comet dust trails were first detected directly. The thick, bright band cutting diagonally across the image consists of debris from asteroid collisions (possibly a large, single collision). The Encke dust trail is seen to the north of the ecliptic. The Tempel 2 dust trail is seen at a distance of 1.3 AU, having just passed through perihelion. Both comets are located several degrees off the left of the image.

4.10 Evolution of Comets

The Source of the Comets

The comets we see are temporary residents of the inner solar system. Many are traveling on nearly parabolic orbits and will not return for millions of years. The short-period comets, however, have limited lifetimes. After approximately a thousand orbits — less than 10,000 years for most of them — their volatile ices will be exhausted and they will no longer be comets, although the dead nucleus might remain as an Earth-approaching asteroid. They are also dynamically unstable. Like all objects with planet-crossing orbits, they run the risk of either impacting a planet or being gravitationally expelled from the solar system. Their average residency in the solar system is no more than a few million years. Therefore, it is important to determine the source of the comets.

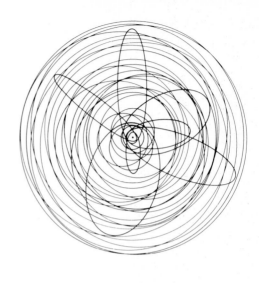

100,000 AU

FIGURE 4.27 A schematic representation of a tiny fraction of the orbits of comets in the Oort comet cloud. The dot in the center is considerably larger than the diameter of Pluto's orbit at this scale.

The Oort Comet Cloud

The first such satisfactory theory for the origin of the comets was proposed by the Dutch astronomer Jan Oort in 1950, the same year as Whipple's model for the nucleus. Oort noted that in all cases where the orbits of new, nearly parabolic comets had been carefully determined, the orbits indicated an aphelion at a distance of approximately 50,000 AU, a thousand times more distant than Pluto. Very few comets seemed to come from greater distances, and none showed evidence of originating outside the solar system in interstellar space. He therefore suggested the existence of a comet cloud associated with the Sun but existing far beyond the known planets. This grouping of comets is now called the **Oort comet cloud** (Fig. 4.27).

At any given time virtually all of the comets in the Oort cloud are too far away for detection, and their ices will be preserved indefinitely in the cold of space. Even if the orbits of these comets bring them in as close as Neptune, they remain frozen and invisible to us. However, occasionally some of these comets can be perturbed by the slight gravitational tugs of nearby stars to bring their perihelia into the inner solar system. Only then will they be recognized as comets.

In order to account for the several new comets discovered each year, Oort calculated that the comet cloud contained about 100 billion (10^{11}) comets. More recent calculations have suggested that the true number is 10 times greater. If there are a trillion comets in the Oort cloud, and ten of these are lost each year by falling too near the Sun, the total number of comets lost to the cloud is still only about five percent of the original population.

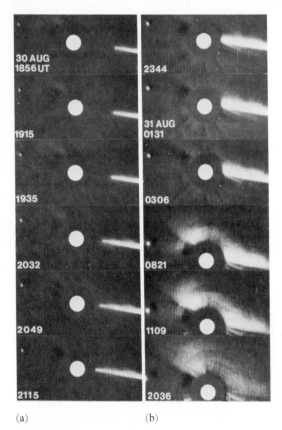

(a)   (b)

FIGURE 4.28 The probable collision of Comet 1979 XI with the Sun on 30 August 1979, photographed with an orbiting coronagraph telescope. The white disk shows the size of the Sun, which is actually hidden behind a disk in the coronagraph to mask the Sun's glare. The comet approaches the Sun from the lower right in the sequence of frames (a). The bright spot in the upper left is the planet Venus. The series of frames (b) shows the dissipation of material from the comet after impact with the Sun.

The Fate of Comets

Although a comet may have rested peacefully in deep space for billions of years, once it is diverted into the inner solar system its life expectancy is limited. There is some chance that it will not survive even its first plunge toward the Sun; several comets have been seen to impact the Sun or to pass so near its surface that they are destroyed by the heat in a single pass (Fig. 4.28). Other comets die even before reaching the Sun; in 1906 and 1913, for example, new comets that had been expected to put on a good show simply faded to invisibility as they neared perihelion.

More likely, a new comet will pass near enough to some planet, usually Jupiter, to suffer a change in orbit. This change can either increase its energy, ejecting it completely from the solar system, or it can lose energy, becoming a short-period comet. The reason most short-period comets have aphelia near the orbit of Jupiter is that they were captured by the gravitation of this giant planet. But the comet has now gone out of the frying pan and into the fire, for it still runs a significant risk of impact with one of the terrestrial planets. Meanwhile the heat of the Sun is also consuming its icy substance at a rapid rate.

There is considerable debate among astronomers concerning the aging of comets. As a comet suffers repeated heating and loss of gas and dust each time it travels around the Sun, does it change or simply grow smaller? If it is of uniform, homogeneous composition, and if the dust is stripped away together with the evaporating ices, then we can expect the nucleus to shrink without otherwise changing much over time. Alternatively, if comets have a layered structure, the successive stripping away of the surface (typically about a meter each orbit) will expose differing materials as the comet ages. To date there is no evidence for such inhomogeneities, since the spectra of old comets are not different from those of young ones. A third possibility is that the nucleus is homogeneous, but the non-volatile debris that builds up on the surface ultimately insulates the interior under a thick crust of dark material. So far the jury is out on this question.

Not all comets achieve old age. Some have their lifetimes cut short by accidents. They may be perturbed into orbits that result in impact with planets or with the Sun or they may also break apart, for reasons not always understood. A recent example of one such breakup is provided by Comet West, which apparently split three times within a few days early in March 1976. The four components drifted apart, extending more than 10,000 km by the end of the month (Fig. 4.29). The smallest fragment survived only a few days, but the other three components of the nucleus still retained their individual identities as the activity level declined with increasing distance from the Sun.

Origin of the Comet Cloud

The idea of the Oort comet cloud provides a framework for understanding the continuing supply of comets to the inner solar system, but it begs the question of ultimate origin of the comets. How did the comets get into the Oort cloud in the first place? Might they have been formed in place, tens of thousands of AU from the Sun? We have already mentioned this possibility, but at such distances in the solar nebula it seems difficult to imagine solids condensing, except possibly when the solar system was passing through a dense molecular cloud. Most astronomers believe instead that the comets were formed in the realm of the planets, at the same time as the other members of the planetary system.

In this scenario, the comets represent bodies that condensed in cooler parts of the solar nebula. The presence of a variety of ices in comets suggests formation temperatures in the range of 50 K to 100 K, corresponding to the region of space now occupied by Uranus and Neptune. Subsequently, some of these proto-comets must have been ejected into the Oort cloud, while others were gravitationally dispersed throughout the solar system or expelled from the system entirely. They are representative of the original condensates from the outer part of the solar nebula, just as the primitive asteroids are remnants of the materials that condensed in the inner part of the nebula.

FIGURE 4.29 A sequence of pictures taken over 16 days shows the head of Comet West breaking up in March 1976. This breakup means that Comet West will never repeat its spectacular display when it returns again. Future generations (a million years hence) may see several dim comets (or nothing at all!) instead.

8 March 76   12 March 76   14 March 76   18 March 76   24 March 76

———————◇———————

Summary

Asteroids and comets are interesting in their own right, as small members of the planetary system, and as possible parent bodies for the meteors and meteorites. Because they are much smaller than the planets, the comets and asteroids have remained relatively unmodified since the origin of the solar system.

The distinction between a comet and an asteroid is relatively easy upon observation. An asteroid is a small rocky world; through the telescope it looks like a tiny, slowly moving star. The term asteroid, in fact, means "small star." A comet, however, does not appear starlike (except when far from the Sun); it develops a thin but extensive atmosphere as its ices evaporate under the influence of solar heating. It is this transient atmosphere that gives a comet its characteristic head and tail and renders the nearer comets easily visible to the unaided eye. For the same reason, comets have been known since antiquity, while the asteroids were not discovered until the beginning of the nineteenth century.

In physical terms, we can outline the main distinctions between comets and asteroids as follows: (1) In composition, asteroids are made of rocky and metallic materials, similar to the inner planets, while comets contain a substantial fraction of volatile ices. (2) The orbits of most asteroids are roughly similar to those of the planets, located primarily between Mars and Jupiter. Comets, in contrast, have highly elliptical orbits that generally take them from within the orbit of Earth to beyond Jupiter. (3) Main belt asteroids are on stable orbits and have probably been in their present position since the formation of the solar system. Comets, however, are on unstable orbits and must originate elsewhere, probably the Oort comet cloud far beyond the orbit of Pluto. It should be noted that Earth-approaching asteroids are also on unstable orbits.

From telescopic studies, the asteroids appear to be similar in composition to many of the major classes of meteorites, and scientists believe that they are in fact the parent bodies of most meteorites. Note that the asteroids, like the meteorites, include both primitive bodies (which make up the majority) and differentiated objects. The best known differentiated asteroid is Vesta, which has a basaltic surface and is very likely the parent body of the eucrites.

The two satellites of Mars, Phobos and Deimos, are probably captured asteroids, with compositions similar to the carbonaceous meteorites. Each has, however, been modified by the unique environment of Mars, so that we must be cautious in drawing conclusions about the larger asteroid population from these interesting objects.

Earth-approaching asteroids are a separate class, sharing many orbital properties with the comets while physically resembling the smallest main belt asteroids. Since their present orbits are unstable, they must also originate elsewhere. Most are probably derived from the main belt, but some may also be now extinct comets that have lost most of their volatiles.

The comets can best be understood in terms of the dirty-snowball model, in which the nucleus is composed of approximately equal quantities of ice, primarily water ice, and silicate and carbonaceous material, largely in the form of small dust grains. Rapid evaporation of the ice under the influence of solar heating leads to the production of the head and the plasma tail, with dust carried along to form the dust tail. Preferential evaporation from the Sun-facing hemisphere of the nucleus also gives rise to jets which produce the non-gravitational forces that alter the orbits of many comets. The gas released from the nucleus dissipates into space, while the dust contributes to meteor showers when the Earth intersects the orbit of a comet.

Although we have not yet sampled cometary material directly, except possibly for the dust collected at high altitudes, it seems reasonable to conclude that it is even more primitive than that of the asteroids and meteorites, in the sense that it more closely represents the composition of the solar nebula. Presumably, the comets are bodies that formed very early in the evolution of the solar system, aggregating from icy grains that condensed in the outer parts of the solar nebula. We may think of comets, then, as messengers from the distant past, preserved for us in the deep freeze of space for 4.5 billion years.

Key Terms

asteroid family

C-type asteroid

coma

comet head

comet nucleus

dust tail

Earth-approaching asteroid

long-period comet

main asteroid belt

meteor

meteor shower

Oort comet cloud

plasma tail

resonance

resonance gap

S-type asteroid

short-period comet

zodiacal dust cloud

PART TWO
REVIEW QUESTIONS

1. Distinguish among meteors, meteorites, and meteoroids, in terms of their properties, appearance, and origin. What relationships exist among these three classes?

2. Distinguish among comets, main belt asteroids, and Earth-approaching asteroids, in terms of their orbits and their physical and chemical properties. What relationships exist among these three classes of small bodies?

3. What is meant by primitive bodies? Which objects are the most primitive? What kinds of information do they provide on the origin of the solar system?

4. Explain the techniques for radioactive age dating. What are the limitations of this technique? How might you check to see if it yields the correct answers?

5. How are meteorites classified? Relate the different classes to more fundamental properties, such as age and probable origin.

6. What makes the Antarctic meteorites special? Explain how they are found, and why they might represent a different population of objects from other meteorite finds.

7. Where did the various kinds of meteorites originate? How can we use the properties of meteorites to determine their parent bodies? Justify the cases for the origins of some meteorites from the Moon, Mars, and Vesta.

8. Why do you think there are differentiated meteorite parent bodies? What might these bodies have been like? Explain how they might have been heated enough to cause differentiation. Consider Vesta as a specific example in addressing these questions.

9. Summarize the types of basic information that can be obtained about an asteroid from telescopic studies. Compare this information with that available for various planets and satellites in the planetary system.

10. Describe the size-frequency distribution expected for a population of fragments. In what solar system environments might we expect this kind of size distribution to apply?

11. Explain the phenomenon of resonance gaps in the asteroid belt. Where else in the planetary system would you expect similar resonance effects to be seen?

12. Trace the typical life history of an Earth-approaching asteroid. Compare this with the life of a short-period comet. Does one last longer than the other? Do you think that short-period comets could evolve into Earth-approaching asteroids?

13. Describe the dirty-snowball model for the nucleus of a comet. Compare this model with the observations. Explain in particular how it accounts for the non-gravitational forces that disturb the orbits of comets.

14. Imagine yourself riding on a comet. Describe what you would see as the comet approached the Sun. Do you think that an instrumented lander on a comet nucleus would survive long? What kind of measurements might such a lander make?

15. Describe how the atmosphere and tail of a comet are formed. How do the phenomena we see in the head and tail relate to the properties of the underlying nucleus?

16. What are the relationships between comets and meteors? If the shower meteors come from comets, isn't it likely that the meteorites come from comets too? Explain.

17. Describe the Oort comet cloud. While a comet is in this cloud, how might it change over time? What is the orbit of a comet in the cloud like? Under what circumstances will a comet in the Oort cloud be able to enter the inner solar system?

18. Could the meteorites, comets, and asteroids have originated in an explosion of a planet between the orbits of Mars and Jupiter? Give arguments for and against.

19. In what ways are the martian satellites, Phobos and Deimos, like the asteroids? In what ways have they been influenced by their environment orbiting Mars?

ADDITIONAL READING

Brandt, J.C. and R.D. Chapman. 1981. *Introduction to Comets*. Cambridge: Cambridge University Press.

Calder, N. 1980. *The Comet Is Coming!* London: British Broadcasting Company.

Chapman, C.R. 1982. *Planets of Rock and Ice* (Chapter 3). New York: Scribners.

Chapman, C.R. 1982. "Asteroids." In *The New Solar System*, 2nd ed., ed. J.K. Beatty, B. O'Leary, and A. Chaikin. Cambridge, MA: Sky Publishing Corp.

Chapman, R.D. and R.L. Bondurant, Jr. 1985. *Comet Halley Returns* (NASA EP-197). Washington: U.S. Government Printing Office.

*Delsemme, A., ed. 1977. *Comets, Asteroids, Meteorites*. Toledo, OH: University of Toledo Bookstore.

*Dodd, R.T. 1981. *Meteorites: A Petrologic-Chemical Synthesis*. Cambridge: Cambridge University Press.

---

*Indicates the more technical readings.

Dodd, R.T. 1986. *Thunderstones and Shooting Stars*. Cambridge, MA: Harvard University Press.

*Gehrels, T., ed. 1979. *Asteroids*. Tucson, AZ: University of Arizona Press.

Heide, F. 1964. *Meteorites*. Chicago: University of Chicago Press.

Hartmann, W.K. 1982. "Small Bodies and Their Origins." In *The New Solar System*, 2nd ed., ed. J.K. Beatty, B. O'Leary, and A. Chaikin. Cambridge, MA: Sky Publishing Corp.

Hutchinson, R. 1983. *The Search for Our Beginnings*. Oxford: Oxford University Press.

*Nature: Encounters with Comet Halley. 1986. *Nature 321*: 6067, Supplement.

Planetary Report: Special Asteroid Issue. 1983. *Planetary Report 3*:4.

Planetary Report: Special Comet Halley Issue. 1987. *Planetary Report 7*:2.

Sky and Telescope: Special Comet Issue. March 1987. *Sky and Telescope 73*:3.

*Wasson, J.T. 1985. *Meteorites*. New York: W.H. Freeman.

Whipple, F.L. 1985. *The Mystery of Comets*. Washington: Smithsonian Institution Press.

*Wilkening, L.L., ed. *Comets*. Tucson, AZ: University of Arizona Press.

Wood, J.A. 1968. *Meteorites and the Origin of Planets*. New York: McGraw-Hill.

PART ◆ THREE

Two Battered Worlds:
Moon and Mercury

The objects in our planetary system are remarkably varied; even the smallest members of the planetary family encompass an impressive range of objects. The larger planets and satellites are still more complex, a result of their more active geologic and chemical histories.

As a general rule among planets, size results in complexity, since internal heat can be retained by larger objects to power an active surface geology. It is therefore reasonable to expect that smaller planets will be less active, and hence probably simpler to understand, than larger ones. This is why we began, in Part II, with the smallest bodies. And it is why we now take up the study of the Moon and Mercury before we tackle the larger and more complex Earth, Venus, and Mars.

A relatively simple geologic history is not the only reason for beginning a detailed study of individual planets with the Moon and Mercury. The fact that both lack an atmosphere also considerably eases our task: there are no obscuring clouds or vapor, no wind or precipitation to erode the surface, and no oxygen or

◀ The lunar roving vehicle used by geologist-astronaut Harrison Schmitt next to a huge boulder in the Taurus-Littrow Valley. Schmitt himself and the other end of this boulder are visible in Plate 6. The first humans reached the Moon two decades ago. Since that time, no astronauts have ventured beyond Earth orbit, and the equipment from the Apollo missions remains untouched on the lunar surface where it will be preserved for millennia in the vacuum of space.

water to weather the rocks or alter their chemistry. Change is a slow process; a million years from now, the footprints of Apollo astronauts in the lunar soil will still be fresh. Unlike the Earth, where the surface geology is often hidden under vegetation, and rock and soil chemistry are altered by local conditions, on the Moon and Mercury what you see is what you get.

The Moon and Mercury also provide an appropriate introduction to the other planets because of the dominant influence impacts have in shaping their surfaces, producing mountains and abundant craters. The other planets have all been subjected to a similar rain of impacting debris. Such impacts provide a common thread, weaving together the histories of different worlds. By first studying impacts on objects where such processes are dominant, we can better interpret planets where both impacts and internally driven geological processes are important.

Although the Moon and Mercury are simpler worlds to understand than Venus, Earth, or Mars, both are complex planets, and we should not delude ourselves into thinking we have solved all of their mysteries. It is sobering to remember that even with all the missions that have studied the Moon and the hundreds of kilograms of returned samples we are still not certain where and how our satellite originated. How much more uncertain may be the conclusions drawn about other, more distant worlds?

CHAPTER FIVE

Our Ancient Neighbor: The Moon

◆

5.1 The Face of the Moon

Our Nearest Neighbor

Although large in comparison to the comets and asteroids discussed in Chapter 4, the Moon is a small planet. With a diameter of 3,476 km, its total surface area is about the same as that of North America. A flight from Los Angeles to New York traverses a greater distance than the lunar diameter. The mass of the Moon is less than 2% of that of the Earth, although it is still substantially greater than that of all of the asteroids combined.

Chemically the Moon differs from both the Earth and the primitive objects introduced in Part II. The Moon is not primitive in the sense of comets and most asteroids; it has been heated and differentiated and lacks water, ice, carbon, and organic compounds. The Moon has its own life history, one very different from that of the Earth.

The Moon is the only planetary body that can be distinguished with the naked eye as a globe, and even without a telescope we can see that its surface is not uniform. Many cultures have associated names and myths with the markings familiarly known as the "Man in the Moon," but it was not until Galileo Galilei turned his first small telescopes on the Moon in 1610 that it became clear that the surface of our satellite was rugged and mountainous like that of the Earth. His observations provided the foundations for considering the planets as other worlds, and thus indirectly gave birth to the science of planetary geology.

Resolution of the Surface

A sharp-eyed observer can see features on the Moon as small as $1/15$ of its apparent diameter, approximately 200 km across. This is the **resolution** of the human eye. Resolution is sometimes expressed in angular terms, but here we are more interested in linear resolution, the dimensions of the smallest features of a planetary surface that can be seen or imaged. At 200 km resolution, the larger light and dark markings can be distinguished, but topographic features, such as mountains and craters, are undetectable (Fig. 5.1). You can demonstrate this conclusion to your own satisfaction by sketching the Moon at different phases and using these sketches to produce a naked-eye map. This exercise is especially instructive when you note that the best Earth-based telescopic views of other planets, such as Venus and Mars, have about the same resolution as naked-eye maps of the Moon.

A telescope, with its higher resolution, is able to reveal lunar surface topography. Galileo's early telescope was barely sufficient to distinguish the largest craters and mountains. You can

FIGURE 5.1 Seen at full phase, the Moon shows little topographic detail even through a telescope. The dominant features visible under these lighting conditions are the dark volcanic maria and the bright rays associated with young impact craters.

do as well today with a good pair of 7 power or 8 power binoculars, which actually perform better than Galileo's best 30 power telescope.

Today's modern astronomical telescopes are far more powerful. As we noted in Section 2.5, however, telescopes on the surface of the Earth are limited in resolution by the atmosphere. For the Moon, this limiting resolution is about 1 km. It is interesting to note that, even after initial spacecraft exploration, most of the other bodies in the solar system have been imaged with resolutions of only one or two kilometers. Thus our current photography of most other worlds is no better than that of the Moon before the start of the space age.

The angle at which sunlight illuminates a surface plays an important part in limiting our ability to distinguish topographic detail. When the Sun is low, features cast shadows, while at moderate angles of illumination slopes and contours are revealed by shadings that vary as the surface is tilted toward or away from the Sun. When the sunlight streams down from directly behind the observer, details of topography become indistinguishable. Thus an image of the full moon emphasizes differences in reflectivity between light and dark areas but suppresses topography. Craters and mountains are best studied near first and last quarter, where we see the border between night and day. Here the early morning or late afternoon Sun is low in the sky, and every topographic detail is sharply etched (Fig. 5.2).

What does the Moon look like at a resolution of 1 km? And what can geologists determine about the processes at work on its surface from images of this quality? The answers not only provide an introduction to the study of the Moon, but also illustrate the limitations imposed upon our understanding of other worlds where we have not yet landed and explored their surfaces directly.

FIGURE 5.2 The nature of the lunar surface is better revealed at quarter phase, when the sunlight strikes obliquely, highlighting topographic features such as craters and mountains.

Lunar Highlands and Maria

The most obvious conclusion that can be reached about the Moon's surface, even from an examination through binoculars or a small telescope, is the presence of two different kinds of surface terrains. The predominant surface type is relatively light (reflectivity about 15%) and extremely rugged, with craters of all sizes stacked one upon the other. Since these lighter, heavily cratered regions also generally lie at higher elevations, they are called the lunar **highlands.** The second surface type is darker (reflectivity about 8%) and smoother, with relatively few large craters. These regions, which make up the features of the Man in the Moon, are called **maria** (singular: **mare**). Mare is the Latin word for sea, and when the term was first applied to the Moon in the seventeenth century these darker regions were thought to be water oceans. Fig. 5.3 contrasts the appearance of these two regions.

Most of the maria are found on the side of the Moon that faces the Earth. The opposite, or far side, had never been seen until 1959 when the Soviet Luna 3 radioed back to Earth the first rudimentary photos. To the surprise of almost everyone, no major maria appeared on the lunar farside, which was almost completely highland terrain. Any explanation for the existence of highlands and maria must address this basic asymmetry in mare distribution. Subsequent mapping of the Moon reveals that only 17% of the surface consists of mare material.

Maria and highlands differ in many ways. Most obvious is the distinction in color and reflectivity, implying that their chemical compositions differ. Second, a difference in cratering suggests different geologic histories. Finally, a difference in elevation exists, which is related to the largest scale forces that have molded the lunar surface. The two types of terrain are representative of different chemical and geological re-

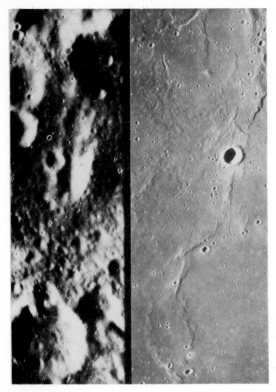

FIGURE 5.3 The contrasting nature of the lunar highlands and maria is clearly shown in these Apollo orbital photos of the two types of terrain. The crater density is 10 to 20 times greater on the highlands (left) than on the mare (right).

Lunar Craters

The prevalence of the characteristic circular features called **craters** provides the second striking feature of the lunar surface. The word crater, derived from the Greek for cup or bowl, refers to the shape of the feature. Note that a crater is a depression; the popular application of the term to a volcanic mountain, such as Vesuvius in Italy or Haleakala in Hawaii, is incorrect. A volcano frequently has a crater at its summit, but the mountain itself is not a crater.

Even binoculars or a small telescope reveal the larger lunar craters, which are hundreds of kilometers in diameter. At the best Earth-based resolution of 1 km, it is estimated that 30,000 craters can be identified on the Moon. These craters range from sharp, new-looking depressions of nearly perfect bowl shape to barely detectable, old, battered craters. In the highlands, the craters are packed shoulder to shoulder, and it seems obvious that new additions must occur at the expense of those craters already present. In some highland areas, the surface is so battered that individual craters almost become lost in the accumulation of jumbled, mountainous debris. In contrast, the craters on the maria are rather widely spread, indicative of a younger surface on which there has been less time for craters to accumulate.

Although the interiors of lunar craters are depressions, their rims are raised above the surrounding surface. Fig. 5.4 illustrates a cross-section through a typical fresh lunar crater. For a mare crater several kilometers in diameter, the rim typically rises hundreds of meters above the surrounding plain, while the interior drops two or three times lower below the mare surface. On the Moon, craters in this size range usually have a depth, measured from the rim crest down to the lowest part, that is between $1/10$ and $1/5$ of their diameter. Larger craters are relatively shallower and more complex, as we will describe in Section 5.3.

gimes, providing windows into two different periods of solar system history.

It is interesting to note in passing that this clear division of the Moon into two terrain categories has had a negative effect on the interpretation of other, more complex planets. For instance, progress in understanding Mars was impeded by the assumption, derived from lunar analogy, that the light and dark areas there must represent regions of differing elevations. Later it was learned that these martian markings have no systematic relation to elevation.

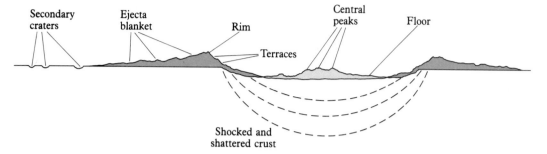

FIGURE 5.4 A typical large lunar impact crater is illustrated here in cross-section.

At the largest scale, the face of the Moon is dominated by a few ringed **impact basins,** which are roughly circular in outline. On the side of the Moon facing the Earth, these basins tend to be the sites of the darker mare materials. Examples readily seen with binoculars or on a photograph of the full moon are Mare Crisium, Mare Serenitatis, and Mare Imbrium (Fig. 5.1). Even more striking would be the bull's-eye feature Mare Orientale (Fig. 5.5), nearly a thousand kilometers in diameter, but Orientale is just barely visible from the Earth. If the Moon were rotated 90° so that this feature faced us, we can imagine the myths that might have developed in an attempt to explain this "eye" staring down on us from space. The major lunar mountain ranges, such as the Carpathians and the Apennines, define the margins of these basins. These mountain arcs are entirely different in origin from those on Earth, having been thrown up as part of the basin-forming process.

For a long time, the origin of the lunar craters and basins was a major issue of debate, as important an issue as that of the distinction between maria and highlands. Largely by analogy with the Earth, most geologists in the nineteenth and early twentieth centuries believed the lunar craters had a volcanic origin. Now we recognize that the great majority, including all of the larger ones, were produced by meteoroidal impacts (Section 5.3).

When the Moon nears full phase and the sunlight strikes it directly, a few craters become the most prominent features on the Moon (Fig. 5.1). In contrast with the maria or even the lighter highlands, these craters and their surroundings appear brilliant white. Radiating from them are long, light **crater rays** that stretch for

FIGURE 5.5 The Orientale basin is the youngest of the great lunar impact basins. Because it has escaped flooding by mare volcanism, Orientale basin shows most clearly the double ringed or "bull's-eye" form of large impact structures.

hundreds and even thousands of kilometers across the surface, passing right across other craters or mountains. Among the brightest of the light ray craters is Tycho, which is marginally visible even to the naked eye. This highland crater, like most on the Moon, is named for a famous scientist. It is 84 km in diameter and is easily spotted in Fig. 5.1. Tycho and the other bright ray craters formed after the mare and highland terrains and appear to be the youngest major features of the Moon; Tycho itself probably formed within the past few hundred million years.

Lunar Stratigraphy

The identification of ray craters as younger than other lunar features is an example of **stratigraphy,** a geological technique for establishing the relative sequence of events that is of great usefulness in studying the planets. It is an application of the principle that younger surface features will generally lie on top of older ones. Thus one crater that overlaps the rim of another is determined to be younger, or one colored deposit that lies on another is concluded to have been laid down more recently. Crater rays, since they overlie all other topography, are concluded to be the youngest. Although no *absolute* ages can be determined by this technique alone, careful stratigraphic mapping of the visible face of the Moon carried out in the 1960s established the main outlines of *relative* lunar ages.

The most important stratigraphic benchmark for the near side of the Moon is the formation of the Imbrium basin. The great impact that excavated this 1200-km feature blanketed much of the Moon with a recognizable debris layer called the Fra Mauro formation. Features that lie below this layer are pre-Imbrium; those on top of it are post-Imbrium in age. After the Apollo missions brought back rock samples to be dated, we learned that the Imbrium event took place 3.8 billion years ago.

Examination of Mare Imbrium itself further illustrates the sequence of events that followed the formation of the basin. There are craters inside the basin that have been flooded by the mare lavas. Obviously, these craters were produced by impacts that postdated the great Imbrium event itself. Equally clear from stratigraphic principles is the fact that the eruption of mare lavas took place after these craters formed. Therefore we conclude that the emplacement of the lavas followed the basin-forming impact by a considerable time, and that these lavas were not produced as a direct result of the impact. This is an example of the kind of lunar studies that were carried out using images with resolutions of a few kilometers in the years before the Apollo program. But the greatest increases in our knowledge have come as a result of the lunar expeditions and from subsequent laboratory studies of the returned lunar samples.

5.2 Expeditions to the Moon
The Lunar Challenge

On 4 September 1957 the first artificial Earth satellite, the Soviet Sputnik 1, launched the space age. Less than a dozen years later, on 20 July 1969, half a billion inhabitants of Earth watched as two Americans stepped onto the surface of the Moon and began the first human exploration of another world. By any standard, the Apollo Moon Program was one of the greatest achievements of history. That we landed on the Moon at all was remarkable; that we did it in so short a time, and without losing a single astronaut in space, is little short of miraculous.

Apollo was the culmination of a decade of lunar research, designed with two goals, that of paving the way for a safe landing by humans and that of providing scientific information regarding the Moon. No one pretends that Apollo or its predecessors were primarily scientific programs. Nonetheless, the Moon program of the 1960s and early 1970s yielded a level of knowledge about our satellite far surpassing what we have learned about any other world beyond the Earth. Even though the last U.S. lunar mission was completed in 1972, scientific results continued to accrue from automatic instruments that operated on the lunar surface until 1978, and analyses of the priceless collection of samples brought back by the Apollo astronauts are still taking place today.

Robot Exploration of the Moon

The USSR dominated initial efforts to explore the Moon by spacecraft; Luna 1, the first spacecraft to escape the Earth, passed the Moon on 4 January 1959. Luna 2, which crash-landed in September 1959, was the first craft to actually reach the lunar surface. A month later, Luna 3 returned the first photographs of the farside of the Moon (Fig. 5.6). Finally, on 3 February 1966 Luna 9 successfully landed and radioed back to Earth the first pictures and other scientific data from the lunar surface.

Meanwhile, the U.S. had begun a three-part effort at unmanned exploration of the Moon in anticipation of the Apollo landings. The first of these, aimed only at hard landings and oriented primarily toward acquisition of close-up photos during final approach to the Moon, was called Ranger. Between 1961 and 1965, nine Rangers were launched, but only the last three attempts of this ill-fated program proved successful. While Ranger was struggling to meet its modest objectives, NASA pushed ahead with two more ambitious programs, Lunar Orbiter and Surveyor.

Five Lunar Orbiter spacecraft were launched successfully during 1966 and 1967. Designed to map the surface of the Moon, with

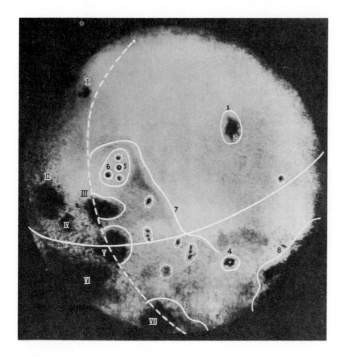

FIGURE 5.6 This first view of the side of the Moon that is always invisible from the Earth was obtained by the USSR's Luna 3 spacecraft in 1959. The continuous line across the chart is the lunar equator; the broken line indicates the border between the hemispheres visible and invisible from Earth. Thus the feature marked II is Mare Crisium, which we can see, while 4 is the crater Tsiolkovsky, which we cannot. Note the absence of maria on the Moon's farside.

special emphasis on the identification of potential Apollo landing sites, the Lunar Orbiters were unique among U.S. automated spacecraft in using film instead of television for their cameras. After onboard processing, the film was scanned electronically and the resulting information transmitted to Earth for reconstruction. The Lunar Orbiters provided coverage of much of the Moon at resolutions as high as a few meters. For many areas not later photographed from Apollo, these images remain unsurpassed today.

Surveyor was the most expensive unmanned program undertaken by NASA until the Viking Mars missions a decade later. Its objective was to achieve the first controlled soft landings on another planet, testing the detailed physical and chemical properties of the lunar surface. Surveyor 1 touched down successfully on 2 June 1966, just two months after its Russian counterpart (Fig. 5.7). While four of the five successful landings were made at potential Apollo landing sites, the final spacecraft was sent to a much more exciting and less accessible highlands area near the prominent ray-crater Tycho (Section 5.1). Tycho is the only large, fresh lunar crater for which we have lander data. Unfortunately, the termination of the Surveyor program prevented the exploration of other scientifically interesting locations. Among the Surveyor results were determinations of the consistency and bearing strength of the finely divided lunar soil and the discovery that the mare materials were basaltic lavas.

FIGURE 5.7 Apollo 12 astronauts visited the Surveyor 3 spacecraft, which had helped pave the way for the first human landings on the Moon. The Apollo 12 Landing Module is visible on the horizon in this photo.

The Apollo Program

The dramatic story of Apollo (Fig. 5.8) is too long and complex to tell here. Following a decade of development work, Apollo 11 astronaut Neil A. Armstrong took his historic first step on the Moon on 20 July 1969. Eleven others followed him in six successful lunar landings, each more ambitious and productive than its predecessors. For a short period, it seemed that humanity had broken the bonds of Earth and had truly begun a new space age. But at the peak of its success, Apollo lost its political appeal, and the program was abruptly terminated. The last human footprints were planted in the lunar soil by Harrison H. Schmitt, the only scientist-astronaut to reach the Moon, before he re-entered the Apollo 17 spacecraft on 14 December 1972.

Several aspects of the Apollo exploration deserve mention even in a brief overview of the scientific results of the program. Perhaps most important is the role of the lunar samples returned to the Earth. Sample collection was a primary scientific objective of every Apollo landing. The first thing each astronaut did upon alighting on the surface was to scoop up a contingency sample of soil and seal it in a pocket of his spacesuit, thus assuring the return of some material even if the moonwalk had to be abandoned. Special tools were designed to pick up and store samples of different kinds. In the later missions, rocks were carefully documented — measured and photographed in place before being picked up — and placed in individual labeled bags for the return trip to Earth. Core samples more than 2 meters long were obtained by forcing specially designed hollow tubes into the surface (Fig. 5.9). All samples were transported back to the Houston Lunar Receiving Laboratory in sealed containers, where they were inspected and cataloged under extremely sterile conditions, exposed only to filtered dry nitrogen and thus protected from contamination or corrosion by oxygen or water vapor in the Earth's atmo-

FIGURE 5.8 Launch of one of the Apollo missions to the Moon (compare Fig. 5.13).

FIGURE 5.9 Among the samples of lunar material returned to Earth in the Apollo program were cores, obtained by drilling into the lunar soil. This Apollo core (lower left) is being inspected in the laboratory at NASA's Johnson Space Center, Houston.

sphere. Most of the 382 kg of returned samples are still in Houston (Fig. 5.10) and are made available to scientists from all over the world.

On the lunar surface, each Apollo landing after the first left behind an automated surface

laboratory called **ALSEP** (Apollo Lunar Surface Experiments Package, Fig. 5.11). Powered by its own nuclear generator, it contained instruments to measure the solar wind, analyze any trace of thin lunar atmosphere, and measure heat flow from the deep interior. Perhaps most important, each ALSEP included a seismometer to measure moonquakes. Because the lunar environment is much quieter than that of Earth, with no winds, waves, truck traffic, footsteps, etc., these seismometers could be much more sensitive than their terrestrial counterparts. With the installation of the third ALSEP by Apollo 15, a network of stations became available to triangulate on the sources of moonquakes. In the later missions, the spent Saturn upper stage rockets and the Lunar Landing Modules themselves were deliberately sent crashing into the Moon to generate artificial moonquakes to be tracked by the ALSEP seismic instruments. The ALSEPs continued to collect data until 1978, when they were turned off by NASA as another cost savings measure.

A third important aspect of Apollo involved the scientific study of the Moon from the orbiting Command Modules. These spacecraft car-

FIGURE 5.10 NASA stores lunar samples (and some meteorites) at its Extraterrestrial Materials Curatorial Facility in Houston. The samples are studied in this laboratory, and small quantities are also made available to qualified scientists throughout the world.

FIGURE 5.11 Automated instruments called ALSEPs (Apollo Lunar Surface Experiment Packages) were deployed on the Moon by the crews of Apollo 12 through 17. Shown here is the Apollo 16 ALSEP, the only instrument package in the lunar highlands.

ried a variety of instruments, including highly sophisticated mapping cameras flown on the last three missions. Film from these cameras could be returned directly to Earth, yielding much better images than had previously been transmitted by radio from a robot craft. Orbital maps were also made of magnetic fields, chemical composition, and surface radioactivity as the Command Modules passed over the Moon. Unfortunately, since all of the Apollos flew on nearly equatorial orbits, this intensive mapping did not extend to higher latitudes.

The Apollo Flights to the Moon

Fig. 5.12 shows the locations of the six Apollo landing sites. The following is a brief summary of each of the Apollo missions that went to the Moon (numbers in the series not appearing here represent Earth-orbital tests).

Apollo 8: December 1968. First human circumlunar flight and lunar orbit. Excellent photography from hand-held cameras as the spacecraft orbited the Moon ten times.

Apollo 10: May 1969. First rendezvous in lunar orbit, in a dress rehearsal of the Apollo 11 landing. LM (Landing Module) maneuvered to within 14 km of the surface.

Apollo 11: July 1969. First lunar landing, in Mare Tranquillitatis (Tranquility Base), a site chosen for its smoothness and freedom from hazards. Astronauts spent 23 hours on the Moon. First lunar samples returned (22 kg).

Apollo 12: November 1969. Landing at another mare site, in Oceanus Procellarum, near a ray from Crater Copernicus. First ALSEP deployed. Landed 200 m from Surveyor 3 spacecraft, retrieved parts of it for return to Earth. 34 kg of samples collected.

Apollo 13: April 1970. Mission aborted after explosion of oxygen tank in Command Module. Astronauts returned safely after harrowing flight around the Moon in their disabled spacecraft.

Apollo 14: January 1971. Landing on the Fra Mauro ejecta from the Imbrium basin.

First use of lunar rickshaw to haul equipment and samples; astronauts traversed 5 km on foot. 43 kg of samples collected.

Apollo 15: July 1971. Landing at the edge of the Imbrium basin, near Mt. Hadley at the foot of the Apennine Mountains. First in new phase of missions with extended scientific capability. Carried improved Command Module instruments, launched subsatellite into lunar orbit. First use of lunar rover vehicle permitted 24-km traverse on surface, including visit to Hadley Rille, an ancient lava channel. First measurement of lunar heat flow. 77 kg of samples collected.

Apollo 16: April 1972. Only landing in a highland site, near Crater Descartes. 95 kg of samples collected.

Apollo 17: December 1972. Final Apollo landing, in the Taurus-Littrow Valley on the margin of Mare Serenitatis. Included only scientist-astronaut to visit Moon, geologist Harrison H. Schmitt. 111 kg of samples collected (Plate 6).

FIGURE 5.12 The landing sites of the six Apollo and three Soviet Luna sample return missions are shown on this NASA map of the nearside of the Moon.

While the Soviets never joined a "space race" to land people on the Moon, they did continue a vigorous unmanned exploration during the early 1970s. Three robot sample return missions, Lunas 16, 20, and 24, brought back 300 g of material, including one core sample. And in 1970 and 1973, two mobile vehicles named Lunakhods were successfully operated on the lunar surface.

It is ironic that more than a decade after Apollo, neither the United States nor any other nation has the capability for manned lunar exploration. Tourists gawk at the giant Saturn rockets, which now lie rusting on the grass at Cape Canaveral and Houston (Fig. 5.13). Leftover Apollo spacecraft, built at costs of hundreds of millions of dollars, take the place of honor in museums instead of resting where they were intended, on the surface of the Moon. No forecaster of the future or writer of science fiction had ever predicted that humans, having once attained the Moon, should have abandoned it so quickly. The scientific legacy of Apollo continues, but what of the exploration potential? Where has that legacy gone?

FIGURE 5.13 The giant Saturn V rockets of the Apollo program were the largest rockets ever built (as of 1986). Unfortunately, Apollo was terminated before its completion, and several Saturn Vs, such as this one on public display in Houston, were left to rust on Earth rather than being sent to the Moon.

◆
───────────────────────────

5.3 Impact Cratering

Craters are the dominant geological feature on the Moon, readily visible to generations of telescopic observers, yet their origin, in the impacts of meteoroids, was not widely recognized until about fifty years ago. It is interesting to examine why this fundamental principle of planetary science remained hidden for so long, and to see what finally provided the convincing evidence in favor of an impact origin for the lunar craters.

Volcanic or Impact Origin?

We learned in Section 3.2 that it was not until the beginning of the nineteenth century that scientists accepted the extraterrestrial origin of meteorites. At about the same time geologists began to recognize that the Earth was at least tens of millions of years old. Both of these ideas serve as prerequisites for understanding that lunar craters resulted from impacts.

Throughout the next 100 years, however, the idea that the lunar craters were volcanic in origin dominated. The argument was really fairly simple. No impact craters had been recognized on Earth, and the largest projectiles known to strike our planet were meteorites. In those days prior to the discovery of Earth-approaching asteroids, no evidence existed for large meteoroids in the inner solar system that could impact either the Earth or the Moon. Volcanoes, on the other hand, were well known to terrestrial geologists, and do exhibit craters, although they are usually small and located at the summit of conical peaks. Therefore, the craters of the Moon, by analogy with the Earth, must also be volcanic.

The first detailed arguments against the volcanic crater hypothesis were presented in the 1890s by G.K. Gilbert of the U.S. Geological Survey (Fig. 5.14). Gilbert was among the few geologists who were interested in the Moon, and he was a strong proponent of quantitative measurements rather than subjective descriptions of what the Moon looked like through a telescope.

Assembling data on the sizes, shapes, and distribution of lunar craters and of terrestrial volcanoes, Gilbert pointed out the many differences between the two. Most significant was the fact that lunar craters do not appear at the summits of mountains. Gilbert was among the first to emphasize that the floors of lunar craters actually lie well below the level of the surrounding plains. Using these and similar arguments, he developed a strong case for the dissimilarity between lunar and terrestrial craters. With the apparent similarity demolished, the argument by analogy crumbled.

Through elimination, meteoroidal impacts emerged as the most probable cause of lunar craters, and Gilbert presented as strong a case for this hypothesis as was possible in the 1890s. He did, however, perpetuate a fundamental misconception about the impact cratering process. In his mind, the formation of an impact crater on the Moon was similar to that of small craters produced by throwing stones into mud or sand. Craters formed in this manner are only round if the projectile strikes the surface from above; an oblique impact produces an elongated crater. Since virtually all lunar craters are circular, Gilbert tried to find a way out of this predicament by hypothesizing peculiar circumstances in which the meteoroids would fall from almost directly overhead. His efforts were not very convincing.

The Process of Impact Cratering

Gilbert's fundamental problem lay in his failure to understand that the high-speed impact of a meteoroid on the lunar surface generates an ex-

FIGURE 5.14 Grove Karl Gilbert, U.S. Geological Survey scientist who correctly argued in the 1890s that the craters of the Moon were caused by impacts rather than volcanism.

plosion similar in many ways to that of a bomb. (Actually, there are significant differences between impact craters and explosion craters that are apparent to the trained geologist, but we will not probe that deeply into the crater-forming process.) It is the explosion, not the mechanical force of the impact, that digs the crater. Once the explosion is underway, it obliterates any evidence of the direction from which the projectile struck the surface. To cite a modern analogy, shell and bomb craters are always circular, independent of the direction from which the bombardment occurred.

Why does the impact of a meteoroid produce an explosion? Not because fragments of asteroids or comets are inherently explosive, but because of the tremendous energy acquired by the meteoroid as it falls to the surface. For the Moon, the minimum speed of impact (which is equal to the escape velocity) is 2.4 km/s, and for Earth, with its stronger gravitational pull, the minimum impact speed is 11 km/s. To these values must be added the original orbital speed of the projectile, relative to the target. Such speeds endow the meteoroid with energy greater than that of an equivalent mass of TNT, so that it explodes upon impact, regardless of its own composition or the nature of the target.

Imagine a large meteoroid striking the Moon at a speed of several kilometers per second. Its energy is so great that it penetrates two or three times its own diameter below the surface before it stops. The force of the blow shatters the surface and generates seismic waves — moonquakes — that rapidly spread throughout the Moon. Meanwhile, most of the energy goes into heating the projectile and its immediate surroundings. The material forms a pocket of superheated gas, and the expansion of this hot, high pressure gas creates the crater.

Fig. 5.15 illustrates the stages of crater formation. At the speeds associated with lunar im-

FIGURE 5.15 This diagram illustrates the formation of a large lunar crater from the explosive impact of an asteroid or comet.

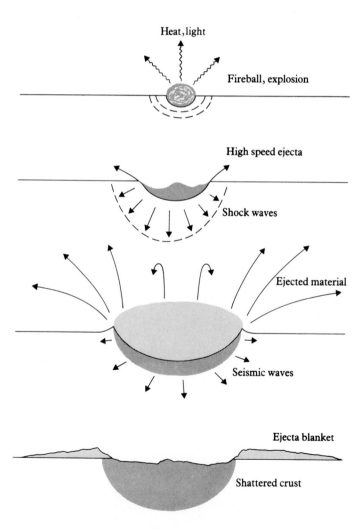

pacts, the energy released is sufficient to excavate a crater with a diameter about ten times that of the projectile and a depth, even after much of the ejecta falls back into the crater, of about ⅕ the crater diameter. The material removed from the crater, which consists of the original projectile plus several hundred times its own mass in excavated rock, is thrown upward and outward. Part of it falls back into the hole, partially filling it, while the rest spreads out over the surrounding area. On the Earth, the higher speed of impact results in a crater about twenty times the diameter of the projectile, and an excavated mass thousands of times greater than that of the impacting body.

The Ejected Material

The ejected material from a cratering explosion consists primarily of broken and shattered rock fragments mixed with gas and liquid droplets generated by the heat of the explosion. The bulk of this mass falls within one crater diameter of the rim, where it produces a rough, hilly deposit known as an **ejecta blanket** (Fig. 5.16). A surge of back-falling debris that flows outward from the explosion center may also occur. On the Earth, similar hot fluid surges are a characteristic and dangerous aspect of some volcanic explosions, such as that of Mt. St. Helens (Washington, 1980). Impact craters formed on Mars, where the subsurface is (or was) saturated with ice, have a unique form of ejecta blanket, as we will see in Chapter 9.

Large fragments expelled by an explosion rain down on the surrounding lunar surface, generating their own small craters, called **secondary craters.** Sometimes a group of these fragments strikes the surface together to produce a chain of secondary craters.

A third type of crater ejecta consists of high speed streams of material spewed forth from large impacts, similar to the splash made by a careless diver. Such streamers can arc for hundreds or even thousands of kilometers across the lunar surface. Where they strike the ground, they produce many small secondary craters and generally disrupt the surface, lightening its color. These long streamers of ejecta produce the lunar crater rays.

FIGURE 5.16 King crater, a 75-km crater on the lunar farside, is one of the freshest lunar craters in its size range. It clearly shows the features usually associated with large impact craters, including central peaks, terraced walls, and a well-formed ejecta blanket that extends outward approximately two crater diameters.

FIGURE 5.17 This spectacular view of the Imbrium basin with the crater Copernicus on the horizon illustrates the ejecta patterns from a large lunar impact. Extending toward the viewer are many rays and patterns of secondary craters formed by the Copernicus impact about a billion years ago.

The nature of crater rays is especially well illustrated in Fig. 5.17, which shows an oblique Apollo view of the Imbrium basin with the 91-km crater Copernicus in the distance. Apollo 12 landed near one of these Copernicus rays. Several rays stretch toward the observer across the dark mare surface. Near the foreground, we can see the association between the light ray material and clusters of irregular secondary craters. Since the impact speed of secondaries is rather low (less than 1 km/s), they do not generate true explosions. Therefore, these craters can be irregular or elongated, just like the pits produced by throwing stones in a sandbox.

Evolution of the Crater

Meanwhile, back at the crater, the lunar crust is trying to adjust in the aftermath of the explosion. For a small crater, up to perhaps 10 km in di-

ameter, little adjustment is necessary, and the crater retains approximately its original bowl shape. Larger craters, however, cannot be supported for long once the original explosion is over. Under the force of gravity, sections of the crater wall collapse and slide downward, partially filling the center and creating a series of step-like terraces along the walls. Near the center of the crater, where a great weight of overlying material has been removed by the explosion, the crust may rebound to create a **central peak** or group of peaks (Fig. 5.16). Occasionally rock melted by the heat of the impact will form pools of lava, called impact melts.

Large craters tend to have flat floors, often flooded by later lava flows. Let us return to Tycho, where you may imagine yourself landing in its interior. Standing on the floor you would not have the sensation of being in a bowl-shaped depression. Instead, you would find yourself on a rather level rocky plain, with the distant rim visible as a low line of mountains serrating the horizon. The central peak would appear as a huge pile of rubble. Craters larger than 100 km in diameter might more appropriately be referred to as mountain-ringed plains, although their true crater form is readily apparent to the observer looking down from above.

In Section 5.5 we will return to the formation of the lunar basins. First we will explore further the origin of the meteoroids that created all of these lunar features, from the largest basins down to microscopic pits in surface rocks. Could enough impacts really have occurred for this process to be the dominant one in shaping the surface of the Moon?

5.4 Dating Cratered Worlds
Crater Densities and Surface Ages

The number of impact craters on a planetary surface serves as a measure of the age of that surface. Here we define the **crater retention age** as the time over which the surface has been sufficiently stable to preserve craters once they are formed. On an active planet like the Earth, many erosional and geologic processes degrade and destroy craters. But on a small, airless, rocky world like the Moon, practically the only events that can destroy craters are later impacts, especially those that create large basins, and lava flooding during periods of large-scale volcanism. For the Moon, therefore, the crater retention ages generally represent the time elapsed since the formation of large basins and their subsequent flooding by lavas.

When we observe differently cratered regions (for instance Fig. 5.3), we naturally interpret the differences as the result of different crater retention ages. After all, we know the impacting meteoroids came from outside, and

there is no reason to think they preferentially struck some regions, the highlands for instance, while sparing others. The situation resembles that of a city street in the midst of a long, windless snowstorm. As you walk along, you will find the sidewalk in front of some houses covered deeply with snow, while in other places the depth is less, with a few areas of sidewalk nearly clear of snow. Do you conclude that different amounts of snow have fallen in different areas of the street? No; you attribute the differing depth to the time that has passed since that section of walk was shoveled. The less snow, the shorter the "snow retention age." It is just the same with craters.

To determine a crater retention age, we must first count the number of craters that can be seen on the particular terrain being studied. The number of craters on a given area is called the **crater density.** This density has nothing to do with the material density introduced in Section 1.6 and used to characterize the bulk composition of an object. It is simply the expression of the number of craters of a given size on a well defined area of the surface, usually taken as one million square kilometers, about the size of the state of Texas on Earth or of Mare Imbrium on the Moon.

Crater Size Distributions

Look at the photographs of the Moon throughout this chapter. You will note that there are always many more small craters than large ones. The reason is obvious: as we saw in the last chapter, small meteoroids are much more abundant than large ones. Fragments in space, from asteroids down to pebbles, tend toward a size distribution in which there are more than a hundred 10-km meteoroids for each one that is 100 km in diameter, more than a hundred 1-km objects for each 10-km object, etc. Since lunar craters have diameters approximately ten times that of the impacting meteoroids, they will display the same sort of size distribution. Indeed, scientists have turned the argument around. By counting lunar craters in different size ranges, they have determined the distribution of crater sizes and thus the size distribution of the meteoroids.

Fig. 5.18 plots the frequency of impacts of different size meteoroids. The smooth curve, spanning the size range from a ton, similar to the larger meteorites, to basin-forming asteroidal impacts, is an attempt to summarize a great deal of evidence acquired from the Earth, the Moon, and studies of comets and asteroids. We will return to this subject in Chapter 7, when we consider catastrophic impacts on the Earth.

The measure of crater density commonly used is a cumulative one, that is, we count all craters larger than a given minimum size, frequently 10 km. For example, the 10-km cumulative crater density on the average mare is 50 per million square kilometers, while that on the lunar highlands is 1000 per million square kilometers. Cumulative crater densities for a number of areas of the Moon are listed in Table 5.1.

For the lunar crater size distribution, the average size crater is about twice as large as the lower limit counted. In our case, where we are considering craters down to a 10-km diameter, the average area is about 400 square km. A quick multiplication shows that on the maria, where the crater density is 50 per million square km, only about 2% ($50 \times 400 / 1,000,000$) of the surface is covered with craters 10 km or larger. On the highlands, in contrast, nearly 50% of the surface is cratered. You can verify these calculations by examining the photos of highland and mare terrain in this chapter and counting the craters yourself.

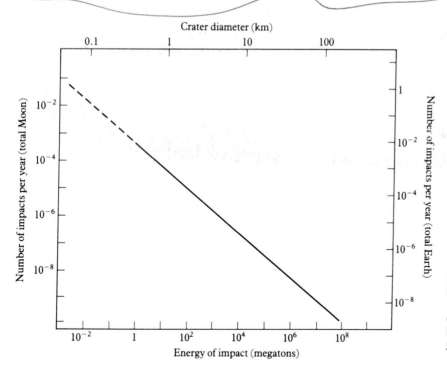

FIGURE 5.18 Estimated present rate of impacts by asteroids and comets on the Earth (total land area) and on the Moon. The lunar and terrestrial scales differ because the land area of the Earth is greater than that of the Moon, and a given size meteoroid will make a larger crater on the Earth than on the Moon. Note that a 100-km crater (like Tycho) is formed on the Moon about once per billion (10^9) years, while a 2-km crater (at the resolution limit of Earth-based telescopes) should be formed about every 100,000 (10^5) years. The cratering frequency on the Earth is about ten times greater.

TABLE 5.1 Cumulative crater densities on the lunar surface

Region	Crater Density (10-km craters per million km²)	Age (by)
Highlands	1000	> 4.0
Fra Mauro (Apollo 14)	130	3.9
Apennine (Apollo 15)	95	3.9
Tranquillitatis (Apollo 11)	50	3.7
Fecunditatis (Luna 16)	20	3.4
Putredinis (Apollo 15)	25	3.3
Procellarum (Apollo 12)	20	3.3
Crater Copernicus	10	0.9
Crater Tycho	2	0.2

Absolute and Relative Ages

The greater the crater density, the older the surface. But how much older? Only for the Moon, where scientists have measured both the crater densities and the associated radioactive ages from returned samples, can this question be answered with any precision. On Earth, there are too few craters, while on other planets there is no absolute age scale defined by returned samples.

All of the lunar maria have roughly similar crater densities, and all therefore appear to have comparable ages. This fact was recognized before the Apollo missions, but without any lunar samples there was a lively debate among lunar scientists concerning the absolute age of the maria and, hence, the period since the cessation of major lunar volcanism. If one assumes that the rate of impacts has remained constant throughout solar system history, then the fact that the mare crater densities are only about 1/20

of the highland densities suggests that the maria are relatively young. If the highlands are as old as the Earth, 4.5 billion years, the maria would be only about 200 million years old. But is this conclusion consistent with what we know about the current impact rates from comets, Earth-approaching asteroids, and other meteoroids?

In the 1960s, many scientists thought not. Starting from the known numbers of objects in near-Earth space, they calculated that it would require closer to 4 billion years, rather than 200 million, to accumulate the observed mare crater density. In other words, the flooding of the maria by lava flows was restricted to the early period of lunar history. They therefore argued for an inactive Moon, on which the volcanic fires cooled billions of years ago.

This explanation poses a problem, however. If the maria and the highlands are both billions of years old, how can the crater density in the highlands be twenty times higher than that on the maria? In order for these parts of the Moon to have accumulated so many craters, a much higher early impact rate would have been required between the formation of the highland crust and the period of lunar volcanism represented by the maria.

When the first lunar samples were returned to Earth in 1969, the most eagerly awaited scientific result was the radioactive age determination for mare materials. The age, for a variety of Mare Tranquillitatis rocks and soil samples, was 3.7 billion years. Subsequent measurements on samples from other mare sites confirm that the mare solidification ages are between 3.2 and 3.9 billion years. Highland rocks, in contrast, have ages greater than 4 billion years.

A Unifying Concept of Cratering on the Inner Planets

The combination of lunar crater counts, dated lunar samples, and measurements of the numbers of meteoroids in near-Earth space today, has given rise to the current concept of early planetary history. Judging from the lunar evidence, the impact rates on all of the inner planets have not changed greatly during the past 3.8 billion years. Presumably the impacting bodies have been the comets and asteroidal fragments that we recognize today. If similar numbers of meteoroids have struck Mercury, Venus, or Mars, then the lunar chronology can be applied to crater densities on these planets as well.

According to this theory, however, quite different conditions prevailed during the first 800 million years of solar system history. Before about 3.8 billion years ago, the rate of impacts must have been much higher; otherwise the high crater densities on the highlands cannot be understood. From lunar research, it appears that the impact rate was a thousand times greater at 4.0 billion years ago than at 3.8 billion years. Earlier, it may have been even higher, but the evidence concerning lunar history before 4.0 billion years ago is limited.

The period around 3.9 billion years ago is variously called the late heavy bombardment or the terminal bombardment of the Moon, and is presumed to be the result of a different source of impacting bodies from those present today. Because the impact rates were very high at this time, most scientists suspect they were the result of a unique event such as the disruption of a large asteroid that scattered fragments throughout the inner solar system. Alternatively, the terminal bombardment may represent the final stages of planetary accretion, before the leftover debris from the solar nebula had been swept away. In either case, this period of heavy cratering has largely obliterated direct evidence of the first half-billion years of lunar history.

The idea that the inner solar system experienced a late heavy bombardment ending 3.8 billion years ago followed by nearly constant impact rates extending to the present constitutes one of the paradigms, or generally accepted theories, of modern planetary science. One of the many scientists contributing to this paradigm is Eugene Shoemaker of the U.S. Geological Survey branch of astrogeology (Fig. 5.19). Shoemaker has dealt with impact craters all of his professional life, beginning with his doctoral dissertation on the geology of Arizona's Meteor Crater. In the 1960s, he was one of the Apollo geologists who guessed the wrong lunar chronology, predicting relatively young mare surfaces. After Apollo, Shoemaker began a major observational and theoretical program to understand the orbits and size distributions of meteoroids in Earth-Moon space today. This program has yielded the recent discovery of many Earth-approaching asteroids (Section 4.5). Thanks to this work, we now recognize that the number of Earth-crossing meteoroids is consistent with the lunar impact rates deduced from the mare crater counts. More recently, Shoemaker has tried to extend this cratering paradigm to the satellites in the outer solar system, a subject we will return to in Chapter 13.

FIGURE 5.19 Eugene Shoemaker of the U.S. Geological Survey has played a leading role in developing a theory of the role of impacts from comets and asteroids over the history of the solar system.

Application of the cratering paradigm to other planets, particularly Mercury and Mars, has yielded many important insights concerning solar system history. All such unproven theories must be treated with some caution, however. We know the lunar chronology, but its application elsewhere carries a certain risk since we do not know the origin of the late heavy bombardment. The Moon may have experienced special, local events related to its formation. We will try to keep this reservation in mind when we begin to interpret crater counts on other planets.

5.5 Lunar Catastrophism
The Lunar Highlands

The lunar highlands (Fig. 5.20) are the oldest surviving part of the lunar crust. From their heavily cratered surfaces we conclude that they formed during another era, at a time when much more debris was impacting the surfaces of the inner planets. This early period, from the formation of the crust through the age of basin formation, was the era of lunar catastrophism.

FIGURE 5.20 This view of the heavily cratered lunar highlands was taken from the Apollo 17 Command Module.

Catastrophism is a geological term referring to sudden or violent events that substantially modify the landscape. Conversely, **uniformitarianism** seeks explanations of the world around us in terms of very slow processes acting over vast time spans. Large impacts and massive volcanic eruptions are examples of catastrophic events, while the gradual accumulation of ocean sediments is the product of uniformitarian processes. Although proponents of the two concepts have been in conflict in the past, we now recognize that both types of geological processes are important.

On the Moon, most of the catastrophic geological events took place early in its history. The degree of highland cratering far exceeds that on the maria. Since the highland craters are packed virtually shoulder to shoulder, we really have no way of determining how many impacts took place in the past. The crater density has apparently reached a state of saturation, in which new impacts do not create additional craters; they simply replace old craters with new ones.

An indication of the magnitude of highland cratering is provided by the behavior of moonquakes studied by the Apollo ALSEP instruments. The reverberation of seismic waves in the highlands indicates that the crust has been shattered to a depth of 25 km, and that the fragmented surface layer of rubble has an average depth of hundreds of meters. In contrast, the fragmented surface layer on the maria is only about 10 meters deep.

As might be expected from this history, the rocks of the lunar highlands are breccias, composed of recemented pieces broken and scattered by previous impacts (Fig. 5.21). Many highland breccias are extremely complex, indicating three and even four generations of shattering and reforming of the rocks. Often the highland breccias contain frozen droplets of impact-melted material. By comparison, the meteoritic breccias discussed in the last chapter are relatively simple, consistent with their formation on smaller bodies.

FIGURE 5.21 Many lunar rocks are breccias, like this sample from the highlands. These rocks, consisting of individual fragments cemented together, display a long history of impacts. Most lunar breccias formed during the high bombardment period of the first half-billion years of lunar history.

Highland Samples

The oldest dated highland samples are not whole rocks but tiny individual fragments within breccias. In general, these show solidification ages not greater than 4.2 billion years, although a few go back to 4.4 billion years. The apparent age limit of 4.2 billion years does not necessarily mean that the surface remained molten up to that time. More likely, the radioactive clock was reset for highland samples by heating and intense shock from multiple impact events.

The dating of Apollo samples indicates that the period of lunar catastrophism extended up to about 3.8 billion years ago, when the rain of impacts subsided to more nearly its present low rate. As lunar scientist Stuart Ross Taylor has observed, "The meteoritic bombardment has drawn a curtain across the landscape through which we peer dimly to discern the earlier history. The [original] structure of the lunar crust is obliterated, and there is no vestige of a beginning."

One way to try to pull this curtain aside is to look at the bulk composition of the highland lunar crust. This material, like all lunar samples, is severely depleted in volatile elements, that is, in those elements that can be evaporated or vaporized at relatively low temperatures. Ices are extremely volatile, but geologists also consider any element that evaporates at the temperature of lava, about 1000 C, to be volatile. On the Moon water is absent; the lunar minerals include no clays or other familiar terrestrial forms that incorporate chemically bound water. Other common volatiles that are depleted include nitrogen, carbon, sulfur, chlorine, and potassium.

The primary material of the highland crust is a class of silicate rocks called **anorthosites.** The lunar anorthosites consist mainly of mineral oxides of silicon, aluminum, calcium, and magnesium, in descending order. Relative to basalts, they are enriched in aluminum, calcium, and magnesium, and are depleted in iron. Anorthosites are rocks that might form out of an originally molten Moon, and their presence argues for an early differentiation before the lunar crust solidified. Battered though these fragments of the crust may be, they still provide some information on this earliest stage of lunar history.

During the hundreds of millions of years of heavy bombardment that followed the formation of the original crust, many episodes of remelting triggered by impacts probably also occurred. We can imagine large projectiles breaking through the crust to release floods of still liquid rock from the interior. At some point after the crust had stabilized and thickened, more conventional volcanic activity also began. Although now nearly obscured by subsequent events, indications exist of the formation of the first basalts as early as 4.2 billion years ago.

Lunar Basins

The surviving lunar impact basins were formed during the final stages of the heavy bombardment of the Moon. We use the term basin to refer to any impact feature more than about 300 km in diameter. The impacting meteoroids that blasted out these features must have been up to a hundred kilometers in diameter. Possibly many basins were formed from a cluster of asteroidal impacts that occurred during the terminal bombardment, in the relatively brief interval from about 4.1 to 3.9 billion years ago. Today lunar scientists have identified about thirty ancient basins, including many on the lunar farside.

The youngest of the great basins are Imbrium and Orientale (Figs. 5.17 and 5.5). The Imbrium event can be precisely dated using samples from the Apollo 14 mission which landed on its ejecta blanket, and from Apollo 15 which explored the rim of the basin at the base of the Apennine Mountains. The time of the impact was just over 3.9 billion years ago, and the extensive ejecta, which overlies older features on much of the nearside of the Moon, provides a reference marker for lunar stratigraphy. The ejecta from Orientale is found on top of Imbrium features, indicating that the Orientale event occurred later. Although no Apollo landing took place on Orientale ejecta, indirect evidence places this last great impact at between 3.8 and 3.9 billion years ago. Both Imbrium and Orientale are mountain-ringed circular features about the size of the state of Texas. Today they are different in appearance primarily because the Imbrium basin was later flooded by lavas that produced Mare Imbrium, while the Orientale basin escaped major volcanic modification.

The outer mountain ring of Imbrium, consisting of the lunar Carpathians, Apennines, and Caucasus, rises as high as 9 km above the present mare material, similar to the height of Mt. Everest above sea level or of the Hawaiian volcano Mauna Loa above the ocean floor. The diameter of this outer ring is 1200 km. Two inner mountain rings apparently once existed but have been largely destroyed by subsequent lava flooding. The Fra Mauro formation, which is the ejecta blanket from Imbrium, extends outward nearly a thousand kilometers and varies in thickness from more than a kilometer to a fraction of a meter (Fig. 5.22).

The Orientale basin is defined by a 900-km outer mountain ring, the Cordilleras, and a 600-km inner ring, the Rook Mountains. Because it has not been flooded like Imbrium, we can still see the unmodified floor of a giant basin in the interior of Orientale, with its rough and hilly contours and patches of impact-melted rock. The primary ejecta from Orientale extends about 500 km beyond the Cordillera Mountains.

The process of formation of basin-rim mountains is not entirely understood, but it apparently involved a combination of uplift, produced by the blast itself, and later subsidence along concentric cracks as the lunar surface adjusted following the impact. A cross-section of the Apennine front is shown in Fig. 5.23, which reveals layers in the uplifted basin rim. Isolated mountains are probably rootless piles of ejecta.

FIGURE 5.22 The Fra Mauro formation, shown in this view of the Apollo 14 landing site, is the hummocky material extending through the center of the frame. This material is ejecta from the impact that formed the Imbrium basin, about 600 km to the north of the location shown here.

FIGURE 5.23 One of the best views of mountains on the Moon was obtained by the Apollo 15 crew, who landed near the base of the Apennine Mountains. Mt. Hadley, shown here, was formed by uplift associated with the impact that created the Imbrium basin.

Basins on Other Planets

Other planets have their own counterparts of the Imbrium and Orientale basins: Caloris on Mercury, Hellas and Argyre on Mars, Valhalla and Asgard on Callisto. The Earth also could not have escaped impacts by 100-km asteroids at the same time that the lunar basins were formed. In fact, the terrestrial basins must have been substantially larger than those on the Moon, as a result of higher impact velocities caused by the Earth's greater gravity. It is staggering to contemplate the consequences of such impacts on the crust of the young Earth. Yet all traces of this early history have been lost, as we will see in Chapter 7.

Compared to conditions today, the era of heavy bombardment was a time of great violence, with impact rates tens of thousands of times greater than those experienced now. Thinking back to that time, we tend to picture a sky filled with projectiles or a surface pockmarked with fresh craters. A bit of arithmetic will quickly convince us otherwise, however. Today a 1-km crater is formed over a million-square-kilometer area about once per million years. When the impact rate was ten thousand times greater, such a crater was still formed only once per century. An individual witness is only aware of events to a radius of about 100 km, or about 4% of a million square kilometers. Thus a hypothetical observer standing on the Moon or Earth during the period of heavy bombardment would have witnessed the formation of a 1-km crater only once in 2500 years! Events that seem catastrophic on a planetary time scale are still rare from our human perspective.

5.6 Lunar Volcanism

Even before the period of heavy bombardment had ended, volcanic vents were erupting on the lunar surface. The major period of lunar volcanism apparently did not begin until shortly after the formation of the Imbrium and Orientale basins, about 3.8 billion years ago. Over the following half-billion years, repeated outpourings gradually filled the nearside basins with dark basalt to create the familiar pattern of lunar maria.

The Lunar Maria

The story of lunar volcanism is essentially the story of the maria. Although there were isolated examples of volcanic activity elsewhere, they were of negligible significance compared to the great outpourings of mare lava. Fortunately, several landing sites on the flat mare surfaces were selected by the safety-conscious Apollo planners, thereby providing us with many observations and large quantities of returned mare rocks and soil.

The mare basalts resemble their terrestrial counterparts and are believed to have originated in a similar way, from subsurface melting and eruption of lava. The compositional difference is that the lunar basalts have a higher iron content, including metallic iron, and are free of water and its effects. Since all lunar rocks are igneous and formed in an environment lacking both water and atmospheric oxygen, they have simpler compositions than terrestrial rocks. Only about a hundred separate minerals have been identified on the Moon, as opposed to more than 2000 on the Earth. There is no clay or other water-bearing rock and very little carbonaceous material.

The oldest lunar basalts, obtained from the Apollo 14 and Apollo 17 sites, have solidification ages of 3.8 to 3.9 billion years. The flows on the margin of Mare Imbrium sampled by Apollo 15 are dated at 3.3 billion years, which presumably fixes the late stages of volcanism in this basin.

Apollo 12, which landed on Oceanus Procellarum, yielded the youngest mare basalt, with a solidification age of just under 3.2 billion years.

The Maria-Forming Eruptions

Unlike most of the volcanic activity with which we are familiar on Earth, the lunar eruptions did not normally create volcanic mountains. Instead, the eruption of large volumes of fluid lava from long fissures yielded thin, flat flows of enormous extent. On Earth, similar flood basalt eruptions are responsible for the lava plains of eastern Washington State and of central India.

Most of the lunar lava flows have been buried by subsequent eruptions, but the final large outpourings can still be identified. On Mare Imbrium, the last major eruptions appear to have originated from a fissure about 20 km long and to have spread more than a thousand kilometers from the vent, covering a total area of 0.2 million square kilometers, about the size of the state of Utah. Along the flow fronts the thickness typically ranges from 30 to 50 meters. A hundred flows of this thickness are required to account for the total depth of the mare, estimated at about 5 km. As much as a million years might have elapsed between individual flows.

After the eruption of the mare material, some settling and subsidence took place, perhaps as a result of shrinkage as the lava cooled. In places, "high water marks" show that lava once reached up to 100 m higher than the current level. Many cracks have formed around the margins of the maria, often concentric with the center of the basin. In other areas the lavas seem to have been compressed to form wrinkle ridges (resembling the wrinkles formed in a rug by pushing it against a wall) (Fig. 5.24).

Lava Valleys

Among the most remarkable geologic features of the maria are the curving valleys or sinuous

FIGURE 5.24 The Aristarchus Plateau, seen in this Apollo 15 view, shows the many wrinkle ridges that characterize the lava plains that make up the lunar maria.

rilles. Before Apollo, many scientists thought these broad, meandering channels, which resemble terrestrial rivers, had been carved by running water. We now know that the Moon is, and always has been, dry. Since the sinuous rilles are found in the maria, it seems likely that they, too, represent some kind of volcanic phenomenon.

Apollo 15 was targeted to visit one of the largest and best known of these features, Hadley Rille, near the edge of Mare Imbrium (Fig. 5.25a). When the astronauts drove their rover to the edge of the channel, they found a curving valley 1.2 km wide and 370 m deep, littered with fallen boulders, some as large as a small house. On the far wall they could make out an exposed 50-meter-thick layer of solidified lava (Fig. 5.25b).

The consensus from these and other data is that Hadley Rille is the channel scoured by an ancient lava river, perhaps the remains of an underground lava tube with its roof collapsed. Similar features, but of a smaller scale, occur in terrestrial volcanic areas. Why the lunar lava tubes should be so much larger than their terrestrial counterparts remains a mystery, but no reasonable alternative has yet been suggested.

Although volcanic mountains are extremely rare on the Moon, they may not be completely absent. A group of domes in Oceanus Procellarum, the Marius Hills, are thought by some geologists to be true volcanoes. The Marius Hills were to have been visited by a later Apollo flight, but when the program was cancelled in 1972, this mission was among the casualties.

Sources of Lunar Eruptions

Why are the maria confined primarily to the side of the Moon facing the Earth? The main reason seems to be that the average surface elevation is lower on the Earth-facing hemisphere. The subterranean pressures that forced lava to the surface appear not to have been great enough to produce major eruptions in the highlands, including most of the farside. In this respect the maria do indeed resemble seas, filling low-lying areas just as the oceans of the Earth occupy the basins between the continents.

Substantial information on the source regions of the lunar eruptions has been derived from a detailed study of the composition of the lunar basalts. This field of planetary science has been very active since the Apollo program, and the results provide a great deal of insight into the history of the Moon. Basically, the conclusion reached from this work shows that the mare samples represent material that has been three times chemically separated or fractionated from the original material of the solar nebula. First, there was the loss of water and other volatiles that is characteristic of the entire Moon. Second, a differentiation, in which the lunar interior separated into core, mantle, and crust occurred. Finally, a process of "partial melting," in which the more easily melted minerals of the mantle were separated out to become the lunar lavas, took place. This partial melting must have occurred below the crust. From a variety of arguments, lunar scientists have identified the lava source as 100 to 400 km below the surface.

The major lunar volcanic activity ceased a little more than 3.0 billion years ago. A few eruptions, possibly including the Marius Hills volcanoes, may have continued for a time, but their products are not represented in our collection of lunar samples. For most of the Moon, at least, no significant internal activity has taken place during the past 3 billion years. This means that if we saw the Moon 3 billion years ago it would look much as it does today, whereas the Earth would have been essentially unrecognizable even a few hundred million years ago.

FIGURE 5.25 One of the largest of the ancient lava rivers on the Moon is Hadley Rille, visited by the Apollo 15 astronauts: a) the curving valley, about a mile in width, is seen from orbit; b) the view from the surface.

(a)

(b)

5.7 The Surface of the Moon

The Lunar Regolith

When the first Apollo astronauts stepped onto the lunar surface they found themselves in a stark but beautiful world. The airless sky was deep black, the surrounding plains a dark brownish grey. The force of gravity was just one-sixth that at the surface of the Earth, making their bulky spacesuits seem relatively light. With no haze to obscure the view, distant details stood out as sharply as those in the foreground. The mare plain itself was flat and covered with scattered irregular rocks of all sizes and shapes. Some distance away the low profiles of small craters could just be identified, but at first sight there was little evidence to establish that they stood on a heavily cratered planet (Fig. 5.7).

Once they began to move about, the astronauts quickly became aware that the surface of the Moon was layered with a fine dark dust. Every step raised clouds of this material, which soon coated their spacesuits and eventually found its way into every item of their equipment. In spite of this omnipresent dry dust, they found the lunar soil to be firm underfoot. Their boots sank only a few centimeters into the soil, and the bootprints they made were crisp and sharp-edged, as if they had been made in damp dirt or crunchy snow (Fig. 5.26).

Although later Apollo flights landed in rougher and more scenic locales than that first site in Mare Tranquillitatis, the basic character of the immediate surroundings was the same everywhere. The entire Moon appears to be covered with fragmented rock and dust, which represents the ejecta from impact craters, far and near. This continuous debris blanket has been named the lunar **regolith.** The finer component is generally referred to as "lunar soil," while the term regolith includes the full range from fine dust to house-sized blocks.

FIGURE 5.26 This astronaut's bootprint in the lunar soil illustrates the strength and cohesiveness of the fine upper layer of the lunar regolith.

The depth of the regolith varies with the age of the surface. On the maria, it is typically ten meters thick. In the older Littrow Valley area visited by Apollo 17 (page 122) the measured thickness ranged from six to forty meters, while in the highlands the regolith thickness may be as great as hundreds of meters. As shown by core samples, the regolith is composed of overlapping layers of ejecta, each typically a few centimeters thick. One Apollo 17 core, which penetrated nearly three meters of regolith, revealed forty-two distinct layers, each representing ejecta from a single impact.

The lunar regolith accumulates at an average rate of about two millimeters per million years, or two meters per billion years. In comparison with terrestrial erosion and deposition processes, this is slow change indeed. Along with the buildup caused by the addition of ejecta layers, the upper part of the regolith is also frequently disturbed by impacts of very small meteoroids which stir or garden the top few millimeters of material and help to maintain the loose fine dust.

Studies of the processes that form the regolith indicate that most of the ejecta at any one site is derived from relatively nearby impacts. If this were not the case, the sharp boundaries between mare and highland materials would have been blurred by horizontal mixing. Only about 5% of the material comes from as far away as 100 km, and only 0.5% from 1000 km. These calculations apply to the normal cratering process, not to ejecta from the major basins, which has blanketed substantial fractions of the lunar surface to depths of many meters.

In addition to original lunar material, the regolith includes meteoritic fragments from billions of years of cosmic bombardment. Typically, between 1% and 2% is meteoritic, apparently of about the same composition as the primitive meteorites collected on Earth (Section 3.4). Most of the small amount of carbon in the lunar soil is derived from impacting carbonaceous meteorites.

Lunar Soil

When examined under a microscope, all lunar soils are found to contain large quantities of **glass** in the form of spherules up to a millimeter in size (Fig. 5.27). Glass is simply silicates that have been melted and then rapidly cooled so that no crystals can form. These glasses, which are characteristic of the lunar soil, result from the melting of lunar rock during meteorite impact. Most of the glass-forming impacts are very small. On Earth, the atmosphere protects us from these micrometeorites, and our soil does not contain similar glass spherules, except at a few sites associated with terrestrial impact craters. Most of the lunar glass is dark, and the

presence of these tiny spherules contributes to the dark color of the lunar surface.

The tiny glass spherules can themselves be pitted by even smaller micrometeorites when they are exposed on the lunar surface (Fig. 5.28).

In addition to the micrometeorites, the lunar surface is also exposed directly to charged subatomic particles of the solar wind and **cosmic rays,** which are not rays or radiation at all, but high-speed atomic and subatomic particles, carrying tremendous energies. Cosmic rays penetrate into rock and soil particles, leaving tracks that can be studied on Earth. The existence of cosmic ray tracks, and of the similar implantation of solar wind particles in lunar materials, has given rise to a new kind of astronomy. In effect, the lunar surface has served as detector and recorder of solar and cosmic particles over geological time scales. For example, studies have shown that the flux of cosmic rays from the galaxy has remained constant over the past billion

FIGURE 5.27 Viewed under a microscope, the lunar soil is composed of a variety of small rock fragments mixed with glass spheres produced by impact melting.

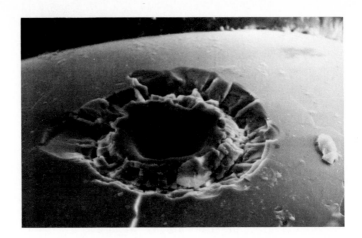

FIGURE 5.28 This photograph of a droplet of lunar glass, made with a scanning electron microscope, shows a beautiful microcrater produced by the high-speed impact of a tiny fragment of interplanetary dust.

years; the rate of major solar flares has not changed greatly over the past 100,000 years; but the isotopic abundances in solar wind nitrogen have changed substantially over the past 2 billion years. Thus the study of returned lunar samples provides information never before available on long-term variations in solar and galactic energetic charged particles.

Surface Conditions and Erosion

In the absence of moderating air or oceans, the surface of the Moon experiences much greater temperature ranges than does the Earth. The contrast between day and night is further magnified by the fact that each lasts two weeks. At the near-equatorial Apollo sites, the maximum surface temperature is about 110 C, higher than the boiling point of water. During the long lunar night, the surface temperature drops to − 170 C, only about 100 degrees above absolute zero. It is a tribute to the design of the Surveyor spacecraft and the ALSEP instruments that they were able to operate over this huge temperature range. The astronauts themselves were never subjected to such extremes, since they always landed in the lunar morning when the temperatures were similar to those on the Earth.

The last three Apollo flights visited mountainous areas: the Apennines, the Descartes highlands, and the Taurus Mountains. Notable in each of these landscapes were the gentle, rounded contours of the lunar mountains and hills (Fig. 5.23; Plate 6). The illustrations in science-fiction stories had always depicted the mountains of the Moon as extremely steep and spiky, but the reality was otherwise. A partial explanation for the soft contours of lunar mountains can be found in the ejecta that blankets all lunar features. But the primary reason that no sharp peaks or steep cliffs are found on the Moon is that no water or ice erosion exists to sculpt them. Water and ice erosion on Earth do not simply wear down mountains; they also cut the valleys and shape the peaks. In the absence of such natural forces, the mountains on any planet will be as gentle as those of the Moon.

Summary

The Moon is the best studied planetary body after the Earth, as the result primarily of the Apollo expeditions and their treasure of returned lunar samples. Unlike the Earth, however, our satellite is no longer geologically active, nor does it possess an atmosphere. Therefore, the Moon is a much easier place to understand, although still complex enough to leave us with a number of fundamental questions unanswered.

The Moon is a differentiated planet, with a highland crust of anorthosite that solidified at least 4.4 billion years ago, much older than any rocks that are still present on the surface of the Earth. During its first half-billion years, the Moon (and the other inner planets as well) was subjected to an intense meteoroidal bombardment, which saturated its surface with craters and created a deep regolith of broken rock. The major basins, including Orientale and Imbrium, were formed near the end of this period. About 4 billion years ago a final burst of impacts occurred, after which the bombardment dropped to approximately its current rate. The highland crust preserves a record of this early period, while the much less cratered lunar maria represent younger surfaces.

The maria, which cover 17% of the surface, are composed of very fluid basaltic lava flows that erupted between 3.9 and 3.2 billion years ago from deep beneath the lunar crust. They generally occupy low-lying areas on the nearside of the Moon, primarily within impact basins. Features of lunar volcanism include the formation of the sinuous rilles, large lava rivers, and the absence of volcanic mountains.

There is no evidence of any major activity on the Moon during the past 3 billion years, a period when the lunar surface has been modified only by impact cratering, including the creation of a fine, glass-rich soil from the eroding effects of many small impacts.

The dominant landform of the lunar surface is the impact crater, ranging in size from basins with diameters of 1000 km to microscopic pits in lunar rocks. We examined the craters in some detail, since they are a common phenomenon on other planets; their characteristic shapes including central peaks, ejecta blankets, secondary craters, and rays resulted from violent explosions caused by the sudden impacts of fast-moving meteoroids. The density of craters is determined by the rate at which meteoroids strike the surface and by the age of the surface. Crater counting provides a good measure of relative surface ages, but it must be used with caution when trying to estimate absolute ages.

The Moon differs greatly from the Earth. Its bulk chemical composition is unlike the Earth's in spite of their proximity in space. This remains a major mystery which we will return to in Chapter 6. In addition, its small mass, only 1% of that of the Earth, has led to much lower levels of geological activity. With its relatively cool, solid interior, the Moon has been unable to form the continents, mountain ranges, volcanoes, and other characteristic geological features of our own planet. Also because of its low mass, the Moon has no atmosphere, creating a very different surface environment of extreme temperatures, with constant bombardment by micrometeorites and cosmic rays. Although unlike the Earth, the Moon is rather similar to one other planet, Mercury. In the next chapter we compare the Moon and Mercury and probe further into the origin and evolution of the Moon.

Key Terms

ALSEP	ejecta blanket
anorthosite	glass
catastrophism	highlands
central peak	impact basin
cosmic rays	mare
crater	regolith
crater density + diagram	resolution
crater ray	secondary crater
crater retention age	stratigraphy
	uniformitarianism

+ Apollo program
Apollo 11

CHAPTER SIX

Strange Relatives: Moon and Mercury

6.1 Radar and the Rotation of Mercury

Radar Astronomy

Because of its proximity to the Sun, Mercury is a difficult object for the astronomer to study. This small grey world is a disappointment to any telescopic viewer, and only the lucky, and persistent, astronomer has ever succeeded in seeing any markings at all on its surface. For this reason, Mercury was a neglected object until the development of radar astronomy in the early 1960s.

Radar is a powerful tool of the modern planetary scientist, the principles of which remain essentially the same whether applied to a distant planet or a nearby airplane on its approach to a foggy airport. An intense pulse of radio radiation is emitted by a transmitter in the direction of the target, and a very, very much weaker echo bounced from the target is picked up by a sensitive radio receiver (Fig. 6.1). The transmitter and receiver can both use the same antenna to focus and collect the radiation. In essence, the technique is similar to illuminating a target with a flashlight (the transmitted radio signal) and then examining the target by means of the light it reflects to our eyes (the radio receiver).

The most powerful planetary radar system in the world uses the 1000-foot antenna suspended in a bowl shaped valley near the town of Arecibo in Puerto Rico (Fig. 6.2) as both a transmitter and a receiver. The radar echo from Mercury is typically only one millionth of a trillionth (10^{-18}) as strong as the transmitted pulse, and the round trip time at the speed of light is more than 10 minutes.

The Doppler Effect

By employing a radio pulse of carefully controlled duration and frequency, a radar system can reveal much more about a target than would be achieved by simply illuminating it with ordinary light. The time required for the radio pulse to cover the two-way path to the target and back provides a measure of the distance to the target. In addition, the motion of the target toward or away from the observer can be measured using the Doppler effect.

The **Doppler effect** or Doppler shift is one of those physical laws of great practical significance. We experience it most commonly in the realm of sound, which is propagated, like light and radio, in the form of waves. A sound source moving toward a listener has a higher pitch (shorter wavelength), while motion away results in lower pitch (longer wavelength) (Fig. 6.3). We hear these effects daily in the sounds of police and firetruck sirens or in the familiar "whoosh" as trains or cars pass by us, their pitch suddenly dropping as their relative motion changes from approaching us to moving away.

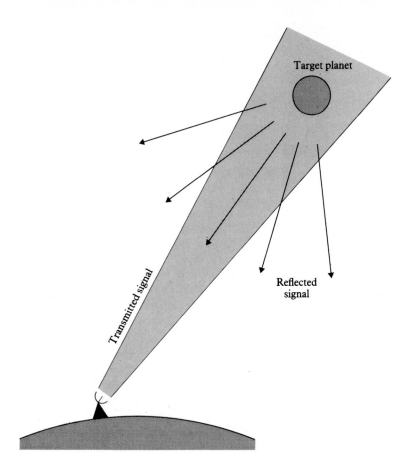

Target planet

Transmitted signal

Reflected
signal

FIGURE 6.1 A radar transmitter
sends a powerful signal into the
Earth-based antenna (lower left).
The antenna then collimates this
radiation into a parallel beam and
directs it toward a planet (upper
right). The planet in turn reflects
the beam, and a small fraction of
this reflected radiation reaches the
transmitting antenna, which is also
equipped with a receiver in order
to detect this returning signal.

The same changes in pitch or frequency take
place when sound is reflected from a moving target.

When either sound waves or radar waves are
reflected from a moving target, the Doppler shift
in frequency is proportional to the relative speed
toward or away from the transmitter/receiver
system. In the case of planetary radar systems,
the frequency of the transmitted pulse is controlled very precisely, so that even tiny changes
in the frequency of the echo can be detected and
analyzed.

The first results from planetary radar were
the determination of the radar reflectivities of the
Moon, Venus, and Mercury, and precise measurements of their distances. Careful tracking of
the motion of Venus, in particular, has been essential in defining the distance scale of the solar
system and thereby making possible the accurate
navigation of spacecraft on complex interplanetary trajectories. The planetary radar observations of the 1960s were an essential prelude to
the Mariners, Pioneers, and Voyagers of the
1970s.

FIGURE 6.2 The 300-m Arecibo radio telescope in Puerto Rico is the most powerful radar instrument on our planet; it is this telescope that first determined the true rotation period of Mercury.

FIGURE 6.3 The Doppler effect. Suppose a source of sound or light is moving from position 1 to position 4. Motion *toward* the observer at left crowds the waves of sound or light together, so the wavelength is shorter and the frequency is higher (light becomes bluer). Meanwhile, the motion *away* from the observer at the right spreads the waves, increasing the wavelength and decreasing frequency (light becomes redder). The observer at the top sees no change in wavelength or frequency, since the source moves neither toward nor away from him.

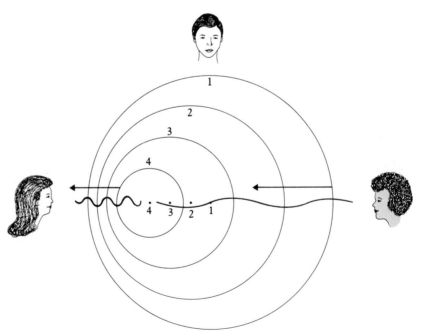

Planetary Rotation Rates

The second phase of radar exploration of the planets yielded our first measurements of the rotation rates of Venus and Mercury. How radar has aided in the penetration of the clouds of Venus and the exploration of its unseen surface will be discussed in Chapter 8. The early radar discoveries about Mercury were equally dramatic, since Mercury was almost as mysterious as Venus as long as astronomers remained dependent on observations with conventional telescopes.

The determination of a planet's rate of rotation by radar utilizes the Doppler effect in a clever way. Since the incident radar beam illuminates an entire planetary hemisphere, the rotation of the planet as well as its orbital motion alters the frequency of the reflected signal (Fig. 6.4). As seen from the Earth, the effect of rotation is to make one side of the planet approach us while the other recedes, relative to the overall motion. The echo from the approaching side of the planet is shifted toward higher frequency, and that from the receding side is lowered in frequency. As a result, the single frequency of the transmitted pulse is broadened in the returned echo, which includes reflections from a full hemisphere. The faster the rotation, the broader the frequency range in the echo (Fig. 6.4).

Mercury's Surprising Rotation Rate

When the Doppler radar technique was applied to Mercury in 1965, the frequency width of the echoes did not conform to astronomers' expectations. Ever since nineteenth century maps had been made showing faint markings on Mercury, it was believed that Mercury always kept the same face toward the Sun, just as the Moon does with respect to the Earth. Thus Mercury, unlike

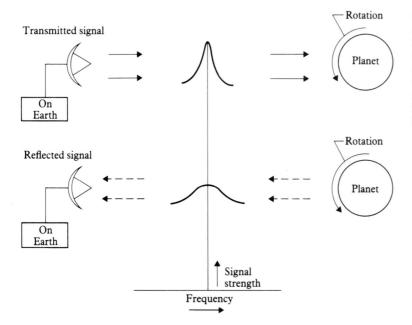

FIGURE 6.4 The broadening of a radar signal by planetary rotation. The side of the planet approaching the radar shifts the signal toward higher frequencies, while the side receding shifts toward lower frequencies. The net effect is to broaden the transmitted signal.

the Moon, really seemed to have a dark side, a hemisphere never illuminated by the Sun. Mercury was thought to be simultaneously the hottest and coldest place in the planetary system. (We now know it is neither.) All maps of the visible markings on the planet were of the supposed Sun-facing side only.

The radar results did not yield the expected 88-Earth-day rotation period, equal to the period of revolution. Instead, the indicated rotation period was approximately 59 days. Subsequent data have demonstrated that the rotation period of Mercury is exactly two thirds of its period of revolution, or 58.65 Earth days.

It is surely no coincidence that the rotation period of Mercury is an exact fraction of the orbital period. Section 6.2 further examines the probable reason for this curious spin-orbit coupling. First, however, we will investigate the consequences it has for Mercury itself. In the process, we must consider just what we mean by the terms day and year as applied to another planet. The year on a planet is simply its period of revolution about the Sun, the time necessary to complete one orbit. The term day is more ambiguous, at times referring to the rotation period, that is, the time for one complete spin relative to the fixed stars. Here, we shall adopt the other, more familiar, meaning of day as the time interval between successive sunrises. This is also called the **solar day.**

A year on Mercury lasts approximately 88 Earth days, during which time the planet rotates one and a half times on its axis. Since the orbital motion and the spin are in the same direction, however, one partially cancels the other, and the Sun appears to move only halfway through its diurnal cycle during this span of time (Fig. 6.5). The solar day on Mercury is equal to two orbits and three complete rotations, or about 176 Earth days. Thus on Mercury the solar day is longer than the year.

The situation for Mercury is further complicated by the large eccentricity of its orbit (Sec-

FIGURE 6.5 Mercury rotates on its axis once every 58⅔ days and orbits the Sun every 88 days. Hence the planet rotates three times (3 × 58⅔ = 176 days) for every two revolutions about the Sun (2 × 88 = 176 days). Starting at position 1, the planet completes one full rotation when it reaches position 5. At 7, the point on the surface that originally faced the Sun now points directly away from it. After the second orbit, this point will again face the Sun.

tion 2.1), which takes it from 0.308 AU away from the Sun at perihelion to an aphelion distance of 0.467 AU. The corresponding range in apparent size of the Sun is from 1.6 degrees to 1.1 degrees, compared to an apparent size of the Sun as seen from the Earth of about 0.5 degrees. Each solar day on this planet includes two perihelion and two aphelion passages.

Strange Days and Seasons

Imagine the appearance of the Sun to a race of hypothetical Mercurians, beginning with a group living at a longitude where the Sun is overhead at perihelion. There are two such hot longitudes, on opposite sides of the planet. At sunrise and sunset the Sun will be at its most distant. The Sun, therefore, rises small and increases in size as it approaches the zenith. At the same time, the combination of rotation and orbital motion causes the apparent motion of the Sun to slow, until it hangs nearly motionless overhead while the planet races through perihelion. Speeding up as it shrinks, the Sun then dips toward its setting point. Meanwhile the stars will be moving through the sky about three times as fast as the Sun. A star that rises with the Sun will set before noon and will rise again before sunset.

Now consider an observer situated at a longitude 90 degrees away. For her, the Sun will be small and rapidly moving at noon but large and nearly stationary at sunrise and sunset. Will such a Mercurian wonder about the favored lands 4000 km to the east or west where the Sun stands still at noon, or will she be grateful that the Sun lingers at rising and setting when its light is most desired? What interesting conversations two Mercurians separated by 90 degrees in longitude would have were they to compare their cosmologies!

In addition to their different cosmological perspectives, the two longitudes that face the Sun at perihelion clearly have a very different thermal history from the longitudes that face the Sun at aphelion. The perihelion longitudes are called the hot longitudes, because they receive extra heating as a result of both the proximity of the Sun at noon and the way it lingers overhead for many Earth days. In contrast, the maximum surface temperatures at the warm longitudes are about 100 C lower, although still a sizzling 550 K.

◆

6.2 Tides and the Spin of Planets
The Nature of Tides

Both Moon and Mercury have rotation periods closely linked to their orbital motion. For the Moon, the two periods are equal and the same side always faces the Earth. Mercury has a

unique two-thirds resonance between the two periods. Many outer planet satellites also experience spin-orbit couplings. How were these linkages forged, and what are the consequences for the evolution of planets and satellites?

The most effective forces that can alter the spins of planets are **tides.** A tide is a distortion in the shape of one body induced by the gravitational pull of another nearby object. The tides with which we are most familiar, of course, are those generated in the Earth's oceans by the pulls of Moon and Sun.

If a planet were completely rigid, the gravitational attractions of other bodies would act on it as if its total mass were concentrated in a point at its center, the so-called center of mass. However, there must be a *differential* gravitational force that results from the fact that planets are not point masses, and that the part of one body facing toward another is more strongly attracted than the part that is turned away. This tendency to pull harder on nearer parts and thus to distort the shape of one planet when it is near another is a **tidal force.**

The tidal force of the Moon on the Earth creates two bulges of ocean water, one approximately facing the Moon and one on the opposite hemisphere. In effect, we may think of the Earth in three parts: the water nearest the Moon is most strongly attracted; the solid body acting at its center of mass is pulled with an intermediate force; and the water away from the Moon feels the smallest gravitational tug (Fig. 6.6). Similar but smaller tidal bulges result from the pull of the Sun, and the highest tides occur twice a month when the two bulges are aligned.

Tidal forces are proportional to the mass of the perturbing body and to the inverse cube of its distance (as opposed to an inverse square law for gravity, Section 1.4). Thus tides most strongly affect satellites in close orbits around their primaries, and tides raised by the Sun are much more important for Mercury than for any more distant planet.

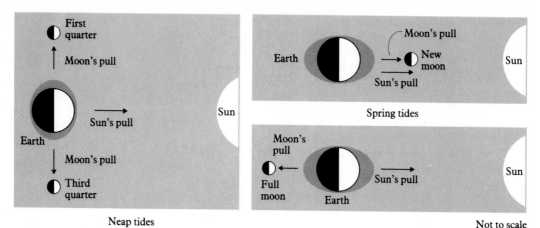

FIGURE 6.6 When the Sun and the Moon pull on the Earth at right angles to one another, the tides are smaller than when they pull together or opposite one another.

Tidal Friction and Orbital Evolution

If the Earth did not rotate with respect to the Moon, the lunar induced tidal bulges would align directly with the Moon. Similarly, if there were no friction between the oceans and the land, the bulges would also maintain their lunar alignment while the solid Earth rotated beneath. This is approximately the case, and an observer on the shore watches the ocean tide rise and fall twice a day as the rotation of the Earth carries it past the nearly fixed tidal bulge. Friction occurs, however, as the restless seas wash against the shores of the continents, and the result is to induce a lag in the tidal bulges. This lag, which varies from one part of Earth to another, explains why high tide does not take place when the Moon is overhead (or underfoot) but usually follows by several hours.

The constant friction between the ocean tides and the land has several other important consequences. First, it slows the rotation of the Earth. Ocean tides dissipate about 2 billion horsepower, lengthening the day by about 5 hundred millionths (5×10^{-8}) of a second per day. Though seemingly small, such a change accumulates in a manner similar to compound interest, and after just a century the terrestrial clock has lost a total of thirty-three seconds, an easily measurable amount. Some of the energy lost by the spinning Earth results in a very small frictional heating of the ocean and our planet's surface, and some alters the orbit of the Moon.

As the Earth slows its spin, the Moon gradually is forced away into a larger orbit, another example of the conservation of angular momentum (Section 3.1). The total angular momentum of the Earth-Moon system includes the spin of both objects and the revolution of the Moon around the Earth. If the Earth's rotation is slowed, losing angular momentum, some other part of the system must gain equivalent momentum. The Moon absorbs this additional momentum by moving outward.

Long ago, the Moon was much closer to the Earth and the length of the month much shorter. Calculations suggest that at one time the Moon may have circled the Earth in as little as a week, while at the same time the solar day on the Earth would have been as short as six hours. Barring unforeseen events, the coupled evolution of the lunar orbit and the rotation of the Earth will continue until the Earth has slowed to the point where it keeps the same side always toward the Moon. With the tidal bulges stationary relative to the solid planet, tidal friction will cease.

We noted in Section 1.1 that total eclipses of the Sun are possible on Earth because the Moon is just large enough in apparent size to cover the face of the Sun. We now see that this situation applies only briefly on a cosmic time scale. In the past, the Moon was nearer and solar eclipses were more common, while in the future the Moon will be too far away to block the Sun completely.

Everything we have said about the effects of the lunar tides on the Earth is equally applicable to the tidal effects of the Earth on the Moon. To be sure, the Moon has no oceans, but a significant tidal bulge can be formed even in a solid planet by gravitational distortion of the rock. On Earth, this body tide results in a bulge a few inches high, and the effects of the larger Earth on the Moon are substantially greater.

Although no friction occurs in the sense that the oceans rub against the land as on the Earth, body tides can also dissipate energy. As an object rotates with respect to the bulge of its body tide, it is subjected to forces that compress and bend its rocky crust and mantle. Tidally induced earthquakes release some energy, while more escapes through heating of the rocks. As in the case of ocean tides, this release of energy slows the object's spin until the tidal bulge faces its companion.

In the distant past the effects of energy dissipation on the Moon from Earth-induced tides slowed the rotation of our satellite until it reached a rotation period precisely equal to its period of revolution. This **synchronous rotation** is a highly stable situation since it minimizes frictional energy loss. Once synchronous rotation is reached, it will be maintained; if the orbital period should change slightly, we can be sure the rotational period will adjust itself accordingly. Most of the satellites in the solar system are in synchronous rotation about their primaries.

Mercury and the Effects of Solar Tides

If synchronous rotation is favored and Mercury is the planet most strongly affected by solar tides, why does it not keep the same face toward the Sun? How did it get onto its present peculiar rotational state? The answer lies in its large orbital eccentricity, which forced this planet to be trapped into a two-thirds synchronous rotation.

Imagine that Mercury started its existence with a relatively rapid rotation and that early in its history it was slowed down by solar tides. At the same time its eccentric orbit carried it first closer to the Sun, then farther away. Since tidal forces are so sensitive to distance (varying as the inverse cube), most of the tidal dissipation took place near perihelion. As Kepler's second law (Section 1.3) tells us, however, the orbital velocity of the planet near perihelion is much higher than the average, and in Mercury's case the orbital speed at perihelion matches a spin period near two thirds of the orbital period. In other words, near perihelion when the tidal forces are strongest, Mercury finds itself turning on its axis in apparent synchronism with its orbit for a rotation period near 60 days. This is the same effect that causes the Sun to appear to stand still in the mercurian sky near perihelion.

Once Mercury's tidal evolution carried it into the two-thirds synchronous rotation period, it was trapped. Given the orbital eccentricity, this spin minimizes tidal dissipation near perihelion when it is most effective. Therefore, the planet remains in a stable rotational state with two fixed tidal bulges that face the Sun at alternate perihelion passages.

What would happen if we were to intervene somehow and place Mercury into a 1 : 1 synchronous rotational state? It could not speed up to return to the two-thirds state, since friction can only slow a body's spin and not increase it. In spite of keeping the same face toward the Sun on the average, however, Mercury would still experience significant tidal heating. Each time it neared perihelion, the increased force of solar gravity would swell its tidal bulge, while the changing direction to the Sun would twist and distort it.

Jupiter's satellite Io finds itself in precisely this situation today, where the resulting **tidal heating** causes a high level of volcanic activity (Chapter 13). Past tidal heating has also been implicated in the thermal histories of many bodies in the planetary system, especially following the Voyager spacecraft discoveries of indications of past internal activity on the satellites of Jupiter and Saturn.

6.3 The Face of Mercury
The View from Mariner 10

Almost everything known about the geology of Mercury was learned from a single spacecraft, Mariner 10, which made three flybys of the planet in 1974 and 1975. For a few months this small planet held center stage in the planetary exploration program, before settling back again into relative obscurity.

The Mariner 10 cameras revealed a planet (Fig. 6.7) that looked remarkably similar to the

FIGURE 6.7 Two mosaics produced by the Mariner 10 spacecraft in 1974 show the visible hemisphere of Mercury as the spacecraft approached (a) and as it receded (b) from the planet. Mariner 10 was able to examine about 45% of Mercury's surface. The remaining 55% remains unexplored.

(a)

(b)

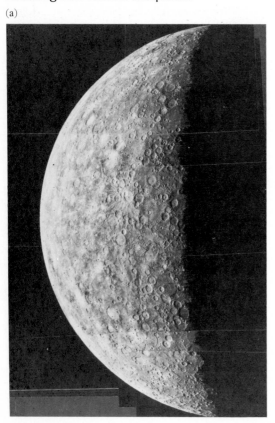

Moon. Although an abundance of impact craters had certainly been expected for this small, airless world, the degree of similarity of the mercurian surface to the lunar highlands was surprising. Most geologists had anticipated greater evidence of internal geological activity, but the Mariner photos revealed only one lava-flooded impact basin. Generally the new data did not support the idea of widespread volcanism on Mercury like that which occurred on the Moon.

While the composition of the surface material on Mercury has not been measured, the general colors and reflectivity of this planet match those of the Moon. The surface therefore consists of some kind of igneous silicate rock; whether there are basaltic lavas such as those of the lunar maria remains unknown, however. With no returned samples to study, there is no way to ascertain if the crust of Mercury is depleted in volatiles as is the Moon's, or if it shares any of the other chemical peculiarities of our satellite. Nor is there any way to measure the age of the surface, beyond noting that the heavy cratering probably implies great antiquity.

On a small scale, we know that Mercury has a regolith and that its surface soil is texturally similar to that of the Moon. These similarities can be deduced from the identical way the two objects reflect sunlight and emit thermal radio energy. With our post-Apollo knowledge of the Moon as a benchmark, we can interpret these astronomical studies of Mercury with confidence. Surface temperatures have also been measured, ranging from a high of about 675 K at noon at the hot longitudes to a low of about 90 K over most of the night side hemisphere. The lowest temperatures on Mercury nearly equal those on the Moon, since the higher daytime temperatures are compensated by a longer cooling period during the 88-Earth-day mercurian night.

Geology of Mercury

The ubiquitous impact craters on Mercury bear many resemblances to their lunar counterparts.

Where differences occur, they are probably caused by the higher surface gravity on Mercury (0.38 of the Earth's value, as opposed to 0.21 for the Moon). As a result, ejecta blankets are smaller, rarely extending more than one crater radius beyond the rim, and the secondary craters are similarly confined in their distribution.

The density of craters on Mercury varies from values near that of the lunar highlands down to about one tenth of this value (from approximately 1000 to 100 10-km craters per million square kilometers). The lower values suggest that Mercury, like the Moon, has experienced some destruction of craters since the period of heavy bombardment. If Mercury had also undergone a period of volcanism that produced a counterpart to the lunar maria, we would then be able to develop a history for the planet that could draw on our more detailed knowledge of the Moon. Alas, life is not that simple. The less cratered regions of Mercury (Fig. 6.8) simply do not look like lunar maria.

FIGURE 6.8 An example of intercrater plains on the surface of Mercury. The large ringed structure in the lower center — with many craters inside it — has been named Dostoyevsky.

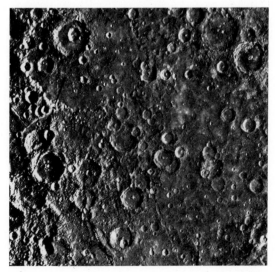

They do not differ from their surroundings in brightness or color, as do the distinctive dark basalts of the Moon; nor are they as flat as maria. Rather, they appear as gently rolling plains. No flow fronts or other distinctively volcanic features have been identified in the Mariner pictures. Thus geologists are left with a fundamental question: what destroyed some older craters to produce the rolling plains? Was it volcanism that took place during the latter stages of heavy meteoritic bombardment but did not leave the characteristic marks of the lunar maria? Or was

some other process responsible, such as blanketing of the surface by impact-produced ejecta? No one knows, although a vote among geologists would probably favor some kind of volcanic process.

The largest impact basin in the half of Mercury seen by Mariner is located near one of the two hot longitudes and is appropriately named the Caloris (Latin term for heat) basin. With a diameter of 1300 km, the Caloris basin (Fig. 6.9) resembles Orientale and Imbrium on the Moon in many ways. The most important difference,

FIGURE 6.9 This mosaic of images obtained by Mariner 10 shows the half of the Caloris basin that the spacecraft was able to see. It is interesting to compare this feature with Mare Orientale on the Moon (Fig. 5.5).

however, arises in the distinctive patterned interior floor of Caloris. If this material is basalt, why does it lack the smooth surface and dark color of Mare Imbrium and other ancient lunar basalt flows? If it is not basalt, what is it? One hypothesis proposes that the floor of Caloris is composed of material that melted as the direct result of the impact, rather than being filled in later as in the lunar maria. It is also possible that the rocks of Mercury are fundamentally different from those of the Moon; perhaps only high-temperature materials were present among the building blocks of this planet (Section 3.1).

Geologically, the most remarkable features on Mercury are compressional scarps or cliffs that have no lunar counterpart (Fig. 6.10). These scarps or wrinkles may have formed in the crust if the interior of the planet shrank slightly. Since they formed after most of the craters (as seen in Fig. 6.10), they must represent an internal event on Mercury that took place many hundreds of millions of years after the solidification of the crust. The slowing of the planet's spin to achieve the present 3 : 2 resonance has been suggested as a possible cause.

FIGURE 6.10 The scarp named Discovery stretches more than 500 km across Mercury's surface. In places it is 3 km high, about the same height as Mt. Olympus in Greece. This scarp is younger than the craters it crosses. It probably formed when Mercury's crust contracted as the planet cooled early in its history.

6.4 Interiors and Atmospheres of Moon and Mercury

The Significance of Density

Planetary interiors remain mysterious places, forever inaccessible to direct measurement. What little is known about them must be indirectly obtained and is inevitably subject to misinterpretation. Yet the effort must be made, for without some understanding of the bulk of a planet that lies beneath the visible surface, an intelligent discussion of planetary evolution is impossible.

The most important clue to the bulk composition of a planet is provided by its density (Section 1.6). The Earth has a density of 5.5 g/cm³; the Moon, 3.3 g/cm³; and Mercury, 5.4 g/cm³. The fact that these densities differ certainly suggests differences in composition. But the average density is also affected by the size of a planet; big planets with stronger gravity are able to compress their interior materials and increase their densities. For example, the iron in the Earth's core has a density of about 15 g/cm³, while ordinary iron on the surface of the Earth has a density of only 9 g/cm³. A better starting point for a chemical interpretation would be an **uncompressed density**; that is, the density corrected for the effects of self-compression in a large planet.

The uncompressed densities of the Earth, the Moon, and Mercury are, respectively, 4.5 g/cm^3, 3.3 g/cm^3, and 5.3 g/cm^3. Note that the Moon because of its relatively small size has experienced little internal compression. Mercury has the highest uncompressed density of any planet, with the Earth second. Therefore, Mercury must have the largest proportion of high-density constituents, the metals. We will discuss the internal structure of Earth in more depth in Chapter 7, concentrating here on the Moon and Mercury.

The density of the Moon is only slightly greater than that of ordinary rocks. Furthermore, the density within the Moon must increase with depth, since it is hard to imagine the alternative in which heavier materials could have floated above lighter ones when the Moon was young and molten. Therefore, the interior must consist primarily of rocky materials only slightly denser than the surface rocks, and there simply is no place to hide a substantial metallic core. One of the fundamental properties of the Moon is its lack of metals relative to the Earth. If Earth and its satellite formed together from the solar nebula, this basic difference in composition is hard to understand.

In contrast, the only plausible model for a planet as dense as Mercury requires a lot of metal, roughly 60% by mass. If 60% of Mercury consists of an iron-nickel core similar to that of the Earth, this core must have a diameter of 3500 km and extend to within approximately 700 km of the surface. Mercury can then be thought of as a metal ball the size of the Moon covered with a 700-km-thick crust of dirt.

The proper way to analyze the situation is to realize that Mercury must be deficient in silicate rocks, rather than enriched in metals. Evidently when this planet formed, temperatures in the solar nebula were so high that only metals and some high-temperature silicates could condense and form solid grains. Thus Mercury may not have the sort of silicates that form basaltic lavas and may also be deficient in radioactive elements which are usually associated with low-temperature rocks. We can speculate in this manner on a number of reasons why Mercury may have experienced a very different geological history from that of the Moon.

Moonquakes and Heat Flow

In the case of the Moon, considerable additional insight into interior conditions has been derived from the Apollo ALSEP experiments. The lunar seismometers measured the response of the whole Moon to impacts, such as those of spent Apollo rockets (Fig. 6.11), while an additional experiment on Apollos 15 and 17 directly measured the rate of heat flow from the interior. Relative to the Earth, the Moon is a very quiet place. No moonquake measured by any of the Apollo instruments was large enough to have been detected by a person standing on the Moon, and the total energy released each year by moonquakes is a hundred billion times less than that of earthquakes. Internally generated moonquakes arise primarily at a depth of approximately 1000 km, apparently near the base of the rigid layer called the lithosphere. Below this depth, temperatures may be high enough to produce some melting of rocks (Fig. 6.12a).

The heat flow from the interior of the Moon also suggests that temperatures are higher at great depths. Most of the heat being generated today results from radioactive elements in the Moon, with only a small remnant of any early, high-temperature phase remaining. If a similar heat flow measurement were made on the surface of Mercury, it would be possible to determine if the metal core were liquid, but in the absence of a lander no such data exist.

Another important line of evidence on interior structure comes from the measurement of a planetary magnetic field. Although the exact mechanisms for generating such a field are not

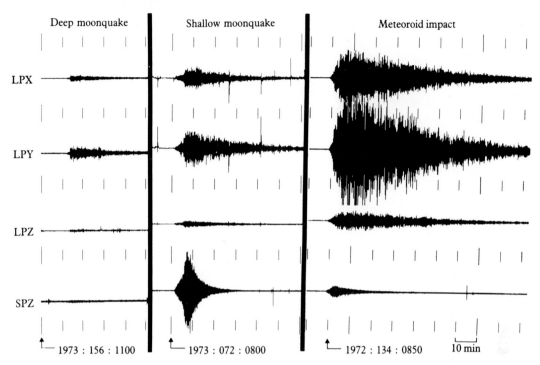

FIGURE 6.11 A comparison of seismograms obtained from various types of seismic events. The long "ringing" of the Moon in response to meteoroid impacts is especially noteworthy.

understood, it is generally believed that a strong field requires motions in a liquid metallic core. It was not surprising, therefore, when early space missions showed that the Moon lacked such a global magnetic field. No one really expected that our small satellite had a liquid metal core. Even today, with all of the collected Apollo data available, the question of whether the Moon has any metallic core at all remains open (Fig. 6.12a).

The Magnetic Field of Mercury

One of the most exciting Mariner 10 discoveries at Mercury was the presence of just such a planetary magnetic field after none had been detected for the Moon, Mars, or Venus. Although only about 1% as strong as the field of the Earth, that of Mercury was similar in character and clearly seemed to indicate the existence of a liquid core. Of course, the presence of a massive iron-nickel core on Mercury had long been inferred from the planet's high density. What was unexpected was that this core might still be hot enough to be liquid, or that sufficient motion within the core could be produced on a relatively slowly rotating planet. A decade after the Mariner flybys this situation remains unclear, since we still have no data on the state of the planet's interior. But the presence of a magnetic field on Mercury is undeniable, and most scientists conclude that the metal core must therefore be at least partially molten (Fig. 6.12b).

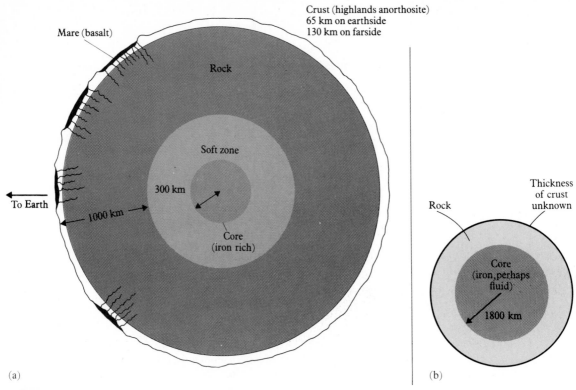

FIGURE 6.12 a) A model of the interior of the Moon derived from seismic data. It is still uncertain whether or not there is a small liquid core at the Moon's center. b) A model of the interior of Mercury based simply on the planet's high density and its requirement for a large, iron-rich core. (Note different scales in (a) and (b).)

Mercury thus presents a paradox to challenge planetary scientists. Its interior is in many ways like the Earth, with a large metal core and a planetary magnetic field. Yet its surface is like the Moon, with little evidence of any internal geological activity for at least several billion years. If the interior is hot enough to generate a magnetic field, why has it not also reworked more of the surface? Or is it perhaps that we have incorrectly estimated the age of the surface, and that the cratering on Mercury was produced more recently than on the Moon? Or perhaps the magnetic field does not indicate a molten core? These are important questions, but without additional missions to Mercury they may remain unanswered.

Atmospheres of Mercury and the Moon

We have spoken at length about the surfaces and interiors of these two similar battered worlds, but what about their atmospheres? Because both objects are small, their gravity is not strong enough to hold on to any substantial atmosphere over the lifetime of the solar system. Their high surface temperatures, up to 400 K on the Moon and 675 K on Mercury, further accelerated the escape of atmospheric gas. For most purposes, therefore, we may consider both to be airless worlds.

Nevertheless, Mercury and the Moon have extremely tenuous, transient atmospheres produced by the slight buildup in density of the solar wind as it flows around them. These flickering envelopes of gas consist mostly of hydrogen and helium, the primary atoms in the solar wind. Since they are partly ionized, the main effect of these thin atmospheres is to provide a degree of electrical conductivity around the object, thus influencing the interaction between the planet and the solar wind.

Mercury's tenuous atmospheric envelope also includes atoms of sodium and potassium that have apparently been dislodged from the planet's rocky surface by impacts from ions in the solar wind. The spectroscopic observations that discovered the sodium showed bright emission lines from this element superimposed on the normal, dark sodium lines present in reflected light from the Sun (Fig. 2.6b).

6.5 The Histories of Moon and Mercury

Reading the Geological Record

The geological history of a planet is written in its rocks and surface topography. This record can be read using the techniques of stratigraphy supplemented by detailed studies of selected rock samples. Because both Mercury and the Moon are relatively inactive internally and lack atmo-

spheric weathering and erosion, the span of solar system history available for study on their surfaces covers a great expanse of time. Approximately 4 billion years of geologic evidence is exposed on the Moon, and the surface of Mercury most likely preserves a similar record although in the absence of dated samples we cannot be sure. Yet there is a limit beyond which we see dimly if at all. As noted in Section 5.5, heavy meteoroidal bombardment during the first half-billion years of the solar system has destroyed most of the evidence from the earliest period of its history.

Models of Planetary Evolution

Direct geological evidence from its surface can be augmented by theoretical calculations of the probable thermal evolution of a planet. To trace this life history, we must begin with what we know today and try to calculate backward and then forward in time.

Such calculations require the capabilities of the most technologically advanced computers if they are to be truly representative of the laws of physics and chemistry governing planetary evolution. To be successful, the theorist intent upon such a calculation must also make certain assumptions concerning the initial state of a planet, especially concerning its chemical composition. Then, if these assumptions are reasonable and the computer program works properly, an outline or **scientific model** of planetary history is produced.

A planetary model builder requires a computer program powerful enough to trace the life story of a planet. Equally important, however, is an understanding of the behavior of materials at the unfamiliar pressures and temperatures of planetary interiors. Near the center of the Earth, for instance, the pressure is about 4×10^6 bars (or 4 million times the sea-level pressure of the

Earth's atmosphere), and the temperatures are measured in thousands of degrees. Under these conditions, the properties of each mineral, such as its melting temperature or strength, may be very different from those observed in the laboratory. The calculations will yield accurate results only if these material properties are properly considered.

The output of a planetary model calculation consists of many columns of numbers, each representing the pressure, temperature, and chemical and physical properties at various points inside a planet. All of these values are calculated for a series of time steps, separated by perhaps a few million years. From these numbers, it is possible to trace the changing conditions at each interior point over a span of billions of years. Such a record serves as a quantitative model of planetary evolution; whether a particular model represents the actual history of a real planet remains a more difficult question.

Lunar history is comparatively well defined by the wealth of evidence obtained by the Apollo missions, while Mercury's past remains more speculative. It can, however, be interpreted using the lunar based paradigm for the cratering history of the inner solar system (Section 5.3). If Mercury experienced impact rates similar to those of the Moon, including a terminal heavy bombardment about 3.9 billion years ago, a consistent picture emerges. Of course, consistency does not guarantee truth, but it does suggest that planetary scientists are moving in the right direction.

Origins and Early History

Obscurity clouds the origin of the Moon (Section 6.6). But we do know that it formed sometime shortly before 4.4 billion years ago and that its original chemical constituents were depleted in

iron and other heavy metals and in some of the more volatile elements. We also know that early in its history the upper layers of the Moon were hot enough to melt. Global differentiation may have taken place. If the Moon has a small iron core, it formed at this time.

By 4.4 billion years ago the highlands crust had begun to form by the freezing of an ocean of melted rock. The oldest anorthositic highland rocks provide us samples of a much earlier period of planetary history than is preserved on the Earth. During this period the heavy rain of debris still impacting the lunar surface repeatedly shattered and remelted the crust, destroying any detailed evidence of its early structure. The entire surface of the Moon came to consist of overlapping ejecta from innumerable impact explosions. By 4.0 billion years ago the magma ocean of the Moon had probably cooled and solidified, while at the same time a terminal bombardment occurred, either as a burst of impacts or as the last stages of a process that had continued since the formation of the Moon.

Mercury's earliest history differed from that of the Moon primarily because of its different composition. Mercury was depleted in silicate rocks rather than in metals, giving it the highest metal concentration of any planet. Differentiation must have taken place early, resulting in a magma ocean and rocky crust similar to that of the Moon. Early heating and expansion of the planet may have triggered widespread volcanism near the end of the period of heavy bombardment, producing the smooth, gently rolling plains that distinguish the surface of this planet from that of the Moon.

From the Terminal Bombardment to the Present

On the Moon, the first 600–700 million years are called the pre-Imbrian era, ending 3.8 billion years ago with the impact formation of the Imbrium basin and widespread blanketing of the Moon with ejecta from this event. Shortly thereafter the formation of Orientale basin marked the last of the large asteroidal impacts. As the meteoritic impacts rapidly declined, large scale volcanism began to fill in the deep impact basins on the lunar nearside. The sources for this lava were located hundreds of kilometers below the surface, in regions being heated by internal radioactivity. Slight expansion triggered by changes in internal temperature of the Moon may have opened fractures produced at the time of the basin-forming impacts, permitting basaltic lava to rise and flood the basins. Imbrium was flooded in this way between 3.4 and 3.3 billion years ago.

On Mercury at approximately the same time the formation and melting of a huge iron-nickel core apparently led to a slightly different evolutionary path. Instead of expanding, calculations show that Mercury began a slow contraction. Cracks in the crust were squeezed shut, inhibiting large scale volcanism. In fact, the compressional forces in the crust became great enough to form the wrinkled scarps we see today on the surface. Shrinkage rather than expansion of Mercury may explain the apparent paradox that this planet seems to have had less surface volcanic activity than the Moon in spite of interior temperatures that must surely have been higher.

Large scale volcanic activity on the Moon ceased about 3.0 billion years ago, and slow cooling of the interior has continued until the solid surface extends down to 1000 km depth. If there remain any partially molten zones, they are still deeper in the interior. Nothing much happens on the surface except for rare impacts of comets or small asteroids and continuing slow erosion by smaller impacts. Mercury is similarly winding down. Although it may still have a partially liquid iron core, the outer parts of the planet have cooled, and major geological activity ended long ago.

It is interesting to contrast these histories with that of the Earth. Except for very rare rocks in the most ancient parts of the continents, the oldest terrestrial rocks have all formed since the end of the period of lunar volcanism. Even the young (billion year old) lunar crater Copernicus formed before the appearance of hard shelled life forms in our oceans. The youngest major lunar crater, Tycho, formed about 200 million years ago, about the time the outlines of the major terrestrial continents were established. In many ways the lunar and mercurian records are the perfect complement to that of the Earth, with all the action on the smaller bodies ending just about the time our own geologic story begins to unfold.

6.6 The Problem of the Origin of the Moon

Moon and Earth: Similarities and Differences

"All explanations for the origin of the Moon are improbable." — H.C. Urey

In some ways the Moon is like the Earth, and in other ways it is different. Reconciling these facts in a self-consistent theory of the origin of the Moon has proved difficult. Before lunar exploration began, the origin of the Moon was a major question for planetary scientists; twenty-five years later it remains largely unanswered.

Study of the Apollo lunar samples revealed a fundamental similarity between the Earth and the Moon in the detailed isotopic composition of their rocks. Not only are many of the lunar minerals similar to many of those found on our own planet, but the relative proportions of different isotopes of oxygen and other elements are also the same. Recall from Chapter 3 that these isotopic ratios differ among the meteorites and that these differences are used to infer incomplete mixing of material in the early solar nebula. Thus the close similarity of the Earth and Moon in this respect strongly suggests that these two bodies were formed together, perhaps out of the same mix of materials condensing from the solar nebula.

The Moon is also different from the Earth in its bulk composition. We have repeatedly noted the absence of a large metallic core on the Moon and its consequent low density. In many ways the composition of the Moon resembles that of the terrestrial mantle, which also has a density of about 3.3 g/cm^3. But significant differences remain between the bulk composition of the Moon and that of the Earth's mantle. Water and other volatiles are severely depleted on the Moon. There is five times less potassium, while the radioactive element uranium has about the same abundance on both bodies. Other elements, such as calcium, aluminum, and titanium, are more abundant on the Moon. If the Moon and Earth formed together, it is difficult to see how these major differences in composition can be explained.

Three Theories of Lunar Origin

Three traditional theories of the origin of the Moon exist. While none of them is now believed to be entirely correct, they do illustrate the range of alternatives considered. These theories are sometimes called the daughter theory, the sister theory, and the capture theory. Let us look at each and confront it with the evidence cited above concerning the similarities and differences between Earth and Moon.

The **daughter theory** or fission theory supposes that the Moon formed from the Earth. It was first proposed in 1880 by astronomer and mathematician George Darwin (1845–1912) — the son of the famous biologist Charles Darwin — who calculated that a rapidly spinning Earth could split to form a double planet. If the alleged split took place after the differentiation of the Earth, the smaller Moon would be formed purely from mantle material and thus be depleted in metals. Furthermore, the idea seemed consistent with the tidal evolution of the Moon, which showed that our satellite is slowly moving away from the Earth.

There are serious flaws in the daughter theory, however, many of them associated with the differences in composition between the Earth's mantle and the Moon. There are additional mechanical problems in the theory as originally proposed by Darwin, and detailed calculations indicate that even if a splitting of the Earth could occur, the "daughter" would be unable to pull itself together to form the Moon and would instead break up into a ring around the Earth.

The **sister theory** suggests that the Earth and Moon formed at close proximity from a spinning cloud of dust, in much the same way that many scientists believe the large satellite systems of Jupiter, Saturn, and Uranus originated. The problem here is simple: the sister theory provides no explanation for the compositional differences between the Earth and the Moon. At least the daughter theory suggests how the Moon might have formed without much metal, but sisterhood does not seem to be at all compatible with the chemical evidence.

The third possibility is the **capture theory,** which proposes that the Moon formed elsewhere in the solar system as an independent body and was subsequently captured into orbit around the Earth. The compositional differences are thus understood as representing different condensation conditions in separate locales in the solar nebula. Again two problems arise, however. First, the isotopic ratios seem to indicate that the Earth and Moon formed from the same pool of material. Second, the Earth could not have captured the Moon in a stable orbit unless a third body intervened, and no helpful go-between has been identified. Capture by atmospheric drag, the method proposed in Section 4.5 for the capture of the martian satellites, does not apply for an object as large as our Moon. Today, the capture theory has few if any defenders.

A Possible Synthesis

If the Moon is neither daughter, sister, nor interloper, what is it? Many scientists feel that the three basic theories described above are too simple, and that we must seek a more complex scenario. Perhaps elements of more than one of these theories were involved, but what is required is a mechanism that permits the Moon or its precursor materials to form initially in the same part of the solar nebula as the Earth and then to undergo some process or processes that remove most of the metals and volatile elements before solidification.

The most widely supported theory of lunar origin in 1987 is a variant of the daughter theory. As understood today, the proto-lunar material did not separate from the Earth through fission as Darwin had suggested, but from one or more extremely large impacts taking place during the late periods of planetary accretion. A catastrophic event of this magnitude could eject huge quantities of iron-deficient mantle material, at the same time heating it sufficiently for most of the volatiles to escape. However, this process would preserve the basic isotopic similarities of the Earth and the ejected material. Subsequently, some fraction of the ejecta could aggregate in the Earth's orbit to form the Moon. As we have seen, the Moon was initially in a closer orbit to the Earth, but tidal forces have since driven it to its present position. Fig. 6.13 outlines this scenario as currently envisioned.

Perhaps we are converging on the correct answer, but the problem of lunar origin has been a difficult one. The most frustrating aspect of the problem arises from the fact that we are dealing, in the Earth-Moon system, with the two most familiar planetary bodies. Yet it is precisely these two bodies that do not conform to our more general theories about the formation of planets and satellites. When we learn more about other bodies in the solar system, will they also prove to be exceptions?

---◇---

Summary

Mercury, like the Moon, is a small airless world heavily scarred by impact craters. Its surface actually looks very much like that of the Moon, although the indications of past volcanic activity are less obvious. However, Mercury differs from the Moon in its bulk properties. Its great density, the highest of any planet, indicates that much of its interior is made of iron, similar to the core of the Earth, a hypothesis further supported by the presence of a small but quite real planetary magnetic field. In general terms, then, Mercury has an interior like the Earth and a surface like the Moon.

The lunar interior is quite different. Our satellite has lost not only volatile elements but also most of its expected allotment of iron and other metals. As a result, the Moon has at most a very small iron core and no global magnetic field. The compositional differences between the Earth and the Moon are great enough to frustrate simple theories in which the two objects form together out of the same raw materials in the solar nebula.

We looked at three simplified theories of the origin of the Moon, and found difficulties with each. Apparently the final answer is more complex, perhaps involving the ejection of terrestrial

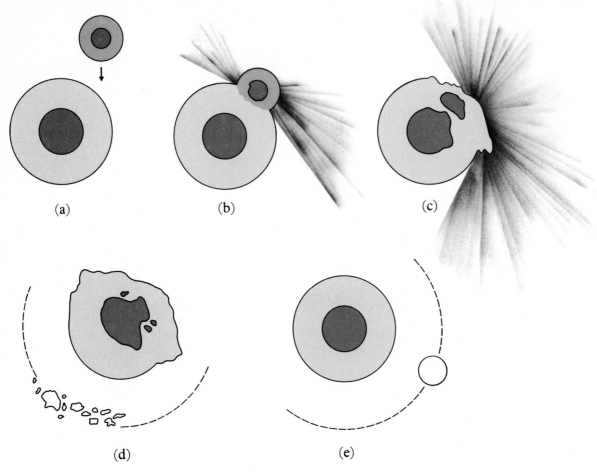

(a)   (b)   (c)

(d)   (e)

FIGURE 6.13 In the impact model for the origin of the Moon that is currently gaining favor, a Mars-size body strikes the Earth and fragments from this collision form an orbiting ring of material from which the Moon accretes.

mantle material in a giant impact and subsequent aggregation of the iron- and volatile-depleted ejecta into the Moon. Such an approach looks promising, but many details remain to be worked out.

Because Mercury is small and near the Sun and has only been visited by one spacecraft, Mariner 10, our data base on Mercury is limited. One important source of information, however, has been planetary radar. In particular the power of using Doppler shifts in the frequency of the transmitted radio pulse to determine planetary rotation has come to play a major role in providing new data. In Chapter 8 we will further see how radar can penetrate through clouds to map the surface of Venus.

Radar has revealed the unusual rotation period of Mercury, which is exactly two thirds of its 88-day period of revolution around the Sun. We examined how tides raised by one body on another can slow planetary rotation and modify orbits, providing an explanation for the fact that the Moon always keeps the same face toward the Earth as well as for the more unusual two-thirds synchronous rotation of Mercury. Tidal forces are important for many planets, satellites, and rings, as we will see later in this book.

---◇---

Key Terms

capture theory

daughter theory

Doppler effect

radar

scientific model

sister theory

solar day

synchronous rotation

tidal force

tidal heating

tide

uncompressed density

+ tidal stress

PART THREE
REVIEW QUESTIONS

1. Describe the appearances of the Moon and Mercury. For each, note what types of features can be seen at different levels of resolution. How does our knowledge of Mercury compare with that of the Moon? Can we make meaningful comparisons? How does our knowledge of both objects compare with that of the Earth?

2. Impact cratering is an important process in planetary geology. Describe how craters are formed and note which features are characteristic of impact craters. How does an impact crater differ from a volcanic crater? Why did it take so long for scientists to recognize that the lunar craters are of impact origin?

3. Determining the age of a planet's surface is another basic part of planetary geology. Discuss what is meant by stratigraphy, with illustrations from the history of formation and flooding in the Imbrium basin. Discuss the concept of crater density and crater retention age. How are these relative ages translated into absolute ages for the lunar surface?

4. Describe the standard paradigm for the cratering history of the inner solar system. Note how our detailed knowledge of the Moon is used to define the stages in this history. What role is played by the comets and asteroids, and is this theory consistent with what we learned in Part II?

5. Use what you now know about Mercury to critique the standard paradigm for solar system history. How would you expect the meteoroidal bombardment history of Mercury to compare with that of the Earth and Moon? Is there any evidence for a late heavy bombardment on Mercury? What are the likely ages for the craters, basins, and plains on Mercury, according to the standard model?

6. Discuss volcanism on the Moon. When and where did the lavas form? Why do the flows occupy the nearside basins almost exclusively? Why are there so few volcanic mountains? Why did the volcanic activity cease billions of years ago? And what evidence exists for similar volcanic activity on Mercury?

7. Discuss the soil on the Moon. What erosional processes are at work? How do large impacts create and modify the surface layers? What are the roles of micrometeorites and cosmic rays? What forms the lunar glass?

8. Describe and contrast what an astronaut would see and feel on the surfaces of the Moon and Mercury. Could you easily tell whether you were standing on Mercury or the Moon?

9. Describe how radar is used in the study of the planets. In particular, make sure you understand how the rotation of a planet broadens the returned signal because of the Doppler effect even though the radar signal is reflected from an entire hemisphere.

10. Discuss how tides are formed. Why are there two high tides every day on Earth? Why does high tide not generally correspond to the time when the Moon is overhead? What sort of tides are raised on the Moon by the Earth?

11. Compare the tidal and rotational histories of the Earth, the Moon, and Mercury. Can tidal theory account for their different rotational periods today? For each planet, is the present rotational state stable or is evolution continuing? What will the final rotation period be?

12. Compare the geology of the Moon and Mercury. What do they have in common? Consider in particular the craters, impact basins, and plains (or mare) areas.

13. Compare the histories and interiors of the Moon and Mercury. What does the geology of each tell us about the interiors? How useful are interior models? How would you go about distinguishing among several models to see which is most likely to correctly represent the thermal history of a planet?

14. Describe three scenarios for the origin of the Moon. How does each compare with the data we have on the Moon? Show how an appropriate combination of elements from these theories may be consistent with what we know about the Moon. Do you believe this synthesis is likely to be correct? Can you think of a way to test it?

15. Contrast the bulk chemical compositions of the Moon and Mercury. Can these be understood in terms of the temperatures in the solar nebula, as described in Section 3.1? If a planet had formed even closer to the Sun than Mercury, what would you expect its bulk composition to be?

ADDITIONAL READING

Chapman, C.R. 1982. *Planets of Rock and Ice* (Chapters 2, 3, 5, and 8). New York: Scribners.

Cooper, H.S.F. 1970. *Moon Rocks*. New York: Dial Press.

Cortwright, E.M., ed. 1975. *Apollo Expeditions to the Moon* (NASA SP-350). Washington: U.S. Government Printing Office.

Dunne, J.A. and E. Burgess. 1978. *The Voyage of Mariner 10: Mission to Venus and Mercury* (NASA SP-424). Washington: U.S. Government Printing Office.

*El-Baz, F. 1975. "The Moon after Apollo." *Icarus* 25, 495.

French, B.M. 1977. *The Moon Book*. New York: Penguin Books.

*Guest, J.E. and R. Greeley. 1977. *Geology on the Moon*. London: Wykeham.

Hartmann, W.K. 1977. "Cratering in the Solar System." *Scientific American* 236:1, 84.

*Kaula, W.M., M.J. Drake, and J.W. Head. 1986. "The Moon." In *Satellites*, ed. J.A. Burns and M.S. Matthews. Tucson: University of Arizona Press.

Masursky, H. *et al.*, eds. *Apollo Over the Moon: A View from Orbit* (NASA SP-362). Washington: U.S. Government Printing Office.

Murray, B.C. 1975. "Mercury." *Scientific American* 233:3, 48.

*Murray, B.C., M.C. Malin, and R. Greeley. 1981. *Earthlike Planets*. San Francisco: W.H. Freeman.

*Roddy, D.J., R.O. Pepin, and R.B. Merrill, eds. *Impact and Explosion Cratering*. New York: Pergamon Press.

Shoemaker, E.M. 1982. "The Collisions of Small Bodies." In *The New Solar System*, 2nd ed., ed. J.K. Beatty, B. O'Leary, and A. Chaikin. Cambridge, MA: Sky Publishing Corp.

*Strom, R.G. 1979. "Mercury: A Post Mariner 10 Assessment." *Space Science Reviews* 24, 3.

*Taylor, S.R. 1975. *Lunar Science: A Post-Apollo View*. New York: Pergamon Press.

*Taylor, S.R. 1982. *Planetary Science: A Lunar Perspective*. Houston, TX: Lunar and Planetary Institute.

*Indicates the more technical readings.

PART ◆ FOUR
Fraternal Twins: Venus and Earth

It is hard to think of the Earth as a planet. We humans are so intimately involved with our world that we have difficulty establishing a planetary perspective. Yet the Earth is quite an ordinary planet in terms of location and size, and it is subject to many of the same processes that shape the other worlds of the solar system. One of the achievements of the space age is that humans have begun to appreciate the close relationships among the planets, and to use information about one to gain insight into another. The other planets have much to teach about the Earth; and even more since we know it so well, the Earth can teach us a great deal about other planets.

In order to study the Earth the way we do other planets, we must step back from the wealth of detail that surrounds us and try to address the same kinds of questions that motivate our study of other worlds. Where and how did the Earth form? What is its composition and internal structure? What sources of energy maintain its geologic activity, and how have these energy sources varied over time?

◀ *The Birth of Venus* by Sandro Botticelli (1445–1510). Aphrodite ("born of the foam") is being gently blown ashore at the island of Cythera, in Greece. Botticelli was born some thirty years before Copernicus, but their lives did overlap in time. One could imagine the two of them discussing the origin of Venus over a glass of wine, from rather different points of view!

Why does the Earth alone have oceans of liquid water and an atmosphere rich in oxygen? What maintains the circulation of the oceans and atmosphere, and how has each evolved with time? Perhaps most intriguing of all, why is the Earth home to living things, and how do the myriad of life forms on the Earth interact with the long-term geological and chemical evolution of the planet?

Much can also be gained from a close comparison of the Earth and its closest neighboring planet, Venus. It is not just proximity in space that makes this comparison appropriate. Earth and Venus are truly twins, more so than any other pair of planets. They have essentially the same diameters and densities and presumably very similar bulk compositions. Yet the twins differ in important ways. Earth has a relatively large Moon; Venus has none. Earth rotates directly in 24 hours; Venus rotates retrograde in 243 days. Earth has extensive oceans of liquid water; Venus is dry. Perhaps most dramatically, Earth has a climate that can support abundant life, while Venus has developed an oppressive atmosphere, sulfurous clouds, and blistering surface temperatures. Twins these two planets may be, but certainly not identical twins. Understanding the reasons for their divergent atmospheric and surface evolution is one of the outstanding problems of planetary science, with important implications for the future of the Earth.

CHAPTER SEVEN

Our Home Planet: Earth

7.1 Earth as a Planet

The View from Space

Let us begin our study of the Earth by imagining ourselves visitors from another solar system, making our preliminary reconnaissance of the planets. Undoubtedly our first attention is directed to the giant planets with their spectacular systems of rings and satellites. Later, perhaps, we will check up on the smaller worlds of the inner solar system. Among these the largest would be Earth, the only inner planet with a major satellite.

Even from a distance, Earth would appear unique (Fig. 7.1). It is the only planet with a partially blue surface, and seen from the correct angle sunlight glints brightly off its oceans. Ice eternally blankets its south polar continent, as well as Greenland and other isolated land areas in the northern regions. The polar caps show large seasonal changes; in winter sea ice covers the polar seas and a transient snow cap invades the temperate land masses, extending halfway to the equator. These surface changes could be seen only dimly through a shifting canopy of brilliant white clouds that reflects about one third of the sunlight striking the Earth. A few tawny brown land areas would be normally clear of clouds, while other regions, primarily near the seas, are eternally clouded. Enigmatic hazes of dust sometimes obscure even the cloud-free areas, and hints of changes in surface color or reflectivity might appear with the passage of the seasons.

More remarkable than its surface appearance would be the chemistry of the terrestrial atmosphere. Nitrogen, the primary gas, makes up seventy-eight percent of the atmosphere, but the real surprise is oxygen, the second most abundant gas at twenty-one percent. On the other inner planets, oxygen, a highly reactive gas, is trapped in minerals in the surface rocks. Perhaps equally surprising is the absence of more than a trace of carbon dioxide, which is the primary constituent of the atmospheres of Venus and Mars. Thus the atmosphere and oceans of Earth represent a triple enigma: too much oxygen, too much water, and not enough carbon dioxide.

To compound the mystery, a trace of methane is present in this oxygen-rich atmosphere. Methane is so easily converted to carbon dioxide in our atmosphere that its continued existence requires an active source. Something must constantly renew the atmospheric methane; is this perhaps related to the production of oxygen?

Although invisible from a distance, one of the most remarkable aspects of the Earth at close range is its abundant and varied life. From mountain tops to ocean depths, from frozen poles to equatorial deserts, even in such extreme environments as boiling hot springs or the dry valleys of Antarctica, living things have found a home. With their extraordinary ability to evolve and adapt, life forms have occupied every possible ecological niche, keeping up with changing conditions and surviving even world-wide catastrophes such as impacts from asteroids.

Terrestrial Geology

The oceans of the Earth contain enough water to cover the entire globe to a depth of 3 km. Yet nearly one third of the surface of the planet rises above this watery layer. Most of this land area is concentrated in the six massive continents, much larger than can be explained by individual volcanoes or upthrust mountain ranges. Something on Earth produces and maintains these continents, which are strikingly different from the ocean basins that cover the other three fifths of the planet. Near the edges of some continents, high mountains rise nearly 9 km above sea level, while in places the sea floor, also near the continental margins, can drop to as low as 11 km below sea level.

A closer examination reveals to the space visitor many additional indications of geological activity: erupting hot springs and volcanoes; frequent earthquakes; youthful mountain ranges with sharp ridges and peaks sculpted by the forces of water and ice; subtle variations in gravity revealing that parts of the crust are not in equilibrium but are currently undergoing uplift or being drawn down by internal forces. On a smaller scale, everywhere the processes of ero-

FIGURE 7.1 The only planet in the solar system with oceans of water, blue skies, and abundant life. We are looking down on the South Atlantic with the west coast of Africa revealed in evening twilight to the right and the west coast of South America peeking out from under a continental cloud cover at the left (see Plate 4).

sion are tearing down the mountains, weathering the surface minerals, and laying down vast beds of sediment under the seas.

To learn something about the interior of this planet, the visitor might look at space surrounding the Earth. There, electrons and ions from the solar wind become trapped in the planetary magnetic field, forming a magnetosphere and setting up currents of electricity that gird the planet. The magnetic field of the Earth is produced in the deep interior, and its existence indicates the presence of a liquid metal core within which electrical currents can generate a magnetic field. Noting the high density of the Earth, which is considerably greater than the density of the surface rocks, the space visitor might conclude that the Earth had differentiated and that its core consists primarily of iron, a cosmically abundant metal.

Cyclic Processes and Planetary Evolution

The next step in the initial investigation of the Earth might be to examine the processes taking place. Earth is a remarkably active planet in many ways, in dramatic contrast to the stable Moon and Mercury. How should we look at this activity? One possibility is that we are seeing evidence of rapid evolution and change: the Earth is in transition from one state to another in much the same way we viewed the history of the Moon and Mercury, although their transitions were slow, not fast. But a little thought quickly shows the fallacy of this perspective when applied to Earth. Consider the rivers flowing into the sea. Do they indicate that the sea is filling up and that soon the land will be submerged? Certainly not, because the water evaporates from the ocean as fast as it is added from the land. What we have here is a cycle, called the water cycle (Section 7.5) in which input and outflow balance.

Many other processes on Earth can also be best understood as cyclic. Volcanoes deposit lava on the surface, but the new rock is eventually drawn down again to the interior to be recycled through new volcanic eruptions. The land is eroded into the sea, but the sea does not silt up; instead the sediment is used to create new continental crust. Marine animals trap carbon dioxide to form vast deposits of carbonate rocks, but eventually some of these rocks also are reprocessed, and their carbon dioxide is released again into the atmosphere. The Earth is almost like a giant machine, with many interlocked cycles turning at different rates.

The idea of the Earth as a machine is useful for understanding many processes but not all. True evolution does take place. Life, for instance, has clearly evolved, and along with it the atmosphere has changed greatly. There are long-term trends in geology that can be separated from the cyclic events. The Earth itself is slowly running down, as its interior cools and its supplies of radioactive heat are gradually exhausted. One objective in looking at Earth as a planet is to distinguish the long-term evolution from the more short-term cyclic processes in order to see the many ways in which our planet's history resembles that of other planets.

7.2 Journey to the Center of the Earth

Probing the Interior

Few places in the planetary system are less accessible to direct study than the Earth beneath our feet. Few people have traveled even a mile down into the upper crust of our planet, and only the very deepest bore holes have penetrated below 6 km, a mere one tenth of one percent of the distance (6378 km) to the center. Determining the structure and composition of the interior of the Earth is a difficult scientific puzzle, dependent on indirect evidence and careful logic.

Many uncertainties exist concerning even the basic question of the exact composition of the Earth's core.

One of the most fundamental properties of any planetary body is its density as we noted in Section 1.6. The uncompressed density of the Earth, 4.5 g/cm³, is relatively high, certainly much higher than that of the crustal rocks (2.6 to 3.0 g/cm³). Such a density strongly suggests that the Earth contains a fraction of a heavy material such as iron. In order to balance the lower densities measured for the crust and inferred for the mantle of the Earth, there must be a core of density greater than 10 g/cm³.

Since it is observed that temperatures in deep mines increase with depth, it also seems clear that the interior of the Earth is warmer than the surface, a conclusion supported by the presence of volcanism at many places on the planet. The interior, therefore, seems to be a place of high temperature and pressure, containing substantial quantities of dense material, probably iron. How is the scientist to probe such an environment and learn the true nature of the center of the Earth?

A powerful technique for exploring the interior of the Earth is provided naturally, in the form of earthquakes. Whenever some shift or slippage of the brittle crust produces an earthquake, low frequency vibrations are generated. These vibrations (Fig. 7.2) are analogous to sound, but with much longer wavelengths, typically several kilometers. The shaken Earth responds like a giant bell, and its interior structure determines the tones that will be detected at different places on the surface. By measuring the vibrations of the Earth from many locations, geophysicists (scientists who study the physical properties of the Earth) can reconstruct the temperature and pressure of the deep interior. They can also distinguish liquids, which transmit only one kind of vibration, from the stronger solids, which respond to the full range of tones pro-

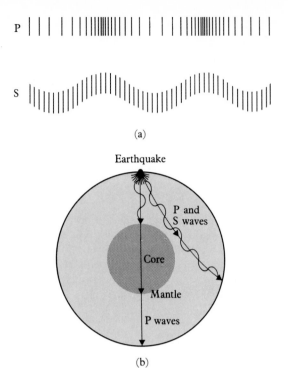

(a)

(b)

FIGURE 7.2 a) A schematic diagram of P (pressure or compression) waves and S (shear) waves. b) Paths of such waves through the Earth. S waves cannot propagate through the core, indicating that it must include a liquid component.

duced. The waves generated by earthquakes are called **seismic waves,** and instruments that measure them are seismometers.

The second important source of information on the interior of the Earth is provided by measurements of the chemical and physical properties of rocks. Most of the crustal rocks of the Earth have been derived, like the lunar basalts, by partial melting of subsurface layers, and they retain some information on the composition and nature of their source regions. In addition, there are places on Earth where material from as deep as 200 km has reached the surface. These occur in a few older parts of the continents, where a rock type known as kimberlite has been erupted under great pressure from deep in the mantle. Kimberlite, named for Kimberly in South Africa, is best known as the source of most of the world's diamonds. Diamond, formed from carbon at great pressures, is an artifact from the mysterious world just a few hundred kilometers beneath us.

A third source of information is provided by the magnetic field of the Earth. Although no one understands all the details of the mechanisms that generate a planetary magnetic field, it is clear that a metallic, turbulent liquid is required. The presence of a strong magnetic field therefore demonstrates that at least some part of the interior is both metallic and liquid. Combining all of this information allows us to undertake a scientific journey into the interior of our planet.

There are six major and several minor landmarks along the path to the interior. The six major divisions (Fig. 7.3) into which we conveniently divide the Earth are, beginning from outer space:

1. The magnetosphere: the region of trapped charged particles that extends from the upper atmosphere, roughly 200 km high, out to the boundary with the solar wind and into interplanetary space at an altitude of about 100,000 km.
2. The atmosphere: the layer of gas (primarily nitrogen and oxygen) that extends from about 200 km altitude down to the surface.

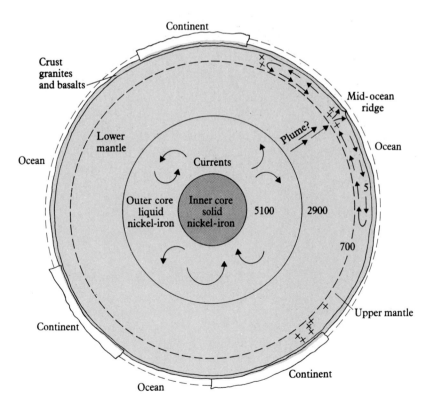

FIGURE 7.3 The interior of the Earth, according to current thinking. A small solid core is surrounded by a liquid region which in turn gives way to the mantle. Rapid convection in the liquid core generates the planet's magnetic field. Convection in the mantle causes the slow motions of crustal plates, moving the continents, forming mountains, and subducting eroded material. (Highly schematic.)

Intimately related to the atmosphere is:

3. The ocean or hydrosphere: the layer of liquid and frozen water that covers about three fourths of the surface of the Earth.

4. The crust: the solid surface of the Earth bounded on the top by the oceans and atmosphere. Under the continents it extends to a depth of tens of kilometers, but under the oceans to just 5 or 6 km.

5. The mantle: a huge region making up two thirds of the mass of the planet, consisting of solid but plastic rock extending from the bottom of the crust down to a depth of 2900 km, almost halfway to the center of the planet.

6. The core of the Earth: composed of dense metal, divided into an outer liquid core and an inner solid core.

The Crust

The **crust** of the Earth is defined as the uppermost solid layer, distinct in composition from the mantle beneath it and derived from it by partial melting. Two forms of crust exist, distinct in both composition and structure. Under the ocean basins, the crust is thin, averaging only 6 km in thickness, and composed of basaltic rocks, generally similar to the lunar and meteoritic basalts. This crust, which covers 55% of the surface, is all relatively young, having solidification ages of less than 200 million years.

Although apparently analogous to the basaltic maria of the Moon, the origin of the terrestrial basalts is rather different. The ocean crust is formed from lavas rising from the interior along well defined rifts or spreading centers, whose position is marked by several mid-ocean ridges. Formed at these rifts, the ocean crust persists until it is destroyed by sinking into the mantle to be reheated and recycled. This process of formation and destruction of the oceanic crust is described in detail in Section 7.3.

The continental crust is thicker, older, and of lower density than the oceanic crust. It covers 45% of the surface and constitutes 0.3% of the mass of the Earth. Its primary constituents are rocks called **granites.** Granitic rocks, like basalts, are igneous, but their composition is different, and their solidification took place *below* the surface under great pressure.

The thickness of the continental crust varies from about 20 km to more than 70 km. Although there are exceptions, normally the continental crust can be thought of as floating on top of the denser mantle. The crust is thickest beneath the highest elevations and thinnest under lower areas. In addition to the igneous granites, the continental crust includes a great deal of sedimentary and metamorphic rock accumulated over hundreds of millions, and even billions, of years. All of the oldest rocks on the Earth are found in the continental crust.

The Mantle

Below the crust lies the **mantle,** composed of silicate rock with elemental composition similar to that of the crust but occurring in the form of higher-density minerals. Heated by its own radioactive elements and by heat leaking from the hotter core below, the mantle is insulated by the crust above and is more plastic than brittle. Within the mantle, three distinct layers can be identified.

The upper part of the mantle, down to a depth of about 100 km, is mechanically attached to the crust. Together, this upper mantle and the crust are called the **lithosphere.** The boundary between the crust and upper mantle represents a change in composition; the boundary between the lithosphere and the mantle underneath represents a change in mechanical properties. The lithosphere is not strongly attached to the mantle beneath, and it can slide over it like an ice cube on a smooth tabletop. For an understanding of the dynamics of the Earth, which are discussed in Section 7.3, the distinction between the lithosphere and the rest of the mantle is more important than that between crust and mantle.

Below the lithosphere, and extending to a depth of about 700 km, lies the part of the mantle that generates most of the geological activity seen on the surface. In this region the properties of the mantle materials permit it to convect, transporting heat upward by a slow mechanical turnover of mantle material. These massive convection currents provide the motive force that moves the plates of the lithosphere, creates mountains and continents, and generates the heat to power volcanoes.

The lower mantle extends from a depth of 700 km down to the core boundary at 2900 km. Within it, the increasing pressure causes the minerals to alter to a denser phase in which convection cannot take place.

The Core

The boundary between mantle and core marks a major discontinuity in the interior composition of the Earth. Below 2900 km the primary constituent is iron, probably containing nickel, sulfur, and other cosmically abundant elements. At the temperature and pressure corresponding to this boundary — about 4500 K and 1.3 million bars, respectively (where one **bar** is defined as the atmospheric pressure at the surface of the Earth) — iron is liquid, so the core-mantle boundary is also a region of transition from solid to liquid phase. At these pressures, the iron has a density of at least 10 g/cm^3. The diameter of the core is almost 7000 km, substantially larger than the planet Mercury, and it accounts for one third of the total mass of the Earth.

Still deeper, at 5200 km, the increase of pressure to 3.2 million bars triggers another phase change, with iron reverting to solid form. This solid inner core continues all the way to the center of the Earth, where the pressure is nearly 4 million bars and the temperature is about 5000 K.

Within the liquid outer core the iron can move rather easily, driven either by heat (causing convection currents) or by chemical changes. Because the Earth is spinning, these motions lead to turbulent flows, which in turn generate the magnetic field that extends through the surface and into the magnetosphere. Thus the force that aligns a compass needle is the result of a special combination of temperatures, pressures, composition, and rotation deep in the interior of the planet.

Periodic magnetic field reversals demonstrate just how turbulent the motions are within the core. We are accustomed to having our compass needles point approximately northward, but there have been many periods in the relatively recent geologic past when our compasses would have pointed south. As a matter of fact, the strength of the Earth's field has been steadily dropping for the past few decades, and if this trend should continue we might experience a field reversal within the next 200 years. More likely, however, we are seeing a short-term fluctuation only. In addition to these major reversals, the magnetic field shifts direction continuously in a phenomenon known as the wandering of the magnetic pole.

The interior structure of the Earth, with its layers of increasing density down to the center, is clearly the result of planetary differentiation. Presumably this differentiation took place very early in the history of the planet, but unfortunately no way exists to probe such distant times. We must use theoretical means, reconstructing the story through comparative studies of other solar system bodies.

7.3 The Changing Face of the Earth
The Geologic Time Scale

There is only one planetary body, Io, more geologically active than the Earth. All around us we see the processes of erosion, and road cuts reveal the strata, often tilted and twisted, of sedimentary rocks deposited by erosion in the past (Fig. 7.4). The mountains and hills are worn down, yet there are still high plateaus and jagged peaks; therefore, invisible processes must be acting to raise new mountains. Volcanoes erupt, and earthquakes testify to movements in the Earth's crust. There is an almost total absence of impact craters, signifying destruction of craters as fast as they are formed.

The geologic history of the Earth (like that of other inner planets) is read in its rocks, which record the events of the past four billion years. The strata deposited in successive layers are like the pages of a history book. Unfortunately, however, the pages have frequently been shuffled or even lost in subsequent upheavals. To read the record, therefore, the geologist must find a way to number the pages, independently of their present sequence.

There are two ways to order the record: by stratigraphy to deduce a relative sequence, often aided by noting the fossils embedded in the rock, or by radioactive dating, to produce an absolute or chronometric age scale. The standard **geologic time scale** (Fig. 7.5) was first determined by stratigraphy, primarily by nineteenth century geologists, who also identified the fossils associated with each stratum. Specific dates were established only during the past half century.

The simplest major division of the geologic record is into two time periods, the **Precambrian,** before fossils of multicelled organisms became widespread (about 590 million years ago), and the **Phanerozoic,** which extends from the Precambrian to the present. The next level of detail involves the division of geologic time into eras: the Archean (from the origin of the Earth to 2500 million years ago); the Proterozoic (the era of one-celled creatures, extending to 590 million years ago); the Paleozoic (an era that saw the development of fish and the occupation of the land by plants and early reptiles, ending 248 million years ago); the Mesozoic (dinosaurs, early mammals, flowering plants, ending 65 million years ago); and finally the Cenozoic (from the extinction of the dinosaurs 65 million years ago

FIGURE 7.4 Examples of layers of rock folded and lifted by the slow but inexorable motions of the Earth's crust. (left) Thin-bedded limestone in Texas. (right) Sandstone in Maryland.

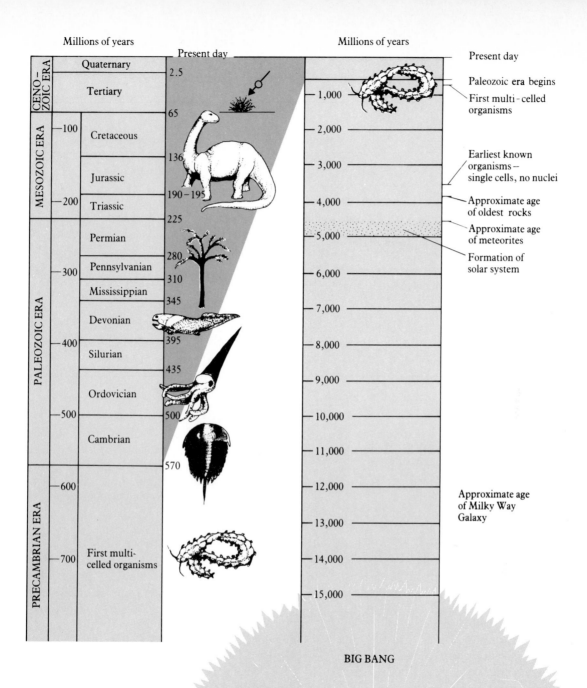

FIGURE 7.5 The geologic time scale. The full scale is given at the right. The left-hand scale covers just the last 800 million years. Note that the exact boundaries in time between periods differ from the values adopted in the text. These differences illustrate the uncertainties that still exist in the chronometric time scale for times older than 200 million years. (Adapted from Foster.)

TABLE 7.1 Comparison of geologic time scales for Moon and Earth

Time Before Present (billion yrs)	Earth Era or Eon	Events on Earth	Events on Moon
0.03	Cenozoic	Mammals	
0.2	Mesozoic	Reptiles	Crater Tycho
0.4	Paleozoic	Land plants	
1.0	Proterozoic	Early multi-celled fossils	Crater Copernicus
2.0	Proterozoic	Abundant oxygen present in atmosphere	Scattered craters
3.2	Archean	One-celled life forms	End of volcanism
3.5	Archean	Oldest evidence of life	Peak of mare volcanism — filling of mare basins
3.8	Priscoan	Oldest rocks	Late heavy bombardment ending
4.4	Priscoan	(no record)	Oldest highland rocks
4.5		(formation of Earth and Moon)	

to the present). In Table 7.1 this time scale for Earth is compared with that of the Moon, the only other planetary body for which precise radioactive dates can be assigned to geologic events.

Each of the four geologic eras of Phanerozoic time is subdivided into periods, and each period further subdivided into epochs. These divisions are based on the fossil record, later calibrated by radioactive age dating. In contrast, there are few subdivisions of the Precambrian eras, reflecting both the absence of fossils and the scarcity of rocks from so long ago.

Using both the fossil-based geologic time scale for sedimentary rock and radioactive dating for igneous rock, geologists can date geological events of the past. A third technique has also proved very useful, especially in applications to the past few million years. As we noted in Section 7.2, the magnetic field of the Earth reverses direction every few hundred thousand years on average. This magnetic shift, of course, has no

effect on the rotation of the Earth, but it does leave a record in the magnetism of lavas. When molten rock cools, its crystals preserve a record of the direction and strength of the Earth's magnetic field. Thus the magnetic orientation of a rock (if it has not been moved) can be used to identify the time when it solidified.

For the most recent time periods, even more accurate dating techniques are available. Tree rings can be counted stretching back about 5,000 years, and ice cores taken from Greenland and the Antarctic now permit year-by-year measurements of temperature and atmospheric dust extending back 20,000 years. For periods up to about 15,000 years, the radioactive isotope C-14 yields ages for organic material and has proved very useful for tracing the early history of human civilization.

Much of the geologic structure that we see around us is comparatively young. The jagged peaks of the Rocky Mountains or the Alps have been sculpted by ice within the past hundred thousand years; the volcanic islands of the Hawaiian chain are at most a few million years old, and new land is still added through volcanic eruptions (Fig. 7.6). Even a feature as large as the Grand Canyon in Arizona has been cut by the Colorado River within the past 10 million years. Thus many of the major landforms around us have ages comparable to the short span of human existence on this planet.

Volcanoes

Volcanism is one of the most spectacular forms of geologic activity. On our planet, heat from within the mantle drives a variety of types of volcanic activity. The largest but rarest types of eruptions lay down vast plains of fluid lavas very similar to the lunar maria; examples, each about 60 million years old, are the Columbia River flood basalts that cover eastern Washington State (400,000 cubic kilometers in volume) and the even larger Deccan lavas that make up the central part of the Indian subcontinent. More familiar, however, are the eruptions that build the distinctive mountains that we call volcanoes.

Although geologists distinguish among many types of volcanoes, we will consider only two basic classes, both of which have been found on other planets as well as our own. First are the **shield volcanoes,** exemplified by the volcanoes of Hawaii (Fig. 7.6). These are low dome-shaped mountains built up by the repeated eruption of relatively fluid lavas from a single vent or fis-

FIGURE 7.6 a) The largest active volcano on Earth, Mauna Loa, on the island of Hawaii. The gentle slopes of this volcano result from the low viscosity of the lava that formed it and the relatively subdued style of Hawaiian volcanism. The summit of Mauna Kea, adorned with observatories, is in the foreground. b) The summit of Mauna Loa showing several circular depressions called calderas. These calderas are not impact craters like those on the Moon, but volcanic features like those found on the huge volcanoes of Mars (Fig. 9.17). Mauna Kea is in the background.

(a)

(b)

sure. Although the individual layers are typically only a few meters thick, the mountains can reach heights as great as 9 km (Mauna Loa and Mauna Kea in Hawaii). Typical shield volcanoes have slopes of under 10 degrees and are topped by large, shallow, flat floored craters called **calderas** (Fig. 7.7). The eruptions that produce shield volcanoes are nonexplosive, and the summit calderas are the result of subsidence rather than explosion (Plate 3).

At the opposite extreme are a variety of explosive eruptions that build and often then destroy steep sided **cinder cone volcanoes.** Cinder cones are formed by the fall-back of plumes of viscous lava fragments ejected to great heights by the force of the eruption. Their slopes are typically about 40 degrees. Perhaps the best known example of a cinder cone volcano is Mt. Fuji in Japan (Fig. 7.8). Stromboli, in Italy, is another (Plate 2). The most violent kinds of eruptions, such as the 1980 explosion of Mt. St. Helens in Washington State (Fig. 7.9) or the eruption of Vesuvius in AD 79 that destroyed Pompeii, can eject many cubic kilometers of rock dust into the atmosphere as well as devastate the surrounding countryside.

Other Geologic Activity

Another easily recognized form of geologic activity is erosion by ice. In glacial regions, the movement of ice sheets can gouge out deep U-shaped valleys in times as short as a few thousand years. It is primarily glacial action that is responsible for the sharp peaks that characterize the best known mountains of the Earth. As we have seen

FIGURE 7.7 A close-up view of the summit caldera of Mauna Loa as it appeared in 1966 (compare Fig. 9.17).

FIGURE 7.8 Painting of Mt. Fuji by the Japanese artist Hokusai. The slopes of this volcano are not as steep as the artist has painted them, but they are much steeper than those of Mauna Loa. The volcanism here is of a different type than that in Hawaii.

on the Moon, mountains tend to be low and smooth in profile on a world without ice erosion.

Finally, **sedimentation** is a process that can alter the landscape rapidly. Shallow lakes can silt up in a few decades, and the recession of the sea due to sedimentation has left the docks of once great ports like Ephesus, Miletus, and Pisa dry and deserted. Since most sediment is deposited in the sea, the fact that the seas have not filled with silt is one of the properties of the Earth that requires an explanation.

By mapping the distribution of rock types and landforms, and determining ages from the fossil record, radioactivity, and rock magnetism, scientists have begun to read the book of geological history of the Earth, although events earlier than the past few hundred million years remain sketchy. Beyond two billion years ago, entire

FIGURE 7.9 The spectacular explosive eruption of Mt. St. Helens in 1980. This event was yet another reminder of just how active the geology of our planet still is.

chapters are missing, and of the earliest periods of terrestrial history not a trace remains.

The story is one of a dynamic, changing planet. The ocean floors that make up 55% of the surface change the most rapidly, with rock continuously being created and destroyed. The continents in contrast are relatively stable with some continental rocks dating back nearly 4 billion years. Yet on the continents also, change is everywhere. The great mountains of the Earth, such as the Himalayas or the Andes, are only tens of millions of years old. Volcanoes are created and destroyed on an even shorter time scale. The continents themselves constantly shift position in response to forces of mantle convection hundreds of kilometers below the surface.

7.4 Plate Tectonics: A Unifying Hypothesis

A Revolution in Geology

In the 1960s the science of geology underwent a dramatic change that altered the basic assumptions of the field. Until then, almost all geological science had been based on the idea that the major features of the Earth — the continents and ocean basins — remained in a fixed location. This was not an unreasonable hypothesis, given that the mantle of the Earth was thought to be solid and unyielding. Adherence to this view was thus as natural as accepting the evidence of our senses that the Sun revolves around a stationary Earth. Within this framework, geology was interpreted in terms of periods of slow uplift to form mountains in competition with the forces of erosion gradually tearing them down again. This outlook is still reflected in older books and museum displays that explain the landscape in terms of alternating uplift and subsidence, as though the crust were part of a seesaw, always moving up and down but never sideways.

The first serious challenge to the orthodox geological viewpoint came from Alfred Wegener, a German meteorologist and amateur geologist. Early in this century Wegener became convinced that the strong similarities between rocks on either side of the Atlantic could best be understood if the east coasts of North and South America had once fitted against the west coasts of Europe and Africa like the pieces of a puzzle. Even casual inspection of a world map suggests that this fit is rather good, but proof was harder to establish.

Wegener was also concerned about the mechanism by which the continents might be moving apart, and in Iceland he studied the active volcanoes, concluding that this mid-Atlantic island was located on the rift where the new world and the old were still separating. For many years he developed his theory of drifting continents, seeing in it ways to explain the mysterious forces that caused the uplift of mountains. But his theory had a major stumbling block: it violated what was then understood about the rigidity of the mantle and crust.

Today geologists realize that Wegener was correct about **continental drift** and about many other things as well. The critical evidence came from measurements of the age of the oceanic basalts, particularly from the symmetric pattern of ages revealed in the Atlantic on either side of the mid-ocean ridge where new rock is being formed and gradually moving outward as the continents pull apart. Soon many other pieces of the puzzle fell into place. The result is a unifying geological concept called **plate tectonics,** which explains much of what happened to form the Earth's topography. As we shall see, it is not yet clear that this remarkable unifying theory can be applied to any of the other planets. It may be another of Earth's unique attributes.

Formation of Ocean Crust

The surface of the Earth is now recognized to consist of six major and about ten smaller plates that float like sheets of ice upon the plastic mantle beneath (Fig. 7.10). The plates constitute the

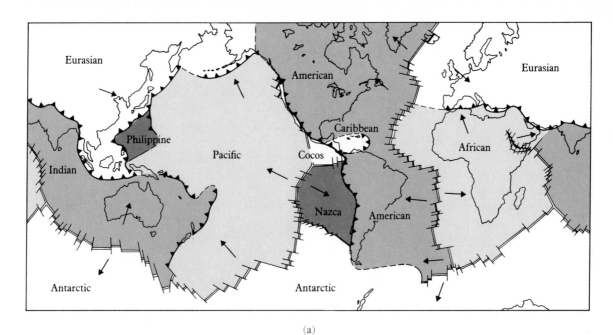

(a)

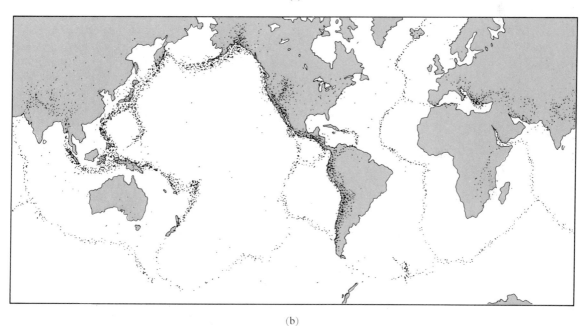

(b)

FIGURE 7.10 a) The main crustal plates on the Earth. b) The boundaries of these plates are delineated by the source points (epicenters) of earthquakes observed over a seven-year period.

lithosphere, about 100 km thick and made up of both the crust and the upper mantle. Driven by mantle convection, these lithospheric plates move at the rate of a few centimeters per year. Since there is no free space on the surface of the Earth, the plates bump into each other as they move.

Three kinds of boundaries between plates are possible. First there is the rift zone or spreading center, such as the mid-Atlantic ridge, where the plates are separating. Second is the subduction zone, such as the deep trenches off the east coast of Asia, where plates collide and one is forced under the other. Third is a fault where one plate scrapes along parallel to another, such as the San Andreas fault in California.

There are 60,000 km of active **rifts,** where new crust is formed by basalts rising from below. Most are in the oceans, and the few on land, such as the great rift valley that extends through central Africa, probably represent places where a continent is being torn apart to be replaced by new ocean. The average spreading rate is a few centimeters per year, corresponding to the creation of about two square kilometers of new oceanic crust each year. Since the total area of oceanic crust is 260 million sq km, it is easy to see that the lifetime of a given piece of oceanic basalt is only a little more than 100 million years

(260 million divided by two). Otherwise the entire planet would be covered by oceanic crust.

Oceanic crust is destroyed by **subduction** when convection pulls a plate into the mantle. As two plates are forced together, one generally slides under the other (Fig. 7.11). The sinking slab grinds down into the mantle, generating deep earthquakes, which permit its descent to be traced. At a depth of 200 to 300 km it succumbs to increasing heat and melts. Some of the melt can rise to the surface in the form of volcanoes, which also mark the position of the subduction zone, but most is recycled back into the upper mantle to await eventual release at another rift zone.

Many of the most violent earthquakes occur along **faults,** where one plate is forced to move parallel to another (Fig. 7.12). Along the San Andreas fault, for instance, the Pacific plate (which includes the southern coastal regions of California) is forced northward at a rate of a few centimeters per year with respect to the main part of the North American continent. At this rate, Los Angeles will be next to San Francisco in about 20 million years.

Instead of sliding smoothly, however, the jagged edges of the plates tend to stick until the strain becomes so large that the restraints suddenly give way. The resulting motion is sudden

FIGURE 7.11 Here a slab of oceanic crust is being subducted. As it slides down under the granitic continental plate, earthquakes are generated and volcanoes form at the plate boundary.

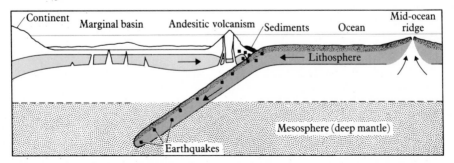

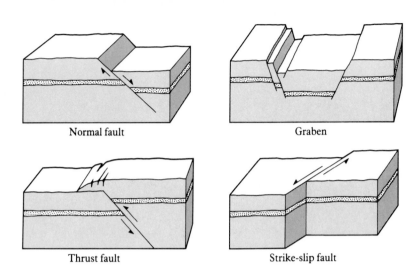

Normal fault

Graben

Thrust fault

Strike-slip fault

FIGURE 7.12 Diagrams of the common types of fault structures. We will encounter faults again on the surfaces of other planets and satellites.

and violent. In California, the interval between major plate motions is about a century, resulting in typical movements of several meters when the strain is finally released. The last major earthquake in the part of the San Andreas near San Francisco took place in 1906, while the southern part of California has not moved significantly since the great Ft. Tejon earthquake of 1857. The 130-year buildup of tension along the southern part of this fault is obviously a matter of concern for the residents of southern California.

Plate motions produce volcanoes along both rift zones and subduction zones. In addition, volcanoes can erupt in the middle of lithospheric plates above mantle hot spots. These hot spots are thought to represent rising plumes that heat the crust from below to induce eruptions. The Hawaiian Islands are an example of volcanoes produced by a mantle plume. Mauna Loa, located over the plume today, is the highest mountain on Earth, rising more than 9 km from its base to its summit (Fig. 7.6). Since the Pacific plate is moving with respect to the plume, the volcanic center has traced out an island chain over a period of tens of millions of years (Fig.

7.13). It is not the plume that moves, however, but the plate.

There are three main ways that the interior heat of the Earth reaches the surface and escapes. First there is heat conduction through the crust, which is thought to be the dominant process on most other planets. Second is the volcanic activity associated with hot spots. But the most important way heat is released from our planet is at the rifts and subduction zones produced by tectonic activity. Without plate tectonics, the heat balance of the Earth's interior would be very different.

The Continental Crust

The continental crust, composed mostly of granite, has a different history from that of the oceans. Like a frothy slag floating on the top of the plates, it is rarely subducted. Continental granite is produced much more slowly than oceanic basalt, but it also lasts longer. The creation and destruction of the ocean floor is a cyclic

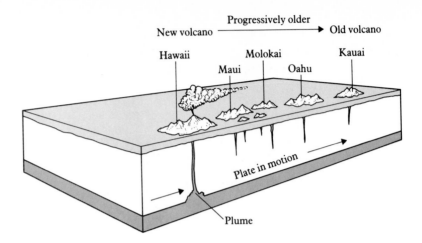

New volcano → Progressively older → Old volcano

FIGURE 7.13 The gradual motion of the Pacific plate past a long lasting "hot spot" (representing the top of a mantle plume) accounts for the formation of the chain of Hawaiian Islands.

process, but the continents evolve, slowly increasing in total area as more new granite is added by volcanism than is destroyed by subduction.

Jostled by mantle convection, the floating continental masses are sometimes pushed together and other times pulled apart. The last time nearly all of the land mass of the Earth was jumbled together was about 200 million years ago, near the transition from the Paleozoic to the Mesozoic eras. This super-continent has been named Pangaea. As illustrated in Fig. 7.14, this land mass has gradually been pulled apart and rearranged since then by continental drift. Not everything is coming apart, however. Some of the land masses are colliding with each other, most notably the Indian subcontinent with Asia. At such points great mountain ranges are folded and uplifted by the pressure of the moving plates. Before the existence of Pangaea, there were other configurations of the land and oceans, but so much of the evidence has been lost that it is difficult to reconstruct the map of Earth any farther back than the Phanerozoic. The Ural Mountains in the Soviet Union, for example, may represent a boundary between ancient plates, but the evidence is fragmentary. The last half billion years — just 10% of the age

of the Earth — is a practical limit for detailed geological study, although individual continental rocks almost as old as 4 billion years have been identified.

The geological processes outlined here are highly dynamic, with much of the landscape of our planet produced by forces deep in its interior. Because the Earth is much larger than either the Moon or Mercury our planet has not cooled as much, and these forces continue to act today. Most of the processes discussed here are cyclic, involving repeated acts of creation and destruction. The time scales, 100 to 200 million years, for both the cycling of oceanic crust and the reconfiguration of the continents, are very short compared to the 4.5-billion-year age of the Earth. In contrast, the production of continental rock takes place more rapidly than its destruction, and the continents thus evolve with time.

7.5 Ocean and Atmosphere

Floating on top of the solid Earth are two layers of great importance to us: the hydrosphere and the atmosphere. The liquid water that covers 70% of the surface, and the oxygen-rich atmosphere that shields us from solar ultraviolet light and replenishes the land with life-giving moisture, are unique to our planet. Without them there would be no life. And in return, both atmosphere and ocean are shaped by the evolution of the very life they make possible.

The Composition of the Oceans

The hydrosphere of the Earth is defined to include the freshwater lakes and streams, the ice caps that blanket the polar land masses and high mountains, and the sea ice that floats on much of the arctic and antarctic seas. Its main compo-

nent, however, is the oceans. The oceans, with volume of 1.3 billion cubic kilometers and a mass of more than one billion billion (10^{18}) tons, are the great reservoir of water on our planet.

Ocean water is not pure H_2O. As we all know, ocean water is salty. These salts include, in addition to common NaCl, a variety of other water-soluble compounds, which together make up about 3.5% of the mass of the oceans. Water is an excellent solvent, breaking down many compounds into their individual atoms. Six elements constitute more than 90% of the salts dissolved in the ocean: chlorine, sodium, magnesium, sulfur, calcium, and potassium.

FIGURE 7.14 Motions of the continents with respect to the South Pole (cross) through geologic time. The dashed circle represents the position of the Earth's equator.

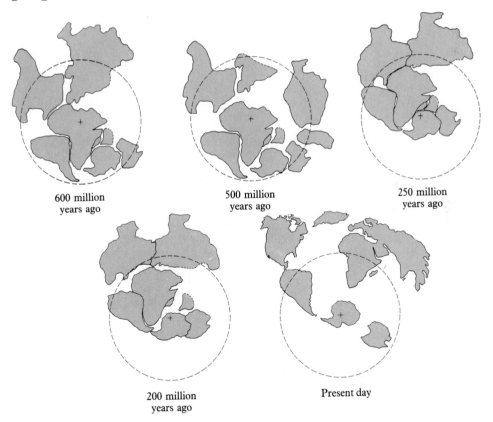

600 million
years ago

500 million
years ago

250 million
years ago

200 million
years ago

Present day

The concentration of each of these elements in seawater represents a balance between inflow — primarily from weathering and erosion of continental rocks — and precipitation into ocean sediment. Calcium provides an example. About 100 million (10^8) tons of dissolved calcium enters the ocean each year, yet the total calcium content of the ocean is only a million times larger, 10^{14} tons. This does not mean that the oceans have existed for only a million years as was once thought. Rather, it means that the average residence time of a calcium atom in seawater is a million years. On the average, then, it takes a million years for calcium to combine with carbonate atoms and become incorporated into limestone. This cycle of calcium from surface rocks to ocean to limestone sediment is completed when the ocean sediment is melted in a subduction zone and recycled back into the continental crust (Fig. 7.11).

Gases as well as salts are dissolved in the ocean water. The most important of these are oxygen and carbon dioxide. The oxygen in the water makes possible the existence of advanced forms of marine animal life, such as the fish, which use gills to extract the oxygen they require from the water. However, the total oxygen content of the ocean is less than that of the atmosphere. In contrast, about sixty times as much carbon dioxide is dissolved in the oceans as is present in the air. As a result, the role of the oceans is critical in determining the amount of carbon dioxide in the atmosphere, and therefore in establishing the climate of the Earth.

The Ocean Floor

While examining the geology of the Earth, we noted that the ocean crust is distinct from that of the continents in composition and origin. The boundary between the granitic continental crust and the thinner basaltic crust of the ocean does not, however, coincide with the current shorelines of the Earth. At present ocean levels, these continental boundaries are drowned under about half a kilometer of water and generally lie about 75 km off shore.

The continental shelves, as the submerged parts of the continents are called, are the best known part of the ocean and the richest in marine life. At their edges, the surface drops off more steeply (with an average slope of 4 degrees) into the true ocean deeps. With the present configuration of the continents there are four primary deep basins, containing the Atlantic, Pacific, Indian, and Arctic Oceans. At other times, of course, the ocean basins were different in size and shape, but these four have dominated for about the past 100 million years.

Sediment washed down from the land is deposited first on the continental shelves but gradually moves out and down into the ocean proper. Great underwater canyons carry sediment into the deep ocean, where it typically accumulates to a depth of two or more kilometers. Much of the ocean floor consists of vast, gently rolling plains deeply buried in sediment. These deep ocean plains are the flattest major parts of the surface of the Earth.

Ocean Temperatures

Because of its huge mass and thermal capacity, the ocean serves as a great heat reservoir. The average temperature of the ocean is 3.9 C, only a little above the freezing point. Although the surface temperature varies from about 30 C near the equator to below freezing on the polar ice caps, the temperature a kilometer or more below the surface changes very little with location or season.

It is convenient to divide the ocean into two regions that interact differently with the air and land. The upper part, less than a kilometer in thickness, is generally warmer and circulates through great current systems that heat or cool adjacent land masses. Here the sunlight deposited in the water is absorbed and photosynthetic life flourishes. Beneath it lies a dark, static world of denser, colder water still largely mysterious to us that extends to the deepest parts of the ocean floor.

The recent discovery of ocean thermal vents (hot springs) illustrates that surprises still await us in the ocean depths. In a number of areas near plate boundaries and regions of volcanic activity, heated water containing many dissolved minerals has been found issuing from massive underwater vents. These remarkable thermal springs support unique life forms. Creatures live here without sunlight, deriving both nutrients and energy from the hot, mineral-laden water. In this dark, mysterious world, giant sea worms, clams, and other exotic creatures thrive in an environment where no life was thought possible. Perhaps there is a lesson here in looking for life on other worlds: the variety of environments that can support life forms is very broad.

The surface of the ocean is heated by the Sun and stirred by the air. The wind generates waves and helps to drive the major ocean currents. Even the most violent storms, however, affect only the upper tens of meters of the ocean. Although this fact is of no comfort to the mariner caught in a wild storm, we should remember that the most fearsome hurricane or typhoon is a superficial event, of little consequence to the placid reservoir of water beneath the surface.

Composition of the Atmosphere

Above the ocean is the atmosphere, which consists primarily of molecular nitrogen (N_2) and molecular oxygen (O_2), which make up 78% and 21% respectively. The remaining 1% of dry air is primarily argon (Ar), with carbon dioxide (CO_2) the fourth most abundant gas at 0.034% (or 340 parts per million). In addition, the atmosphere contains considerable water vapor,

ranging from 3% in wet, warm regions down to much less than 1% in cold, dry locations. Table 7.2 summarizes the compositional measurements.

The atmosphere exerts a pressure on the surface equal to the weight of the gas. On the Earth, the force of gravity and the amount of gas generate a sea-level pressure of about 1,000,000 dynes per sq cm, or the same pressure that would be exerted by a 10-meter-thick layer of water. This pressure is just one bar, the unit of atmospheric pressure used when studying the planets. Note, however, that the pressure is not a measure of total amount of gas in an atmosphere. Pressure is the product of the amount of gas multiplied by the force of gravity. On Saturn's satellite Titan, for instance, about ten times as much gas is required to generate a pressure of one bar as would be the case on Earth, since the surface gravity of Titan is only about one-tenth that of our own planet.

Nitrogen is a chemically lazy gas, little inclined to react with other substances. Even more inert is argon, one of the noble gases discussed in more detail in the next chapter (Section 8.7).

TABLE 7.2 Composition of Earth's atmosphere

Gas	Fraction (percent by number)
N_2	78.1
O_2	20.9
H_2O	3–0.1 (variable)
Ar	0.93
CO_2	0.034 (increasing)
Ne	0.0018
He	0.0005
CH_4	0.0002 (increasing)

Nitrogen and argon are also sufficiently massive to resist rapid escape from our planet's upper atmosphere. The present abundance of argon therefore reflects the total accumulation released into the atmosphere over the history of the Earth. In contrast, both oxygen and carbon dioxide are highly reactive chemically, so the abundances of both must represent an equilibrium between competing reactions that produce and destroy them. Nitrogen is also removed from the atmosphere, primarily by living organisms. As we will see below, the primary sources and sinks of both oxygen and carbon dioxide involve life, and their atmospheric abundances have evolved along with life itself.

Both water and carbon dioxide are under-represented in the current atmosphere. Both are in the atmosphere only as trace gases. Consider, however, the difference that a slightly higher temperature would make. If the oceans warmed only a few degrees, both water vapor and dissolved carbon dioxide would be released into the atmosphere. If the warming continued up to 100 C, the boiling point of water, the entire ocean would vaporize and become part of the atmosphere.

How much water vapor would be produced if the ocean boiled away? We have noted that the average depth of the ocean is about 3 km, at which depth the weight of the water is 300 bars. (Remember that a 10-meter thickness of water exerts a pressure of 1 bar.) The weight of this water would be the same if it were in vapor form rather than liquid; therefore the pressure of the atmospheric water if the oceans boiled away would also be 300 bars. Water vapor would thus be by far the major constituent of the atmosphere. The carbon dioxide content would also rise, particularly if the carbonate minerals in the ocean sediment were broken down and their CO_2 was released into the atmosphere. We will return to the question of reservoirs of carbon dioxide later.

If the Earth were hotter, it might have a massive atmosphere consisting primarily of water vapor and carbon dioxide. If it were just a little cooler, the oceans would freeze and there would be no liquid layer on the surface. Only within a fairly narrow temperature range can the Earth maintain liquid water oceans and a nitrogen-dominated atmosphere.

Structure of the Atmosphere

The density of air at sea level on Earth is about 0.001 g/cm^3, and the pressure is, by definition, 1 bar. This pressure drops off rapidly with elevation, reaching one-half its sea-level value at 5.5 km (Fig. 7.15). Water vapor is even more strongly concentrated near the surface, and it declines to half its sea-level value at 2 km; this is why mountain air is dry and skiers and mountain hikers suffer so much from chapped skin.

For the astronomer, operating telescopes above the lower atmosphere offers great advantages in terms of clarity and steadiness of the air. Mauna Kea Observatory (Section 2.5) is at an altitude of 14,000 ft (4.2 km), above one third of the atmosphere and nearly 90% of the water vapor. Even more dramatic gains, especially in infrared transparency, are obtained with the telescope of the NASA Kuiper Airborne Observatory, which flies at 40,000 feet (12 km) above the surface. At this altitude, the airplane is above 80% of the air and 99% of the water vapor.

The lower 10 to 15 km of the atmosphere is called the **troposphere**. This region, which contains 90% of the mass, is the location of nearly all of the phenomena we call weather. The main characteristic of the troposphere, which sets it off from other regions of the atmosphere, is the convection of air within it. Convection is caused by the heating of the ground and lower atmosphere by sunlight. The warmer air near the surface expands and rises under its own buoyancy,

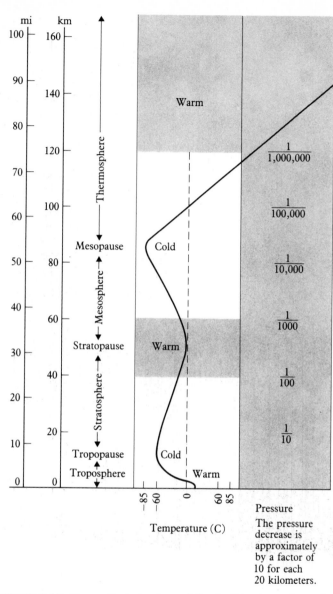

FIGURE 7.15 The vertical structure of the Earth's atmosphere showing the approximate altitudes of the principal regions and the change in pressure and temperature with altitude.

to be replaced by downdrafts of cooler air from above. The convection in the troposphere generates storms and other effects of weather and also maintains a steady decrease in temperature with altitude. This temperature gradient of 6 degrees C per kilometer results directly from the vertical circulation of air. The top of the troposphere, called the tropopause, is at a temperature of about -60 C or 213 K.

Above the troposphere is the **stratosphere,** in which very little vertical mixing takes place and the temperature is nearly uniform at about -60 C. In the stratosphere water is almost entirely frozen out and the atmosphere is very dry. In the upper part of the stratosphere the absorption of solar ultraviolet light creates a layer of ozone (O_3), an unusual form of oxygen with three atoms per molecule instead of two. The ozone layer, sometimes called the ozonosphere, extends from approximately 20 to 50 km. Without it, solar ultraviolet light would penetrate to the surface, doing irreparable harm to most living organisms, both plant and animal.

At altitudes above 90 km the temperature rises again due to the absorption of sunlight, while above 80 km some atoms lose electrons, producing the **ionosphere.** We will return to these tenuous upper parts of the atmosphere in Section 7.9.

One of the most important cycles linking the ocean and the lower atmosphere is the hydrologic or **water cycle.** Water in the atmosphere forms clouds and precipitates as rain and snow on both land and sea. Some of the precipitation is stored temporarily in the ground or in lakes, but most finds its way to the ocean. Evaporation back into the atmosphere completes the cycle. In the process of evaporation and condensation, moreover, the water can store and release vast amounts of solar energy. This energy as well as direct heating by sunlight drives the winds and generates weather. Without the water cycle, the Earth would be a much less dynamic place.

7.6 Life and the Atmosphere
Origin of Life

Surely the existence of life on Earth is the most extraordinary characteristic of our planet. Earth is apparently the only inhabited planet in the solar system, and terrestrial life is the only life we know. Obviously there are some underlying properties of this planet that make it especially suitable for the origin and development of life. Certainly the nature of the oceans and atmosphere, as well as an appropriate and relatively stable temperature, are requirements which the Earth meets. But to allow life is not to ensure that it develops. One of the many remarkable properties of life is its close coupling to the evolution of the atmosphere and its apparent modification of the environment of the planet in ways that are conducive to its survival and further development. In this section we consider the intertwined histories of the life and the atmosphere of Earth.

Life is a remarkable natural phenomenon through which the complex genetic material — the DNA in the cells of all living tissue — organizes the chemical reactions that permit it to reproduce itself and maintain the continuity of its existence from its origin billions of years ago to the present. While all of the steps that might have originally led from inanimate matter to animate have not been reproduced in a laboratory, it is clear that these events took place very early in the Earth's history and that the necessary organic building blocks for life could only have been produced in abundance in an environment free of atmospheric oxygen. If oxygen, a highly reactive gas, had been present in the primitive atmosphere, the compounds necessary to life would have rapidly been converted into oxides. Life has since developed the capacity to protect itself against oxygen and, more importantly, to use it to extract energy.

The primordial solar nebula, described in Section 3.1, must have been rich in simple compounds of carbon, hydrogen, oxygen, and nitrogen, the very elements that are the predominant constituents of life. In addition, we know from the primitive meteorites that substantial amounts of more complex organic compounds also formed early in the history of the solar system (Section 3.4). Some of these materials must have been present on the surfaces and in the atmospheres of Mars and Venus as well as the Earth at the time these terrestrial planets were forming.

Additional organic compounds, including amino acids and other biologically important compounds, form naturally from gas mixtures that contain methane (CH_4), carbon monoxide (CO), ammonia (NH_3), nitrogen (N_2), and water (H_2O). Laboratory experiments first carried out by Harold Urey and Stanley Miller at the University of Chicago in the 1950s show that when these gases are irradiated with ultraviolet light or bombarded with electrons, this complex chain of reactions takes place naturally. If all of the terrestrial planets once had the kind of primitive atmosphere we expect the Earth to have developed, they too might have acquired these chemical building blocks of life. However, the detailed chemical steps that led from these organic compounds to the much more complex self-replicating molecules of DNA remain obscure.

The Earliest Records of Life

It would be immensely helpful if we could examine the rock record to see what was happening on Earth at the time life was first developing. But as we saw in Section 7.3, this record has been irretrievably lost. The oldest rocks currently known solidified 700 million years after the Earth formed. The oldest record of life preserved in the fossil record dates from still later,

3.5 billion years ago, by which time life had already progressed to the level of complex colonies of photosynthetic microorganisms that form fossils called stromatolites (Fig. 7.16). The bacteria that made the stromatolites are quite sophisticated life forms, compared with a self-replicating molecule. Most likely it took hundreds of millions of years for life to evolve to this level of complexity.

We gain an idea of how much time is required by noting that the further evolution from the stromatolites to multi-celled creatures required another 2.7 billion years. During this immense stretch of time the most important developments were cells with nuclei and specialized subsystems for locomotion, which first appear in the fossil record about 1.5 billion years ago. After that, evolution occurs at a more rapid pace. By the beginning of the Phanerozoic, 590 million years ago, sexual reproduction had evolved along with hard skeletons and other specialized tissues. Another hundred million years later the backbone evolved, and shortly thereafter the colonization of the land began. The time that passed from colonization of the land to the evolution of Homo sapiens was just a short interval on the cosmic calendar (Fig. 7.5).

Further searches on Earth may reveal older rocks that will shed some light on the early history of life, but it is clear that such rocks will be extremely rare if they exist at all. If other planets underwent an early history similar to Earth's, but without developing our subsequent destructive geological activity, perhaps the early history of life can be read someday in their rocks. As we shall see in Chapter 10, Mars looks particularly promising in this regard.

Earth's Early Atmosphere

Although we are fairly sure that the stromatolite-forming bacteria of 3.5 billion years ago were deriving energy from the Sun through photosyn-

(a)

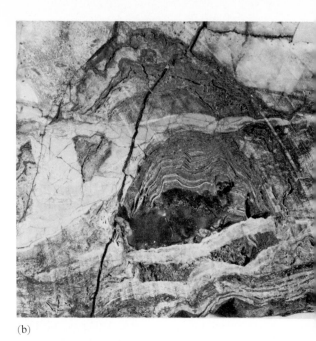

(b)

FIGURE 7.16 a) Modern stromatolite beds on the western coast of Australia are exceptional now, but they represent what was once the most sophisticated form of life on Earth. b) The oldest evidence of life on Earth (discovered so far!). This is a 3.5-billion-year-old stromatolite found in Western Australia. It has been cut in half to show the characteristic layered, domal structure produced by colonies of microorganisms that were presumably carrying out photosynthesis. The rock is about eight inches across.

thesis, they were probably not producing oxygen in the process. Even today, there are bacteria that liberate sulfur (from H_2S) instead of oxygen (from H_2O) using sunlight as a source of energy. Hence the atmosphere that existed 3.5 billion years ago probably still lacked free oxygen, since any small amounts of this reactive gas that might have been produced would quickly have recombined with surface rocks.

With or without life, however, the early atmosphere of the Earth would have evolved with time. If many hydrogen-bearing compounds were present (something we're not sure about), ultraviolet light from the Sun would break them apart, and the resulting light hydrogen gas would escape from the atmosphere. Thus the at-

mosphere experienced an inevitable transition from methane (CH_4) to carbon dioxide (CO_2), and from ammonia (NH_3) to nitrogen (N_2). Alternatively, some scientists have concluded that the atmosphere was dominated by carbon dioxide and carbon monoxide from the beginning. In either case, we are sure no free oxygen was present in the original atmosphere of the Earth.

The fact that the conditions for the formation of life were relatively short-lived has a rather awesome implication. If at any later time all life had been destroyed on the planet by some cosmic catastrophe, it could not have started again. The Earth would have become totally sterile. The fact that we are here means that such a catastrophe did not occur.

The Effects of Life on the Atmosphere

Our human presence on this planet is actually a gift from the green plants, since it is oxygen-producing photosynthesis that created the environment required for the development of warm blooded animals. It is not clear just when the transition to an oxygen-rich atmosphere occurred. Rocks much older than 2.5 billion years seem to have formed in the absence of free oxygen, and 2.5 billion years — a little more than half the lifetime of the planet — is usually cited as the approximate time of transition, although this estimate remains controversial. Certainly the atmosphere has had approximately its present composition for the past billion years.

Life has also modified the atmosphere by removing carbon dioxide. Marine shell-forming creatures, most of them microscopic, manufacture their shells from calcium and carbon dioxide dissolved in the ocean. As these shells are incorporated into sediment, great deposits of **carbonate** limestone are created at the expense of the atmospheric carbon dioxide.

The quantity of CO_2 represented by these carbonate deposits today is truly staggering. If all this gas were put back into the atmosphere at once, the pressure of CO_2 alone would be 70 bars, or seventy times the total atmospheric pressure today. As we will see when we discuss the atmosphere of Venus, the amount of CO_2 in the atmosphere can have an important influence on the surface temperature through the greenhouse effect (Section 8.2). Thus the history of CO_2 in our atmosphere is closely tied to the past climate of the Earth.

The Gaia Hypothesis

Oxygen and carbon dioxide are two atmospheric gases that have been strongly influenced by the evolution of life. A careful examination of the interplay between life and the atmosphere reveals a variety of other less obvious relationships. The intricacy of these interactions has led two scientists, Lynn Margulis and James Lovelock, to suggest that life on Earth actually regulates the composition of the lower atmosphere. They have given the name Gaia (after the Greek goddess of the Earth) to the total system of organisms, air, and oceans. Their **Gaia hypothesis** suggests that Gaia will strive as though it were a conscious being to maintain that particular equilibrium within which life can survive most successfully.

As an example of how Gaia would work, consider a situation in which the average input of solar energy to the Earth decreases, perhaps as a result of changes in the luminosity of the Sun. Gaia could respond by promoting the growth of organisms that liberate more CO_2 into the atmosphere. The additional CO_2 would result in stronger atmospheric warming by trapping more thermal radiation from the Earth's surface. The surface would then return to its original temperature.

The concept of Gaia is certainly intriguing and adds another dimension to our growing awareness of the interdependence of life on Earth. As we now realize, the growth of human population and its effects on the environment risk more than eliminating a few endangered species and producing esthetically unpleasant effects such as smog and dirty rivers. We face the potential of disturbing a global balance, and the ramifications of such a disturbance are impossible to predict.

7.7 Climate and Weather

Weather and climate are both aspects of the closely coupled systems of air, land, and water on our planet. Weather refers to the state of this system at a given place and time and its short-term variations, while climate is concerned with average conditions and their long-term trends. While no sharp demarcation line exists between the two, climate generally refers to changes occurring over a period of ten years or more, while more rapid changes are considered a part of the phenomena of weather.

Heating by the Sun

The source of energy that drives motions in both the atmosphere and the water is the Sun. Of the energy incident on the planet, about 30% is reflected back to space from the atmosphere or surface. The principal reflectors are water and ice clouds, atmospheric dust, surface snow and ice, and unvegetated deserts. Thus the basic energy balance of the planet is affected, through changes in reflectivity, by the amounts of dust in the atmosphere, the degree of forestation of the land, and the size of the polar deposits of snow and ice.

The 70% of incident sunlight absorbed by the planet corresponds to an average power of 240 watts for each square meter of surface, or a total input of more than a hundred billion megawatts $(1.2 \times 10^{17}$ W). (Imagine the consequences of more efficient ways of harnessing the incredible amount of solar energy.) If the climate is to be stable, an exactly equal amount of energy must be radiated from the Earth back into space, primarily at infrared wavelengths. Part of this infrared radiation escapes directly from the surface, but most is absorbed and reradiated by the lower atmosphere.

The average surface temperature of the Earth is about 10 C, a value about 25 degrees higher than would be the case without an atmosphere. This increase in surface temperature is the direct result of the blanketing effect of the atmosphere, primarily caused by the carbon dioxide and water vapor greenhouse effect, which is discussed in the next chapter.

Circulation of the Atmosphere

The Earth is heated more near the equator than in polar regions, and therefore the average surface temperature decreases with increasing latitude. However, this temperature decrease is less dramatic than it would be if there were no oceans or atmosphere to redistribute the solar energy. Air that is warmed near the equator tends to rise and move toward the poles, carrying much of this energy with it (Fig. 7.17). Transport of energy from warm regions to cooler ones generates the most important forces that drive the circulation of the atmosphere.

Because the rotation axis of the Earth is tilted by about 23 degrees, we have seasons on our planet (Section 1.4). Each year the familiar cycle is repeated, as the latitudes of greatest solar heating move north and south, bringing with them the succession of spring, summer, autumn, and winter. If the axis were not so tilted, the atmosphere might be able to sustain a single pattern of atmospheric circulation. The constantly shifting deposition of energy complicates the whole system immensely, however, encouraging the circulation pattern to break up into the cyclones and storm fronts and other phenomena we call weather.

The most stable weather occurs near the equator. Between 35° N and 35° S latitude the large-scale winds blow toward the equator at the surface and away from it at higher altitudes, carrying warm moist air upward near the equator and recirculating cooler, dryer air back to complete the cycle. The rising air generates a region of high precipitation within 10 degrees of the equator, producing the tropical rain forests of the Earth. In contrast, most of the great deserts of the Earth are at latitudes between 15 and 35 degrees, where the dry air sinks and flows back toward the equator.

Poleward of 35 degrees, in the temperate and polar regions of the Earth, the basic character of the circulation changes. Here the predominant winds are east-west, rather than north-

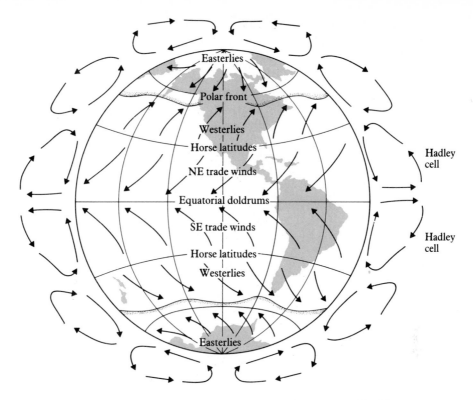

FIGURE 7.17 The global circulation of Earth's atmosphere in simplified form.

south, in direction. Instead of a smooth flow of warmer air toward higher latitudes and cooler air back toward the tropics, a series of large wave-like patterns develops, with the primarily east-to-west winds veering north and south. As the waves develop, masses of cool air are carried southward, and masses of warm air penetrate toward the poles. At any one time, about a dozen of these waves, each a couple of thousand kilometers across, are likely to exist in the temperate regions of each hemisphere.

Planetary Rotation and the Coriolis Effect

The rotation of a planet as well as thermally-driven atmospheric motions is an important fac-

tor in determining its weather. If it were not for the relatively rapid spin of the Earth, we would not have the rotating weather systems that are variously called cyclones, hurricanes, and typhoons (Fig. 7.18). All of these phenomena are generated when the motion of the air is deflected by the Coriolis effect.

To understand the **Coriolis effect,** we must imagine ourselves riding along on a north or south moving wind, perhaps as passengers suspended from a hot air balloon. Suppose we start our trip moving north from a point at latitude 30° N, and that the wind is initially blowing northward. The air, of course, shares the rotation of the surface, which at 30° N corresponds to an eastward speed of 1200 km/h. As we move

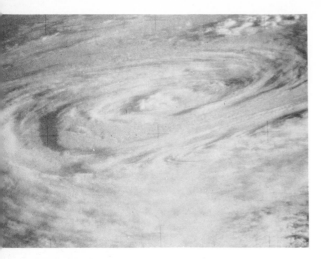

FIGURE 7.18 A typhoon southeast of New Zealand as seen from Skylab in 1974; it exhibits the spiral pattern of a typical tropical storm on Earth.

wise spin in the northern hemisphere and a clockwise spin in the southern. This sense of motion, corresponding to a low pressure region, is called **cyclonic.** A cyclone is a circular storm system moving around a low pressure region. Diverging air spins the opposite way and is called **anticyclonic.** Because high pressure regions do not generate clouds and rain, however, we rarely refer to an anticyclone when discussing weather.

A low pressure cyclone has the potential to become a major storm. When the inward-moving air is moist, it may cool as it nears the center of the system and is carried to higher altitudes. The resulting condensation of water vapor not only produces rain, it also releases energy to increase the speed of the wind. (Since it requires the addition of energy to water to produce steam, it follows that the opposite phase change, when water vapor condenses back to liquid, must release energy.) A situation can then develop in which the growing storm feeds upon local sources of water, drawing in more and more moist air and releasing more and more energy. The result is a tropical storm, more commonly known as a hurricane or typhoon. Only on Earth, where water is abundant, can self-sustaining storms of this type be formed. When such storms move over land, they immediately begin losing their strength since their source of water is cut off.

north, we continue going east at this speed, but soon notice that the landscape below us is moving east more slowly, since it has less distance to go to complete one rotation in twenty-four hours. Thus our extra eastward momentum causes us to turn toward the right. In a similar way, a balloonist moving south from a starting point north of the equator is deflected to her right because her initial eastward speed is lower than that of the ground nearer the equator. If both are deflected toward the right, the result is a counterclockwise circular motion (Fig. 7.19).

The same line of reasoning will convince you that in the southern hemisphere the direction of turning is reversed and that air diverging from a point is deflected toward the left. A similar mental exercise shows that air blowing inward toward a center is deflected in the opposite direction from that blowing outward. You may even have experienced the Coriolis effect directly, if you have ever tried to walk quickly from the outside edge to the center of a carousel.

As a result of the Coriolis effect, air converging toward a center sets up a counterclock-

Changes in Climate: The Ice Ages

Weather on Earth is an extremely complex phenomenon, difficult to understand and almost impossible to predict with any high degree of accuracy. Climate presents fewer problems, in that changes are small. However, change in the climate of the Earth has much more far-reaching consequences than any storm.

We know that the climate on Earth has changed, even during the relatively brief period that human records have been kept. At the time civilization was developing in the fertile crescent of the Middle East, rainfall in that region was higher than it is now, and vast forests supported lions and other animals now confined to sub-Saharan Africa. About a thousand years ago, when Norse seafarers founded colonies in Greenland and present-day Canada, the climate in these regions was warmer than it is today, and when a global cooling took place around the year 1400 these colonies did not survive. Even more recently, there is evidence of recurring droughts in the plains of North America at approximately 22-year intervals.

The most dramatic climatic changes for our planet are those associated with the great **ice ages** of the past million years. At intervals of about 100,000 years the average temperature of the Earth has dropped by 2 or 3 degrees, sufficient to produce vast ice sheets up to 3 km thick over much of the northern hemisphere land masses (Fig. 7.20). During an ice age, the sea level drops and atmospheric circulation patterns alter significantly. The last such glacial period ended only about 10,000 years ago, and the Earth today is in an unusually warm period. It is so warm, in fact, that the sea ice that still covers the Arctic Ocean and the thick ice cap on Greenland are in some danger of melting. If this should happen, the sea level would rise still further and perhaps submerge our coastal cities.

One reason that a change as small as a degree or two in average global temperature can have such major climatic consequences is that climate changes tend not to be self-correcting. During an ice age, the average reflectivity of the surface increases and less sunlight is absorbed, magnifying the cooling effect. Similarly, the melting of the Arctic sea ice would greatly increase the absorption of sunlight in the polar regions, generating further temperature increases. The climate of the Earth is not as stable as it seems.

FIGURE 7.19 The Coriolis effect on air moving north or south on a rotating planet. You can experience this same force by trying to walk toward or away from the center of a moving carousel.

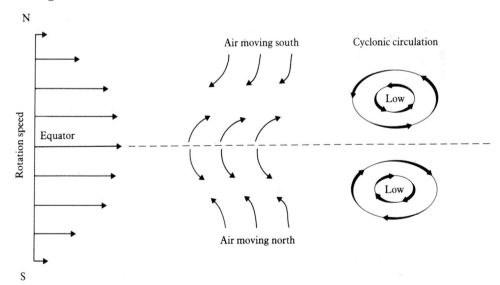

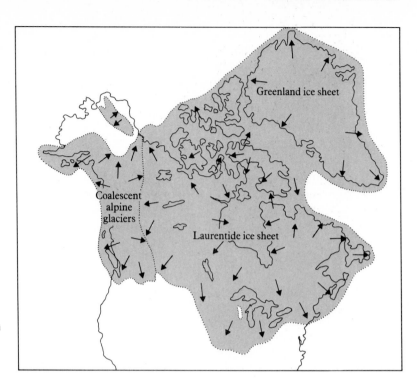

FIGURE 7.20 The area of North America covered by the great ice sheets during the Pleistocene ice age.

The Causes of Climate Change

Changes in the orbit of the Earth and the tilt of its rotation axis are now believed to be the primary cause of the great ice ages. Astronomers have calculated the changes in orbit and tilt expected from the gravitational influence of other planets over the past million years, and the pattern of ice ages follows these changes closely. Even though the resulting variations in solar heating are small, they seem to be sufficient to shift the Earth from an interglacial equilibrium, such as we have at present, to an ice age condition, with about equal intervals of time spent in each climatic state.

Other changes in climate are probably related to the amount of dust in the atmosphere. The primary source of dust is large volcanic explosions (Fig. 7.9), which can eject several cubic kilometers of fine silicate dust and sulfuric acid droplets into the stratosphere. Calculations show

that such quantities of dust result in lowering the average temperature of the planet by 1 to 3 degrees. Such a change would be sufficient to trigger an ice age if it persisted long enough, but fortunately most stratospheric dust falls back into the lower atmosphere within a few months. Even so, the effects of even a single volcanic explosion can be detected, as when the 1982 eruption of the Mexican volcano El Chicón apparently triggered the 1983 weather condition called the El Niño, which altered the equatorial ocean circulation pattern and produced severe storms in temperate latitudes. A sustained period of high volcanic activity could have more severe climatic consequences. Impact of an asteroid or comet on the Earth would cause even greater problems. A 10-km asteroid would inject more than 10,000 cubic kilometers of dust, darkening the planet for months and utterly disrupting both atmospheric and ocean circulation.

7.8 Impacts and Other Catastrophes

Impact Craters on Earth

In this chapter, many features of terrestrial geology have been discussed, but conspicuously absent have been the craters that dominate the surfaces of the Moon and Mercury. Why does the Earth have so few impact craters relative to the Moon? Both objects occupy the same region of space, and Earth with its greater surface area and stronger gravity should receive many more impacts than does the Moon. The presence of an atmosphere does not explain the absence of craters, for while the air effectively filters out the smaller debris, any projectile capable of digging a large crater will punch through the atmosphere as easily as a fist through a cobweb. Not even the oceans could protect the solid surface against a huge impact. How then did the Earth avoid being heavily cratered? The fact is, it did not avoid impact cratering. The difference between the Earth and the Moon lies not in the rate at which craters are formed, but in the rate at which they are destroyed. We are deluded into thinking we are immune to large impacts because the evidence of past catastrophes has largely been destroyed. Remember that most of the craters on the Moon were formed more than 3.9 billion years ago, while the most ancient rocks found on Earth are only 3.8 billion years old.

The primary agents that obliterate impact scars are tectonic and erosional. The surface of our planet is constantly being reworked. New crust is formed along rift zones, and old crust is destroyed in subduction zones. Elsewhere, collisions between plates squeeze the crust into mountain ranges, while volcanism periodically floods sections of the continental blocks. In addition, the processes of erosion and sedimentation, aided by the living creatures of our planet, fill in and wear away the craters generated by impacts. Only recently have geologists learned to recognize the healed scars of a few ancient impacts on the continental crust of our planet.

As we noted in Section 4.1, the debris orbiting the Sun has a size distribution that greatly favors small objects, making small crater-forming impacts much more frequent than large ones. Fortunately, the atmosphere shields us from the intense rain of sand or pebble-size material, which contributes hundreds of tons of dust to the upper atmosphere each day. Only objects with masses greater than about a kilogram have any hope of surviving intact to the surface, and only objects weighing hundreds of tons can dig craters when they land.

Perhaps once every decade a small meteorite crater is formed. In terms of significant effects on the surface of our planet, however, we must consider the much rarer impacts of objects that weigh from hundreds of thousands to billions of tons. Four case studies follow.

Tunguska River

A remarkable and still somewhat mysterious event took place above the wilderness of Russian Siberia on 30 June 1908. Witnessed by only a few, an explosion of unprecedented magnitude occurred, registering its shock wave on atmospheric instruments around the world. A brilliant fireball in the sky briefly rivaled the Sun in brightness, followed by explosion and fire. Over a region of a thousand square kilometers the forest was flattened, the trees stripped of leaves and branches and toppled in parallel rows radiating from the blast center (Fig. 7.21). Thousands of caribou died, and a man in a trading post 60 km away was thrown from his seat and knocked unconscious. Yet, in spite of all this violence, no crater was formed, and the Siberian forests are already erasing the last evidence of this explosion.

No one knows exactly what caused the Tunguska event. Since no crater was formed, some believe it may have been a fragment of cometary material that broke up at an altitude of 8 km, creating what nuclear weapons strategists call an

FIGURE 7.21 A small part of the forest destroyed by the 1908 explosion of a cometary or asteroidal fragment in the Earth's atmosphere near Tunguska, Siberia.

air burst. It has been suggested based on orbital grounds that it might have been a small fragment associated with Comet Encke. Another suggestion is that the incoming object was a stony meteoroid of low strength. From the radius of destruction, we can estimate that the explosion released the energy of a ten megaton bomb, similar to the larger weapons presently in the nuclear arsenals of the United States and the Soviet Union.

Ten megatons of energy corresponds to an impacting body with a mass of more than 100,000 tons. For an icy body this implies a diameter of about fifty meters — the size of a large city office building. It is estimated that such an impact should occur every few centuries somewhere on Earth. We must hope that the next such event does not take place near a populated area, and even more ominous, we must guard that it could not be mistaken for a nuclear attack, and thereby accidentally trigger a nuclear holocaust.

Meteor Crater

The scene is the Painted Desert of northern Arizona, near the volcanic region of the San Francisco Peaks. The time is 50,000 years ago, before there were humans living in the region to witness the impact of a million tons of iron with the Earth. There was no chance that this solid mass of metal would break up in the atmosphere. Rather, the force of its impact carried it several hundred meters below the surface before it finally came to a halt, transferring its energy into an explosion that destroyed the impacting body and blasted a hundred million tons of pulverized rock into the atmosphere, leaving a crater a kilometer in diameter. Around the edges of the crater the thick layers of sediment were twisted up and folded back (Fig. 7.22). Droplets of metal from the impacting object were spread over hundreds of square kilometers around the crater.

The force of the Meteor Crater explosion is estimated at nearly 20 megatons, only about twice that of the Tunguska event. The primary difference between the two is that Tunguska was an air burst due to a fragile projectile, while Meteor Crater was an underground explosion caused by an iron meteorite. Impacts on land of iron meteorites of this size probably take place on an average every hundred thousand years.

Today, Meteor Crater is the best known and most obvious impact crater on Earth. Tens of thousands of tourists take the short detour from Interstate Highway 40 each year to visit it. Although erosion has partly filled the interior with sediment and has washed away much of the ejecta, the characteristic uplifted and folded-over strata around the edges clearly testify to the explosive origin of this feature. Tiny metal fragments can still be found in the surrounding desert. Yet even here, the recognition that this is the scar of an extraterrestrial object has only come during the past fifty years; previously, most geologists thought it was an unusual volcanic crater associated with the nearby San Francisco Peak volcanic field.

(a)

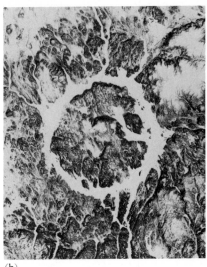

(b)

FIGURE 7.22 a) Meteor Crater, near Winslow, Arizona, is 50,000 years old, only yesterday on the geological time scale. Thus a similar impact could occur anywhere on Earth at any time. In fact a near miss was reported just a few years ago when a large meteor streaked over Wyoming. Meteor Crater is about 1.3 km across and 200 m deep. A similar crater would be a tiny pit on the lunar landscape as viewed from Earth, but quite a nuisance if it occurred in downtown Chicago. b) The Manicougan Lakes Crater in northern Quebec. This circular feature, 70 km in diameter, was caused by an impact in the late Triassic period, about 200 million years ago.

No equally large impact crater has been located on Earth younger than Meteor Crater, but recent searches have found three craters with diameters of 7 to 18 km, two in the Soviet Union and one in Ghana, that may have been formed within the last three million years. Craters much older than this become very difficult to identify, especially if they are less than a few kilometers in diameter (Fig. 7.22b).

Ries Crater

For a much larger impact event, we must go back 15 million years, before the appearance of humans on Earth. The target this time was the southern plains of present-day Germany, and the projectile was neither ice nor metal, but rocky in composition. Its diameter was more than a kilometer, its mass more than a billion tons, and the explosion it generated far outstripped anything humans can produce (yet!) even with our most horrible instruments of war. Presumably the projectile was similar to the Earth-approaching asteroids observed today.

The Ries crater is 27 km in diameter, and when first formed must have been 5 km deep. Although the crater is not deep in comparison with the continental crust, a similar impact in a region of thin oceanic crust could penetrate into the upper mantle. As the impacting body plunged through the atmosphere, it left a cylindrical vacuum behind it, permitting some of the explosion debris to escape into space. Millions of tons of this debris re-entered the atmosphere, where friction molded it into teardrop-shaped particles of glass. These strange frozen droplets, called tektites, are found today scattered over much of central Europe (Fig. 7.23). The billions of tons of dust blasted from the crater spread around the Earth, darkening the sky for weeks and perhaps disrupting the global climate for years.

Today the Ries crater is virtually unrecognizable from the ground. The ancient town of Noerdlingen on the German "Romantic Road" is a popular tourist destination, but not because of the surrounding impact structure. The force of the explosion can be recognized, however, from the characteristic shock patterns in the underlying rock and from the tektites strewn over half of Europe. We can only imagine what the effects of such a catastrophe must have been on the climate — and on the ecology of the planet.

The K/T Mass Extinction

One of the most important catastrophes of the last few hundred million years was caused by the impact of an asteroid about 10 km in diameter having a mass of more than a trillion tons. The

FIGURE 7.23 These curiously rounded tektites are fragments of rock created by an impact on Earth and melted during their double passage through our atmosphere.

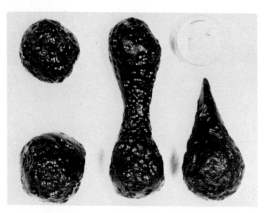

location of the impact is unknown, but the time is well defined in the geologic record as the boundary between the Mesozoic and Cenozoic eras, 65 million years ago. This is better known as the boundary between the Cretaceous and Tertiary periods, and the event is known as the **K/T event.** (If this seems a strange choice of letters, note that "Cretaceous" is spelled with a K in German.)

The K/T boundary has long been recognized by geologists as one of the major breaks in the history of life on Earth. It corresponds to what is called a **mass extinction,** in which many species became extinct almost at once, to be succeeded by new species in subsequent strata. In the K/T mass extinctions, more than half of the major marine species were destroyed. At about the same time the dinosaurs — those great reptiles that had dominated the Earth throughout the Mesozoic era — also died. Some scientists believe that the two extinctions happened at the same time, and they have looked for a common cause for both.

The key to understanding the nature of the K/T event was the discovery in 1980 of a thin sedimentary layer of remarkable composition exactly at the boundary where the fossil record indicated the mass extinctions (Fig. 7.24). In this boundary layer, which has subsequently been recognized all over the globe, the quantities of a number of rare elements, notably the metal iridium, are dramatically enhanced.

Iridium is very rare on the surface of the Earth because it dissolves readily in iron, and most of our planet's allotment is thought to be locked up in the core. The relative concentrations in this layer of isotopes of the element osmium are also peculiar, being typical of meteoritic materials rather than terrestrial. Both of these anomalies can be understood by the sudden addition of about a trillion tons of typical stony meteorite material to the Earth's atmosphere, which would include 200,000 tons of iridium. The impact of a 10-km stony asteroid would do just that.

The energy of the explosion as the asteroid struck the Earth is almost beyond imagining, equalling a hundred million megatons or 5 billion bombs of the size that obliterated Hiroshima and Nagasaki. The crater was nearly 200 km in diameter and deep enough to puncture the crust even in continental regions. Because they have found no remnant of a crater of this size that is 65 million years old, geologists suspect that the impact took place in the ocean. Since about a third of the oceanic crust of 65 million years ago has subsequently been subducted, no record may still exist. If it did occur in the ocean, the impact would have generated waves kilometers high and inundated the shorelines of half the world.

Almost surely any living thing within a thousand kilometers of the impact would have been destroyed by the blast. Even more devastating for the biota of Earth would have been the dust cloud raised by the explosion, the same dust that is responsible for the iridium-enriched layers that mark the K/T boundary. With its mass of a hundred trillion tons, this cloud would have blotted out the Sun from the surface of the Earth for a period variously calculated from sev-

FIGURE 7.24 A layer of clay just a few inches wide marks the boundary between the Cretaceous and Tertiary eras on Earth (see Fig. 7.5). Here we see it exposed on a cliff wall near Trinidad, Colorado. It was during the time that this clay was deposited that massive extinctions of various types of life occurred on Earth. The clay is strongly enriched in iridium and other metals suggesting an extraterrestrial origin for part of this material.

eral weeks to several months. Either way, the temperatures would have dropped drastically, plants would have died from lack of sunlight, and the climate of the planet would have been seriously disrupted perhaps triggering the onslaught of an ice age. Probably 99% of all living things were killed over the whole Earth.

Horrible as this event may sound, it had profound effects for us. The new species that rapidly evolved to fill the ecological vacuum of this mass extinction gave rise to most of the life on Earth today. The tiny mammals that succeeded the dinosaurs were our ancestors. Had it not been for the K/T impact, evolution of life on Earth would have taken a very different course.

There are half a dozen mass extinctions recognized in the Phanerozoic (Fig. 7.25), but at this writing no other clear identification has been made between an impact and a mass extinction. Important as such impacts can be in influencing conditions on the Earth, they may not be the only catastrophes that occasionally disrupt the flow of evolutionary history. Immense volcanic explosions, for instance, can also produce enough dust to darken the skies of our planet. On the other hand, we can be pretty sure that the K/T impact was not unique. The crater densities on the lunar maria indicate that several impacts of objects 10 km or more in diameter have taken place on Earth during the Phanerozoic (Fig. 5.18). The large crater Tycho, which was caused by an asteroid about the size of the asteroid that triggered the K/T extinction, is the most recent example from the nearside of the Moon. Surely these other Earth impacts have also placed their mark upon the planet and its life forms.

Nuclear Winter: Human Threats to Earth

All of the violent events discussed above resulted from naturally occurring impacts of interplanetary debris with the Earth. However, human technology has now reached the level where we are capable of inflicting comparable destruction

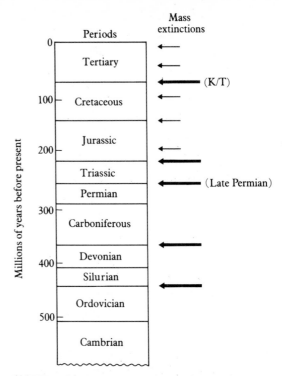

FIGURE 7.25 Major and minor mass extinction events (marked by arrows) during the past 500 million years. Some scientists are convinced that these mass extinctions are periodic, coming at intervals of approximately 30 million years, but this is a difficult suggestion to prove or disprove with existing data. (Data after Raup.)

on ourselves through nuclear war. In recent years the study of the consequences of nuclear war for the Earth and its delicate ecology have become topics of major interest. These studies draw upon discoveries about the K/T event and other natural catastrophes of the past, as well as comparisons with the other terrestrial planets.

At first you might think that a nuclear war, though catastrophic for human civilization, would not threaten the Earth as a whole. Consider, for example, an exchange involving 1000 warheads of one megaton each. The total explosive energy released — 1000 megatons — is only a small fraction of the energy of the Ries impact, and a mere 0.001 percent of the energy release that triggered the K/T mass extinction. Yet the destruction of life from such an exchange would probably follow the pattern of the K/T event. The reason for this magnification of destructive power is to be found in the concept of **nuclear winter,** a term introduced by Richard Turco in 1983 and widely publicized by Carl Sagan and others since that time.

The global impact of a nuclear war is amplified primarily by the nature of the targets, which would inevitably include major cities and military and industrial centers. These targets represent tremendous concentrations of flammable wood, hydrocarbon fuels, plastics, and other products of human civilization. The burning of these materials would generate huge quantities of smoke and soot in addition to the dust lofted by the explosions themselves. Since smoke and soot particles are much more efficient than silicate dust particles at absorbing sunlight, the blanketing effect of a worldwide pall of smoke in the stratosphere would be equivalent to that of the much larger mass of dust that would result from a natural impact.

Many studies are being made using computer models of the Earth's atmosphere to evaluate the effects of nuclear wars of various sorts, and the answers are not all in. Most investigators agree that the results of a large-scale nuclear war would include a period of global darkness lasting weeks to months before the dust and soot settled out of the atmosphere, the period of the nuclear winter. Depending on the season when the nuclear winter occurred, it could destroy vast amounts of food crops and animals and result in the death of many of the people and other species of life who survived the initial attack.

The threat to life on Earth posed by the presence of humans goes beyond the possible effects of an all-out nuclear war. Scientists estimate that the destruction of our tropical rain forests will cause the loss of as many species as the K/T extinction in the next thirty years.

The knowledge that humans are capable of such severe damage to the ecosystem of the Earth should have a profound effect on our thinking. The concept of a nation winning or even surviving a nuclear war appears to be obsolete. The preservation of species on the only inhabited planet in our system is a weighty responsibility. Such realizations could represent one of the most important results from our recent progress in the exploration of the planetary system and the study of the Earth as a planet.

7.9 Probing Near-Earth Space
The Upper Atmosphere

Until the advent of space exploration, the limits of the Earth were thought to be at the ionosphere, a few hundred kilometers above the surface, about the altitude limit of the instrumented rockets of the 1940s and early 1950s. From our perspective on the surface, the atmosphere above these altitudes is virtually undetectable, and the gas density is so low that Earth satellites can remain in orbit for centuries without being forced down by atmospheric friction. In spite of the low density of matter above a few hundred kilometers, however, there is a vast region of space surrounding the Earth called the magnetosphere that is very much a part of our planet.

A convenient if somewhat arbitrary altitude to mark the beginning of the tenuous upper atmosphere is 100 km. This altitude represents the highest level at which meteors are sufficiently heated by atmospheric friction to become visible, and the lowest level at which an Earth satellite can complete even a single orbit. It is also about the level at which solar ultraviolet light breaks apart molecules more rapidly than they can recombine, so that the composition of the atmosphere changes from primarily molecules to primarily atoms. Finally, and perhaps most important, 100 km represents approximately the lower boundary of the ionosphere of the Earth.

The Ionosphere

The ionosphere is the region of the upper atmosphere in which many of the atoms are ionized, that is, broken apart into positively charged ions and negatively charged electrons. The atmosphere in this region is therefore a plasma, albeit a weak one, since most of the atoms remain unionized. The ionosphere was discovered early in the twentieth century by its ability to reflect radio waves transmitted from the ground. Before communications satellites, the ionosphere provided the primary means for long distance communications, and it is still critical for the propagation of AM and short-wave radio.

The degree of ionization of the atmosphere is measured by the density of electrons. Below 80 km ions and electrons recombine as quickly as they form, and there is no ionization. The electron density rapidly increases with altitude, forming a first maximum at 100 km of a little more than 100,000 electrons/cm^3. A stronger ionization peak occurs at about 300 km, where the electron density reaches somewhat more than a million e/cm^3. Above 300 km, the electron density falls off gradually with altitude, dropping to 10,000 e/cm^3 at about 2000 km. At each altitude, the electron density represents a balance between ionizing solar ultraviolet and X-ray radiation, recombination of the ions and electrons, and the total amount of gas available. Above 500 km most of the gas is ionized, and the declining electron density with altitude simply reflects the thinning of the atmospheric gas.

The Aurora

Within the ionosphere between altitudes of about 150 and 400 km, one of the most beautiful of natural phenomena occurs: the **aurora** or polar lights. Named for the Roman goddess of the dawn, the aurora is regularly seen at polar latitudes but only rarely graces the night sky in temperate zones. The displays can take on a variety of forms, with the most common being pale rays or curtain-like sheets of green or red color that silently sway and dance across the night sky (Fig. 7.26a). Viewed from space (Fig. 7.26b),

FIGURE 7.26 a) The aurora borealis or northern lights. This shimmering curtain of light with distinct rays is only one example of the various forms the aurora can take. Note trees in foreground. b) The aurora as seen from outside the Earth, photographed by astronaut Bob Overmeyer from the Challenger Shuttle.

(a)

(b)

the aurora is seen to be concentrated in rings centered on the north and south magnetic poles. The positions of these rings of light and the intensity of the glow vary from hour to hour as well as from year to year, with the greatest likelihood of temperate-latitude displays during periods of high solar activity.

The aurora is the result of electric currents flowing through the ionosphere, stimulating the gas to glow much as an electric current produces light from a fluorescent lamp. The green color comes from the fluorescing oxygen, while red is contributed by hydrogen atoms. The currents themselves originate at higher altitudes in the magnetosphere.

The Magnetosphere

The Earth's **magnetosphere** was discovered in 1958 by instruments on board the first U.S. Earth satellite, Explorer 1 (Fig. 7.27).

The scientist who built the high energy charged particle detectors on Explorer 1 was Iowa University professor James Van Allen, and his name has been given to the primary features of the inner magnetosphere, the Van Allen belts. These are regions of space containing large numbers of protons and electrons, trapped in the magnetic field of the Earth.

Unlike the ions and electrons of the ionosphere, the magnetospheric charged particles are highly energetic, spiraling back and forth within the magnetic field at speeds of thousands of km/s. When they strike solid material, such as the skin of a spacecraft, they produce additional subatomic particles and gamma- and x-radiation, leading to the misnomer radiation belts. It is not radiation that is trapped in the Van Allen belts, but energetic electrons, protons, and other ions.

The configuration of the magnetosphere (Fig. 7.28) is the result of interactions between the solar wind and the magnetic field of the Earth. From the direction of the Sun, the

FIGURE 7.27 From right to left: Werner von Braun, James Van Allen, and William Pickering holding aloft a model of the Explorer 1 satellite that discovered the Earth's radiation belts in February of 1958. Von Braun organized the development of the launch vehicle, Van Allen designed and built the instruments, and Pickering was the director of the Jet Propulsion Laboratory, which built and operated the satellite.

charged particles of the solar wind stream toward the Earth at a speed of about 400 km/s. As they near our planet, however, these particles are deflected by the Earth's magnetic field. This deflection distance, which represents the point at which the magnetic field strength is just sufficient to balance the pressure of the solar wind, is at about ten Earth radii, or 60,000 km. The cavity within which the planetary magnetic field dominates over the solar wind is the magnetosphere, shaped rather like a wind sock or a stubby comet tail pointing away from the Sun. Downstream, it extends just about to the distance of the Moon. The outer boundary of the magnetosphere is called the magnetopause.

Sources and Sinks of Charged Particles

Charged particles from three different sources contribute to the magnetosphere of the Earth. First, solar wind particles leak across the mag-

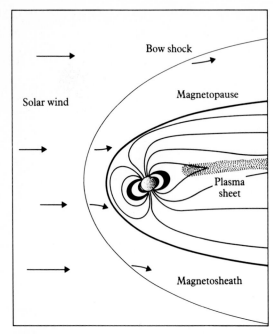

FIGURE 7.28 A diagram of Earth's magnetosphere. The planet's magnetic field forms a shield that stands off the solar wind. The tilt of the field lines close to the planet results from the inclination of Earth's rotational axis. The belts of electrons and protons (discovered by James Van Allen) surround the Earth within the magnetosphere.

netopause and become trapped inside. Second, escaping atoms from the atmosphere contribute most to the population of the inner magnetosphere. Third, particles generated from impacts on the atmosphere of the very high energy galactic cosmic rays contribute primarily to the middle belt a few thousand kilometers up. On other planets, additional sources of charged particles may be important, such as the volcanoes of Jupiter's moon Io or water molecules from the rings of Saturn.

To balance the sources of charged particles, corresponding sinks must exist. For the Earth, two are important. First, particles escape from the downstream side of the magnetosphere back into the solar wind. Second, the impact of charged particles on the atmosphere of the Earth stimulates atmospheric emissions and produces the aurora. Both sinks — escape outward and precipitation into the atmosphere — are triggered when the boundary between the solar wind and the magnetosphere becomes unstable as the result of fluctuations in solar wind pressure. These fluctuations, generated by outbursts on the Sun, produce the bright auroral displays and ionospheric disruptions associated with periods of intense solar activity.

The study of planetary magnetospheres has become a major branch of space science during the past twenty-five years, pursued by physicists like Van Allen using Earth-orbital and interplanetary spacecraft. Many of the plasma-physical processes that take place here are similar to those in the much larger astrophysical systems studied by astronomers, such as quasars and black holes. Magnetospheres represent a marvelous physics laboratory, where conditions are unlike anything that can be duplicated on the surface of the Earth.

———————————◇———————————

Summary

The Earth is the planet we know best, and is, therefore, an appropriate object with which to compare the other worlds of the planetary system. We did not begin with the Earth only because of its complexity, which led us to start with the smaller, simpler planets Mercury and the Moon. Now, with our own planet, we must look at a much wider range of phenomena related to our active geology, our oceans and atmosphere, and the unique influences of life on the evolution of the planet.

The various layers of the Earth, above and below its surface, have given their names to similar divisions on other planets. Thus we must understand how the interior is divided into crust, lithosphere, mantle, and the liquid and solid cores. Similarly, we have the division of the atmosphere into the troposphere, stratosphere, and ionosphere, with the magnetosphere extending still farther out to the edge of interplanetary space.

The complex geology of the Earth can be understood in terms of plate tectonics, the process by which heat released in the interior causes convection currents in the mantle which in turn exert forces on the lithospheric plates floating on the top. These plate motions result in the formation of new oceanic crust at rifts, the destruction of the crust at subduction zones, and the generation of earthquakes and volcanoes along faults where one plate scrapes against another. In this process the ocean crust is recycled on a time scale of about 200 million years, while the floating continental granites are moved about, occasionally compressed and raised up to form great folded mountain ranges.

Above the crust are the oceans and the atmosphere. The short-term variations called weather and the longer-term conditions that constitute the climate both result from varying solar heating coupled with the large-scale motions of ocean and atmosphere. We looked at atmospheric circulation and saw the way Coriolis forces give rise to cyclones and anticyclones in the temperate regions of the Earth. Also considered were the processes that can affect the climate, such as changes in the inclination of the Earth's axis of rotation or the injection of dust from volcanoes or asteroidal impacts.

The role of impacts on the Earth is different from that on the Moon or Mercury because of our oceans and atmosphere. A large impact not only produces a crater, it can also profoundly modify the climate with disastrous effects on life, leading in some cases to mass extinctions. The best documented mass extinction arising from an impact is the K/T event, 65 million years ago.

Atmospheric chemistry is another complex topic. Our present atmosphere is the result of outgassing from the interior, primarily through volcanic eruptions, combined with changes due to interaction of the gas with the crust, the oceans, and especially with life. It is life that has removed almost all of the carbon dioxide (now mostly in the form of carbonate deposits) and has generated free oxygen through photosynthesis. Just as the evolution of life has been influenced by conditions in the atmosphere such as its progression to oxidizing conditions, so has the evolution of the atmosphere been influenced by life. This interaction is the origin of the interesting but speculative Gaia hypothesis.

The Earth is a remarkable planet, like the others in many ways but also uniquely influenced by its liquid water oceans, its oxygen-rich and carbon dioxide-poor atmosphere, and its abundant and varied life forms. As we will see in the next chapter, Venus should be the other planet most like the Earth, yet it has evolved to a strikingly different condition.

Key Terms

anticyclonic	lithosphere
aurora	magnetosphere
bar	mantle
caldera	mass extinction
carbonate	nuclear winter
cinder cone volcano	Phanerozoic period
continental drift	plate tectonics
Coriolis effect	Precambrian period
crust	rift
cyclonic	sedimentation
fault	seismic wave
Gaia hypothesis	shield volcano
geologic time scale	stratosphere
granite	subduction
ice age	troposphere
ionosphere	water cycle
K/T event	

CHAPTER EIGHT

Venus: Earth's Exotic Twin

8.1 Unveiling the Goddess of Beauty

Earth's Twin?

Venus is the nearest planet to the Earth. When we see it glowing brightly in the twilight skies as the morning or evening star, we can understand why this planet was named for the goddess of love and beauty. Venus owes its brilliance both to its proximity to the Earth and Sun and to its

FIGURE 8.1 A picture of Venus taken in visible light with the Palomar 200-inch telescope when the planet was relatively close to us. It shows no detail, indicating the presence of an atmosphere filled with featureless clouds.

layer of clouds that serves as an excellent reflector of sunlight. The Moon, with a cloudless surface of grey rock, reflects only 12% of the sunlight that strikes it, whereas Venus has a reflectivity of about 75%. Being both bright and near us in space, Venus is a relatively easy object for astronomers to study. Nevertheless, it has proved difficult to learn much about this beautiful planet, and scientists are still finding it to be full of mysteries and surprises.

If we simply consider size and density, Venus and Earth indeed appear to be twin planets; both also have substantial atmospheres and brilliant clouds. However, Venus is completely covered by its clouds, unlike the ragged canopy of Earth. The clouds prevent astronomers from seeing the planet's surface, and they frustrated attempts to measure even so basic a property as the rotation rate (Fig. 8.1). The temperature of those clouds, first measured in the 1930s, is 240 K, similar to the temperature that would be measured at the top of our own water ice clouds, and it was generally assumed that the clouds of Venus were composed of water, like those of Earth.

Less than thirty years ago, the only gas that had been detected spectroscopically in the atmosphere of Venus was carbon dioxide (CO_2) (Fig. 8.2). These observations, however, did not preclude the existence of other gases, since observational difficulties could not rule out substantial undetected amounts of nitrogen and much smaller amounts of oxygen and water vapor. Therefore, it was possible that the atmosphere of Venus was composed primarily of nitrogen, like our own, and that carbon dioxide might be a relatively minor component. All of these observations and inferences reinforced the idea of the similarity of Earth and Venus to the supporters of this hypothesis who wished to believe that Venus might even have oceans of water beneath its brilliant clouds and a climate conducive to the existence of life.

Surface Temperature

This optimistic picture changed dramatically in the late 1950s when radio telescopes were used

to measure the thermal radiation from Venus. Unlike visible light, radio waves easily penetrate clouds. (You know this from your experience in listening to radio broadcasts or watching television on cloudy and rainy days.) Furthermore, any object having a temperature not at absolute zero (0 K) radiates some energy at all wavelengths.

It is a basic physical law, known as Wien's law, that the hotter an object is, the more the maximum energy output shifts toward shorter wavelengths. The outer layers of the Sun are at a temperature (5800 K) that puts this maximum in the range of visible light. A planet with a temperature like Earth's is radiating most of its energy in the infrared, at wavelengths near 10 μm (0.001 cm). But some of the thermal energy from

FIGURE 8.2 A comparison of portions of the spectrum of Venus (recorded photographically at the McDonald Observatory in Texas), of Mars, and of the Sun. Carbon dioxide absorptions appear strongly in the Venus spectrum, weakly on Mars, and not at all in the spectrum of the Sun. We conclude that even Mars has more CO_2 in its atmosphere than our planet does, while Venus must have an enormous amount of this gas (see Fig. 2.8).

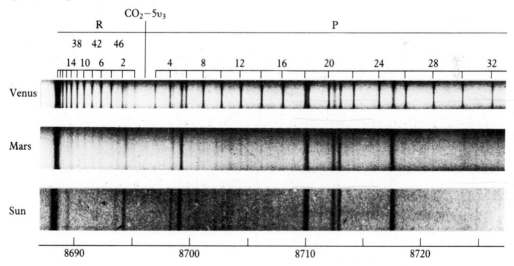

both Sun and Earth is radiated at much longer wavelengths in the centimeter to meter range, and this is the region of the spectrum where radio telescopes operate (Fig. 2.5).

Since Venus is closer to the Sun than we are, astronomers expected its temperature to be higher than the Earth's. On the assumption that both planets rely on solar radiation alone to warm them, it is possible to calculate what temperature Venus should have if it were indeed a planet just like Earth. Its greater proximity to the Sun is countered to some extent by the highly reflective clouds that send most of the incident sunlight back into space; therefore, Venus should be about 15 C warmer than our planet.

In 1958, however, radio astronomers found that the amount of thermal energy Venus was producing implied a surface temperature greater than 600 K. Venus was radiating twice as much energy at radio wavelengths as had been expected. This result was so surprising that at first it was simply not accepted. Alternative explanations were sought to explain the relatively high intensity of the radio radiation; perhaps emission from a dense ionosphere on Venus gave the appearance of a high surface temperature.

One of the primary scientific objectives of the first interplanetary spacecraft, Mariner 2, was to test whether the radio emission from Venus was thermal radiation arising from the surface or might instead be ionospheric in origin. In 1962 the results came back: the radiation really did come from the surface. At about the same time, radar astronomers succeeded in bouncing radio waves off the surface of Venus and found that not only was the ionosphere transparent but the planet was rotating backwards at an incredibly slow rate. One rotation took about 243 days.

The discoveries of the late 1950s and early 1960s changed forever our concept of our sister planet. No longer could the surface of Venus be imagined as a lush jungle populated by exotic creatures. Continuing optical and radio observations from Earth, verified by direct measure-

ments carried out by Soviet spacecraft that landed on the surface of Venus, left no doubt of a temperature hot enough to melt lead, tin, and zinc. We now know that Venus is the hottest planet in the solar system, despite the fact that Mercury is closer to the Sun. Indeed, the picture we now have of conditions at the surface of Venus is reminiscent of traditional concepts of hell.

8.2 The Atmosphere
The Greenhouse Effect

Why is Venus so hot? The answer comes from a phenomenon known as the **greenhouse effect,** which we introduced briefly in Section 7.6. In a typical greenhouse, the glass in the roof allows visible sunlight to enter and be absorbed by the plants and soil within. These objects then heat up and radiate at infrared wavelengths just like the Earth itself. The glass of a greenhouse, however, is largely opaque to infrared radiation. It acts as a color filter, letting short wavelengths through, but limiting passage of longer-wave thermal radiation. Since most of the heat cannot escape, the interior of the greenhouse warms until enough infrared radiation escapes through the glass to balance the energy coming in. A similar effect occurs in a car left out in the sun on a hot day. Once again, sunlight passes easily through the glass in the windows, heating the upholstery and metal of the car's interior. Infrared radiation is trapped inside, and soon the temperature inside the car is much higher than that of the surrounding air.

The gases in a planet's atmosphere can play the same role as the glass in the greenhouse if they have the same property of transparency to visible light and opacity to infrared. A major difference between the greenhouse or the parked car and a planetary atmosphere is that both of these structures limit convection in addition to trapping infrared radiation. The heated air in such enclosures cannot rise and escape, to be replaced by descending cooler air, as it does through circulation in a planetary atmosphere.

Recall that the troposphere is defined as the part of an atmosphere where convective circulation takes place. Tropospheric convection establishes a temperature gradient, with high temperatures near a planet's surface and cooler temperatures aloft. The temperature in an atmosphere is not uniform with altitude, the way it would be in a real greenhouse.

It is still the opacity of the gases to infrared radiation, illustrated in Fig. 2.7 for the discussion of atmospheric windows, that causes the atmosphere to heat up. The reason the infrared opacity is so significant in this process is that the infrared is the region of the spectrum where the bulk of the energy is radiated (Wien's law; see Section 8.1). The amount of energy escaping directly from the hot surface in the form of radio emission, in the long-wave window where the atmosphere is transparent, is too small to affect the energy balance.

The Greenhouse Effect on Venus

Not all gases are opaque to infrared radiation, however. For example, nitrogen and oxygen are virtually transparent at these wavelengths, so they cannot contribute significantly to an atmospheric greenhouse effect. Water vapor and carbon dioxide are very good infrared absorbers, and even the amounts of these two gases in our own atmosphere are sufficient to raise Earth's average temperature some 25 degrees above the value our planet's surface would have if only nitrogen and oxygen were present or if Earth had no atmosphere at all.

The greenhouse effect obviously plays a much more significant role on Venus than it does on Earth. This is clear not only from the high value of the surface temperature, but also from the fact that the temperature is uniform to within a few degrees all over the surface. The entire planet acts as though it were encased in a heavy blanket.

We have already noted the presence of substantial amounts of carbon dioxide on Venus together with a relative lack of water vapor. Carbon dioxide then must be the major source of infrared opacity (Fig. 8.3). The much greater greenhouse effect on Venus compared to the Earth requires both that CO_2 be the major constituent of the atmosphere and that the atmosphere itself must be very massive.

Carl Sagan, the Cornell University astronomer who has become one of the best known scientists of our time, began his research career in 1960 working on theoretical calculations of the Venus greenhouse effect. At that time, the large mass of the atmosphere was not suspected, and early calculations failed to explain the temperatures that were measured by the radio astronomers. During most of the decade of the 1960s, Sagan and his student and eventual colleague, James Pollack, produced theoretical models of increasing sophistication utilizing new measurements of the atmosphere and increasingly more powerful computers for their work. Only after a decade of this effort were Sagan and Pollack able to match the magnitude of the observed effect with their theories.

Mass of the Atmosphere

As Sagan, Pollack, and others struggled to interpret the evidence that our sister planet had an extremely high surface temperature, spacecraft were finding that the atmosphere of Venus is much more massive than ours. In fact, the pressure exerted at the surface of Venus by its atmosphere is 90 bars (ninety times the sea-level pressure on Earth). Venus has nearly the same surface gravity as the Earth, with one bar corresponding to the pressure exerted by a layer of water about 10 m deep (Section 7.2). That

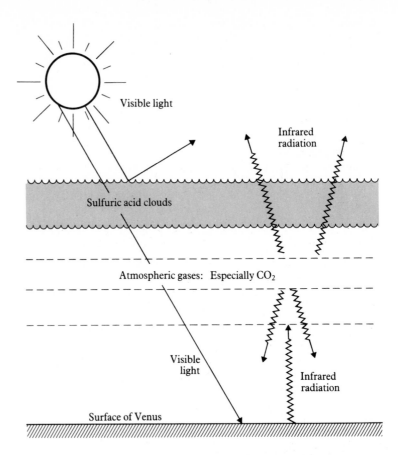

FIGURE 8.3 The atmospheric greenhouse effect. Although about 75% of the sunlight incident on Venus is reflected by the planet's brilliant clouds, some light still reaches the surface where it is absorbed. Infrared radiation from the warm surface then tries to make its way out of the atmosphere but is strongly absorbed by carbon dioxide, causing additional warming of the surface.

means that if you were standing on the surface of Venus (in an air-conditioned, asbestos suit!) you would feel the kind of pressure a deep sea diver deals with at a depth of 900 meters. Heavy armor is required for a human to withstand such conditions. The high surface pressure adds to the infrared opacity of carbon dioxide, making it a much more efficient absorber of thermal radiation than the CO_2 in our atmosphere. Some additional opacity is added by the small amount of water vapor present on Venus.

In order to maintain a greenhouse effect at least some of the solar heat must be deposited at the surface. When scientists first realized the extent of the atmosphere and the thickness of the clouds, some questioned whether any light would filter to the ground. However, experi-

ments on Soviet and American spacecraft that have reached the surface show that some sunlight does get through the cloud layer, providing an illumination about equivalent to a heavy overcast on Earth. This is sufficient to warm the ground and provide the necessary heating.

Composition of the Atmosphere

The discovery that the atmosphere of Venus is ninety times more massive than that of Earth and consists of 97% carbon dioxide surprised many planetary scientists who had expected the atmosphere of Venus to be composed mainly of nitrogen, by simple analogy with the contemporary Earth. They also expected the presence of water vapor, since water is so abundant on Earth. Even oxygen, a gas that might betray the existence of Earth-like life on Venus, was anticipated in this Ptolemaic perspective in which Earth was considered to be the standard planet.

Early searches for spectroscopic evidence of nitrogen, water, and oxygen in the atmosphere of Venus were entirely negative, although carbon dioxide had been discovered spectroscopically in the 1930s (Fig. 8.2). Nitrogen does not have any absorption bands in the part of the spectrum accessible to ground-based observers, and possible evidence of water and oxygen was blocked by the strong absorption by these gases in the atmosphere of the Earth. In the 1950s and 1960s efforts were made to put telescopes on high altitude airplanes and balloons in order to rise above most of the absorptions in our own atmosphere. The results were still negative. Not until the late 1960s were very weak absorption lines caused by water vapor found, followed by carbon monoxide, hydrochloric acid, and hydrofluoric acid. A tiny amount of oxygen was discovered spectroscopically a few years later.

Evidently Venus is much drier than Earth. At its high temperature, all the H_2O on Venus should be present in the atmosphere. However, only traces were found, instead of the equivalent of oceans. The absence of water on our sister planet has been one of the major mysteries of the inner planets, and we will return to its solution later in this chapter.

Keep in mind that all of these observations were made from the Earth and only refer to that portion of Venus' atmosphere above and just within the planet's ubiquitous clouds. But these results were confirmed and extended by the space missions of the 1970s, when probes descended into the atmosphere and landed on the surface. A list of all the gases now known to be present in the atmosphere is given in Table 8.1.

The Clouds of Venus

Ever since astronomers first realized that Venus was covered by a cloud layer, they have also wondered about the composition of those clouds. At first the clouds were assumed to be made of water or ice, similar to those of the Earth. The discovery of very low concentrations of water vapor in the planet's atmosphere led some scientists to seek alternative explanations. Hydrocarbon droplets were briefly in favor, but no proof surfaced to support this conjecture. The liquid or solid droplets of a cloud (like the surfaces of asteroids discussed in Chapter 4) do not display the sharp, diagnostic spectral lines that would permit a definitive identification. Unfortunately for the astronomer, solids and liquids always present a more difficult problem for analysis than do the simpler gases.

The puzzle was finally solved by a series of observations in the 1970s. Improved data obtained from an airplane showed features in the infrared part of the spectrum corresponding to

TABLE 8.1 The composition of the atmosphere of Venus

Gas	Formula	Abundance
Main constituents		
Carbon dioxide	CO_2	96.5%
Nitrogen	N_2	3.5
Trace constituents		
Water vapor	H_2O	150[a] ppm[b]
Sulfur dioxide	SO_2	150[a]
Argon (40)	Ar-40	33
Argon (36)	Ar-36	30
Oxygen	O_2	30
Carbon monoxide	CO	20
Neon	Ne	9
Hydrochloric acid	HCl	0.6
Hydrofluoric acid	HF	0.005

[a]Abundances of these gases vary with altitude and latitude. They are not yet well defined.

[b]parts per million

an unexpected material: concentrated sulfuric acid (H_2SO_4) (Fig. 8.4). The same conclusion had already been reached independently by observers studying the variations in brightness and polarization of sunlight reflected from the clouds in different directions. At last after decades of speculation and years of hard work, we knew the composition of the clouds of Venus, the brightest object in the sky after the Sun and Moon!

Even after the composition of the upper layers of the visible clouds was determined, there remained the question of the origin and structure of the clouds. Unlike the clouds of the Earth, those of Venus reveal no breaks through which to glimpse the surface beneath. Apparently, the clouds of Venus are not simply condensation clouds as are those on Earth. Other processes must be at work to generate and maintain them.

The most important cloud-forming process on Venus is probably **photochemistry,** chemical reactions driven by the energy of ultraviolet sunlight. Photochemical reactions are important in

FIGURE 8.4a Beyond 3 μm, the spectrum of Venus is very dark, meaning that its clouds are poor reflectors. Laboratory spectra of other substances once thought to compose the clouds are shown for comparison; these materials are too reflective beyond 3 μm to match the Venus clouds.

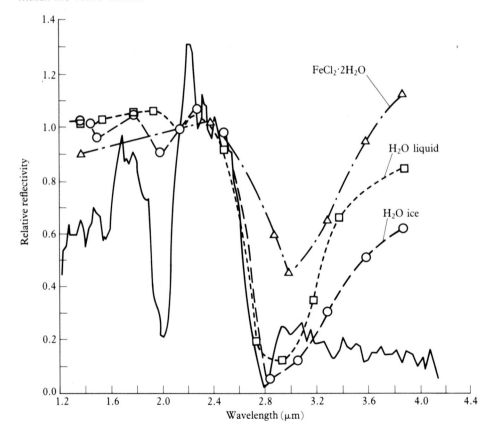

the upper atmospheres of planets, including the Earth, where the production of ozone from oxygen is an example of a photochemical process. On Venus, the H_2SO_4 and the unknown ultraviolet-absorbing clouds are probably both produced and destroyed photochemically (Fig. 8.6).

In addition to photochemical reactions, at least some of the basic cloud material may be supplied from below by active volcanism on the planet's surface. Observations over the past twenty years have indicated that large fluctua-

tions occur in the concentration of sulfur dioxide (SO_2) in the atmosphere of Venus above the clouds. When these observations are combined with indications of volcanic topography and lightning discharges from possible volcanic plumes, the case for erupting volcanoes on Venus becomes rather strong. Although this theory has not yet been proved, at least clear evidence exists for the upward convection of large amounts of SO_2, a fundamental component of the H_2SO_4 clouds.

FIGURE 8.4b A comparison of the reflectivity of solutions of concentrated sulfuric acid and the clouds of Venus. The reflectivities are quite similar, indicating that the clouds of Venus could indeed be made of sulfuric acid.

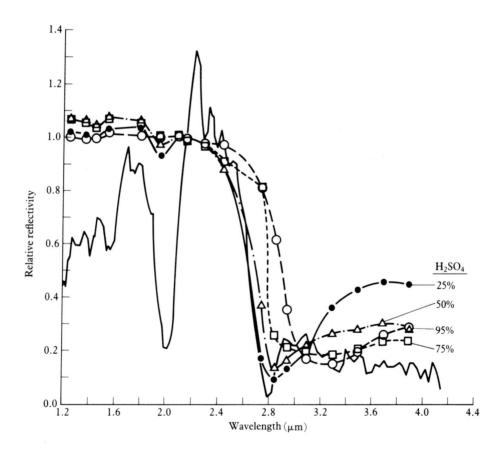

Space probes that have passed through the clouds have given us the picture shown in Fig. 8.5 of a series of discrete cloud layers. Clouds were seen extending from 30 to 60 km above the surface. But what are these various cloud layers made of? Are they all sulfuric acid, as are the topmost layers? Only the Soviet probes have attempted compositional measurements, and their results have been contradictory. Sulfur or possibly chlorine compounds of some sort are indicated, but their exact identities are unknown.

8.3 Weather on Venus

The Upper Atmosphere

Telescopic observations of the clouds of Venus showed that while they lack features in visible light, dusky markings could be seen in pictures of the planet that were taken through filters only transmitting ultraviolet light. Studying such pictures in the 1960s, astronomers found that they could follow the motions of some of these dusky

FIGURE 8.5 The vertical structure of the atmosphere of Venus. The clouds actually consist of several layers with different concentrations of particles. Below the clouds, the atmosphere is clear.

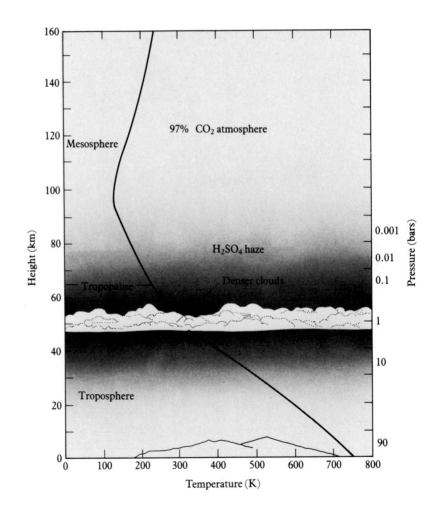

features long enough to see them move completely around the planet and back to their starting position. A complete circuit of Venus required four days, taking place in a retrograde direction (opposite to the direction in which most planets rotate). This motion corresponds to high altitude winds blowing at a speed of 100 m/s (360 km/h) from east to west.

The high altitude, retrograde winds are strongest at the planet's equator, tapering off toward either pole, unlike the pattern of east and west blowing winds found at temperate latitudes on Earth. Subsequently, more detailed imaging from spacecraft, again taken using ultraviolet filters, has permitted scientists to map these high altitude jet streams on Venus (Fig. 8.6). However, we still do not know the chemistry of the thin, ultraviolet-absorbing clouds that make the dark patterns in these pictures. Presumably they have a higher sulfur content than the brighter background, but the compound involved has not been identified. Nor do we have an explanation for the four-day circulation at these high altitudes, although these are current research problems of great interest to atmospheric scientists.

Structure of the Atmosphere

In the case of Venus we are fortunate to have two programs of exploration carried out independently by the United States and the Soviet Union. As a result, numerous probes carrying a variety of instruments have penetrated the clouds of Venus. These probes have revealed the vertical structure of the clouds as well as the temperature and pressure profile of the atmosphere (Fig. 8.7).

The atmospheric temperature profiles of the two planets cross; Venus is colder than Earth at high altitudes and warmer near the ground. A region in the atmosphere of Venus exists where the pressure is near the sea-level pressure on Earth and the temperature reaches a balmy 30 C.

FIGURE 8.6 A picture of Venus obtained with the camera on the Pioneer Venus orbiter spacecraft, using an ultraviolet filter to bring out contrasts in the clouds (cp. Fig. 8.1). The Y-shaped cloud marking, dimly seen from Earth, is clearly visible here.

FIGURE 8.7 The atmosphere of Venus is much warmer than Earth's near the planet's surface, but it is actually colder at high altitudes.

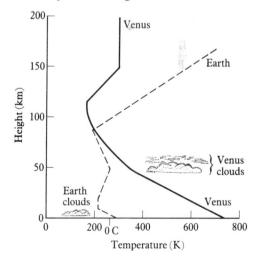

Such conditions would create a shirtsleeve environment for astronauts in the gondola of a balloon floating in the cloudy skies of Venus, were it not for the fact that the clouds are mainly sulfuric acid and the atmosphere unbreathable carbon dioxide.

In June of 1985, the Soviet VEGA spacecraft deployed balloons in the atmosphere of Venus at an altitude of about 54 km. Probes on the balloons measured pressure, temperature, and wind speed as part of a cooperative Soviet-French-American experiment. The results indicated that the winds at this altitude on Venus are far more blustery than had been anticipated. Updrafts as fast as 3 m/s (7 mph), possibly caused by surface topography, were measured over the region known as Aphrodite (see page 243). The balloon experiments found no evidence for lightning discharges, even though previous experiments from orbiters had identified Aphrodite as a region where lightning occurs on Venus.

Near the surface of Venus wind speeds are low, with measured values ranging from 0 to 2 m/s (approximately 0–4 mph). The pressure and density are so great at these levels that the atmosphere behaves more like an ocean than the air we are familiar with on Earth. Like the deep oceans of our planet, the surface of Venus has a nearly uniform temperature, from pole to pole and noon to midnight, resulting from the massive, slowly moving atmosphere.

Atmospheric Circulation

Above the surface of Venus, a pattern of air rising near the equator and traveling north and south to descend near the poles occurs (Fig. 8.8). This simple type of atmospheric circulation is called a **Hadley cell,** after the British scientist who first proposed it as a model for the circulation of the Earth's atmosphere. While not describing our own planet very well, Hadley cell circulation turns out to be a good model for the lower atmosphere of Venus.

At higher elevations, near the main cloud layers about 50 km above the surface, the strong retrograde (westward) super-rotation becomes apparent, with 100 m/s currents seen in the motions of the upper clouds. At this altitude, of course, the atmosphere is as thin as that of our own planet, making high wind speeds possible. At the poles, these winds form a vortex of descending air, rather like the pattern of water swirling down a drain (Fig. 8.9).

Computer calculations suggest that the key to the difference between the circulation of Venus and Earth lies primarily in the slower rotation of Venus. If the rotation of Earth were this slow, the high and low pressure systems that correspond to centers of fair and foul weather would fade away, the mid-latitude jets would disappear, the small Hadley cells now confined to the equatorial zone would expand all the way to the poles, and a globe-encircling wind would begin to blow at high altitudes. The decrease in rotation does two things to produce these changes. First, it lengthens the duration of daylight, increasing the effects of heating in the daytime and cooling at night. Second, the Coriolis effect (Section 7.7) is less strong on a slowly rotating planet, and winds are less likely to be deflected into swirling, cyclonic motion. The combination

FIGURE 8.8 In a simple Hadley cell circulation, warm air rises at the equator of a planet and travels toward the pole where it sinks and returns to the equator along the surface (see Fig. 7.17).

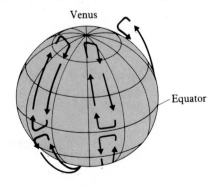

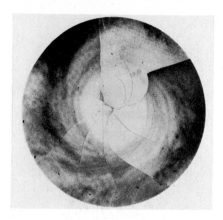

FIGURE 8.9 A mosaic composed of several pictures of the south pole of Venus, showing the spiral cloud structure associated with the polar vortex of descending air.

of longer-term heating and a low Coriolis effect, aided by the larger total mass of the atmosphere, evidently produces a circulation pattern like the one on Venus.

The analysis given above is an example of the ways in which other planets can help us understand the Earth. In order to evaluate the true importance of various forces at work in natural systems, a scientist likes to perform an experiment, to change the forces and then study the effect of these changes. Since we obviously can't slow down the Earth, the next best recourse is a computer than can model the effect. Uncertainty remains, however, since a planetary atmosphere is complex. Perhaps something important is absent from our model. We can test the model by looking at another planet where basic conditions governing circulation are significantly different. Slowly turning Venus is one extreme; the giant planets with rotations more rapid than Earth's provide another. The best assurance we have that our models for the atmosphere of the Earth are accurate is their ability to deal correctly with these other worlds as well, where rotation rate, distance from the Sun, and other basic parameters are fundamentally different from those on Earth.

8.4 The Hidden Landscape
Radar Studies

The dramatic differences between Venus and Earth extend to a comparison of the geology of the two planets. As we cannot photograph the surface of Venus through its clouds, other techniques must be used to map the topography. Thermal radiation from the surface of Venus has no difficulty escaping through the planet's cloud layer at radio wavelengths so we anticipate that it should be possible to send radio waves through those clouds from the outside. This radio signal will then be reflected by the planet's solid surface and pass back through the clouds to our receivers, where we can analyze it for the effect the reflection at the surface has had on the radio transmission. This is another example of the use of radar (Section 6.1) for astronomical purposes. The first radar studies of Venus were made for the same purposes as for Mercury: to determine the rate of rotation.

As usual for this surprising planet, the results for Venus were unexpected. The period of rotation is 243.08 days in a retrograde direction. In other words, Venus rotates on its axis in the opposite direction from the course of its motion around the Sun. On Venus, if only you could see it through the clouds, the Sun would appear to rise in the west and set in the east.

The length of a day on Venus requires further definition. While not quite as peculiar as Mercury, Venus also has a day that is longer than its year. The rotation period measured with respect to the stars is 243.08 days, about 19 days longer than the period of revolution around the Sun, namely 224.7 days. This is the length of time required for Venus to make one rotation about its axis. As we saw for Mercury, another way to determine the length of a day is to measure the time between two successive noons (or sunsets, sunrises, etc.). The time between successive noons — the solar day — is much shorter, amounting to 116.67 Earth days, with the Sun moving across the sky in the wrong direction, from west to east.

In what appears to be simply a coincidence, the true rotation period of Venus is very close to the period needed to align a surface feature with the Earth at successive inferior conjunctions — the times when Venus is between Earth and the Sun (Fig. 8.10) — which is 243.16 of our days. Even though the difference between the two periods is very small (2 hours out of 243 days), it is real, so no tidal lock or resonance exists between Venus and Earth as was once suspected. While probably having no fundamental significance, this coincidence has striking effects on observations of Venus from Earth. It means that over a period of many years, we confront the same side of Venus when the two planets are closest together, with the practical result that we have much better radar coverage of one side of Venus than of the other.

Large-Scale Topography

In addition to determining the rotation period, radar can be used to map the surface terrain on Venus. This mapping can be done for selected regions from Earth, but is an especially powerful technique when used from a spacecraft in orbit about Venus. Being so much closer to its target, the radar on the spacecraft can afford to be much less powerful than its ground-based counterpart and still achieve higher resolution and more extensive surface coverage.

The U.S. Pioneer Venus spacecraft was the first to map the surface of Venus, using a simple kind of radar that measured the altitude of the spacecraft. As the spacecraft orbited from north to south, the planet turned underneath from west to east, building up complete surface coverage over a period of about two years. A resolution of 25 km was obtained in some areas that were beneath the spacecraft's closest approach. The topographic map based on these Pioneer Venus radar observations gave us our first global picture of the landscape on Venus.

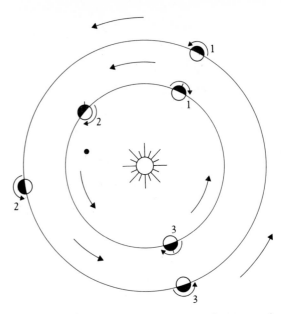

FIGURE 8.10 Venus rotating as it revolves around the Sun. At position (1) Venus is at inferior conjunction, with a feature on its surface pointing toward the Earth. 486 days later the two planets are at the positions labeled (2). Venus has undergone two complete retrograde rotations and slightly over two revolutions around the Sun. Earth has undergone 486 rotations and 1⅓ revolutions around the Sun. Ninety-eight days later the two planets have moved to position (3). Venus is again at inferior conjunction 584 days after position (1), and the surface feature has almost rotated into position to face the Earth.

Plate 7 allows a comparison of the topography of Venus and the Earth at this same low resolution. In looking at the two maps, we are immediately struck by the rarity of large continents on Venus. Evidently the geological processes at work are different. This difference is shown in a comparison of the relative areas of different kinds of topography on the two planets. On Earth, 45% of the crust is continental, while on Venus only 8% consists of highlands that may or may not be similar to terrestrial continents. The rest of the landscape of Venus has been classified as lowlands (27%) and rolling plains (65%), terms descriptive of relatively low and flat terrains. Apparently Venus is not experiencing the same large-scale plate tectonics as the Earth; or if it is, the continental masses are much smaller and the rifts and subduction zones less well defined.

The Search for Impact Craters

If the geology of Venus is unlike that of the Earth, perhaps it resembles Mercury and the Moon with landscapes dominated by old lava-filled basins and heavily cratered surfaces. Venus might have a similar appearance if it cooled rapidly so that the ancient crust was preserved with little geologic activity to destroy impact craters once they had formed. A priori, this alternative seems less likely, since a planet as big as Venus should generate a lot of heat from radioactivity which it will have difficulty releasing because of its large size. Its evolution should therefore be much more like the Earth's than like the small bodies Mercury and the Moon. Perhaps an intermediate case exists, however; our knowledge of planetary evolution is not yet so secure that we can confidently make predictions from simple physical laws.

Turning again to the map in Plate 7, we note little evidence of impact-produced landforms. Hardly a crater or basin can be seen. However, as remarkable as it is the Pioneer Venus map mainly shows surface features with dimensions greater than 100 km, not much better than the naked-eye resolution on the Moon (Section 5.1). We need better resolution than this if we are to find analogs of the narrow rift valleys in our ocean floors that provide clear evidence of sea floor spreading or the large population of small impact craters that would prove that the ancient crust is indeed preserved.

Results from the Soviet spacecraft Venera 15 and 16 are taking us a long way toward this goal. Unlike the Pioneer Venus orbiter, the Venera 15 and 16 spacecraft carried **imaging radar** systems that produce pictures of the surface very similar to photographs made with ordinary cameras. Similar imaging radars are used on Earth to map regions (such as parts of Papua New Guinea) where nearly constant clouds and rain make ordinary photography impossible. They are also of increasing importance for military surveillance satellites, especially to keep track of the motions of ships in all weather conditions.

An example of the results coming from these two spacecraft is shown in Fig. 8.11. Here we

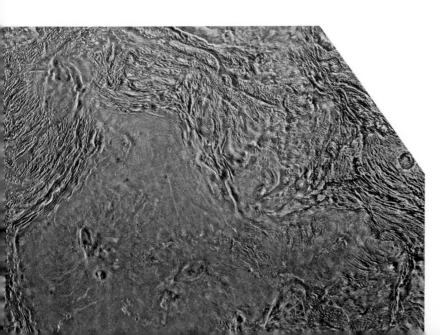

FIGURE 8.11 The plateau called Lakshmi Planum as revealed by Veneras 15 and 16. The plateau is surrounded by "wrinkles," which are actually mountain ranges. At the right center are the Maxwell Mountains, with the crater Cleopatra at the extreme right margin (see Fig. 8.13).

see a mosaic composed of many separate radar swaths that cover the huge plateau known as Lakshmi Planum, centered at 70° N, 330° W (Plate 7). While there are a few impact craters apparent on the smooth surface of the plateau, the crater density appears even less than that of the lunar maria. The single prominent crater is Cleopatra, at the middle right margin. Studies by the Soviet scientists indicate that at the typical resolution of 2 km for these radar maps, the density of impact craters is at most only 10–20% that of the lunar maria.

If the impact rates have been the same for Venus and Earth, this crater density corresponds to a surface about 1 billion years old in the northern regions of Venus that have been surveyed. This crater density is similar to that on the youngest major lunar features, such as the large craters Copernicus and Tycho, and also similar (as we will see in the next chapter) to many areas on Mars. Something must be modifying the surface of our sister planet, but not nearly as dramatically as the processes at work on the Earth where any impact craters and even billion-year-old rocks are relatively rare.

◆

8.5 Geological Processes: Impacts, Volcanoes, and Possible Plate Tectonics

Nomenclature of the Surface of Venus

In order to discuss the processes that have been shaping the landscape of Venus, we must first gain some additional familiarity with the map in Plate 7. It is helpful to know the names of the principal features so we can refer to them in the same way we talk about Africa or Greenland on Earth. For this purpose, we have reprinted the Venus map in black and white as Fig. 8.12.

The features that have been identified by the radar scans have been named after real and mythical women, appropriate to the one planet in our solar system with a female name. Thus the largest upland region is called Aphrodite, using the Greek name for the goddess the Romans called Venus. It stretches along the equator from 60 to 240° in longitude, but is only 20 to 30° wide in latitude on average, making it about the size of Africa.

Ishtar, named after the Babylonian goddess of love and beauty, is the prominent upland in the north, at about the same latitude as Greenland on our own planet (Plate 7). Ishtar is bigger than Greenland but is only a small fraction of the size of North America. It contains the highest elevations on Venus, named the Maxwell Mountains after the nineteenth century (male) British scientist who first formulated the laws of electromagnetic radiation. These mountains rise about 11 km (35,000 ft) above the surrounding lowlands, comparable to the height of the Hawaiian volcano Mauna Loa above the Pacific Ocean floor (Section 7.3; Fig. 7.6). This region of Venus is an excellent reflector of radar waves and has been extensively studied from Earth with the huge Arecibo antenna (Figs. 6.2 and 8.13).

Ishtar and the Maxwell Mountains

The high radar reflectivity of Ishtar and the Maxwell Mountains has allowed the ground-based observers to achieve about the same 1–2 km resolution obtained by the Soviet spacecraft. Fig. 8.13 shows a series of parallel ridges and grooves approximately 15 to 20 km apart. This topography is similar to that exhibited by the Appalachian Mountains on Earth, where compression of the crust has caused buckling and wrinkling, but the Maxwell construct exhibits a much greater altitude difference than the Appalachians. Over a distance of just 50 km, the mountain range on Venus rises 6 km, then gradually decreases in elevation over a horizontal distance of 400 km. This abrupt rise is similar to that of the Rocky Mountains just west of Denver.

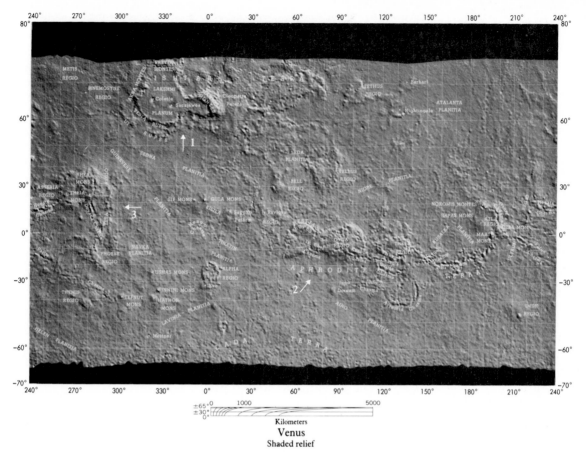

FIGURE 8.12 The surface of Venus revealed by the Pioneer Venus radar. Arrow 1 points to Lakshmi Planum and Maxwell Mountains; 2 is Aphrodite, and 3 is Beta Regio.

Do the Maxwell Mountains provide an example of plate tectonics at work on Venus? There are scientists who think that they do. Just as the collision of the Pacific and North American plates has pushed up the Rocky Mountains, these investigators think that two colliding plates on Venus may have produced the Maxwell Mountains.

A large circular depression called Cleopatra occurs on the slope of these mountains, shown in Fig. 8.13 and in the Soviet Venera 15 radar images (Fig. 8.11). Cleopatra is 85 km across,

about the size of the crater Copernicus on the Moon or the crater Herschel on Saturn's satellite Mimas (Chapter 14). It is still not known if this is an impact crater or a giant volcanic caldera.

Evidence of Volcanism

South of Ishtar, we encounter the two great mountainous regions called Alpha and Beta. Like Maxwell, these areas are excellent radar reflectors and are on the hemisphere of Venus that

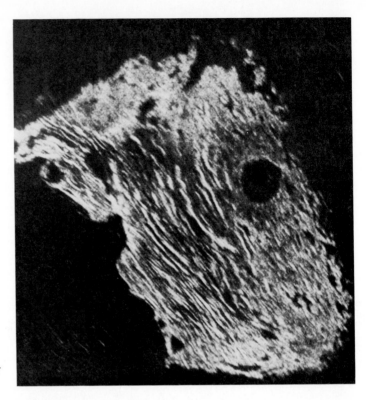

FIGURE 8.13 The Maxwell Mountains with the circular feature known as Cleopatra as imaged by the Arecibo radar (see Fig. 8.11).

FIGURE 8.14 An oblique view of the surface of Venus including the area known as Beta Regio. This regio looks very much like a rift valley whose walls are defined by volcanic peaks. The landing sites of various Soviet spacecraft are labeled V9, V10, etc.

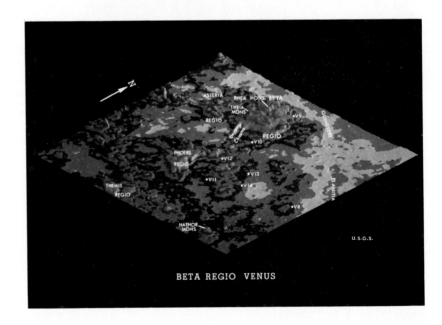

has been well studied from Earth. That is why all three of these regions were given non-female names: they were well known before the present nomenclature convention was adopted.

Both Aphrodite and Beta exhibit intriguing topography, since both are cut by long valleys similar to the great terrestrial rift valley in East Africa. This African valley is one of the places on Earth where two crustal plates are slowly moving apart, the continental equivalent of what is happening along the rift zones of the ocean floors. Is that what is happening in Beta and Aphrodite on Venus? Or are these great valleys of Venus simply examples of **tectonic** fractures — cracking of the crust in response to elevations or depressions elsewhere on the surface? The giant Valles Marineris system on Mars (Section 9.5) is an example of a tectonic valley formed by crustal fractures.

A key to this riddle may lie in the presence of mountain peaks along the sides of the valleys in both Beta and Aphrodite (Fig. 8.14). Some scientists have suggested that these peaks may be volcanoes, since it is difficult to see how they were formed otherwise. If they really are volcanoes, then some plate movement indeed seems to be implied. Recall from Section 7.4 that a rift valley forms in the Earth's crust when new material rises from the mantle, pushing the plates away on either side. Thus molten rock could even now be near the surface under these giant valleys on Venus, and these volcanoes might be active at the present time.

This surprising conclusion has gained credence from observations of bursts of radio and acoustic noise that seem to be produced near Beta and Aphrodite. The noise burst observations made independently by U.S. and Soviet investigators on the Pioneer Venus and Venera spacecraft provide evidence for lightning discharges near the planet's surface, far below the known cloud layers. To explain the presence of lightning in such unusual circumstances, the investigators have suggested that these discharges

occur in the plumes above active volcanoes. Such discharges are often associated with volcanic eruptions on Earth (Fig. 8.15).

We thus have three independent sets of evidence pointing toward active volcanism on Venus: large-scale fluctuations in atmospheric sulfur dioxide, topography suggesting volcanic peaks associated with rift valleys, and lightning discharges apparently coming from regions near the surface in the vicinity of these peaks. (But not all the time; the 1985 VEGA balloons found no lightning.) This is an impressive array of evi-

FIGURE 8.15 A volcanic eruption on Earth sometimes includes lightning discharges. This same combination of geological and meteorological activity *may* be occurring on Venus.

dence, but we still lack the geological equivalent of the famous smoking gun in a criminal case, namely a smoking volcano. Without that, active volcanism on Venus remains an inference, highly probable but still unproven.

Additional Discoveries in the Radar Images

There is also intriguing evidence of possible past volcanic activity on an even larger scale. This evidence is in the form of large flat circular structures with diameters of several hundred kilometers, discovered in 1984 in the Venera 15 and 16 radar images (Fig. 8.16). The Soviet geologists have dubbed these circular features coronas, using the Latin word for crown. They may be examples of the collapsed tops of former mantle plumes which once actively brought heat up from the planet's interior (like terrestrial hot spots, such as the one beneath the Hawaiian Is-

lands). Alternatively, they may also be the result of impacts in which the original crater has been greatly modified by subsequent geological processes. Perhaps more detailed study of these radar images will reveal the answers.

Obviously current radar studies of Venus are revealing a world very different from the others we know. It is fortunate that the Soviet spacecraft have worked so well, mapping most of the northern hemisphere of Venus at a resolution of 1–2 km. In 1990, the United States will launch a new orbiter called Magellan with radar imaging to bring us another step forward in resolution, revealing details as small as 200 m. With these data we should finally be able to identify the geological processes taking place and discriminate among the various interpretations that we now confront. Perhaps we shall be lucky enough to see a volcano in action with an advancing lava flow whose progress can be charted.

FIGURE 8.16 This large (∼150 km) circular feature on the surface of Venus is an example of a corona. Impact craters are visible to the upper right and lower left of the corona in this mosaic of Venera 15 and 16 radar swaths across the planet's surface.

8.6 On the Searing Surface

The Venera Landers

To gain an appreciation for what conditions are like on the surface of Venus, you might begin by looking at the temperature control of a modern kitchen oven. You will find that the highest temperature the oven can achieve is about 500 F. The ground temperature on Venus is over 350 degrees higher: within 10 or 15 degrees of 860 F. And remember that this temperature is coupled with an atmospheric pressure 90 times the value on Earth.

Despite these extremely inhospitable conditions, direct measurements have been successfully carried out on the surface of Venus at seven locations by spacecraft in the Soviet Venera and VEGA series (Fig. 8.17). These landers are suitably armored against the high pressure and equipped with some internal cooling. Some of them have survived nearly two hours, during which a variety of investigations were carried out. The data from these spacecraft even include pictures transmitted back to Earth, four of them in color. These are the only close-up pictures we have of the surface of another planet except for the Viking pictures of Mars and, of course, extensive surface photography of the Moon.

The Venera cameras are designed to provide panoramic views of the surface extending from

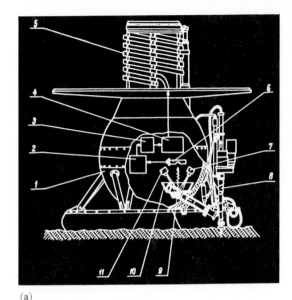

(a)

(b)

FIGURE 8.17 a) The line drawing above shows the principal components of the landing module from the VEGA spacecraft that reached the surface of Venus in June 1985 (cp. Fig. 4.18a). The large, round pressure vessel (1) is surrounded by a coiled antenna (5) for communication with Earth. A drill (8) can obtain samples of surface material which are then brought inside the module for analysis by X-ray fluorescence (7). b) Yuri Surkov, of the Vernadsky Institute in Moscow, holding a proof-test model of the X-ray fluorescence device he built for the VEGA lander.

the soil directly in front of the spacecraft out to the horizon. The cameras are positioned about one meter above the ground, providing a perspective comparable to that of an observer sitting on the surface of Venus. The cameras do not gaze straight out at the landscape as the human eye would. Instead, they are directed at mirrors set at a 45° angle, and the mirrors are rocked back and forth to produce the desired panoramas.

The immediate foreground of each picture shows the view toward the feet of the hypothetical observer, including a part of the spacecraft itself with a scale reference provided by the triangular teeth spaced 5 cm apart. The white

ladder-shaped boom extending out just left of center is about as long as your arm, while the distant white viewport covers (ejected after landing) are about the diameter of a human head (Figs. 8.18 and 8.19).

Images of the Surface

The first pictures of the surface of Venus, transmitted by Veneras 9 and 10 in 1975, showed rough, undistinguished landscapes dominated by loose rocks. Surface illumination, as we noted previously, was similar to that on a heavily overcast day on Earth. The more detailed pictures transmitted by Veneras 13 and 14, which landed

FIGURE 8.18 A view of the surface of Venus obtained by Venera 13. The triangular teeth in the foreground are about 5 cm apart. Note the presence of fine grained, soil-like material in the foreground. The bright crescent-shaped object in the center of the frame is the camera cover. The device to the left of the ejected camera cover measures the surface hardness. The banded boom to the right is for color calibration.

FIGURE 8.19 Venera 14 was surrounded by flat, plate-like rocks resembling pieces of a dried lake bed. One of them was overturned by the shock of the landing, revealing a bright surface (left foreground). In this case, the deployed camera cap unfortunately ended up under the surface-hardness measuring device, showing that Murphy's law also applies on Venus!

in March 1982, represent the best data available on the appearance of the surface of Venus. They provide some exciting new information: evidence of layering and perhaps ripple marks on rock surfaces, especially around Venera 14. Coupling these observations with the measurement of low surface bearing strength and low density derived for these rocks by other instruments on the spacecraft, the Soviet investigators conclude that these might be sedimentary rocks. They were not, however, sediments produced by water; instead some new geological process must be involved that operates under the special conditions existing on Venus.

The Venera 13 pictures show many small rocks and some fine-grained soil, indicating that processes are at work that cause the large blocks to disintegrate. In the absence of water erosion some form of chemical weathering seems the most likely agent. Another possibility is nearby volcanism with concurrent release of dust, ash, and lava. It is also possible that the fine material is ejecta from an impact crater.

Composition of the Surface

Analysis of the chemical composition of the rocks at the Venera landing sites provides further clues. The first Venera landers carried devices to detect gamma rays emitted from radioactive isotopes of uranium, thorium, and potassium. By analogy with the levels of radioactivity of terrestrial rocks, these data suggest igneous rock, probably basalt at the Venera 8 site and possibly granitic rock at the locations of Veneras 9 and 10.

On Veneras 13 and 14, and on VEGA, a more sophisticated technique measured the surface composition. A drill was deployed to gather samples from underneath the surface and bring them inside the spacecraft for analysis (Fig. 8.17). Using an X-ray source to stimulate emission from the sample, the instrument was sensitive to additional elements. The Venera 13 sample indicated a composition typical of oceanic

basalts on Earth, while the result from Venera 14 resembled a much rarer kind of basalt with a high percentage of potassium. The VEGA spacecraft, which landed just north of the Aphrodite continent, again revealed a basaltic composition, this time unusually rich in sulfur.

The igneous nature of the rocks at the Venera landing sites indicates that Venus, like the Earth and Moon, differentiated during the course of its formation. The widespread presence of basalts indicates a history of volcanism, while the possibly granitic rock suggests slow cooling below the surface at higher temperatures and pressure.

Interior Structure

We have no information about the deep interior structure of Venus, since no seismometers have been deployed on the planet's surface yet. Evidence of differentiation at the crust certainly implies internal sorting, with formation of a dense, metal-rich core. But is that core liquid or solid?

The orbiting spacecraft have detected some near-surface variations in the gravity of Venus that appear to be associated with Beta and eastern Aphrodite. Such variations are common on Earth near the margins of crustal plates, especially at rift or subduction zones. Therefore, these **gravity anomalies** have been cited as additional evidence for volcanism on Venus. The argument is that these massive mountains are evidently not resting (floating) on rock of lower density as is true of large mountains on Earth. If they were, there would be no net gravity anomaly. So what is holding them up? Perhaps they are supported by rising plumes from the mantle.

All of these various bits and pieces of evidence seem to point toward a model for the planet that includes a hot interior. No evidence of a planetary magnetic field has been found, arguing against the presence of a liquid core. The argument is that a conducting liquid is required at the center of a planet in order to produce the self-sustaining dynamo that generates the magnetic field. This argument loses some strength because of the planet's slow rotation. It is rotation, after all, that feeds the energy into the core to keep the dynamo running. So there is no clear answer about the state of the interior.

More Differences Between Earth and Venus

Meanwhile, we have found another glaring difference between Venus and the Earth. The absence of a planetary magnetic field means that the solar wind impinges directly on the upper atmosphere of Venus. A continuous aurora keeps the planet's outer fringes glowing, and a tail of ions extends downstream from Venus, balanced by the arriving solar wind ions on the upstream side. Nothing like the huge magnetosphere of Earth with its belts of trapped atomic particles exists on Venus.

Returning to the planet's surface, we would obviously like to know why the topography of Venus is so different from that of the Earth, given the fundamental kinship implied by evidence of differentiation. Some geologists speculate that Venus today resembles the geology of the Earth 3 billion years ago, in the Archean era. At that time the Earth may have been more volcanically active than today, but it had not yet developed the particular type of mantle convection and lithospheric structure that gives rise to plate tectonics, nor did it have the large continents we find today.

One geological theory that Venus may help to evaluate is that water is required to lubricate the movement of the lithospheric plates on Earth and keep the plate motion from freezing up. If this is correct, then the absence of continents and plate tectonics on Venus may result from the absence of water. This question brings us back to a problem posed by the absence of abundant water vapor in our sister planet's overheated atmosphere: why does Venus have so little water while the Earth has so much?

8.7 Atmospheric Evolution: Why Did the Twins Diverge?

The Different Atmospheres of Venus and Earth

The present atmospheres of Earth and Venus are obviously very different, and we would like to understand how evolution led in such divergent directions for the two planets. The most important differences to be explained are the depletion of water on Venus relative to Earth, the depletion of carbon dioxide on Earth relative to Venus, and the unique presence of oxygen in the atmosphere of our planet. As we have seen, this final difference is easy to understand since the Earth's oxygen is constantly being replenished by photosynthesis in the green plants that grace our planet. But the differences in water and carbon dioxide may seem more difficult to explain.

It turns out that the difference in carbon dioxide content between the two planets is also related to the presence of life on Earth. We have already described how carbonate rocks are formed through the action of tiny marine creatures that use carbon dioxide dissolved in the ocean to make their shells, which then become fossilized as ocean sediment. Carbonates are also formed directly by the action of dissolved carbon dioxide on silicate rocks. A much smaller amount of carbon has gone into the creation of deposits of fossil fuels such as coal and oil. All of this carbon now in the rocks was once in the atmosphere as carbon dioxide. Some of it is being recycled when carbonate rocks are subducted.

Suppose there had been no life and no water on Earth. A survey of the materials taken out of circulation shows that if all the carbon were put back into the atmosphere in the form of carbon dioxide, the atmosphere of Earth would have a pressure of about 70 bars, and a composition of more than 98% carbon dioxide and only a little more than 1% nitrogen. The total inventory of carbon dioxide is thus similar for Venus and the Earth (Table 8.2).

Why Is Venus So Dry?

We see that the differences in both oxygen and carbon dioxide content of the atmospheres of Venus and the Earth result from the fact that the Earth is wet and inhabited, while Venus is not. Once again water seems to be the real key, since life would not have developed on Earth if our planet were as dry as Venus is today.

Why is Venus so dry? Until recently, there were two competing explanations. One school held that Venus simply formed without water. We have seen that Mercury has an anomalously high proportion of iron since it accreted in such a hot part of the solar nebula that some of the low-temperature silicates could not condense. Following this train of thought, scientists pointed out that the same models for the solar nebula predict that at the position of Venus, the nebula would have been too hot to allow water to condense. Thus the materials accreting to form the planet would be deficient in water, and Venus would form and remain dry.

The alternative point of view suggested that most of the volatile material on all of the inner planets did not come from the regions of the nebula in which these planets formed, but was brought in from farther out in the solar system in the form of primitive meteorites and comets. In Section 4.7 we noted that comets are about half composed of water ice, and certainly comets impact the surfaces of the inner planets even today. The numbers of these messengers from cooler regions of the nebula would have been much greater in the past than they are now. So the water in our oceans as well as the carbon, nitrogen, phosphorus, and other volatile elements may have been brought to the Earth as a coating rather late in the course of its accretion.

If Earth received many of its volatiles from meteoritic and cometary impacts, the same process should work on Venus and Mars since all three planets would be bombarded to the same extent. Mercury and the Moon would also have received this volatile-rich bombardment. Perhaps their small sizes prevented them from retaining these elements the way their larger siblings did. The fact that the amounts of CO_2 and N_2 in the Venus atmosphere today are so similar to the amounts degassed by the Earth over its history was offered as supporting evidence for this point of view.

If one accepts this model for external acquisition of volatiles, however, one must then be able to account for the absence of water on Venus today. Where did the water go? The answer probably lies in a phenomenon known as a runaway greenhouse.

The Runaway Greenhouse Effect

The **runaway greenhouse** is a process through which a planet can fundamentally alter the state of its surface and atmosphere. Imagine what would happen if we could move the Earth into the orbit of Venus. Our planet would suddenly be closer to the Sun, at 72% of its present distance. Sunlight would be delivering about twice as much energy to every square meter of the Earth's surface. Most of the Earth is covered by oceans, so the immediate result would be an increase in the temperatures of these huge bodies of water. The increase in temperature would lead to increased evaporation. More water vapor would be present in the atmosphere, which

TABLE 8.2 Atmospheres of Earth and Venus

Gas	Earth		Venus
	Now	Total[a]	Now
N_2	78%	1.9%	3.4%
O_2	21	trace	trace
Ar	0.9	190 ppm	40 ppm
CO_2	0.03	98%	96.5%
Water depth	3 km	3 km	trace
Pressure	1 bar	~ 70 bar	88 ± 3 bar

[a]No weathering, no life.

would trap more infrared radiation from the Earth's surface. In other words, we would have increased the greenhouse effect. This in turn leads to a further increase in the planet's surface temperature, resulting in more evaporation of water and a continuation of the cycle. We have established a positive feedback loop, in which the initial disturbance — increasing the Earth's surface temperature — produces consequences which lead to an enhancement of that disturbance. The cycle continues until the oceans literally boil away, and all water is converted to vapor in what is then an exceedingly hot atmosphere. This is the runaway greenhouse effect.

At this point the atmosphere is so hot that water vapor can easily rise to great heights where it becomes exposed to solar ultraviolet light. This is a crucial step. On Earth, water is protected by the natural cold trap in the atmosphere. The air at the top of the troposphere is so cold that water cannot diffuse upward to levels where it could be attacked by ultraviolet light. A runaway greenhouse can raise the temperature throughout the lower atmosphere, giving water free access to high altitudes. Just as in the case of evaporating water molecules from an icy comet nucleus, the water vapor is broken apart into its constituent atoms by ultraviolet light from the Sun. Because of the large mass of Venus, only the light hydrogen atoms escape into space. The oxygen remains behind to combine with rocks on the planet's surface and with other gases that have been produced by the intense heating. The runaway greenhouse leads to the elimination of water from the planet in a perfectly natural way (Fig. 8.20). Quantitative studies of this phenomenon indicate that water simply cannot remain on the surface of a planet at the distance of Venus from the Sun.

A Test of Two Hypotheses

We have indicated two alternative explanations for the absence of water on Venus: that it was never present in significant amounts or that it was once present (probably deposited by late-accreting materials similar to the primitive meteorites and comets) and subsequently lost through a runaway greenhouse effect. How can we decide between them? The customary scientific method in such a situation is to examine the implications of the two hypotheses to see if additional effects can be predicted and then look for these effects.

Let's return to the runaway greenhouse. To remove oceans of water from a planet's atmosphere a huge amount of hydrogen must escape from the planet. As we saw in Section 2.3, the ease with which an atom can escape from a planet at a given temperature depends on the atom's mass. Hydrogen has two stable isotopes, the common form (H) and the rarer deuterium (D), which has twice the mass of H (Fig. 2.1). On Earth, there are about ten thousand atoms of H for each atom of D. Although both of these isotopes can escape from Venus, it is slightly harder for deuterium than for ordinary hydrogen. Therefore, if oceans of water have left Venus, we would expect the hydrogen remaining on the planet to be enriched in D relative to H. This is a prediction we can test. The competing hypothesis — that there was hardly any water on Venus to begin with — does not predict any enhancement of the deuterium abundance.

The test for deuterium was made with one of the instruments on the Pioneer Venus probes: a mass spectrometer that could distinguish between H and D when these two forms of hydrogen are incorporated into water molecules. (The small amount of hydrogen in the atmosphere of Venus today is nearly all bound up in water or sulfuric acid.) The result was positive. There is one atom of D for every one hundred atoms of H on Venus, approximately 100 times the abundance ratio found on Earth. The conclusion that each of the terrestrial planets may have received much of its volatile inventory from the impacts of comets and asteroids seems secure. What is less certain is just how wet Venus was at the beginning. Did it start out with oceans like ours or

Idea experiment

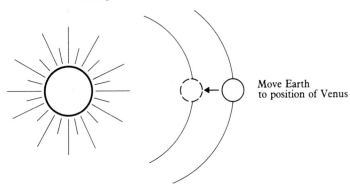

Move Earth
to position of Venus

Consequences:

1. Oceans become warmer.

2. More water evaporates.

3. Increased water vapor in atmosphere blocks infrared radiation from surface, increasing surface temperature.

4. Cycle repeats until:

5. Oceans boil away, atmosphere becomes very hot, full of water vapor, which rises into upper levels.

6. At high altitudes, ultraviolet light breaks water molecules into hydrogen and oxygen.

7. Hydrogen escapes from planet, oxygen remains to combine with rocks, atmosphere becomes dry and full of CO_2. (Carbonates cannot form.)

8. Earth resembles Venus!

FIGURE 8.20 The runaway greenhouse effect.

with a smaller (or larger) endowment of water? This issue is still being argued.

In any case, it appears that the basic reason for the differences between the atmospheres of Venus and Earth is that Venus is just too close to the Sun. At this distance water will not be stable, and without water, the carbon dioxide released from the interior of the planet stays in the atmosphere producing the hellish surface conditions present today. Furthermore, the departure of water may be coupled with the cessation of mantle convection, as suggested in the previous section. This would explain why the crust of Venus seems to be frozen in its development at a stage comparable to the Earth three billion years ago. There is still more evidence we can bring to bear on this problem.

The Message of the Argon Isotopes

An aphorism planetary geologists like to quote states, "Rocks remember but gases forget." It means that it's easier to read an ancient record in a rocky surface than by studying an atmosphere. But this is not always true. We have seen how the hydrogen on Venus remembers the runaway greenhouse effect. Isotope ratios like deuterium/ hydrogen tend to be relatively unaffected by chemical reactions so that large changes in such ratios imply that other processes have been at work.

Probably the must useful gases for such isotope studies are the **noble gases,** which are chemically inert. Because of their atomic structure, atoms of these gases cannot normally join with other atoms to make compounds; they cannot even combine with themselves the way nitrogen and oxygen do. Because they do not become trapped in chemical processes, the heavier noble gases — neon, argon, krypton, and xenon (from Greek words meaning new, lazy, hidden, and strange) — provide an important tool for probing the history of planetary atmospheres.

Argon is particularly revealing. We saw in Section 7.5 that this gas is the third most abundant in the terrestrial atmosphere. One isotope of argon with mass 40 (Ar-40) is produced by the decay of radioactive potassium, as we noted in the discussion of the radioactive age dating in Section 3.3. The other stable isotopes of argon with masses 38 and 36 are primordial, that is, they were created in the interiors of stars along with the carbon, nitrogen, oxygen, and hydrogen of which you are mostly made.

The existence of both radiogenic and primordial isotopes of argon provides a useful tool for studying a planet's history. Once an argon atom of either form is released from the interior and enters the atmosphere it remains, being too heavy to escape and too inert to combine chemically with surface rocks. Radioactive potassium in these rocks is constantly decaying, producing Ar-40 at a predictable rate. If the rocks are melted or heated sufficiently, the newly produced argon will be released into the atmosphere. Making the reasonable assumption that the distribution of potassium is relatively similar from one planet to the next, the amount of Ar-40 in an atmosphere today gives us an estimate of the vigor of crustal degassing over the lifetime of a given planet. Let's apply this analytical tool to Earth and Venus.

We found that the quantities of carbon dioxide and nitrogen degassed from the interiors of these two planets were generally similar, with somewhat greater amounts on Venus. The Ar-40 reverses this trend: six times less of this isotope occurs in the atmosphere of Venus than in ours. Using our previously established logic, we conclude that there has been less degassing on our sister planet than on Earth in relatively recent times. How can we be sure? The carbon dioxide, nitrogen, and primordial isotopes of argon on Venus today could all have been released within the first billion years of the planet's history. But not enough Ar-40 had been produced at that time to equal the amount in our own atmosphere.

The half-life for decay of radioactive potassium is 1.5 billion years. It is a slow process, leading to the gradual accumulation of Ar-40 in Earth's atmosphere, since degassing on our planet has been fairly continuous. The low abundance of the argon daughter in the atmosphere of Venus suggests that early geologic activity, during which the nitrogen, carbon dioxide, and other noble gases were released, has been followed by billions of years of relative inactivity. There must be more Ar-40 trapped in the planet's crust and mantle waiting to escape. The fact that it hasn't suggests that the crust of Venus has indeed become frozen at an early stage of its evolutionary development.

What about the planet's internal heat? The thermal energy generated by the decay of potassium, uranium, and thorium still must find a way out, and by analogy with the Earth we would expect this internal activity to release the recently formed Ar-40 into the atmosphere. Hot plumes rising through the mantle to form huge volcanic constructs on Venus's surface may provide a way for heat to be released without widespread surface melting and release of argon. The high temperature of the surface itself, a product of that immense greenhouse effect, should also improve the thermal conductivity of the crust allowing a more efficient release of heat from below without volcanism. Identifying the processes responsible for determining the heat balance of

Venus promises to be an important area of research on that planet in the years ahead.

————————⬦————————

Summary

Similar to Earth in size and bulk composition, Venus is dramatically different in many large-scale characteristics. It rotates slowly and backwards on an axis nearly perpendicular to its orbit. Its surface temperature is an incredible 730 K, at a pressure of 90 bars. The immense atmosphere that produces this high surface pressure is made mainly of carbon dioxide with a small amount of nitrogen and argon and traces of other gases such as sulfur dioxide, hydrochloric acid, and hydrofluoric acid. This atmosphere maintains the high surface temperature through a greenhouse effect. Some visible light from the Sun penetrates the thick sulfuric acid clouds and reaches the planet's surface, but infrared radiation from the surface is absorbed by the carbon dioxide, which heats the atmosphere and the entire planet.

The circulation of this atmosphere is considerably different from ours, owing to the planet's slower, retrograde rotation and greater proximity to the Sun. Hadley circulation, a simple equator-to-pole exchange, seems to dominate the flow. The thickness of the atmosphere maintains a nearly constant temperature over the entire surface of the planet.

The surface of Venus has been explored on large scales by radar. The topography emerging from these studies is very different from that of the Earth or the Moon. Well defined impact craters and many volcanic features occur. However, no large continents similar to ours exist. It is not yet clear whether large-scale plate tectonics shaped the chains of mountains found on Venus, or whether some other process unfamiliar to us on Earth has been at work. A variety of evidence suggests that active volcanoes may be present on Venus, but again certainty is lacking.

Mineralogical studies of surface rocks have been carried out by the Soviet Venera landers. The results indicate the presence of basalts and granite at those sites that have been sampled. This evidence for differentiation implies the existence of a dense, metal-rich core, but the magnetic field that such a core could be expected to generate does not exist, perhaps because of the planet's slow rotation.

Many of the striking differences between Venus and Earth can be explained in terms of our sister planet's closeness to the Sun. At the distance of Venus, water cannot remain on a planetary surface. A runaway greenhouse effect will occur, ultimately leading to the breakdown of water molecules in the planet's upper atmosphere by solar ultraviolet light. In the absence of water, carbon dioxide will remain in the atmosphere instead of forming carbonate rocks as it has on Earth. Indeed, our own planet would have an atmosphere nearly as dense as that of Venus if all the terrestrial carbonate rocks surrendered their CO_2. Without water, it also seems likely that mantle convection will be inhibited, choking off the development of the crust at an early stage. Independent evidence for this situation on Venus is provided by the relatively small amount of argon-40 in the planet's atmosphere. Evidently Venus has not outgassed as continuously as has the Earth.

————————⬦————————

Key Terms

gravity anomaly

greenhouse effect

Hadley cell

imaging radar

noble gas

photochemistry

runaway greenhouse effect

tectonic

PART FOUR
REVIEW QUESTIONS

1. In simple terms, how does the Earth rate as a planet? Compare its size, density, composition, and other basic properties with those of its neighbors. Would you expect the study of other planets to help us understand our own?

2. Distinguish between evolutionary processes and cyclic processes. Give examples of each. Why are cyclic processes more common on the Earth than on the Moon? What is the relative importance of these two kinds of processes on Venus?

3. Compare the interior structures of the terrestrial bodies studied so far: Earth, Venus, Mercury, and the Moon. Consider also what we know about their thermal evolution. How do structure and evolution depend on the size and composition of a planet?

4. Compare the atmospheric structures of the Earth and Venus. Does the atmosphere of Venus have the same regions as were identified on Earth: troposphere, stratosphere, ozone layer, ionosphere, magnetosphere? Why are there differences between the two planets?

5. Describe how the geologic time scale is determined for the Earth. There are some people today who argue that the Earth is only ten thousand years old; how would you test this claim? Compare the terrestrial geologic time scale with that of the Moon.

6. Explain how plate tectonics works: describe how this process gives rise to mountains, volcanoes, deep-sea trenches, and earthquakes. What does plate tectonics tell us about the fate of deep-sea sediments, including the carbonates that trap most of the carbon dioxide that would otherwise be in our atmosphere?

7. The change in geological thinking that resulted from the acceptance of plate tectonics and continental drift was one of the major scientific revolutions of the twentieth century. Why do you think these ideas were so slow in being widely adopted? You may wish to compare this revolution with others of the past century: Darwin's discovery of the role of natural selection in biological evolution, Pasteur's proof of the role of germs in causing disease, the insights into human psychology provided by Freud, and the revolution in

physics represented by relativity, quantum mechanics, and the discovery of the nature of the atom.

8. Compare the ocean crust and the continental crust, in terms of composition, origin, and evolution. Are these two divisions of the terrestrial crust at all analogous to the division of the lunar crust into highlands and maria? Explain.

9. Discuss the chemical balance that exists between the land, the ocean, and the atmosphere. Imagine what would happen to the other two if any one of these were dramatically changed.

10. Describe the coupling between life and the environment. How has the existence of life produced the special properties of the Earth's atmosphere? Do you find the Gaia hypothesis useful or do you feel it goes too far in its treatment of the Earth as a living organism?

11. Consider the role of impacts in influencing the history of the Earth. What would conditions have been like during the period of terminal heavy bombardment? Could life have existed then? Explain what a mass extinction is. Consider the effect of the impact of a 10-km asteroid on the Earth today, either on land or in the ocean. Would humans survive or would we suffer the fate of the dinosaurs?

12. Compare the atmospheric circulation patterns of Venus and the Earth. What are the roles of solar heating and planetary rotation? How does the presence of water on Earth give rise to weather that would be impossible on Venus? What is the effect on atmospheric circulation of the much larger mass of the Venus atmosphere?

13. Explain the greenhouse effect. How do conditions in a planetary atmosphere actually compare with those in a greenhouse? Do you think the name is a good one, or would you prefer to drop it, as many people suggest? Why is the greenhouse effect so much stronger on Venus than on Earth?

14. Explain the runaway greenhouse. How does it differ from the ordinary greenhouse effect? If Venus once had oceans of water and later lost them through the runaway greenhouse what evidence today might reveal this past history? Consider both chemical and geologic clues.

15. Describe the surface topography of Venus as revealed by radar. Which features most resemble those on the Earth and the Moon? Which ones are unusual or even unique to Venus?

16. What is the evidence in favor of and against an active volcanism and plate tectonics on Venus? How convincing do you find the arguments?

17. At present the resolution of most radar maps of Venus is limited to a few kilometers, but in 1990 the Magellan mission should yield resolutions as high as 200 m. What would you look for in those radar images? What kinds of processes might be revealed that cannot be distinguished from the images we have now?

18. Compare the atmospheric evolution of Venus and the Earth. Are you satisfied that we understand why the two have diverged so strongly? How close did Earth come to following the route of Venus? Is there any possibility that our planet might still follow Venus' course?

ADDITIONAL READING

Chapman, C.R. 1983. "The Vapors of Venus." *Mercury 12*, 130.

Dewey, J.F. 1972. "Plate Tectonics." *Scientific American 227*:5, 56.

Goldsmith, D. 1985. *Nemesis: The Death Star and Other Theories of Mass Extinction*. New York: Walker and Co.

*Harland, W.B. *et al.* 1982. *A Geologic Time Scale*. Cambridge: Cambridge University Press.

*Hunten, D.M. *et al.*, eds. 1983. *Venus*. Tucson, AZ: University of Arizona Press.

Margulis, L. 1984. *Early Life*. Boston: Jones and Bartlett.

Miller, R. 1983. *Continents in Collision*. Alexandria, VA: Time-Life Books.

*Murray, B., M.C. Malin, and R. Greeley. 1981. *Earthlike Planets*. San Francisco: Freeman.

Pollack, J.B. 1982. "Atmospheres of the Terrestrial Planets." In *The New Solar System*, 2nd ed., ed. J.K. Beatty, B. O'Leary, and A. Chaikin. Cambridge, MA: Sky Publishing Corp.

Raup, D.M. 1986. *The Nemesis Affair: A Story of the Death of Dinosaurs and the Ways of Science*. New York: W.W. Norton.

Revelle, R. 1982. "Carbon Dioxide and World Climate." *Scientific American 247*:2, 35.

Russell, D.A. 1982. "The Mass Extinctions of the Late Mesozoic." *Scientific American 246*:1, 58.

Sagan, C. 1983. "The Nuclear Winter." *Parade*, Oct. 30, 1983, pp. 4–8.

Schneider, S.H. and R. Londer. 1984. *The Coevolution of Climate and Life*. San Francisco: Sierra Club Books.

Scientific American, special Earth Science issue (1983), Vol. *249*, No. 3.

*Shoemaker, E.M. 1983. "Asteroid and Comet Bombardment of the Earth." *Annual Review of Earth and Planetary Science 11*, 461.

Siever, R. 1975. "The Earth." *Scientific American 233*:3, 82.

*Silver, L.T. and P.H. Schultz, eds. 1983. *Geological Implications of Impacts of Large Asteroids and Comets with the Earth*. Boulder, CO: Geological Society of America.

Sullivan, W. 1985. *Landprints*. New York: Times Books.

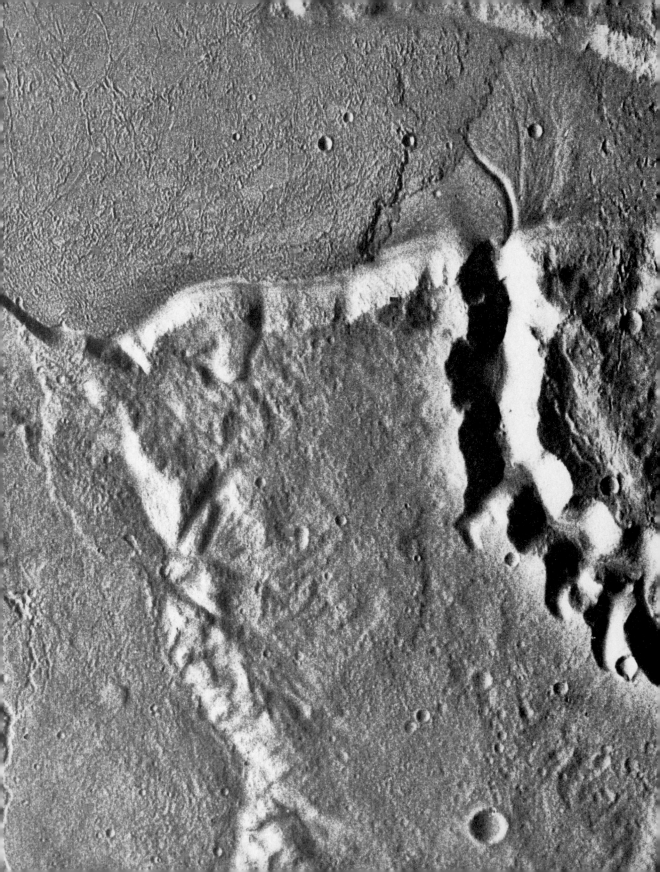

PART ◆ FIVE
Tantalizing Mars

Out of all the planets, Mars has the most romantic appeal. The Moon is closer, Venus brighter, Jupiter larger, and Saturn more beautiful, but Mars remains the one other place with which humans can identify. In the past, this planet was thought to harbor life, even intelligent creatures with civilizations in advance of our own. Now, even though we suspect that Mars may be biologically dead, we recognize its potential both as an exciting target for exploration and as a possible site for human colonization in the twenty-first century.

Today, Mars is a forbidding world. With only one-tenth the mass of the Earth or Venus, it has reduced levels of geological activity and, consequently, less outgassing of volatiles from the interior. Being farther from the Sun, Mars is also noticeably cooler. It is now a planet caught in a terminal ice age, with most of its water frozen in subsurface permafrost. Yet conditions in the past may have been much less hostile.

◀ A potential landing site near Mangala Vallis on the surface of Mars. The lower part of the picture is of older, cratered terrain, cut by a sinuous channel that opens onto a flat plain. A much narrower channel (about 500 m wide) extends out into the plain, where it is buried by a lava flow from the Tharsis volcanoes. At least two flow fronts are visible to the left of the narrow channel, beyond which they have filled the visible plains area, lapping up against the base of the cliffs. A roving vehicle deployed at this site could collect samples from a variety of different terrains for later shipment to Earth.

Its many resemblances to Earth and its evidence of very different past climates make Mars a fascinating planet. Great volcanoes once erupted vast amounts of lava and, presumably, released water and other gases from the interior. River-like channels give testimony of a time when water flowed over the planet's surface and perhaps rain fell from the martian skies. Layered deposits in the polar regions indicate cyclical climatic variations that appear to continue today. Mars is the one place besides the Earth where we see a clear indication of climatic cycles in addition to long-term evolution of the surface and atmosphere. It is the only other planet where there is evidence that liquid water once flowed on the surface.

Although not as many robot spacecraft have been sent to Mars as to Venus and the Moon, the return from these spacecraft may have been higher. The U.S. Viking orbiters and landers that explored Mars in 1976 represented the high point of our planetary program, and most of the material in Chapters 9 and 10 is derived from the Viking mission. Mars is intrinsically a much more interesting place than the Moon, without presenting the formidable environmental challenges of Venus. Today both the United States and the Soviet Union are directing studies toward new missions to Mars, leading perhaps toward human visits early in the next century. Thus Mars could become the object of a renewed space race, or if the nations of the Earth join forces, the exploration of this planet might become a model for international cooperation.

CHAPTER NINE

Mars: The Planet Most Like Earth

9.1 A Century of Changing Perceptions

Mars Through the Telescope

As seen from Earth, Mars follows a twenty-six-month cycle from one closest approach to the next. When the two planets are far apart, Mars looks like an inconspicuous red star, and even the best telescopes reveal no surface features (Plate 2). When they are close together, however, the distance from the Earth to Mars can be as little as 55 million km, and telescopes reveal features as small as 100–200 km across (Fig. 9.1). Note that this resolution is approximately the same as that of surface features on the Moon as seen with the naked eye.

As we noted in Section 5.1, a resolution of 100 km is insufficient to reveal *topographic* features on a planet. In the case of Mars, the situation is made worse by the fact that we always see the planet at nearly full phase. Recall how little topography is visible at full moon, even through a telescope. Thus the features that are seen telescopically on Mars are markings that represent different colors and reflectivities of surface materials. Primarily, we map dark regions and light regions, corresponding to reflectivities near 15% and 30%, respectively.

In addition to these relatively permanent light and dark surface markings, Mars displays bright white polar caps that grow and shrink with the seasons, just as one would expect for deposits of ice or snow. Transient bright clouds

FIGURE 9.1 Telescopic photos of Mars, such as these images obtained in 1971 with a 0.6-m telescope on Mauna Kea, have a best resolution of about 200 km. The major light and dark regions and the polar caps are prominent, but no topographic information is present in such pictures.

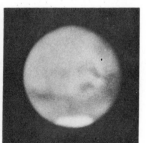

of yellow or white color are also seen. From their presence, the existence of a martian atmosphere was correctly deduced by visual observers in the last century. Not until the 1940s, however, when CO_2 was identified spectroscopically, was the presence of a specific gas detected.

The Canal Controversy

The history of martian studies notes well 1877, the year that the Italian observer Giovanni Schiaparelli recorded the linear markings he called *canali*, or channels. These faint dark lines, glimpsed near the limit of detectability, seemed to stretch for thousands of kilometers across the surface. In English-speaking countries, the name was translated as canals, a term implying construction by intelligent beings.

By the early years of the twentieth century, the conviction that the canals of Mars proved the existence of intelligent life on that planet was widespread. The most vocal advocate of this position in the United States was Percival Lowell, who for two decades dominated the American public image of astronomy. Most astronomers, in fact, could not see the canals, but not until after World War I did their skeptical viewpoint prevail.

The idea that Mars was peopled by technologically advanced creatures faded along with the canals. Mars as an abode of life, however, retained its appeal. The fact remained that many of the phenomena seen on this planet could be interpreted in terms of widespread plant life.

The Seasonal Cycle

Since the tilt of Mars' polar axis (25°) is nearly the same as that of the Earth, it experiences similar seasons, except they are roughly twice as long. In each hemisphere a polar cap forms each fall and winter under an obscuring layer of clouds, reaching its maximum extent (to latitude 65°) at the start of spring. It then recedes with the coming of warm weather, shrinking to a residual cap a few hundred kilometers across by the end of summer, when the cycle begins again. If the caps were composed of water ice, one would expect their retreat to release water, either as liquid or vapor, which would be available to growing plants near the period of warmest weather.

Many twentieth century observers saw changes in the dark surface markings that seemed to be seasonal. Often the regions darkened in summer, as might be expected if life forms were abundant and were growing in response to water released by the cap. Some observers claimed that the darkening was accompanied by a change of color to hues of green. Perhaps most persuasive of all, the dark areas reformed after major dust storms as if new plants were pushing up through the layer of lighter material deposited by the storms. The circumstantial evidence seemed strong, and nothing was known about conditions on Mars that precluded hardy plant life, possibly similar to terrestrial lichens.

The First Flybys

Shortly after the successful Mariner 2 flyby of Venus, a similar spacecraft, Mariner 4, was launched toward Mars. The main difference between the two was that among its instruments Mariner 4 carried a simple television camera. On 15 July 1965, twenty-two close-up pictures of Mars were transmitted to Earth. They showed impact craters, superficially similar to those on the Moon (Fig. 9.2). Today, we expect to find craters on planets, but in 1965 these pictures sent shock waves throughout the scientific world. Geologists had hoped for a more active planet with valleys, mountains, plains, and perhaps a volcano or two. Scientists speculated that Mars was *geologically* dead, but newspapers missed the adverb, and it was widely reported that Mars was a dead world. The public image of Mars would never be the same again.

In 1969, two more advanced spacecraft, Mariners 6 and 7, flew past the planet photographing much of the surface at resolutions sometimes as high as half a kilometer. Again, they saw primarily cratered terrain, and by pure bad luck missed the huge volcanoes, canyons, and channels that we now know are there. Had we stopped after these three flybys, we would never have suspected the true geological complexity of Mars. Fortunately, plans for an orbiter were already well advanced, or the entire Mars program would probably have been terminated.

In addition to photographing the surface, these first flybys made many other measurements. The winter polar caps were determined to be made primarily of frozen carbon dioxide (dry ice) instead of water; an atmospheric surface pressure of less than 0.01 bar was measured; CO_2 was confirmed to be the major constituent of the atmosphere; and the absence of any measurable planetary magnetic field was established.

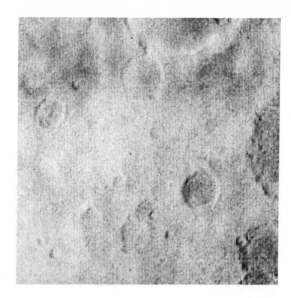

FIGURE 9.2 A small section of the surface of Mars as viewed by Mariner 4. The field of view is about 250 km for each side. Seeing the lunar-like craters in pictures such as this led several scientists to conclude that Mars was as geologically inactive as the Moon.

The Mariner 9 Orbiter

On 14 November 1971, Mariner 9 became the first spacecraft to go into orbit about another planet. Initially the results were very disappointing. The spacecraft had arrived in the midst of a planet-wide dust storm that obscured any view of the surface; had the mission been a flyby, it would have been an embarrassing failure. This time, however, the mission controllers at the Jet Propulsion Laboratory in Pasadena, California, could afford to wait. Over the next weeks the dust cleared, and soon a new world was being revealed to Mariner 9's cameras.

In January 1972 a ten-month program began to map the entire martian surface at a resolution of about 1 km. One remarkable feature after another was discovered: the largest volcano in the solar system, the greatest canyon system, and even vast drainage channels that dwarfed terres-

trial river systems. Repeated radio probes were also made of the structure of the martian atmosphere.

Scientists were ecstatic about the Mariner 9 results; however, it was not easy to communicate their findings to the public. The press, oriented toward instant news, reported only the problems encountered during the dust storm that was raging when the spacecraft arrived at Mars. Newspapers carried featureless pictures that showed only the pall of dust. A few months later, when spectacular pictures were being sent back, most of the U.S. press refused coverage on the grounds that Mariner was no longer news. The other countries, however, did not share this narrow perspective, and many of the Mariner discoveries reached audiences abroad while the people in the United States remained ignorant of the advances being made.

Viking

The next step beyond orbital surveys is to land on the planetary surface. An ambitious Mars lander had been planned since the late 1960s, based on the huge Apollo Saturn V rocket. When the Nixon administration cut back the space budget, this program was scaled down to a modest pair of landers and orbiters, launched by the Titan/Centaur rocket (Fig. 2.17). Arriving at Mars in June 1976, the first Viking had been designated for touchdown on 4 July, the 200th anniversary of the Declaration of Independence.

The Viking orbiters photographed the surface at higher resolution (typically 100 m) than had been achieved by Mariner 9, revealing unexpected hazards in the planned landing site. The landing was postponed day by day while frustrated Viking scientists searched for a smoother place to land. Finally, on 20 July the Viking 1 Lander successfully made its atmospheric entry and settled onto a rock-strewn surface in Chryse Planitia, the Plains of Gold (Plate 10a) exactly seven years after Neil Armstrong took the first human step on the Moon. Two months later the second lander touched down in a region called Utopia (Fig. 9.3; Plate 10b).

All four Viking spacecraft — two landers and two orbiters — were spectacularly successful. This time the public also followed the exciting developments as the first pictures were returned from the surface and the search for life on another world began. The *New York Times*, for example, eventually wrote nine separate editorials in praise of Viking.

After Viking, the U.S. suspended its Mars program. The reasons are complex, with the primary problem being a general decline in funds for planetary exploration. The disappointment that Viking did not find life was also a major negative influence. Although in 1986 a presidential commission recommended that the United States set as a major goal of its space program the eventual human exploration of Mars, at this writing it is not clear if the nation will respond to this challenge. A small Mars orbiter equipped primarily for geochemical mapping has just been approved for launch in 1992. This new mission will possibly become the vanguard of a new United States initiative for Mars exploration. The USSR has already begun a program of missions leading to a proposed sample return from Mars before the year 2000.

FIGURE 9.3 A close-up view of Utopia from the Viking 2 Lander. One of the footpads of the lander is visible in the lower right corner of the frame. Note the many holes in the nearby rocks, indicating that they were full of gases when they were molten.

♦

9.2 Global Perspective

Bulk Properties of Mars

Mars is a mid-sized terrestrial planet. Its diameter of 6787 km is just over half that of the Earth, resulting in a surface area almost exactly equal to that of the continents on our planet. The mass of Mars is 11% that of the Earth, or about nine times that of the Moon. Table 9.1 summarizes the properties of Mars, in comparison with Earth and Moon.

Mars' density is also intermediate, at 3.9 g/cm³. When corrected for the weight of the planet, the uncompressed density of Mars is 3.8 g/cm³. Clearly, this planet must be deficient in metal or more enriched in light elements relative to the Earth, whose uncompressed density is 4.5 g/cm³. Like the Moon, Mars is not expected to have a large iron core. Detailed tracking of orbiting spacecraft has demonstrated, however, that some sort of core is present. It is thought to consist primarily of iron sulfide (FeS) and to have a diameter of perhaps 2400 km, or 40% of the diameter of the planet. If so, the cores on Mars and Earth take up a similar volume relative to the size of the planet, with the main difference being that Mars' core contains a larger proportion of FeS, a lower-density material than metallic iron.

As a differentiated planet, Mars has a mantle and a crust in addition to its iron sulfide core. Little is known about the properties of these layers, however. Although the two Viking landers each carried seismometers in the hopes of probing the interior, no marsquakes were detected. One seismometer experienced mechanical failure, while the other proved incapable of distinguishing between seismic activity and trembling of the spacecraft caused by wind.

Magnetosphere

Calculations suggest that the core of Mars is solid, not liquid. If so, it should come as no surprise that Mars lacks the strong magnetic field that is generated on Earth by convective motions in its spinning liquid iron core.

The only spacecraft that have actually searched for a magnetic field near Mars are Mariner 4 and several early Soviet orbiters. The results have been essentially negative. The maximum surface magnetic field inferred from the Mariner 4 data is less than 0.1% that of the Earth. Thus Mars, like Venus, has an insufficient intrinsic field to repel the charged atomic particles of the solar wind. There is no magnetosphere, except for the very small one created by the solar wind itself as it sweeps past the

TABLE 9.1 Comparison of Earth, Moon, and Mars

	Earth	Mars	Moon
Diameter (km)	12756	6794	3476
Mass (Earth = 1)	1.0	0.107	0.0123
Density (g/cm³)	5.5	3.9	3.3
Uncompressed density (g/cm³)	4.5	3.8	3.3
Surface area (Earth = 1)	1.0	0.28	0.07
Atmospheric surface pressure	1.0	0.006	0.000

planet. Venus, Mars, and the comets are all examples of objects with atmospheres that interact directly with the solar wind.

Surface Nomenclature

A century ago, telescopic observers gave classical names (such as Hellas, Utopia, and Arabia) to the fixed light and dark features on Mars, which they originally thought to be deserts and seas, respectively. Before the first spacecraft visited Mars, scientists engaged in endless debates as to the nature of these features: for instance, whether they represented highlands and lowlands as do the lunar markings. To the surprise of many scientists, most of these markings actually have little to do with elevation or any other topographic property. Thus a new nomenclature was needed to deal with the topographic features photographed by Mariner 4 and its successors.

The spacecraft images revealed flat plains, mountain-ringed basins, canyons, valleys, volcanoes, and a great many impact craters. Following lunar convention, the craters were named for past scientists and others associated with the study of Mars. The great canyon system became Valles Marineris, named for the spacecraft (Mariner 9) that discovered it. (Actually, this feature had originally been seen from the Earth and was called Coprates canal, one of the few canals that turned out to be real.) The river-like channels were given the names for the planet Mars in languages from all over the world: for example, Kasei Vallis (Japanese) and Nirgal Vallis (Assyrian).

Some large-scale topographic features on Mars correspond approximately to spots identified from Earth and given classical names. The largest martian volcano often collects bright clouds around it, and these clouds seen telescopically had been called Nix Olympica, the snows of Olympus. The volcano was therefore called Olympus Mons, Mount Olympus. Similarly, the largest martian impact basin also collects clouds in its interior and is one of the most prominent bright regions seen from Earth: Hellas. This name was applied to the basin itself, once its true nature had been established. A major lowland plain in the northern hemisphere had appeared on some old maps as Chryse Regio, or region of gold; it now became Chryse Planitia, plains of gold. In this way, much of the older nomenclature was preserved.

Surface Elevations

In an overview of a planetary surface, we begin with the largest-scale features, such as the continents and ocean basins of Earth, or the highland and maria regions of the Moon. Thus we wish to know surface elevations. On Mars, these elevations are measured relative to the altitude at which the average atmospheric pressure is 0.0061 bar. This value was picked because it is the triple-point pressure of water, the pressure above which it is possible for water to exist in liquid form. At its triple point, water can exist simultaneously as a solid, liquid, and gas provided the temperature is 0 C. Increasing the temperature even slightly at this low pressure causes the water to boil; lowering the temperature leads to freezing.

The range in elevations on Mars is large, greater than that on any planet so far discussed. Four volcanic mountains rise to a height of 27 km above the reference level, while the lowest region, the Hellas basin, drops to 4 km below the reference. Thus the total range in elevations on Mars is 31 km, compared with about 20 km on both the Earth and Venus (Fig. 9.4).

If you guessed that it is no coincidence that Venus and Earth have the same elevation range, in spite of the differences in their geology, you would be right. The explanation lies in the balance between the gravitational pull of a planet and the strength of its crustal rock. On planets

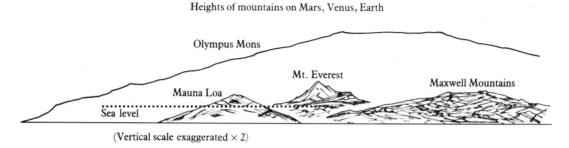

Heights of mountains on Mars, Venus, Earth

(Vertical scale exaggerated × 2)

FIGURE 9.4 A comparison of elevations on Venus, Earth, and Mars. The individual features themselves may be seen in Plates 3 and 12 and Figs. 8.13, 7.6, and 9.16.

the size of Earth and Venus, a mountain can rise only about 20 km before its rock begins to deform under its own weight. A larger structure is simply unable to support itself.

If this self-limiting process is what determines the maximum elevations of Venus and Earth, we can understand why the martian mountains are higher. The surface gravity on Mars is two-fifths that of Earth or Venus, so its mountains can grow higher before they begin to sag under their own weight.

Large-Scale Topography

Fig. 9.5 is a map of Mars on which elevations have been indicated, measured in kilometers above (+) or below (−) the reference level. Two features are outstanding: a north-south hemispheric asymmetry, and the presence of a huge bulge near the equator at longitude 100°.

Most of the southern hemisphere of Mars lies above the + 3 km contour, while much of the northern hemisphere is below the reference level. In many places, the boundary between the two elevations is relatively sharp, with the surface dropping by 4 km in the span of a few hundred kilometers distance. A number of other properties of Mars also differ in the two hemispheres: in particular, the higher southern re-

gions are more heavily cratered (hence older), and most of the dark regions visible in Earth-based telescopes are in the south.

The difference between the northern and southern hemispheres of Mars is one of the most fundamental and mysterious aspects of the planet. If the heavily cratered highlands of the south represent the older crust of the planet, then something peculiar has happened in the north, lowering the elevation by several kilometers and reworking the surface by volcanic or other processes.

The second major feature of the large-scale topography of Mars is the Tharsis bulge, a volcanically active region the size of North America that rises to about 10 km above the reference level. Tharsis straddles the boundary between the southern highlands and the lower northern plains. Being the least cratered, it is the youngest part of the planet's surface. To its east lies Chryse Planitia, which drops to an elevation of − 3 km, while to the west is the shallower depression of Isidis Planitia.

Geologic Features

Near the top of the Tharsis bulge are three great volcanoes, each rising 18 km to a summit at an elevation of 27 km. A still larger volcano, Olympus Mons, rises 25 km above the northwestern

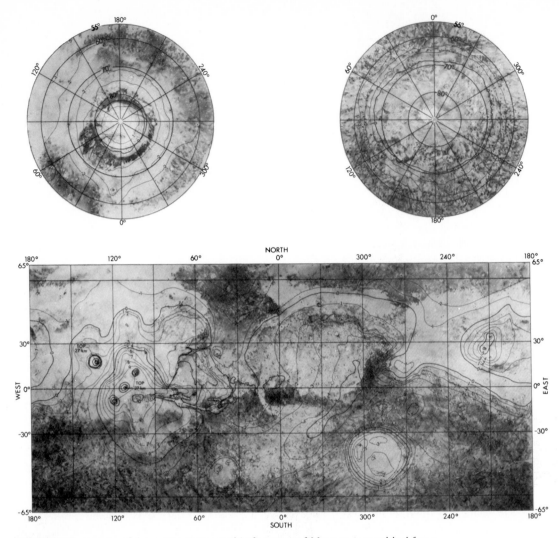

FIGURE 9.5 A map of the major topographic features of Mars as assembled from
Mariner and Viking data. Note especially the Tharsis bulge near longitude 100° and the
prominent boundary between the heavily cratered southern highlands and the smoother
northern plains.

slope of Tharsis to reach the same height of 27
km above the reference level. Olympus Mons is
nearly 700 km across at its base, about the size
of France (Plate 12).

Associated with the Tharsis bulge is the
Valles Marineris, the central part of an intercon-

nected system of east-west running canyons approximately 4000 km long. The individual canyons are about 3 km deep, but in the center of
the Valles Marineris the depression is nearly 7
km and the width of the canyon more than 500
km. Major terrestrial valleys, such as the Grand

Canyon in Arizona, could easily fit into one of the tributaries of this canyon system.

Geologists have concluded that the volcanoes, the east-west canyons, and the Tharsis bulge are all expressions of a major, long-lived center of tectonic activity that has dominated this one side of Mars. We will explore these ideas further in the following parts of this chapter.

The Surface Material

Mars is the Red Planet, but what accounts for its surface color? The Viking landers have confirmed what was suspected for decades: the color of Mars is due primarily to iron oxides in the surface soil. The soil itself is largely fine-grained material, part of which can be raised into the atmosphere and transported for large distances during major dust storms. Chemical analyses carried out by the Viking landers showed that martian soil is similar in composition to some iron-rich clays found on Earth.

Most of the fine dusty material appears to be fairly light in color, like the sand in the deserts of the Earth. During dust storms, this light material is deposited over the surface, obscuring underlying rock, much of which is darker. Thus after a major storm period the overall reflectivity of the martian surface is high. Subsequently, wind redistributes the lighter material, and part of the darker underlying terrain reappears.

This process of wind-blown dust, distributed over the surface by seasonal circulation patterns of the atmosphere, accounts for the seasonal changes in the surface markings recorded by generations of telescopic observers. Among the first scientists to advocate this non-biological explanation for the surface changes were Carl Sagan and James Pollack. It is somewhat ironic that Sagan, who is so closely associated with the search for life on other planets, should have played this role in the history of martian studies.

9.3 View from the Surface

The Viking Landers

The Viking landers were each self-contained laboratories weighing about a ton and similar in size to a subcompact car. Powered by a radioactive electric generator, each spacecraft could communicate with Earth either directly or via one of the orbiter spacecraft. The scientific experiments on board included those aimed toward a general examination of the immediate environment (cameras, chemical analysis devices, meteorological weather station, seismometer, and a 3-m-long mechanical arm to probe and manipulate the soil), as well as a complex biology package to look for evidence of microscopic life on Mars. In addition, a special group of instruments investigated the atmosphere of Mars during spacecraft entry and descent.

The twin cameras on each Viking lander could record pictures in either color or monochrome, and operating together they provided stereoscopic views of the surface. The weather station measured atmospheric pressure, temperature, and wind speed and direction. Chemical analysis of soil was carried out with an X-ray spectrometer. An even more powerful general analysis device, the GCMS (discussed in more detail in Chapter 10), could measure the composition of the atmosphere or of volatile components in the soil, such as organic molecules.

Each Viking lander was operated by controllers on Earth who could command it to poke and dig in the soil, push rocks, pick up and deliver samples to its various instruments, and take a wide variety of pictures of its surroundings. The data these spacecraft returned provide us our best impression of the appearance and feel of the surface of a planet where no humans have yet walked. Unfortunately, however, the Viking landers lacked mobility. Fixed at the spots where they landed, they could not show us the view over the next hill, or even behind a nearby rock.

Thus the choice of the two places on Mars where the spacecraft would land was extremely important.

Selection of the Viking Landing Sites

The one imperative in selecting landing sites on Mars was safety. Therefore, just as had been the case for the first Apollo Moon landings, every effort was made to find dull, flat locations for Viking. This meant avoiding all the remarkable geologic features found by Mariner 9: no sites were considered involving volcanoes, canyons, channels, large impact craters, or anything else of geological interest. In addition, it was necessary to select sites at relatively low elevations, to ensure sufficient atmosphere to support the descent parachutes. The only places meeting these criteria were the lowland plains of the northern hemisphere.

The site selected for the Viking 1 landing was in Chryse Planitia, the depression to the east of the Tharsis bulge. In Mariner 9 images this region appeared flat and featureless, but the higher-resolution pictures from the Viking 1 Orbiter revealed a much more complex topography, including features that appeared to be the result of water erosion. The Viking Project Manager rejected the preplanned site and began a search for a smoother area. It required nearly a month to identify a less rugged looking alternative, still in Chryse but several hundred kilometers to the northwest of the original site (22° N, 48° W).

Viking 2 was aimed for a more northerly latitude of 44° where seasonal effects might be stronger. Again, the site originally selected turned out to be unexpectedly rough, and a search was made for a smoother area at the same latitude. A spot in Utopia, at longitude 226° W, was ultimately found for the September 3 touchdown.

The Plains of Gold

The Viking 1 site in Chryse is illustrated in Fig. 9.6. This orbiter view shows a mare-like plain with a crater density suggestive of an age of 2 to

FIGURE 9.6 The site of the Viking 1 landing in Chryse, a low basin characterized by wrinkled ridges similar to those of the lunar maria.

FIGURE 9.7 The view from the Viking 1 Lander showed a desolate plain with wind-sculpted hills and numerous rocks of all sizes.

3 billion years. Like the lunar maria, this area is thought to be volcanic, but with a surface possibly modified by later floods of water (Section 9.6) as well as by wind erosion.

As seen from the lander (Fig. 9.7; Plate 10a), Chryse is a desolate but strangely beautiful landscape not very different from some terrestrial deserts. The topography is gently rolling, and the surface is thickly strewn with rocks that range in size from golf balls to boulders. The absence of smaller rocks distinguishes this site from those on the lunar maria (Fig. 5.7); the soil between the rocks is fine-grained and has a relatively hard, crusty surface.

The ubiquitous rocks in the Viking 1 site appear to be volcanic in origin, most likely ejecta from impact craters like their lunar counterparts. The scene has been modified by erosion, particularly by wind-blown dust. Erosion takes place slowly, however, and only a sharp-eyed observer can see any differences between photos taken years apart at the Viking 1 site.

Utopia

The landing site of the second Viking is shown in Fig. 9.8. Although the intent was to touch down on an area with a smooth wind-deposited surface, it appears that the spacecraft landed instead in ejecta from the 90-km-diameter crater Mie, located about 200 km to the east. As seen from the lander (Fig. 9.9; Plate 10b), this site is even rockier than Chryse. Indeed, we are fortunate that the lander was not damaged by a protruding boulder as it settled to the surface.

Viking 2 found itself in a very flat area, lacking the low hills that make the Chryse site more attractive. About all there is to see are the rocks. These are interesting enough to the geologist, however, with their distinctive angular shapes and wide variety of colors and textures. Most are heavily pitted, perhaps an indication of their volcanic origin. The consensus among geologists is that the lander is on a debris blanket from Mie. The finer material that must once have been present in this ejecta blanket has been stripped away by wind, leaving the fragmented rocks.

While every effort was made to land both Vikings in flat regions with smooth surfaces, it is apparent from the pictures that this goal was not achieved. Both sites are very rocky, with evidence of wind erosion. Presumably the fine material swept clear from these areas is deposited somewhere else, perhaps near the poles, where large fields of sand dunes have been photographed from orbit (Section 9.7). In any case, it may be that the two Viking sites are more typical than had been expected and that a great deal of the surface of Mars consists of rock-strewn plains similar to these.

FIGURE 9.8 The Viking 2 landing site in Utopia, a more northerly plain that from orbit appeared to be mantled in wind-driven dust.

The Thomas Mutch Memorial Station

Each Viking lander had a nominal lifetime of only ninety days, sufficient to complete the life-detection experiments that will be discussed in Chapter 10. However, scientists and engineers alike hoped the landers would survive much longer on the martian surface, and they were not disappointed.

Viking Lander 2 operated in the Utopia plain until 11 April 1980, spanning two martian years. In spite of its far northern latitude, the vehicle survived two cold martian winters, providing unique meteorological information (Section 10.1).

The first lander performed even better, and at the end of the regular Viking mission it was still in excellent condition. NASA then began a "Viking Survey Mission," in which the lander was placed into an autonomous operating mode, returning data to Earth only about once a month. This lander was designated the Thomas Mutch Memorial Station, in honor of the Viking Imaging Team Leader.

FIGURE 9.9 The Viking 2 Lander touched down in a plain heavily littered with rocks, probably derived from a nearby impact crater. There was no evidence of the mantle of dust that had been expected.

Thomas ("Tim") Mutch (Fig. 9.10), a geologist from Brown University, made the transition from studying the Earth to the Moon (during the Apollo era) and from there to Mars. In 1979 he accepted the highest-level science-management post at NASA, becoming Associate Administrator for Space Science. A little over a year later, Tim Mutch was killed in a mountain climbing accident in the Himalaya Mountains.

It was hoped that the Mutch Memorial Station could continue operations into the 1990s, but it fell victim to a human programming error that misdirected its antenna away from the Earth, breaking the vital communications link with controllers at the Jet Propulsion Laboratory. The last data from Mars were received on 5 November 1982, a total of 6 Earth years and 106 Earth days after landing.

In 1984, NASA formally transferred ownership of the Viking 1 Lander to the Air and Space Museum of the Smithsonian Institution. It is the only museum exhibit on another planet. Someday, perhaps, it will be brought back to a place of honor in Washington, next to the Wright brothers' first airplane and the Apollo 11 spacecraft. Or perhaps it will remain on Mars, for the benefit of the first human colonists and the generations that succeed them.

FIGURE 9.10 Thomas (Tim) Mutch was the highly popular leader of the Viking Lander Imaging Team.

It is appropriate to end this section with two quotes from Tim Mutch, one made at the time of the Viking 1 landing, the other after he became NASA Associate Administrator:

"Touchdown minus seven, six, five, four, three, two, one, zero." There was a long silence. It seemed like minutes. Over the loudspeaker came a muffled prayer, "Come on, baby." Nothing. I looked down at my shoes. I remember thinking, "You always wondered what a failure would feel like. Now you know." I mentally composed some remarks for friends standing nearby. Finally, the silence was broken. "We have touchdown!"

Even if the immediate future is uncertain, I have no doubts about the distant years. Some day man will roam the surface of Mars. Those wonderful Viking machines will be crated up, returned to Earth, and placed in a museum. Children in generations to come will stand before them and struggle to imagine the way it was on that first journey to Mars.

9.4 Craters and the Age of the Martian Surface

The Martian Craters

Like the Moon and Mercury, Mars has thousands of impact craters, concentrated particularly in its southern hemisphere. Generally speaking, these craters look much like their counterparts on airless planets. They have the same sort of raised and terraced rims, flat depressed floors, and often central peaks. Presumably they were formed by the same impact processes.

Although the craters themselves look similar, the patterns of ejecta around many martian craters are unique. As we saw in Section 5.3, the typical large lunar crater is surrounded by a rough, hilly ejecta blanket close to the rim with radial streaks and chains of secondary craters farther out. In contrast, many martian craters have smooth ejecta blankets with well-defined edges. There are no extended rays or streaks of material. These forms have been called **fluidized ejecta** craters or, more informally, splosh craters.

Three Types of Crater Ejecta

The martian crater ejecta falls into three main types. Most common is a multi-lobed pattern, called a flower form. Several ejecta layers are present, with each lobe bounded by a steep, smooth edge (Fig. 9.11a). These *flower craters* are found primarily in the equatorial regions (between 40° N and 40° S latitude). In the second type, illustrated in Fig. 9.11b, the ejecta looks like a relatively smooth pancake of material with a slightly scalloped rim. We call these *pancake craters*, and they occur primarily at higher latitudes. In the third type, most of the ejecta is missing, leaving the crater perched on a relatively small elevated platform. These *pedestal craters* are probably the result of erosive action by wind, which has stripped away the outer layers of ejected material.

Evidently, the martian ejecta blankets were formed by debris that flowed along the ground rather than following explosive aerial trajectories. It seems the material of the martian surface was fluidized in the cratering impact. This hypothesis of fluid flow also explains a number of cases where the ejecta flowed over and around obstacles. The presence of an atmosphere may also have played a role in confining the ejecta so that it flowed along the ground.

The most likely explanation for the fluidized ejecta of martian craters lies in the presence of large quantities of water in the surface at the time the impacts occurred. Probably, if martian conditions then resembled those we see today, the water was in the form of ice until melted by the energy of the impact. In contrast, water was absent from the dry surface materials of the Moon or Mercury. Possibly large impact craters on the Earth would look more like martian, rather than the lunar, craters, but the ejecta from large terrestrial craters have not survived to allow us to test this hypothesis.

FIGURE 9.11 The unusual ejecta blankets of martian impact craters probably result from the presence of frozen water (permafrost) in the soil. a) Yuty, an 18-km crater at 22° N, showing a central peak and several thin sheets of ejecta, with multiple lobes. b) Arandas, a 28-km crater at 43° N, with a thick platform of ejecta terminating in a flow front.

(a)

(b)

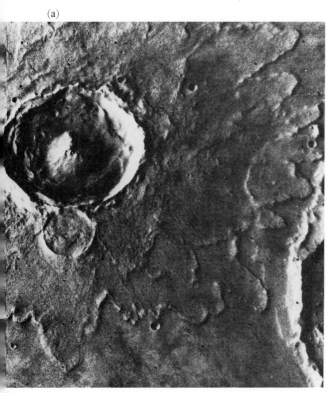 

Large Impact Basins

The largest martian basin is Hellas, about 1800 km in diameter and some 6 km deep. Unlike its lunar counterparts, Hellas has a simple form with a single rim of uplifted mountains. Its interior, which is frequently hidden by hazes of dust or water fog, appears to be relatively smooth, suggesting that the original floor has been buried by wind-blown dust. Parts of the rim mountains are missing. The second largest identified basin, Isidis, approximately 1200 km wide, is eroded even more than Hellas.

FIGURE 9.12 The best preserved large impact basin on Mars is Argyre, about 700 km in diameter.

The best preserved large basin on Mars is Argyre (Fig. 9.12). Its diameter is roughly 700 km. The rim consists of a broad rugged area more than 200 km in width, made up of many individual mountains thrown up in the basin-forming impact. The next largest basins are only about half as large as Argyre, making them really no more than large craters. An example is the 210-km-diameter crater Galle, shown in Fig. 9.12 lying on the rim of Argyre.

Mars has fewer large basins than the Moon, in spite of its much larger surface area. What does this mean? One possibility, reflecting a uniformitarian outlook, is that many of the basins that formed early in martian history have been degraded beyond recognition by subsequent geologic activity, perhaps including extensive erosion. Another way of stating this idea is that the martian crust stabilized later than that of the Moon, near the end of the period of terminal heavy bombardment. An alternative possibility, of course, is that there was no terminal heavy bombardment by large asteroids at Mars.

Age of the Martian Surface

When was the martian surface formed? This basic question is fundamental to understanding the history of the planet. Unfortunately, we have no returned samples with which to measure the solidification age (except perhaps the SNC meteorites, which we will discuss later). Thus any age estimate must depend on a hypothesis or model for the impact history of the planet.

In Section 5.4 we described the standard paradigm for the impact history of the inner solar system, based on the measured solidification dates and crater densities for the Moon. The lunar situation can be extended to Mars *if* the impacting bodies on both objects were members of the same population, represented today by the asteroids and comets. To the degree that we can extrapolate this current situation backwards in time, we can also use crater counts to date the martian surface.

Martian Cratered Terrain

A wide variety of crater densities exists on Mars. Some areas, such as the summits and slopes of the Tharsis volcanoes, have few craters. The crater densities in these areas are substantially less than on the youngest lunar maria. In contrast, the densities on the older cratered uplands are higher than on the lunar maria, although still less than those on the lunar highlands.

One fundamental difference between the Moon and Mars must be considered when interpreting crater densities: erosion. On the Moon, the only important agents that degrade old craters are further impacts. Mars, in contrast, has an atmosphere. Even today, the erosive effects of martian windstorms are significant, and in the past it seems likely that the atmosphere was thicker. At that time, even running water, with its major erosive effects, was present.

The effects of erosion are particularly evident in the heavily cratered uplands (Fig. 9.13), where the large craters are shallow and subdued. In these areas, there are many fewer small craters than would be expected from the numbers of larger ones present, as you can readily see by comparing Figs. 5.3, 6.8, and 9.13. Fig. 9.14 also displays this effect, plotting the density of martian craters of different sizes on both the uplands and the lower northern plains. The shortage of craters with diameters less than 5 km is attributed to enhanced erosion early in the history of Mars. The fact that smaller craters are not in short supply relative to large ones in the younger terrains suggests that this era of high erosion rates terminated billions of years ago.

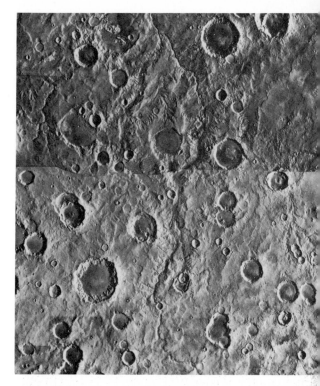

FIGURE 9.13 The cratered uplands that characterize most of the southern hemisphere of Mars. The craters are generally subdued by erosion, with a notable absence of small craters, relative to similar views of the Moon and Mercury. This frame is about 500 km across.

In the northern plains the density of craters is much lower than in the uplands, as shown in the crater counts plotted in Fig. 9.14. The best studied northern plains area is Chryse Planitia, the place where Viking 1 landed. Chryse is a flat basin of apparently volcanic origin (Fig. 9.6). For craters with diameters between 1 and 10 km, the crater density on Chryse is virtually identical to that of the average lunar mare, while for larger crater sizes the density on Chryse is somewhat greater. Similar crater densities also apply to the other northern lowlands.

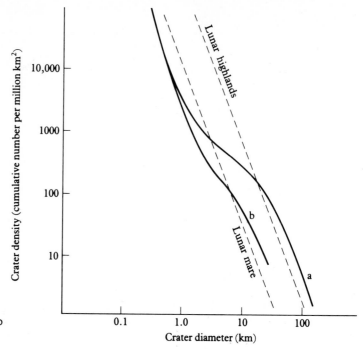

FIGURE 9.14 The distribution of crater sizes in the martian uplands and in the Chryse basin, compared with crater densities on the Moon. Note the depletion of small craters on Mars, presumably the result of erosion. (Legend: a = Average martian uplands; b = Chryse plain.)

Interpreting the Crater Densities

If the craters on Mars and the Moon are primarily caused by impacts with the same population of asteroids and comets, the cratering rates on these two bodies must have been roughly the same. To refine this result, actual impact rates must be calculated and an allowance made for the differing impact speeds on individual planets. Shown in Table 9.2 are results of one such cratering model, calculated in 1981 by William K. Hartmann of the Planetary Science Institute in Tucson. (Cratering rates are expressed relative to the Moon.)

Hartmann's model provides one system for dating different martian terrains from their observed crater densities, if we assume the lunar cratering history, including a terminal heavy bombardment. We will use these age estimates throughout this chapter when discussing various features on Mars. However, note that cratering models developed by other scientists yield ages that differ by as much as a factor of two for the younger volcanic terrains. All investigators agree, however, that the oldest terrains date from about 4 billion years ago. Table 9.3 summarizes the results for some representative surface features.

TABLE 9.2 Relative cratering rates for the terrestrial planets[a]

Mercury	Venus	Earth	Moon	Mars
2.0	1.0	1.5	1.0	2.0

[a]The higher value for Mars is due primarily to its proximity to the asteroid belt, while that of Mercury is a result of the Sun's gravity.

TABLE 9.3 Crater ages for martian surface features

Feature	Crater Density Relative to Lunar Mare	Crater Retention Age (billion years)
Olympus Mons volcano	<0.1	<0.2
Arsia Mons volcano	0.1	0.2
Tharsis plains	0.5	1.6
Elysium plains	0.7	2.6
Chryse plains	1.1	3.2
Alba volcano	1.8	3.5
Hellas basin	1.8	3.5
Cratered uplands	10	4.0

9.5 Volcanoes and Tectonic Features

Shield Volcanoes

The large volcanoes of Mars are concentrated in the Tharsis and Elysium regions, with a few additional older examples located in the southern hemisphere near the Hellas basin. Most of these resemble in shape terrestrial shield volcanoes like those in Hawaii (Fig. 7.6). A shield volcano is formed by the eruption of relatively fluid basaltic lavas, building up a massive dome with gentle slopes of only a few degrees. Most shield volcanoes on both Earth and Mars are also distinguished by large craters at the summit called calderas, produced by collapse of the surface when the underground pressure of the fluid lava is withdrawn.

The four highest volcanoes, all in the Tharsis region, rise to 27 km above the martian reference level. The story of their discovery is an interesting one. As we noted in Section 9.2, Mars was shrouded by a planet-wide dust storm when the Mariner 9 orbiter arrived in 1971. At that time, there was no reason to expect large volcanoes on Mars. As the dust began to clear from the upper atmosphere of Mars, the first surface features to appear were the tops of the four highest volcanoes, each with its large summit crater. Thus scientists faced an essentially featureless planet, with four faintly visible large craters.

On the basis of just this limited information, geologist Harold Masursky of the U.S. Geological Survey correctly reached the conclusion that Mars had volcanoes and that these volcanoes were much larger than their terrestrial counterparts. He reasoned that if this were not the case the coincidence of large craters being found at just the four spots that first emerged out of the dust was too unlikely. Subsequently, as the atmosphere cleared to reveal the great shield volcanoes, Masursky's insight was confirmed.

Tharsis and Olympus Mons

Each of the three Tharsis shield volcanoes is about 400 km in diameter, and all rise to the same height (Fig. 9.5). The southernmost volcano, Arsia Mons, has the largest caldera, 120 km in diameter (Fig. 9.15); the other two calderas are less than half as large. All three calderas are 3–4 km deep. Individual lava flows can be traced on the gently sloping flanks of these volcanoes, and counts of impact crater density on the flanks and in the calderas suggest that these volcanoes have been active within the past few hundred million years.

Olympus Mons, which rises from a lower area on the side of the Tharsis bulge, is the largest volcano on Mars (Fig. 9.16; Plate 12). In fact, it is the largest volcano in the entire planetary system. It has a diameter of 700 km, similar to the Maxwell Mountains on Venus. The entire chain of Hawaiian volcanoes, from Kauai to the Big Island, would fit within it. Olympus Mons

lies to the northwest of the main Tharsis bulge, and its summit is 25 km above the surrounding plains. The volume of this immense mountain is nearly one hundred times greater than that of the largest volcano on Earth, Mauna Loa.

Olympus Mons is a shield volcano, with a multiple caldera about 80 km wide (Fig. 9.17). The lava flows that compose its flanks slope downward at an angle of only 4 degrees. Individual lava channels about 100 m in width can be traced, together with many lava flow features that are virtually indistinguishable from their counterparts in the Hawaiian volcanoes.

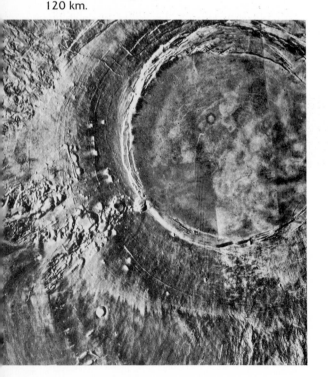

FIGURE 9.15 The west flank and caldera of Arsia Mons photographed by the Viking orbiters. This caldera is the largest on Mars, with a diameter of 120 km.

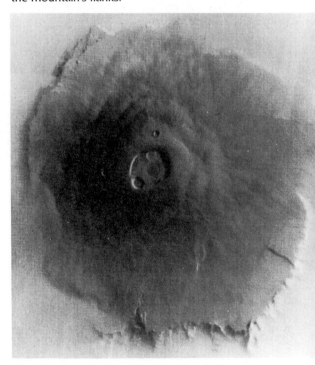

FIGURE 9.16 An aerial view of Olympus Mons, as seen from orbit by the Mariner 9 cameras. Compare this figure with the oblique in Plate 12, showing the summit caldera and clouds around the mountain's flanks.

FIGURE 9.17 The summit caldera of Olympus Mons is 80 km across, with multiple collapse features and surface cracks similar to those found on terrestrial shield volcanoes. Compare this view with the summit caldera on Mauna Loa on Earth (Fig. 7.7).

Other Martian Volcanoes

The four volcanoes we have been discussing are the largest on Mars, but even modest martian volcanoes are huge by terrestrial standards. Fig. 9.18 shows two shield volcanoes in the Tharsis region, each about 100 km in diameter. Hundreds of individual volcanoes have been identified in the high-resolution Viking surveys of the planet.

The older volcanoes on Mars tend not to be very high, perhaps as a result of subsidence of the surface. The best example is Alba, an immense ancient volcano north of Tharsis. Crater counts suggest that this feature may be 3 to 4 billion years old, approximately as old as the lunar maria. Alba is 1600 km across, with four times the area of Olympus Mons and nearly 200 times the area of Mauna Loa. However, it is only a few kilometers in height.

Volcanic Plains

Much of the northern hemisphere of Mars consists of plains, with about the same crater density as the lunar maria. Both of the Vikings landed on such plains, in the Chryse and Utopia basins. Three distinct types of plains can be identified, all apparently the result of volcanic processes.

The lower slopes of the Tharsis bulge and its surroundings are the youngest such features on Mars, judging from their low crater densities. These plains are clearly of volcanic origin, and they exhibit many individual lava flows, some hundreds of kilometers in length. These flows are typically 20–30 m thick. Most of the lava vents were in the plains themselves and not part of the shield volcanoes.

The second class of volcanic plains, which includes the Viking 1 site in Chryse, are called ridged plains. Their low domes and lava wrinkle ridges are very similar to those of the lunar maria.

The plains at higher northern latitudes, above 30° N, are very different in appearance, partly as a result of wind erosion. The most striking of these northern plains show a variety of strange surface patterns: wrinkles, cracks, and low ridges. Viking 2 landed near some of

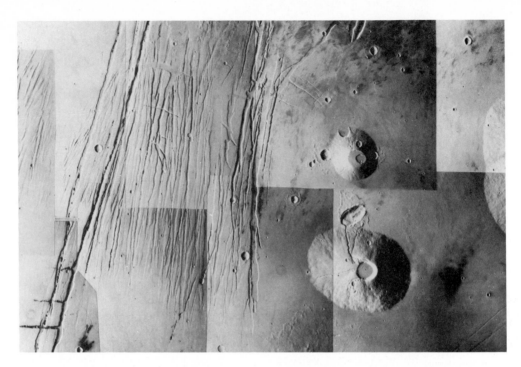

FIGURE 9.18 Two of the intermediate-sized Tharsis volcanoes are shown in this Viking orbiter mosaic of photos, together with a region of extensive tectonic fracturing associated with the formation of the Tharsis bulge. The width of this frame is about 600 km.

these regions. Fig. 9.19 shows a remarkable area with polygonal fractures, spaced a few kilometers apart. The origin of most of these features is poorly understood. Superficially similar polygonal fractures are found in arctic soils on Earth, where they are known as patterned ground. It is very rare on our planet to find this terrain at the large scales exhibited on Mars.

Tectonics and the Tharsis Bulge

In Section 9.2 we noted that the Tharsis bulge is the major surface feature of Mars on a global scale. In part, the Tharsis bulge results from the accumulation of volcanic lavas over billions of years, but in addition it appears to be partly the product of tectonic forces that have bent the crust and uplifted an area that is as large as North America.

Dramatic evidence of the uplift that helped form Tharsis is provided by the extensive fracturing of the surrounding crust (Fig. 9.18). Most of these fractures, which can be hundreds of kilometers long and a kilometer or more in width, point away from the center of the Tharsis bulge. Detailed studies of the apparent ages of these tectonic features, again based on crater densities (Table 9.3), indicate that the uplift began long ago, perhaps 3 billion years, and that the surface fracturing had tapered off by 2 billion years in the past. If these studies are correct, the uplift largely preceded the formation of the large Tharsis volcanoes.

The Martian Canyons

The most spectacular tectonic features on Mars are the great equatorial canyons, especially the Valles Marineris. These canyons, each as much as 100 km wide and up to 7 km deep, stretch a quarter of the way around the planet. The upper parts of the canyon system, near the summit of Tharsis, are called Noctis Labyrinthus (the Labyrinth of the Night), and consist of short intersecting segments that produce a labyrinth (Fig. 9.20). Further down, the Valles Marineris descends the eastern flank of the bulge and extends into the old cratered uplands east of Tharsis (Fig. 9.21; Plate 9).

The term canyon is really a misnomer as the martian canyons exhibit none of the characteristics associated with the term. They were not carved by rivers, do not contain features characteristic of running water, nor do they terminate in lowland basins. They are also distinct from the channels, which are discussed in the next section. The canyons are tectonic in origin; they are huge cracks in the planet's surface, which subsequently have been widened and shaped by forces of erosion. Fig. 9.22 shows a detail in which several landslides tens of kilometers across can be seen. Note that on the far wall the upper part of the slide consists of blocky

FIGURE 9.19 Many of the northern plains areas of Mars have a strange patterned appearance when photographed from orbit. Individual ridges are separated by distances of a few kilometers. These fracture patterns have no analog on Earth and remain poorly understood even a decade after Viking.

FIGURE 9.20 Noctis Labyrinthus, a region of intersecting closed depressions possibly caused by a combination of tectonic fracturing and collapse associated with the melting of permafrost. The frame is 120 km across.

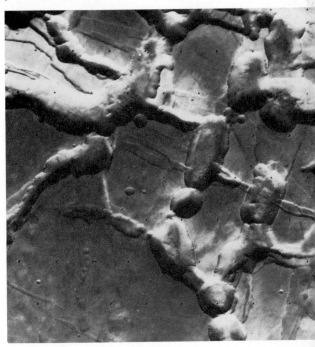

material, while the lower part of the slide shows evidence of a more fluid flow across the canyon floor, probably at speeds greater than 100 km/h. Just as in the case of the fluidized impact crater ejecta, subsurface ice may have helped to lubricate these flows. Such landslides have played a major role in widening the original fractures.

It is unclear how much of the present width of the canyons represents the original size of the crack and how much has resulted from erosion. It seems unlikely to most geologists that fractures 100 km or more in width could have formed on Mars. If most of the canyon width does result from erosional forces, where has the eroded material gone? Perhaps some was water

FIGURE 9.21 Tithonius Chasma and Ius Chasma, both part of the vast Valles Marineris canyon system on Mars. The canyons are about 2–3 km deep in this region, and the huge landslide (lower center) is about 100 km across.

FIGURE 9.22 Landslides in Ganges Chasma in the martian canyonlands. Note the flow patterns in the large fan-shaped slides, and the indication of a harder caprock at the top of the canyon wall. Width of the frame is about 100 km.

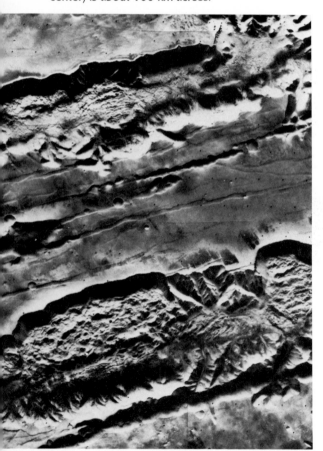

ice, which is believed to make up a sizable fraction of the crust. Alternatively, the fine material may have been carried away by winds whistling down the canyons.

Tectonic Forces

Studies of the stresses in the Tharsis region and the canyon complex show that the crust of Mars is probably thicker than the Earth's but not so thick as that of the Moon: about 100 km is a good estimate. Below lies a mantle of more plastic material of unknown composition. In the absence of any seismic data for Mars, the nature of the planet below the visible surface remains highly speculative.

What produced the Tharsis uplift? No one knows for sure, but the most likely current hypothesis is that some form of mantle convection with a single, upwelling plume or hot spot caused this buckling and swelling of the crust. There is no sign of plate tectonic motion, so the plume has remained in the same location with respect to the crust. Understanding the differences between mantle convection and its expression in tectonics on Mars, the Earth, and Venus remains a profound challenge to planetary scientists.

The Nature of Martian Volcanism

Detailed studies of the martian volcanoes have been carried out by geologists such as Michael J. Carr of the United States Geological Survey, who was leader of the Viking Orbiter Imaging Team. Even without surface measurements of chemical composition, Carr and his colleagues have concluded from the shapes of the volcanoes and the behavior of their flows that they were produced by basaltic volcanism similar to that which formed the Hawaiian volcanoes. Individual flows extend for very great distances, indicating that the lavas had low viscosity, and the rate of flow during eruptions was large.

Estimates of ages of different flows also suggest that the eruptions spread over vast time spans with individual volcanoes active for hundreds of millions or even billions of years. If the SNC meteorites (Section 3.5) come from Mars, they presumably are samples of some of these flows, probably from the Tharsis area.

The concentration of volcanoes in the Tharsis region is consistent with the idea that the Tharsis bulge is the result of a mantle hot spot. In this case, the same forces that raised the bulge provided the heat to generate the volcanoes. It is also easy to see why the martian volcanoes are so much larger than those on Earth. If there was no plate tectonic motion, the crust remained stationary over the hot spot. Thus each eruption was superimposed on the previous one, and the volcanoes continued growing over billions of years.

One fascinating question pursued by Carr and other volcanologists is: Are any of the martian volcanoes still active? No evidence of present day eruptions exists, as it does for the volcanoes of Venus. The probable martian samples, with solidification ages of 1.4 billion years (Section 3.5), could have originated anywhere in the Tharsis area. But what of the youngest martian volcanoes, such as Olympus Mons? Carr and others have determined from crater counting that the youngest lava flows on Olympus Mons are less than 100 million years old. One hundred million years is a very short time on the scale of planetary history, and there is no reason to think a volcano that has been active for much of the past billion years or more will not continue to erupt periodically in the future. Therefore, scientists must conclude that Olympus Mons, at least, probably remains intermittently active, although the intervals between major eruptions could be many millions of years.

9.6 Mysteries of the Martian Channels

Classification of Channels

No geological discoveries on Mars have equalled the channels in either excitement or mystery. Impact craters, volcanoes, tectonic fractures, canyons — all of these might have been expected, and each has counterparts on other planets. The channels, in contrast, speak to us of running water on Mars: a phenomenon once thought to be unique to the Earth. From a study of the martian channels, we hope to illuminate the larger questions of the history of water on Mars and the possibilities that the planet could even have supported water-based forms of life.

There are two main types of channels on Mars, runoff channels and outflow channels. Both were caused by running water in contrast to the tectonic canyons discussed in Section 9.5. Both are, of course, dry today, since current conditions on Mars do not permit the existence of liquid water.

Runoff Channels

The martian features most like terrestrial dry river valleys are the **runoff channels,** found exclusively in the heavily cratered uplands of the southern hemisphere. Over much of this old terrain channels are extremely common, usually consisting of simple valleys tens to hundreds of meters wide and a few tens of kilometers long, often on steep slopes such as crater walls.

More interesting than the singular valleys, because they are more clearly indicative of running water, are the networks of interconnected runoff channels. Dozens of major networks have been identified and are typically several hundred kilometers in length. They have the characteristic form of terrestrial river systems, with small tributary channels connecting into larger chan-

nels to provide drainage for a wide area (Fig. 9.23). Compared with terrestrial drainage systems, however, most of the martian runoff channels show relatively few tributaries.

In some cases, the smallest channels have been obliterated by subsequent erosion, but there are also instances where the water originated primarily from underground springs

FIGURE 9.23 A drainage network of runoff channels in the old upland terrain near 48° S. These finely divided and branching erosional features were almost surely formed at a time when there was rainfall on the surface of Mars. Width equals 250 km.

rather than from direct surface runoff. Similar looking channels are formed on Earth by this process of sapping from springs. The process may begin with water seeping out at the base of a cliff from a subsurface reservoir. As the water escapes, it carries some soil with it, and the cliff face above the flow begins to crumble. Thus the flow eats its way back into the surrounding terrain, forming a channel with a characteristically blunt appearance. The process continues until the groundwater reservoir is depleted, or the spring is stopped by freezing.

Origin of the Runoff Channels

The shapes of the channel networks, and the way they clearly drain from higher areas into lower, point inescapably to erosion by liquid water. Many geologists have sought alternative explanations involving lava, wind, and ice, but none has survived detailed criticism. Unless we are grossly misinterpreting the data, Mars once experienced actual rivers flowing across its upland regions fed by large underground springs and possibly by drainage of rainfall.

When did the rivers flow? An important clue is given by the restriction of these channels to the cratered uplands. There are no such channels in the lower northern plains, which formed at about the time of the lunar maria, and none in any of the volcanic areas such as Tharsis. Apparently conditions have not been suitable on Mars for the formation of runoff channels during the past several billion years.

Detailed crater counts in the areas drained by the runoff channels confirm this time scale. The climatic conditions which permitted rivers to flow ceased before the formation of the northern plains or the emergence of the Tharsis bulge. Standard chronology, then, puts the age of the rivers back to approximately four billion years ago. Thus the channels we now see were active

near the time of terminal heavy bombardment, which is also the period suggested for enhanced levels of surface erosion. As we shall see in Section 10.6, this coincidence is important. Running water requires a thicker, warmer atmosphere than Mars has today, which would also have contributed to greater erosion than is taking place under present conditions.

Outflow Channels

Even larger and more controversial than the runoff channels are the martian **outflow channels.** These tremendous valley systems are confined to the equatorial parts of Mars, primarily leading from the southern uplands into the northern plains. Like the smaller runoff channels, they appear to have been carved by running water. However, in the case of the outflow channels the flow was probably intermittent and catastrophic, and the source of the water remains somewhat mysterious.

The largest and best studied outflow channels lie north and east of the martian canyonlands and drain into the Chryse basin. These include Kasei Valley, Maja Valley, Simud Valley, Tiu Valley, and Ares Valley, together with smaller tributaries. Each major outflow channel is 10 km or more in width and hundreds of kilometers long, and each drops several kilometers in elevation along its length (Fig. 9.24).

Over most of their lengths the outflow channels have cut multiple parallel channels that diverge and interconnect, each showing characteristic patterns of water erosion such as terraced walls, streamlined islands, and sand bars. In places, the flow seems to have escaped these well-defined channels and spread over areas tens of kilometers wide.

Where the outflow channels descend from the uplands into the Chryse basin, they are characterized by teardrop-shaped islands and other

FIGURE 9.24 Several of the large outflow channels that drain from the southern uplands toward the northern basins. These appear to be the product of catastrophic water floods that took place about 3 billion years ago.

sculpted landforms (Fig. 9.25). The islands are remnants of the plateau, shaped by the flood of water rushing out into the lowland plains. Within Chryse, evidence of water erosion extends for hundreds of kilometers from the channel mouths, but not as far as the location of the Viking 1 Lander, so lander images cannot help us to understand the origin of these features.

Catastrophic Floods

The scale of the martian outflow channels and the evidence of massive flows of water dwarf terrestrial rivers. Estimates of the rates of flow in the larger channels are a hundred times greater than the flow from Earth's greatest river, the Amazon. To seek comparable rates of flow on Earth, we must look to catastrophic floods caused by the sudden failure of natural dams holding back large lakes.

The best studied example of such a flood on Earth is to be found in the states of Washington and Montana, where 18,000 years ago an ice dam broke releasing the water in a prehistoric lake called Lake Missoula. Within a few days the entire lake emptied westward across the Columbia Plateau, generating flows up to 120 m deep over a front tens of kilometers wide. The awesome force of this water erased previous surface features in western Washington State, cutting new channels up to 200 m deep through the bedrock. The resulting terrain, called the channeled scabland, is very different in appearance from that generated by slow, steady processes of water erosion. With its multiple parallel channels, teardrop islands, and huge bars, the channeled scabland is the closest analog we have to the martian outflow channels.

Presumably the martian features were formed in a similar way, by brief catastrophic floods arising in the uplands and dissipating in the broad northern basins. But where could such vast quantities of water have come from, and when did these remarkable events take place?

Chaos: The Sources of the Outflow Channels

To find the origin of the martian floods, we must look to the sources of the outflow channels. These are strange landscapes called **chaotic terrain.** They resemble regions where the surface of the old cratered uplands has collapsed into a huge jumble of irregular rugged hills and valleys (Fig. 9.26). Both the chaotic terrain and the great floods could have a common origin in the sudden release of huge reservoirs of underground water, mud, or ice.

When and how did these events take place? From the fact that the outflow channels extend into the northern plains we can immediately see that they represent a more recent era than that of the runoff channels, since the latter are only found in the south. Crater counts in Chryse indicate that the deposits from channel outflow were formed at about the same time the volcanic plains were formed, probably about 3.5 billion years ago. This is also the time when internal forces were producing the Tharsis uplift.

FIGURE 9.25 Flow features, such as teardrop-shaped islands, mark the region where several outflow channels spread out into the lowland basin of Chryse, near the site originally targeted for the Viking 1 landing on Mars. Each island is approximately 40 km long.

FIGURE 9.26 The "chaotic terrain" at the head of an outflow channel system. Evidently a local collapse of the surface was associated with the generation of a massive flood of water.

While most geologists agree that the release of underground water generated the floods and produced the chaotic terrain, there is no consensus concerning the details of the mechanism. One possibility is the melting of subsurface ice by deep-seated volcanic activity. Another suggestion is the chemical release of water loosely bound to the clays of the martian soil. A third invokes the shifting of large quantities of underground water in response to the Tharsis uplift. Possibly all of these processes were involved.

Both the magnitude of the outflow channels and the huge depressions of chaotic terrain left behind indicate that water or ice must have made up a sizable fraction of the martian crust. Large quantities of ice are probably still present below the surface, but the conditions necessary to produce catastrophic floods ceased billions of years ago. We shall examine the fate of that water in Section 10.1.

9.7 The Polar Regions

The geological features of Mars described in the previous sections of this chapter are generally confined to equatorial and middle latitudes. Above 70° north and south the surface takes on a different character, one strongly influenced by the polar caps.

The Seasonal Polar Caps

The polar caps seen through a telescope — the white surface deposits that grow and shrink with the changing seasons — are called the seasonal caps. They are analogous to the seasonal snow cover on the Earth. We do not usually think of the snow that covers the ground in winter as a part of the Earth's polar caps. If we looked at our planet from space, however, the seasonally changing snow cover would be very obvious, and the area involved much larger than that covered by the permanent ice caps. It is the same for Mars.

The substantial eccentricity of Mars' orbit carries the planet farther from the Sun during southern winter, resulting in long cold winters and a large southern seasonal cap, extending down to 55° S. The northern cap, occurring half a martian year later, extends only to about 65°, although some surface frost was seen as far south as the Viking 2 landing site, at 48° N. During spring, each cap retreats at a rate of about 20 km/day, reaching its minimum size in early summer.

Both seasonal caps are composed of CO_2 frost or dry ice. Since CO_2 is the main component of the atmosphere, it is always available to condense on the surface whenever the temperature falls low enough (about 150 K). From changes in atmospheric pressure as the caps form and evaporate, the thickness of the seasonal deposit has been calculated to vary from a meter or more at high latitudes down to a few centimeters near its edge. This is less, but not much less, than the corresponding thickness of the seasonal snow cap on Earth.

The Residual Polar Caps

As the seasonal polar cap retreats during southern spring, it reveals a brighter, underlying permanent cap which persists through the summer months (Fig. 9.27a; Plate 11). The diameter of this residual south polar cap is about 350 km, and the cap is centered a short distance from the true south pole, with its center at 30° W, 86° S. Measurements of its surface temperature made from orbit show that the residual cap remains at the frost point of CO_2, about 150 K, throughout the summer, indicating that it is composed primarily of CO_2. Water ice may also be present, but clearly the temperature of the cap is controlled by the CO_2 ice, and this cold reservoir is sufficient to maintain itself in spite of the continuous summer sunshine.

The residual southern cap has a remarkable appearance, showing a spiral swirl pattern of dark lanes in the bright ice. These lanes are sunward-facing slopes and valley bottoms a few kilometers wide, which heat sufficiently to lose their ice cover. The thickness of the CO_2 ice deposit that makes up the bulk of the cap is unknown, but from the fact that these frost-free areas emerge, we presume that the thickness is only a few kilometers at most.

FIGURE 9.27a The residual south polar cap of Mars, composed of CO_2 (dry ice), is only about 300 km across, much smaller than the northern cap.

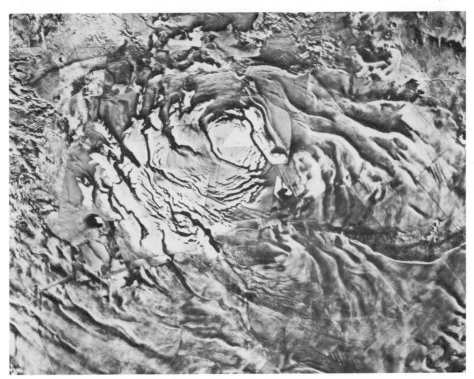

The permanent north polar cap (Fig. 9.27b) is much larger than that in the south, never shrinking below a diameter of about 1000 km. It shows a similar swirl pattern of exposed sunward-facing slopes. In spite of similarities in appearance, the northern cap is fundamentally different from its southern counterpart. Orbital temperature measurements show that its temperature climbs above 200 K, far too warm for CO_2 ice to be stable. At the same time, the concentration of atmospheric water vapor rises sharply above the cap. Thus we see that the northern residual cap does not contain frozen CO_2, but instead is composed of ordinary water ice rapidly evaporating in the summer warmth. With its di-

ameter of 1000 km and an unknown thickness, the residual northern cap may be one of the main storehouses for water on Mars.

Why this major difference between the two martian polar caps? It is not simply a question of the summer temperatures, since the southern summers are hotter. Yet it is the southern cap that stays cold enough to trap CO_2. The difference is probably associated with the seasonal dust storms, which are further discussed in the next chapter. The main, planet-wide dust storms always take place during southern summer, when the northern cap is forming; therefore, the northern cap becomes dusty. Being dustier, it is also darker and therefore it can absorb more sun-

FIGURE 9.27b The residual north polar cap of Mars, composed of water ice, remains nearly 1,000 km across throughout the summer.

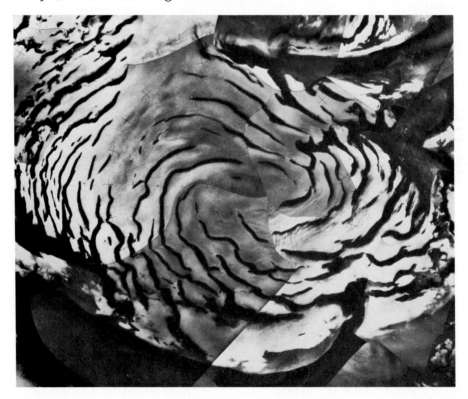

light. It may be just this difference in the reflectivity of the ice that allows the northern cap to heat up and lose its frozen CO_2.

Polar Terrain

So far we have concentrated on the polar caps of Mars, but the underlying polar terrain is equally interesting. At latitudes above 80° in both hemispheres, the ground is covered by deep layered deposits of sediment. Here this term has its geological meaning of finely divided material; sediment does not have to be produced or deposited by water. These deposits lie on top of the old cratered uplands in the south and overlie the volcanic plains in the north. In southern summer, the southern layered terrain is exposed by the retreating polar cap, while in the north the residual cap continues to cover most of these deposits throughout the year.

The polar layered deposits blanket all but the largest underlying topography, indicating a depth of several kilometers. They are sparsely cratered and are, therefore, relatively young. On the defrosted slopes within the residual polar caps, the layering is especially well shown (Fig. 9.28). The individual layers are between 10 and 50 m in thickness, distinguished by darker bands alternating with light and also at times by terraces between the bands.

In the north, there is another special polar terrain, consisting of extensive sand dunes. These dunes form a band about 500 km wide around the pole, between latitudes 70° N and 85° N. They extend to the boundary of the layered deposits near the edge of the north residual cap. The martian dune field (Fig. 9.29) consists of evenly spaced, crescent-shaped dunes a little less than a kilometer long. In the southern hemisphere no similar band of dunes occurs, but smaller dune fields are common inside craters at high southern latitudes.

FIGURE 9.28 A region near the north polar cap that is partially covered with frost. Many layers can be seen on partly defrosted slopes, presumably representing past climate cycles on Mars. Width equals 65 km.

Formation of the Polar Deposits

The polar terrains we have been discussing differ from most other parts of the martian surface in that they are *depositional*; they represent wind-blown material transported from elsewhere. Near the poles we may have located the final resting place of much of the material scoured out of the canyons and other eroded areas of the planet.

The distinct layering of the polar deposits is particularly intriguing. Calculations of the amount of dust raised on Mars during a major dust storm suggest that the maximum annual deposit from this source is less than 1 mm/yr,

FIGURE 9.29 A large dune field at latitude 81° N. The spacing between dune crests is a few hundred meters, and the total width of the frame is 62 km. Such dunes almost completely encircle the martian north pole just south of the layered terrain.

which of course we cannot see; the layers must represent much longer time scales. To build up a thickness of 10 meters would require about 10,000 years at this rate. Thus the polar layering must represent climatic changes over tens of thousands of years. What might these be?

The most likely causes of such climatic changes are to be found in variations in the orbit of Mars caused by the combined gravitational attractions of the other planets. The Earth undergoes similar but smaller cyclical variations, which are probably a major cause of the ice ages that our planet has experienced at almost regular intervals. One of these variations causes the role of the two martian poles to alternate with a period of 51,000 years. So about 25,000 years into the future it will be the north pole of Mars, not the south, that will experience the hottest summers. Another effect is a periodic shift in the tilt of the martian axis of rotation. Calculations show that this tilt varies with a period of about 100,000 years. The eccentricity of the orbit also changes, this time with a period nearer to a million years.

As a result of these orbital and rotational changes, conditions at the martian poles undergo regular cycles of change. At times CO_2 will be retained by either the south or the north residual cap, and perhaps at times by both or by neither. Undoubtedly the polar layered deposits are the result of these changes, but the details of the processes of polar deposition and erosion remain unknown. The study of martian climatic change is a subject that is in its infancy.

Summary

Mars has the most Earth-like surface environment of any planet. In spite of its small size, the thinness of its atmosphere, and the low temperatures of its surface, we can recognize this planet as the closest we shall ever come to another Earth in our solar system. It is not surprising that Mars has been the target of so many spacecraft, or that serious plans for a piloted expedition are being discussed.

At the beginning of the twentieth century Mars was widely believed to have intelligent inhabitants, and even until the 1960s the possibility of widespread plant life was assumed. Spacecraft exploration unfortunately failed to discover life, but the Mariner 9 orbiter and the four Viking craft have revealed a world equally fascinating in other ways.

The topography of Mars is dominated by a north-south asymmetry. Most of the southern hemisphere consists of ancient cratered uplands, probably formed at about the same time as the lunar highlands but substantially modified by erosion. The northern half of the planet, in contrast, is about 4–5 km lower and has a younger, largely volcanic surface, perhaps about as old as the lunar maria.

A second, younger east-west asymmetry is created by the Tharsis bulge, a volcanic area about the size of the continental United States with four huge shield volcanoes that rise to 31 km above the lowlands, creating elevation differences much greater than any on Earth, Venus, or the Moon. Because of its low surface gravity and persistent volcanic activity at a few sites, Mars has the highest mountains in the solar system.

The Viking landers provide detailed information for two surface sites, both rather dull lowland plains. Each is a rock-strewn volcanic basin, apparently stripped of fine-grained material by the powerful martian winds. The soil resembles a terrestrial clay reddened by its content of iron oxides, while the rocks may be basalt fragments ejected from nearby impact craters.

Martian craters differ from those of the Moon primarily in their ejecta blankets, which show evidence of fluidization, presumably by the impact melting of permafrost. Counts of crater densities yield relative crater retention ages, just as on the Moon, but unfortunately we have no absolute chronology based on radioactive rock dating. Only with the aid of a calculated model for the impact history of Mars can we attempt to date the geologic events of the martian past.

The shield volcanoes of Mars look much like their terrestrial counterparts, except that they are vastly larger. Olympus Mons is the largest volcano known in the solar system. In addition to volcanoes, the Tharsis uplift has produced numerous tectonic features, the most spectacular of which make up the 4000-km-long canyon system of the Valles Marineris. These canyons were not cut by running water, although water and wind erosion have both played a role in enlarging the original tectonic features.

The channels of Mars, which were cut by running water, should not be confused with the canyons, which were not. There are two main types of channels: the older runoff channels, which appear to have been produced by rain; and the larger and slightly younger outflow channels, which drain from the southern uplands into the northern basins. The outflow channels appear to have been created in short, catastrophic floods, presumably resulting from the sudden release of groundwater or permafrost.

The martian polar regions offer other fascinating enigmas. The two residual polar caps are quite different from each other; the southern is smaller and is composed of CO_2, while the northern is a thousand kilometers across and is composed of H_2O ice. The underlying topography of both polar regions consists of a unique layered terrain, indicative of periodic changes in global climate on Mars.

Key Terms

chaotic terrain	outflow channel
fluidized ejecta	runoff channel

CHAPTER TEN

Mars: The Atmosphere and the Search for Life

10.1 The Atmosphere of Mars

Early Ideas

Nineteenth century astronomers recognized that Mars had an atmosphere. The existence of polar caps that waxed and waned with the seasons, the presence of white clouds over certain regions of Mars at certain seasons, and the occasional development of giant dust storms, all seemed to require the existence of an atmosphere. It has proved to be much more difficult to determine how thick the atmosphere is and what gases it contains.

The process of discovery on Mars was similar to that for Venus. At first most scientists assumed that the dominant constituent must be nitrogen, even after spectroscopic observations showed that some carbon dioxide was present. The surface pressure was thought to be about one tenth of the Earth's. As spectroscopic methods improved in the early 1960s, it became clear that carbon dioxide must be the major component of the atmosphere and that this atmosphere was very thin. These findings were dramatically confirmed by the first spacecraft sent to Mars in 1964. Mariner 4 found that the surface pressure of the atmosphere was only 0.0065 bars, or $\frac{1}{150}$ the sea-level pressure on Earth. This thin envelope of gas was at least 95% carbon dioxide.

Viking Results

The Viking landers gave us direct atmospheric measurements along two descent trajectories, as well as detailed measurements of atmospheric composition and weather in Chryse and Utopia. One of the authors, Owen, was a member of the team responsible for investigations of atmospheric composition using the GCMS (Section 10.4). After years of studying Mars with telescopes and spectrographs from Earth, it was thrilling to actually measure a sample of the atmosphere, finding gases such as nitrogen, argon, neon, krypton, and xenon that could not be detected by remote sensing.

The composition results are shown in Table 10.1, and the structure summarized in Fig. 10.1. At first glance, the martian atmosphere looks remarkably like that of Venus, but it is a few thousand times thinner. Like both Earth and Venus, Mars has a troposphere and a stratosphere. The thickness of the troposphere varies with location and season, however, and at night it practically disappears.

Because the atmosphere is so thin and there are no moderating bodies of water, the temperatures of the surface and the lower few kilometers of the troposphere experience a wide daily fluctuation. Air temperatures at the two Viking sites reached summer daytime highs of about − 30 C, while the ground itself occasionally exceeded the freezing temperature of water in the strong afternoon Sun. Under these conditions convection can take place, and the troposphere is typically 10–20 km thick. In contrast, nighttime temperatures of both ground and air dropped to nearly 100 C below freezing. Similar low surface temperatures are found over the polar caps. Convection ceases, and the atmosphere becomes highly stable. On Earth, with its much thicker atmosphere, the day/night variations are much less, and tropospheric convection rarely ceases entirely.

As shown in Table 10.1, the major constituents of the martian atmosphere are CO_2 (95%), N_2 (2.7%), and Ar (1.6%). These same gases exist in the atmosphere of Venus. However, the minor constituents of the martian atmosphere are different from those of Venus. No sulfur compounds or acids have been detected on Mars, and the distribution of noble gas abundances is very different. In particular, it seems that Mars has outgassed more argon-40, the isotope produced by the decay of radioactive potassium, relative to argon-36, the primordial isotope. Evidently both the chemistries and the histories of the two atmospheres are not the same.

TABLE 10.1 The composition of the atmosphere of Mars

Gas	Formula	Abundance
	Main constituents	(Percent)
Carbon dioxide	CO_2	95.3
Nitrogen	N_2	2.7
Argon (40)	Ar-40	1.6
Oxygen	O_2	0.13
Carbon monoxide	CO	0.07
Water vapor	H_2O	0.03[a]
	Trace constituents	(Parts per million)
Argon (36)	Ar-36	5
Neon	Ne	2.5
Krypton	Kr	0.3
Xenon	Xe	0.08
Ozone	O_3	0.04–0.2[a]

[a]Abundance varies with season and location.

Martian Weather

The local weather at the two Viking landing sites turned out to be rather dull, reliably repeating from one day to the next. The first weather report from the surface, as given by Viking meteorologist Seymour L. Hess:

> Light winds from the East in the late afternoon, changing to light winds from the Southeast after midnight. Maximum winds were 15 miles per hour. Temperature ranged from −122 F just after dawn to a maximum of −22 F. Pressure steady at 7.70 millibars.

The global circulation of the atmosphere is very different on Mars than on Earth. The absence of large bodies of liquid water greatly simplifies the situation. The main seasonal driver is the exchange of carbon dioxide between the atmosphere and the polar caps. The amount of gas stored in the polar caps is so great that the average atmospheric surface pressure on the planet changes by 20% with the seasons. Such a change is unheard of on Earth, where a severe storm may cause a drop in pressure of no more than a few percent. Strong winds have been observed flowing off the sunlit caps, although no measurements of these winds have yet been made.

The clouds seen by early telescopic observers were confirmed by the Viking cameras. Morning fogs occur in low lying areas (Fig.

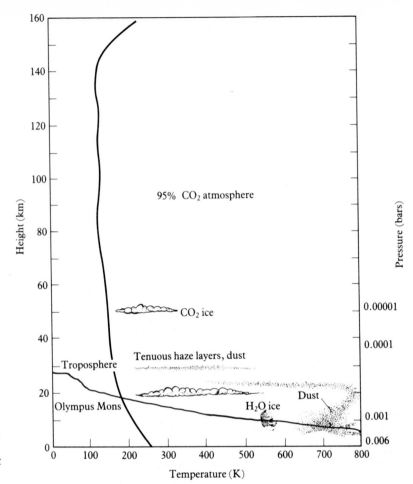

FIGURE 10.1 The vertical structure of the martian atmosphere. Both CO_2 clouds and H_2O clouds can form. Dust from the surface occasionally fills the atmosphere, warming it considerably.

10.2), and extensive clouds are regularly found around the high volcanoes such as Olympus Mons (Plate 12). As air rises over the volcano it cools, and water vapor condenses to form the clouds. This association of mountains and clouds is well known on Earth.

Global Dust Storms

Only during the southern hemisphere summer do conditions change dramatically. Since neither Viking lander was in the southern hemisphere, we are not sure of the details, but apparently the atmosphere becomes unstable, and conditions favorable to the formation of global dust storms develop. Once a large quantity of dust is raised, most of the sunlight is absorbed by the dusty atmosphere and blocked from reaching the surface. As a result surface temperatures fall, but the atmosphere remains unstable and the dust clouds spread until nearly the entire planet is shrouded. It requires about one quarter of a martian year (six Earth months) for the dust to settle and the normal weather patterns to re-emerge.

How can such a thin mixture of gases produce and maintain the great dust storms of Mars? The answer must lie in high speed local winds that are produced by sunlight heating the planet's surface. Something like this occurs in the deserts of our own southwest during hot summer days. The air near the ground gets so hot that it suddenly soars upward, allowing cooler air to rush in. The result is a very local, circular wind that picks up surface dust and carries it aloft in a swirling column called a dust devil.

It is estimated that the local winds on Mars must achieve speeds of 150 km/h in order to raise the dust in such a thin atmosphere. The dust itself must include a substantial amount of very fine-grained material since larger grains would rapidly drop out, clearing the air. The orange skies revealed by the Viking landers give elo-

FIGURE 10.2 Two pictures of the same region on the martian surface show the development of early morning fog in low spots (see arrows in picture at right). Scene at left was photographed first, shortly after local martian dawn. Same scene 30 min. later is at right. Slight warming of the sub-zero surface by the rising Sun has evidently driven off a small amount of water vapor which has recondensed in the colder air just above the surface.

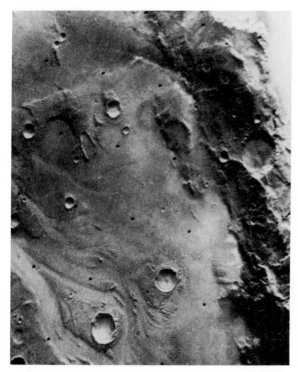
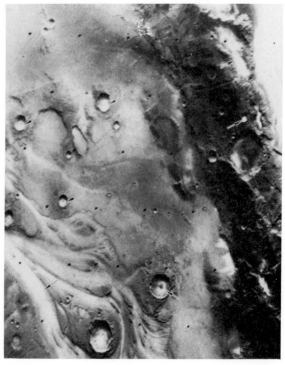

quent testimony to the continued presence of these fine dust particles in the atmosphere, while pictures of the landers obtained late in the mission show how dusty they have become from the coarser particles that settle (Fig. 10.3).

FIGURE 10.3 Frost on the surface of Mars and dust on the surface of the Viking 2 Lander are both visible in this picture taken about one Earth year after landing.

Water on Mars

One of the great mysteries of Mars is the presence of so many indications of water in the past, while present conditions do not admit liquid water. Liquid water is absent because of the low pressure of the atmosphere as well as the low temperatures. The reason water boils in a pan heated to 100 C (or 212 F) in Boston, New York, or Los Angeles is that at this temperature, the upward pressure exerted by water vapor molecules leaving the liquid water in the pan is exactly equal to the pressure of the overlying atmosphere. If the atmospheric pressure is reduced, the boiling point is also lowered. If we carried a pan and a stove to higher and higher altitudes, we would find at a pressure of 0.0061 bars that the water boils at 0 C, which is also its freezing point; at this pressure liquid water can no longer exist. On Mars, this triple-point pressure is reached only at the lowest elevations, with the result that no liquid water is present on the surface of Mars, even on the rare occasions when the temperature is warm enough for it to form. The water would boil away into the atmosphere.

Water on Mars behaves like carbon dioxide on Earth. Dry ice is frozen carbon dioxide. As it warms up, it goes directly from the solid to the gas, without ever becoming liquid. Both H_2O and CO_2 share this property on Mars. So if you stood unprotected on Mars, your blood would boil until it became cold enough to freeze.

What Happened to the Martian Water Supply?

Depending on your point of view, the atmosphere of Mars is either very dry or very wet. Because it is so thin and cold, the atmosphere can hold very little water vapor; in this sense it is exceedingly dry. But it is also true that the martian atmosphere normally holds nearly the maximum amount of water vapor possible for its pressure and temperature. Thus relative to what it could contain it is damp; we say that there is a large *relative* humidity, or that the atmosphere is close to saturation. As a result, it is common for H_2O clouds and fog to form at low elevation during the cold martian night (Fig. 10.2).

On Mars the atmospheric water vapor is continually being broken apart by solar ultraviolet light, with hydrogen escaping into space and oxygen remaining behind, just as on Venus. This destruction of water has been observed directly in the form of a huge thin cloud of hydrogen surrounding the planet.

To compensate for this loss of water, there must be a source or reservoir supplying fresh vapor to the atmosphere. Where is this reservoir? We have already seen that some water ice is frozen out at the polar caps, but this is not enough to maintain the high relative humidity of the atmosphere. In addition, there must be ice buried in the martian soil in the form of permafrost deposits. If we could magically warm Mars in some way the atmosphere would be capable of holding more water vapor, and this reservoir of buried ice would supply it. Future astronauts who explore this planet should have no trouble finding enough water for their needs.

Returning to the puzzles posed by the martian landscape, we can see that this subsurface reservoir of water helps explain the runoff channels cut into the planet's surface in the distant past. If atmospheric pressures and temperatures were higher, the water needed to generate rain would also be present. The widespread permafrost accounts for the peculiar ejecta blankets around the splosh craters. It provides the lubrication for massive landslides and is probably the source of the floodwater that cut the outflow channels.

10.2 Signs of Life
Early Ideas

With all this evidence of water, even in the frozen state, the prospects for finding life on Mars might seem to be worth considering. In fact, the idea that there might be life on this planet is a very old one. As soon as ancient astronomers developed the idea that some of the points of light in the night sky might be other worlds like ours, they began to speculate about the possibility that these worlds were inhabited. This speculation had its ups and downs during the Middle Ages, the burning of Giordano Bruno in 1600 for his heretical belief in other worlds being an obvious low point. After the invention of the telescope, Mars gradually emerged as the focus of the search for other inhabited worlds. Nearly a century ago a feeling of kinship with Mars had developed, supported by observations of the seasonal changes in surface markings and polar caps. Might this be evidence of the growth of vegetation in response to warmer temperatures and an increased water supply? Some astronomers certainly thought so.

How Could Life Be Detected?

More definitive evidence for the existence of life was needed, and the growing sophistication of astrophysical techniques provided more tools for interested investigators. To see how these tools could be used, let us consider some manifestations of life that could be detectable to a remote observer.

1. A mixture of gases in the atmosphere of Mars that requires the presence of living organisms for its maintenance.
2. Large-scale surface features that are clearly artificial in origin, or seasonal changes in

surface markings that could *only* be caused by vegetation.

3. Radio (including radar and television) signals produced by an advanced civilization on Mars.

4. For the non-remote observer, the direct detection of living organisms by experiments deployed on the martian surface.

To see how these manifestations can lead to the detection of life, we can ask what an intelligent Martian could deduce about the Earth following this same approach. Until quite recently, looking and listening would have failed to indicate the presence of life on Earth to a Martian who was using the same technology employed by our scientists who were struggling to find life on Mars. The distorting effects of the Earth's atmosphere would obscure our planet's surface sufficiently that detection of seasonal changes in vegetation or large-scale human constructions would be very difficult to do from Mars. Although a martian astronomer would be able to view the night side of Earth (as we can observe the dark hemisphere of Venus), the nighttime glow of our cities would not have reached the level of easy detection before the widespread adoption of electric lighting. Similarly, no radio transmissions would have been detectable, since humans began to generate large amounts of radio power only in the 1920s. The first test, however — based on the composition of the atmosphere — would have yielded positive results, even if no civilizations had arisen on Earth.

Clues from Atmospheric Composition

Spectroscopic observations made from Mars would reveal the presence of a great number of oxygen molecules in the Earth's atmosphere, along with a small amount of methane (CH_4). This situation would provoke a Martian's curiosity, because methane burns in the presence of oxygen, and even if no fires existed on Earth, ultraviolet light from the Sun would rapidly stimulate the combination of oxygen and methane to form carbon dioxide and water. The continued existence of methane on Earth, therefore, implies the presence of a source that replaces the methane in spite of the natural processes that destroy it. Terrestrial methane is produced primarily by bacteria that live in swampy marshes and in the forward stomachs of grass-eating animals.

In an equally striking way, the large amount of oxygen in the Earth's atmosphere would itself constitute a puzzle for our intelligent Martian, since oxygen is a highly reactive gas that rapidly combines with rocks that are continually being brought up from the interior to the Earth's crust (Chapter 7). This process removes oxygen from our planet's atmosphere, so the continued existence of oxygen also requires a source — in this case, the presence of plants that release oxygen through photosynthesis.

Thus, the first in our list of life indicators would suggest to Martians that life exists on Earth, long before they thought about landing a spacecraft to sample the immediate environment. Likewise, if astronomers on our planet had discovered large amounts of oxygen in the atmosphere of Mars, they would have considered life to be very likely there. But after repeated efforts, they found only a trace of this biological by-product. The tiny amount of oxygen present can be easily explained as the result of photochemical reactions that occur when sunlight strikes the small amount of water vapor in the atmosphere. Methane and other hydrocarbon gases remain undetected on Mars, so the planet shows no signs of the chemical clues that mark Earth as biologically active.

The Canals of Mars

In the decades surrounding the year 1900, there were many astronomers who thought they saw

◆ 1 ◆

The crescent Moon and the planet Venus in the pre-dawn sky above the Canada-France-Hawaii Telescope on Mauna Kea. Sunlight reflected from the Earth illuminates the part of the Moon beyond the bright crescent. This same effect occurs on satellites of other planets.

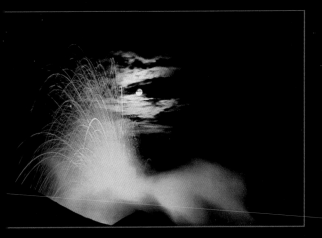

◆ 2 ◆

A typical nighttime eruption of the principal vent at Stromboli, a volcano in Italy. Embedded in the trajectories of the glowing, molten ejecta, the planet Mars appears as a star-like object to the left and just below the Moon. Mars, Moon, and Earth all exhibit the effects of volcanic activity.

◆ 3 ◆

A nighttime eruption at the summit of Mauna Loa in Hawaii, much less violent in style but producing far larger amounts of lava than Stromboli. The constellation to the left is the Southern Cross; stars appear as short trails in this time exposure.

◆ 4 ◆

The Earth as viewed from space, one of the remarkable sights available to travelers who journey to the Moon. South America is peeking through the clouds at left center, while the Sun is setting over the Sahara Desert, visible at the upper right.

a)

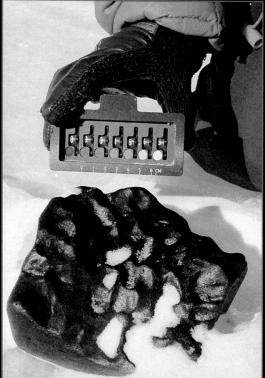

Two visitors from outer space.
a) The recovery of a meteorite from
the ice of Antarctica. b) The dark
lump shown in this picture is a piece
of the Murchison meteorite, the carbonaceous meteorite in which
amino acids were first discovered.

b)

Astronaut Harrison "Jack"
Schmitt examining a huge
Moon rock. The gray colors of
the lunar landscape stand out
against the black, airless sky.

▶ The surface of Venus revealed
by the Pioneer Venus Orbiter
radar (above) shows fewer continental areas than Earth, seen
at the same resolution (below).
Here lowland regions are colored blue, while green, yellow, orange, and red denote
regions above the average
surface.

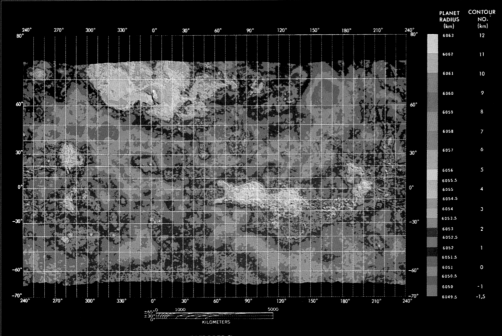

PLANET RADIUS (km)	CONTOUR NO. (km)
6063	12
6062	11
6061	10
6060	9
6059	8
6058	7
6057	6
6056	5
6055.5	
6055	4
6054.5	
6054	3
6053.5	
6053	2
6052.5	
6052	1
6051.5	
6051	0
6050.5	
6050	-1
6049.5	-1.5

KILOMETERS

VENUS

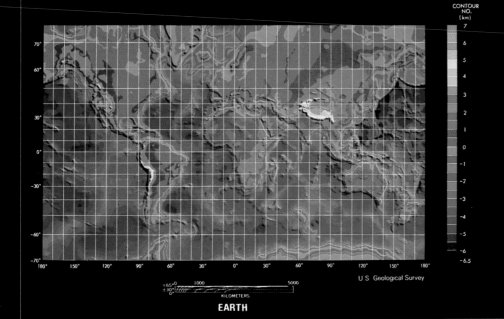

CONTOUR NO. (km)
7
6
5
4
3
2
1
0
-1
-2
-3
-4
-5
-6
-6.5

KILOMETERS

U S Geological Survey

EARTH

◆ 8 ◆

A high-resolution view of the Candor
Chasma region of Valles Marineris
(see Plate 9) on Mars. The frame is
204 km x 236 km; the blue-gray
material at the bottom of the
canyon may be water-deposited
sediment or simply volcanic ash.

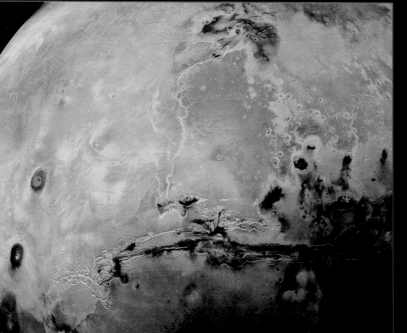

◆ 9 ◆

This enhanced color mosaic of
Viking 1 pictures of Mars extends from the Tharsis ridge,
with its 25-km-high volcanoes
seen at left, across the great
valley system known as Valles
Marineris. This system is
4000 km long, equal to the
coast-to-coast distance across
the U.S.

a) b)

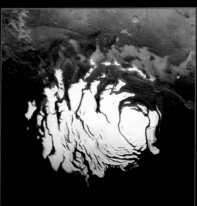

◆ 10 ◆

▲ a) The first Viking lander settled down in Chryse Planitia. The local landscape is dominated by rocks and wind-swept dust.
b) The second lander discovered that in another lowland plain (Utopia), the martian landscape was different in appearance, although still dominated by wind-swept rocks.

◆ 11 ◆

◀ The residual south polar cap of Mars is composed of frozen CO_2 (dry ice) of unknown thickness. The 300-km-diameter cap is seen here surrounded by the ubiquitous red terrain that gives Mars its color.

◆ 12 ◆

▼ The highest volcano in the solar system towers above surrounding clouds. Olympus Mons on Mars rises 25 km above the adjacent plains.

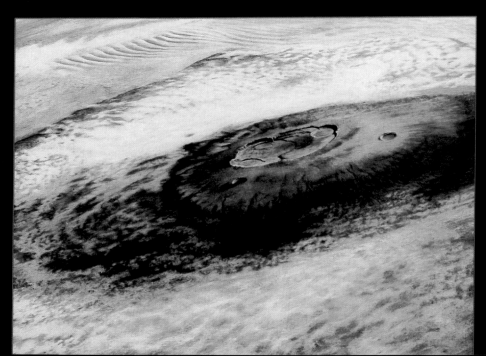

◆ 13 ◆

This global view of Jupiter was obtained by Voyager 1 while the spacecraft was still 30 million km from the planet. The great complexity of the constantly changing, pastel-colored cloud systems is clearly apparent.

◆ 14 ◆

A close-up view of Jupiter's Great Red Spot. This giant storm system could swallow two planets the size of Earth, side by side. The white oval below it has existed for over forty years.

◆ 15 ◆

Saturn looks very differ-
ent from Jupiter, even
in this global view
obtained by Voyager 2.
Aside from its splendid
system of rings, Saturn
is more subdued in ap-
pearance, without the
complex cloud systems
or variety of colors that
Jupiter exhibits.

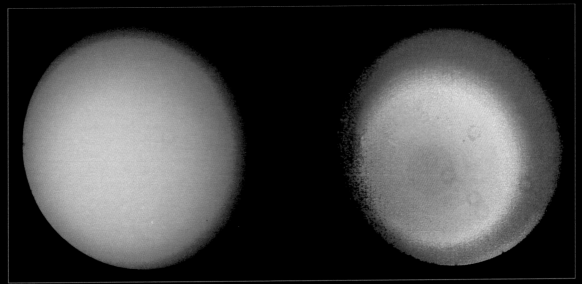

◆ 16 ◆

The left-hand image shows Uranus as it would have appeared to someone riding along on the Voyager 2
spacecraft. The right-hand image has been specially processed to reveal the presence of a banded polar
haze. (The rotational pole is nearly facing us.) The colors in this image are totally artificial.

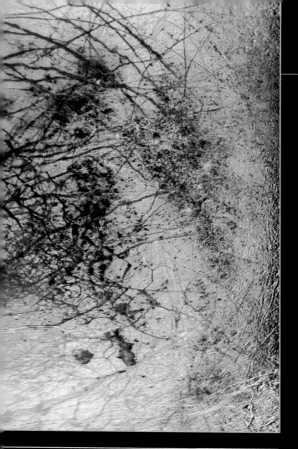

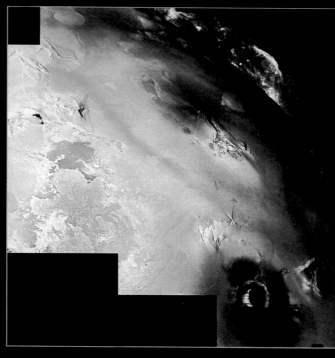

◆ 17 ◆

Jupiter's Europa is the smoothest satellite
we know. This view obtained by Voyager 2
shows almost no impact craters, indicating that the icy surface must be very
young. The streaks and ridges may represent cracks and extrusions in the icy crust
of this satellite.

◆ 18 ◆

This is a view of the large volcanic structure on Io called
Pele, after the supreme Hawaiian volcano goddess. We are
looking through the plume toward the eruptive center on
Io's surface, and we can see turbulent clouds in the plume
projected against the dark sky at the upper right.

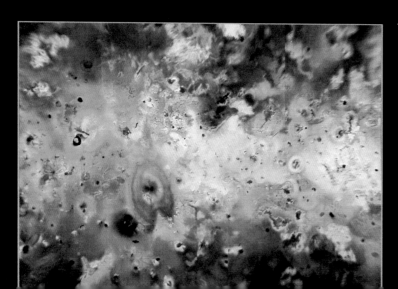

◆ 19 ◆

A map of the surface of Io in
Mercator projection showing the
remarkable variety of volcanic
features produced by the inces-
sant eruptions that wrack this
satellite.

linear markings on the martian surface when they studied the planet with their telescopes, as previously discussed. These lines were the famous canals of Mars, especially championed by Percival Lowell (Fig. 10.4). While Lowell succeeded in implanting the idea of intelligent Martians upon the public mind, his conclusions were disputed by many astronomers. Most were unable to see the canals, although Lowell dismissed their failures as the result of inferior telescopes or observing sites. Some scientists pointed out that an observer can unconsciously interpret faint, discontinuous detail seen near the limit of resolution as straight lines (Fig. 10.5). As new, larger telescopes failed to confirm the canals, the public lost interest. After Lowell's death, the astronomical controversy nearly disappeared, but for the next half-century the mystery of the martian canals was repeatedly resurrected in the popular press and became a part of the standard mythology of science.

FIGURE 10.4 Percival Lowell at the eyepiece of his 24-inch refracting telescope. Lowell built this observatory at Flagstaff, Arizona, especially to study Mars.

FIGURE 10.5 The same small region on the surface of Mars as drawn by G. Schiaparelli (left) and E. M. Antoniadi (right). Antoniadi did not agree with Schiaparelli, Lowell, and others who thought there were canals on Mars.

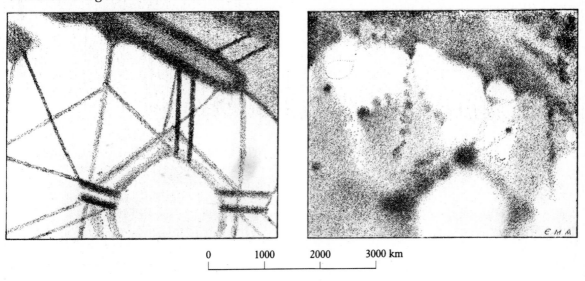

0   1000   2000   3000 km

◆

10.3 The Advantages of Searching for Microorganisms

Since Mars does not exhibit the first three signs of life (no unexpected atmospheric gases, no canals or cities, no radio broadcasts), humans decided in the 1970s to undertake a direct search for organisms on the martian surface. At a cost of approximately 1 billion dollars, the United States sent two Viking landers to Mars, where they carried out a wide variety of experiments, including three separate tests for the presence of martian microbes. Let us now look at those experiments specifically aimed at the search for life.

The Viking Strategy

Although the greatest excitement in a search for life would be the discovery of large, advanced organisms capable of communication, the history and present status of life on Earth demonstrate that microscopic organisms far outnumber large ones and are far more versatile in their adaptations to various environments. The fossil record shows that microbial life was the only kind of life on Earth for billions of years, far longer than larger creatures have existed. Nor have microorganisms decreased in numbers or in adaptive ability. The dirt in your backyard contains more organisms than the number of stars in the galaxy. There is no reason to expect another planet with life to differ from Earth in the overall development of living systems, so we greatly increase the chances of finding life by including a search for microorganisms as a key component of our strategy.

This realization underlay the design of the Viking experiments to find life on Mars. But an ordinary camera cannot photograph tiny organisms, and it is difficult (as design studies showed) to send a powerful microscope to Mars

and have it function properly on the planet's surface. Instead, Viking had to rely on experiments that would detect microorganisms in slightly more complex and indirect ways.

The Contamination Problem

From the beginning the Viking experiments had to deal with the possibility that the carefully planned and marvelously miniaturized laboratory that landed on the surface of Mars might detect Earth microbes that had been carried millions of kilometers to Mars on the lander itself! Concern over this possible **biological contamination** extended beyond the fact that microbes from Earth would render the life-detection experiments invalid. There was also the much more serious possibility that the entire planet might be contaminated by terrestrial organisms. After all, if Mars could support its own life forms, it might also prove hospitable to some terrestrial microorganisms. Thus even if subsequent, better designed spacecraft succeeded in identifying life on Mars, it might be only transplanted terrestrial life, and perhaps we would never know the difference. Much was at stake if the possibility of native martian life forms was to be investigated properly.

The Viking scientists overcame the problem of possible biological contamination through high-temperature sterilization of the entire spacecraft. The two landers were baked in a giant oven at a temperature above the boiling point of water for three days and then sealed in capsules for transport to Mars.

The long-term protection of Mars from possible contamination remains a concern for the planetary exploration community. We certainly don't want to pollute the one other planet in the solar system that might have developed life independently from the Earth. A variety of international agreements exist that are designed to protect Mars from unsterilized spacecraft for the next forty years, in the hopes that a definitive answer to the question of martian life will have been obtained by then.

Experiments to Detect Microorganisms

Suppose all these obstacles are solved and an uncontaminated soil sample has been safely brought inside the sterile Viking spacecraft. What do martian microbes like to eat and drink? Here the Viking scientists relied on basic principles of biology and chemistry, principles that must be at least somewhat biased from the fact that we derive them from one single example of biology, life on Earth.

With this bias firmly in mind, the principles are the following: Life on Mars, if it exists, ought to be based on carbon chemistry and should involve a fluid solvent, some common substance that can exist as a liquid under martian conditions. For Mars, which has no ammonia or alcohol, this means a carbon chemistry with water; that is, a system of life essentially identical in its chemical outlines to life on Earth. This may seem a restrictive conclusion, but the imposition of a chemical similarity at such a basic level permits enormous flexibility in the type of life that might actually exist. For example, it need not have DNA and RNA or use the same twenty amino acids as life on Earth. So these principles don't really provide much help in determining a suitable menu to attract and nourish our prey. To use an Earth-bound analogy, a fly-fisherman intent on catching trout would settle down at this point and study the local environment in more detail. This is the logical next step for a martian microbe hunter as well.

All organisms, even microbes, develop an intimate connection with the places they inhabit. If microorganisms are sufficiently abundant, they will modify the chemistry of their surroundings, thus producing evidence of life that should be detectable through analysis of either the atmosphere or the soil of Mars. This is the philosophy behind the Viking life-detection experiments.

Design of the Experiments

After considering many possibilities, the Viking project selected three experiments to search for microorganisms on Mars. These experiments, called the gas exchange (GEX), the labeled release (LR), and the pyrolytic release (PR), reflect scientific experience with life on Earth. For example, all the organisms that we know derive their energy from two basic processes, oxidation (removal of hydrogen, combination with oxygen) and reduction (removal of oxygen, combination with hydrogen). Both of these processes deserve investigation on Mars, but how? What would the Martians "eat" for energy, and under what conditions?

The experimenters were forced to return to the general principles we discussed above, tempered by their knowledge about conditions on the planet. Thus they judged it reasonable to offer the martian microbes some suitable mixture of organic (carbon-based) nutrients dissolved in water, to see what — if any — reactions would occur with this diet. The GEX and LR experiments followed just this procedure.

The GEX Experiment

The **gas exchange (GEX)** experiment put a soil sample in contact with several dozen likely nutrients (Fig. 10.6). These nutrients have such broad appeal for Earth-based life that the GEX quickly became known as the chicken soup experiment. Once the soil sample came in contact with the chicken soup, the question was: Would the gas chromatograph attached to the chamber detect any changes in the composition of the gas above the soil? Such changes could arise from the life processes of the martian microbes. On Earth, the chicken soup approach would reveal the presence of life through changes in the amount of oxygen, carbon dioxide, or hydrogen in the chamber caused by the metabolic activity of organisms in the soil.

The LR Experiment

The **labeled release (LR)** experiment aimed at checking on biological activity more directly, using a set of compounds that were labeled by substituting radioactive carbon atoms for some of the ordinary carbon atoms in the compounds (Fig. 10.6). This labeled mixture dripped onto the soil sample, and the gases above the sample

were tested to see whether any radioactive compounds, such as carbon dioxide or methane, had been released by the life processes of martian organisms. On Earth, the LR would be viewed as a respiration experiment, designed to see whether organisms are releasing gases into the atmosphere.

The GEX and LR experiments have two basic drawbacks. First, liquid water cannot exist on Mars today, so martian microbes may be completely unused to a watery medium. Instead of drinking the water, they might drown, dissolve, or undergo a destructive chemical reaction. Second, organic nutrients found delicious by terrestrial organisms may be found poisonous or simply inedible by martian microbes.

FIGURE 10.6 A schematic representation of the GEX, the LR, and the PR experiments shows that each of them analyzes the martian soil in a different way to test for the presence of life. The GEX experiment exposed a broth of nutrients to a few grams of soil and then looked for changes in the gas above the soil and nutrient mixture. The LR experiment tagged carbon-rich compounds with radioactive carbon-14 atoms in place of some of the usual carbon-12 atoms. These labeled compounds then dripped over the soil sample. Any biological processes should have caused some tagged compounds to appear in the gas above the sample. The PR experiment replaced the normal Mars atmosphere with an equivalent set of gases labeled with radioactive carbon atoms. Any organisms that ingested some of these labeled molecules would produce a radioactive signal when the soil in which they lived had been roasted.

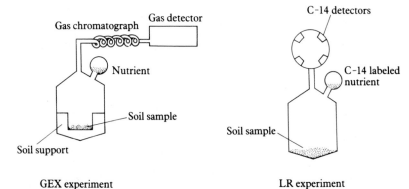

GEX experiment

LR experiment

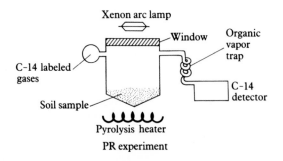

PR experiment

The PR Experiment

The third biology experiment, the **pyrolytic release (PR),** dealt with these drawbacks. In this experiment, martian soil was maintained in an environment virtually identical to that on the planet's surface, but the atmospheric gases in the test chamber were labeled by adding carbon dioxide and carbon monoxide that had been tagged with radioactive carbon (Fig. 10.6). After allowing any organisms to live for a while, the soil was heated to 750 C, and the volatile gases released by this heating passed into a vapor trap and then into a counting device.

Only carbon dioxide and carbon monoxide could pass completely through the vapor trap. Once these two gases had left the system, the trap was heated to drive off any organic vapors that might have been produced. These vapors could be recognized because they would contain some of the radioactive atoms from the added carbon dioxide or carbon monoxide. In other words, the PR experiment incinerated the remains of martian microbes to release any carbon atoms that the microbes had incorporated through biological activity. The most important terrestrial activity of this kind consists of plant photosynthesis, in which carbon dioxide in the Earth's atmosphere is converted into organic compounds by green plants. Baking the plants vaporizes these organic compounds. In the PR experiment these vapors would be radioactive, and could therefore be detected by charged particle counters.

10.4 Viking Results

The Viking GCMS

Once the first Viking lander descended onto the surface of Mars and the first soil sample rode the scoop into the test chambers, the results quickly came back to Earth. No evidence of organic compounds appeared in the martian soil, and the atmosphere of Mars showed only the compounds expected if life did not exist.

The Viking lander made its soil analyses with a highly sophisticated instrument called a gas chromatograph-mass spectrometer, or **GCMS.** An example of the ability of the GCMS to detect organic matter is shown in Fig. 10.7, which illustrates the results of an analysis of Antarctic soil on Earth with a laboratory version of the Viking instrument. Although this soil contains barely enough living organisms to give weak positive results in the three Viking life-detection experiments, the GCMS found lots of organic compounds. In other words, the fingerprint of life can be seen more easily than the living creatures themselves. For instance, the Antarctic soil sampled in Fig. 10.7 contains 10,000 times more carbon in organic molecules than in microorganisms.

We must realize, however, that not all organic compounds require living creatures to produce them. Fig. 10.7 also shows a GCMS analysis of a carbonaceous meteorite that contains amino acids, the building blocks of proteins. Again carbon-laden molecules are present in abundance, but these organic molecules were not made by life. Thus, the detection of organic molecules in general would not show that life exists on Mars, for the simpler kinds of organic compounds could have been brought to Mars by meteorites or might even have arisen from photochemical reactions in the martian atmosphere.

The Martian Soil

In practice, these potential ambiguities did not plague the GCMS experiment, because it found no organic compounds whatsoever in the soil of Mars. On the scale of Fig. 10.7, the martian soil results would be indistinguishable from the zero horizontal line of the graph. These results apply to both landing sites and to two different samples at each site, including one from underneath a

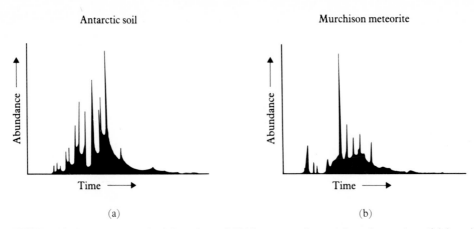

FIGURE 10.7 A test model of the Viking GCMS was used to analyze Antarctic soil (a) and a piece of the Murchison meteorite, a carbonaceous chondrite, (b). The graphs indicate that a rich variety of organic substances exists in each of these samples. Each peak in the graphs represents one or more organic substances. On this scale, the GCMS analysis of martian soil would be a straight horizontal line at zero, indistinguishable from the horizontal axis itself.

rock that might (so it was thought) have sheltered organisms from deadly ultraviolet light. The only compounds found in the soil by the GCMS were water and carbon dioxide, hardly surprising since these gases exist in the planet's atmosphere. The upper limits on all likely organic compounds in the soil fall at a few parts per billion. This is the kind of sensitivity that would allow detection of the proverbial needle in a haystack.

These negative results set powerful constraints on any models of martian biology: How could life exist on Mars without leaving any trace of its presence? Could martian organisms be such efficient scavengers that all traces of their wastes, their food, or their corpses could not be found, even with the great degree of sensitivity that the Viking landers brought to Mars? The biologists involved in the Viking project considered this an extremely unlikely possibility.

In view of these negative findings, the prospects for finding life on Mars with the experiments designed for precisely that purpose seemed poor. (The trout fisherman would be inclined to move on to another stream.) So the Viking scientists received quite a shock when all three of the experiments designed specifically to search for life showed positive results! Something very peculiar was happening on Mars, but was it truly biological?

Initial Results from the Biological Experiments

On 28 July 1976, the eighth day after the first landing on Mars, the Viking 1 Lander scoop dug a trench in the soil and distributed samples to the various experiments. The GEX placed about a gram of soil into a tiny, porous container positioned above the nutrient medium. Two days later, the first analysis of the gas in the container showed an exciting result: A large quantity of oxygen had appeared in the chamber, fifteen times the proportion in the planet's atmosphere! The simple exposure of martian soil to the humidity in the test chamber produced by the nutrient-laden fluid had apparently been sufficient to liberate oxygen from the soil. Was this an indication of life on Mars? After months of testing, the biologists concluded that they were observing not biological activity but merely the chemical interaction of martian soil with a higher pressure of water vapor than had been present on Mars for millions of years. In other words, not the chicken soup itself, but the humidity it generated, had led to oxygen-releasing chemical reactions in the martian soil.

The day after the first GEX data, the LR experiment also reported a positive result! But the sequence of events that led to this report was once again followed by a controlled experiment that diminished its impact. After checking to be sure that the background radioactivity level was low, the LR added about two drops of the radioactive nutrient material to the soil that had been brought into its chamber. A sudden rise in the radioactivity of the gases above the soil sample appeared, a more dramatic reaction than biologists had found in their comparison experiments with many life-bearing soils on Earth.

Unfortunately for those who had hoped for proof of life, the Viking scientists soon realized that the radioactive gas, almost certainly carbon dioxide, could arise from simple chemical reactions that involve oxygen compounds called peroxides. If, for example, hydrogen peroxide (H_2O_2) exists in martian soil, it could easily react with an organic compound in the nutrient medium, such as formic acid (HCOOH), to form water and carbon dioxide. A second wetting of the soil showed no increase in the amount of radioactivity in the gas in the test chamber. In fact, the additional nutrient apparently absorbed some of the radioactive carbon dioxide that was originally released. Hence, the scientists concluded not that life exists on Mars, but rather that the martian soil may contain chemicals such as peroxides that release carbon dioxide when exposed to simple organic compounds.

PR Experiment Results

Since the first two experiments yielded information that seemed ambiguous, the Viking team eagerly awaited the results from the PR experiment, which did not use a water-based nutrient and thus avoided one of the primary agents (water) that the biologists suspected of causing purely chemical reactions in the soil. The PR experiment required an incubation time of five days, during which radioactive carbon monoxide and carbon dioxide stayed in the sample chamber. Analysis of the initial experiment revealed that radioactive carbon had indeed become part of compounds in the soil (Fig. 10.8). Weak as this signal was, it seemed clearly positive: to the PR experiment, martian soil behaved much like an Antarctic soil on Earth, nearly sterile but not entirely so. Thus, this experiment, apparently the most difficult to fool by nonbiological reactions, yielded an undeniably positive result. Yet

FIGURE 10.8 The PR experiment showed a rise in the amount of radioactivity in the gases above the soil being tested for living organisms once C-14-labeled carbon dioxide and carbon monoxide were introduced. This suggested initially that living organisms had incorporated some of the radioactive carbon atoms from the simulated martian atmosphere.

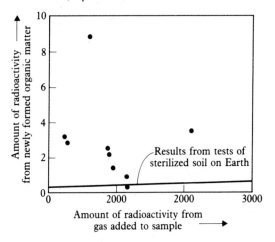

even in this case the Viking scientists were skeptical that they had found life on Mars.

The reason for this skepticism comes from the fact that when the scientists arranged to heat the martian soil in separate experiments to 175 C and to 90 C respectively for several hours before the radioactive gases were injected, they still found positive results from the PR experiment. The higher temperature reduced the reaction by 90%, but a positive signature still emerged. The lower temperature had no effect. Since Mars' surface temperature is never more than 30 C, the Viking scientists did not believe that any martian life could have adapted to survive three hours at 175 C, which no terrestrial organisms can do. Furthermore, because the GCMS showed that the soil lacks any organic material down to a level of a few parts per billion, the likelihood that a tiny amount of the (supposed) heat-resistant microorganisms could provide the positive results of the PR experiment seemed minuscule. Effects arising from the nature of the soil itself are much more likely.

The Chemically Active Martian Soil

The consensus among the Viking biologists is that martian soil contains loosely bound oxygen in various compounds, such as peroxides. In the absence of a shielding amount of atmospheric ozone, solar ultraviolet light reaches the surface and apparently enhances the chemical activity of the iron-rich soil. Because the temperature on Mars is low and because liquid water is entirely absent from the planet's surface, these active compounds can remain outside the state of chemical equilibrium for long periods of time. If even a small amount of water is added to the soil, however, it produces chemical reactions that mimic the effects of biological activity.

The Viking missions have revealed that martian soil is chemically active, but probably contains no carbon-based living organisms, at least at the two landing sites. Could the landers have

killed any microbes by landing nearby or by bringing them inside the spacecraft? This seems unlikely, since any such microbes must have adapted to daily temperature variations of almost 100 C, and the temperature inside the spacecraft (27 C) did not exceed the maximum temperature on the planet's surface. The PR experiment, in fact, was designed to simulate the actual martian environment as closely as possible.

The Variety of Martian Environments

What about life elsewhere on Mars? The similarity between the soil analyses at the two landing sites, separated by more than 2000 kilometers, argues against great variations from one place to another. This argument gains strength from current models of the planet's surface, which suggest that the soil is turned over (gardened) by meteorite impacts, volcanic eruptions, and winds to depths of one to several meters on time scales ranging from tens of thousands to tens of millions of years.

Strong evidence also exists that the soil particles are scattered across the entire surface of Mars during the periodic global dust storms. Thus every part of Mars' surface seems likely to come in contact with every other part, so that sampling the dusty material at one location should be equivalent to sampling at all locations, as was indeed the case for the soil analyses made at the two Viking landing sites.

A possible exception to this rule arises in the polar regions (Section 9.7). Since the north pole of Mars has a permanent cap of water ice, specialized environments could exist at the edges of this polar cap where conditions would be more favorable to life during the summer months. A relatively abundant source of water, one that we know enters the atmosphere and then freezes again, lies in the polar cap. The ground temperatures there, as measured by the Viking orbiters, reach -33 C, nearly as warm as the highest temperatures measured near the second landing site.

They are not, however, subject to extreme changes during summer as Mars rotates.

Simply because the polar regions stay cold at all seasons organic material might concentrate there, since it could never warm up, vaporize and be blown away. The south polar region of Mars, not as well observed as the north polar cap, seems to be still colder; thus the north pole of Mars seems to offer the most hospitable high latitude environment at the present time.

However, most hospitable may not be good enough; color pictures of Mars show that even near the pole the bare ground has the same reddish color as the rest of the planet. Here, too, the soil may be gardened and blown about by the winds, and it is certainly sterilized by ultraviolet light from the Sun. This soil should therefore be as thoroughly oxidized as the soil tested by the biology experiments at lower latitudes.

Models for the wind circulation on Mars show that the dust that becomes airborne in middle latitude regions will be carried to the poles, where layers of material have (presumably) built up as the result of repeated cycles of erosion, airborne transport, and deposition of fine dust particles. Even here, it seems likely that no living organisms exist in the soil, which should resemble the soil already tested at the two Viking sites. Perhaps under the ice at the pole there exist some protected carbon compounds, but here the continual low temperatures seem to rule out the existence of active life. Remember that in winter temperatures fall to 150 K, a rugged environment for the survival of life.

Dogmatic statements have fared poorly in the history of science, and we must be reluctant to state categorically that no life exists on Mars. All of the evidence accumulated so far points toward the absence of life, though some discovery may one day convince biologists that Mars does harbor some living organisms. Comparing Mars as we see it now with our own planet, it seems clear that no known terrestrial organisms, including the toughest and best adapted microbes, could grow in the present martian environment.

On the other hand, we cannot exclude the presence of life on Mars in the past, when its atmosphere was more substantial and water flowed on its surface. Thus we turn again to the question of past climates of Mars.

10.5 Evolution of the Atmosphere and Surface

Early Geological History

In Chapter 9 we discussed many aspects of the geology of Mars, including the results of efforts to determine the ages of surface features such as the channels, and in this chapter the chemistry of the atmosphere and soil have been described. Let us now see if we can combine this information with what we know of the histories of the Earth and the Moon to outline the development and evolution of Mars.

Presumably Mars formed along with the rest of the planetary system 4.5 billion years ago. This date of origin has been measured only for the Earth, the Moon, and the meteorites, but no one seriously questions its applicability to the rest of the system. From the existence of heavily cratered uplands on Mars, we also know that the crust of the planet had formed in time to record the late heavy bombardment. It seems highly probable that this period of higher impact rates coincided with the late bombardment of the Moon about 3.9 billion years ago. It also seems clear that during the first few hundred million years of its existence Mars differentiated, and that outgassing of volatiles created a substantial atmosphere.

Detailed examination of the cratered terrain suggests that the period of late bombardment was also a time of enhanced erosion. Sometime not long after, the extensive run-off channels of the southern uplands were formed. For a period of perhaps hundreds of millions of years, Mars may have been blessed with blue skies, rivers, and even the miracle of rain.

Volcanic Upheaval

Two great events soon altered this state: the loss of much of the atmosphere, and a nearly simultaneous transformation of the northern hemisphere due to an internal thermal evolution of the planet. Over nearly half the surface, the old cratered terrain was destroyed, the surface collapsed, and vast lava flows spread across the land. At about the same time, catastrophic floods caused by the melting of permafrost originated in the southern uplands and carved immense channels as they deposited their flows into the northern volcanic basins. If our cratering chronology is correct, the formation of the northern lowlands and the activity of the outflow channels took place at the same time the lunar maria were forming.

The continuing evolution of the interior of Mars now created another asymmetry, as a massive uplift produced the Tharsis bulge separate from the north-south division between the uplands and the volcanic basins. A similar, smaller uplift took place in the Elysium region. As the land rose it also split, forming networks of tectonic cracks thousands of kilometers long; many of these joined to form the Valles Marineris. Great volcanoes formed with the repeated eruption of lava from the same vents. By two billion years ago most of these volcanoes had fallen silent, but the four largest in the Tharsis area persisted and may be intermittently active even today.

We can assign approximate dates to these geological events from crater counts. We can also estimate the much shorter time scales involved in producing the layered polar deposits from periodic changes in the orbit and rotation axis of Mars. But we can only venture a guess as to the dates when the atmosphere and climate of Mars shifted to their present state.

History of the Atmosphere

To trace the past history of the martian atmosphere, we need to examine isotope ratios and noble gas abundances, the same clues that were so helpful on Venus and the Earth. Deuterium has just been detected on Mars, so we will soon be able to use this isotope to help determine the history of water on the planet. Meanwhile, recall that the hydrogen escaping from Mars can be observed directly. Knowing how much hydrogen is escaping from Mars today, we can calculate how much has left the planet in the past. This amount is equal to the hydrogen contained in a layer of water that would cover the entire surface of Mars to a depth of about 4 meters.

At least this much water has escaped, but how much was present to begin with, and how much carbon dioxide, nitrogen, etc.? On Earth, we attacked this problem by estimating the total amount of carbon dioxide in carbonate rocks. Unfortunately, none of the Viking experiments was designed to detect carbonates, so we have no idea how much gas may be chemically bound in this way. Instead, it is necessary to proceed indirectly, returning to isotopes and noble gases.

The nitrogen on Mars is highly enriched in its heavier isotope, providing a clue similar to the deuterium enrichment on Venus: a lot of nitrogen had to escape from Mars to produce such an effect. Knowing the escape process and the present value of the isotope ratio, it is possible to calculate the original nitrogen abundance. The conclusion is that Mars must have started out with about ten times the nitrogen we now find in its atmosphere.

The noble gases tell a similar story. As we saw in the discussion of the atmosphere of Venus (Section 8.7), the relative amounts of neon, argon, krypton, and xenon are the same on Mars and the Earth, suggesting that other volatiles, such as carbon and nitrogen, should also be present in similar relative abundances on the two planets. Once the noble gases are released into the atmosphere, they remain, whereas the other volatiles can escape or combine chemically with the soil. Knowing the ratio of total nitrogen and carbon dioxide to atmospheric neon on Earth and knowing the amount of neon on Mars, we can calculate how much nitrogen and carbon dioxide should be present on the red planet.

The result for nitrogen obtained from the noble gas abundances is similar to that from the nitrogen isotope study. Buoyed by this agreement, we can examine the results in Table 10.2. Evidently Mars outgassed an atmosphere great enough (if it were all present at the same time) to produce a surface pressure on that planet equal to at least twice the sea-level pressure of our own atmosphere. The composition of this reconstructed atmosphere is similar to the present atmosphere of Venus and the reconstructed (no life, no rock weathering) atmosphere of the Earth.

This denser, carbon dioxide-rich atmosphere could produce a greenhouse effect that would lead to an average surface temperature above 0 C. These are exactly the conditions that are needed for the development of the runoff channels discussed in Section 9.6: a high surface pressure and a warm surface temperature. In this environment, liquid water can flow freely over the martian surface.

Catastrophic Climate Change

What happened to this early atmosphere? Without detailed chemical information on the composition of surface rocks, we cannot be sure. One likely possibility is that the liquid water created its own demise. Dissolving atmospheric CO_2, it would weather the ancient rocks and form deposits of carbonates. Evidently this chemically bound CO_2 was not replaced or recycled.

As the amount of CO_2 in the atmosphere diminished, nitrogen was also escaping into space. The total surface pressure gradually decreased. As the greenhouse effect became weaker with less CO_2 and less total pressure, the planet began to cool. Such cooling reduced the amount of water vapor in the atmosphere, thereby further decreasing the greenhouse effect. This positive feedback is similar to that established for the

TABLE 10.2 Atmospheres of inner planets[a]

Gas	Venus	Earth	Mars
N_2	3.4%	1.9%	1.7%
O_2	trace	trace	trace
Ar	40 ppm	190 ppm	850 ppm
CO_2	96.5%	98%	98%
Water	> 9 m	3 km	30 m
Pressure	88 ± 3 bar	~ 70 bar	~ 2 bar

[a]No weathering, no life, no escape.

runaway greenhouse on Venus only with opposite results, and we are left here with a runaway refrigerator.

◆

10.6 Goldilocks and the Three Planets

Comparing Mars with Earth and Venus

Let us reflect on the reasons for the dramatic differences between Venus, Earth, and Mars. Of the three inner planets, only Earth has abundant liquid water on its surface and only Earth has abundant life. These two facts are obviously related, since life as we know it is totally dependent on water. Liquid water requires that the planetary temperature be kept within certain limits. Like the bowls of porridge confronting Goldilocks, one planet is too hot, one is too cold, and only Earth is just right!

Planetary temperature is primarily determined by the distance from the Sun. As we saw in Section 8.7, if the Earth were closer to the Sun, it would develop a runaway greenhouse and become much like Venus. However, moving the Earth farther from the Sun than its present orbit won't necessarily produce the runaway refrigerator we have just described for Mars. Our planet could stay warm if the composition of its atmosphere were changed, by introducing larger quantities of carbon dioxide.

Thus the problem with Mars is not just its greater distance from the Sun; Mars must be in trouble for some other reason. Looking at Table 10.2 again, we notice that even when we attempt to estimate how large the reconstructed atmosphere of Mars might have been, we find a total surface pressure thirty-five times smaller than the comparable reconstructed atmosphere for Earth, which in turn is approximately equal to the bulk of the present atmosphere of Venus. This is surprising, since we might reasonably expect that Mars would be more rich in volatiles than either Earth or Venus, because it formed farther from the Sun in a cooler region of the solar nebula where volatile substances could condense more readily.

In fact Mars may once have had an atmosphere comparable to those of Venus and the reconstructed Earth. With the current popularity of the massive impact theory for the formation of the Moon (Chapter 6), studies have suggested that large impacts must have occurred on the other inner planets as well. Such events would cause the loss of any existing atmospheres, which would have to be replaced by subsequent bombardment by volatile-rich meteorites and comets. Because of its small size, Mars would be likely to suffer more atmosphere-removing impacts for a longer period of time than its two larger sister planets. By the time these large impacts had ceased on Mars, the flux of small bodies that could *add* volatiles to the planet was greatly diminished also. Hence the subsequent atmosphere, which is the one we reconstruct from the nitrogen isotopes and the noble gases, was never very massive.

If Mars had been a bigger planet, it presumably would have retained a greater supply of the volatile elements, such as carbon and nitrogen, that are essential for the existence of a dense, stable atmosphere within the inner solar system. A larger planet would also have released (outgassed) more of these retained volatiles from its interior because of its more vigorous tectonic and volcanic activity, which would be driven by its greater internal heat. It could thereby replenish some of the gases lost to chemical compounds in the soil. This hypothetical Mars would no longer show a surface pockmarked by craters, for the primitive crust would have disappeared, just as it has on Earth.

Would the thicker atmosphere have managed to maintain itself over billions of years? This would depend on how large a greenhouse effect our hypothetical Mars would have established to counteract the lower temperatures that arise at its distance from the Sun.

An Ancient Eden?

Even on the actual Mars, early conditions were evidently much more clement than the harsh, cold climate we find there today. While the huge, flood-formed outflow channels could occur in the presence of a low-pressure atmosphere, the branching drainage systems of runoff channels seem to require a warmer, thicker atmosphere. This requirement in turn is entirely consistent with the reconstructed atmosphere described in Table 10.2. A surface pressure of 1 to 2 bars suffices, given a high proportion of CO_2 in the atmosphere.

This table also suggests that all three planets had an equal proportion of the major gas-forming elements produced by outgassing. Only on Earth was the resulting atmosphere changed forever by the origin of life. If the list of elements was so similar, then presumably the early conditions on all three planets could have been similar as well. If Earth had an early atmosphere of hydrogen-rich compounds, perhaps Mars and Venus did also. On Mars, at least, some of the same chemistry that took place on Earth might have occurred, given the existence of liquid water on Mars at that time. In our present stage of ignorance, we must even consider the possibility that life actually originated on Mars, only to die out as the climate changed to its present state (Fig. 10.9).

In this context, it is interesting to point out that the most ancient life found on Earth so far (the 3.5-billion-year-old stromatolite shown in Fig. 7.16b) was contemporaneous with the formation of some of the large channel systems on Mars. During the first billion years of the history of our planet, life began and evolved to the point where it could leave a visible trace of its existence. It was this same billion years in which liquid water was present, at least sporadically, on the surface of Mars. Might there be traces of martian life, waiting for some future expedition to discover them?

Future Missions: Searching for Signs of Ancient Life

The timing of the existence of water and the speed with which life is capable of originating are critical factors. Cold as it is today, Mars is still warm enough for water vapor to be present in its atmosphere. This would have been even more true in the past. The breakup of H_2O by solar ultraviolet light provides a ready source of oxygen atoms that will combine with surface materials, organic compounds, and other atmospheric gases. Thus any inner planet will inevitably progress toward an oxidized state, regardless of whether or not life evolves there.

The red sands of Mars show us the end result. The present surface of this planet is directly attacked by solar ultraviolet light and active oxygen compounds, while organic compounds are totally destroyed. This bleak landscape gives us an even greater appreciation of the remarkable balance that life maintains on Earth — between its own hydrogen-rich chemistry and the oxygen-rich environment in which it lives. So the question becomes: did this remarkable system evolve quickly enough on Mars — before solar ultraviolet radiation sterilized the planet for all time?

Finding the answer to this question will require a much more sophisticated mission than any we have sent to Mars so far. Consider the

FIGURE 10.9 A flood channel intersecting martian craters. Evidently water from the channel has poured into the large crater at the left-hand side of the breach in the crater wall. Craters may have been the martian equivalent of ponds or tide pools on the early Earth.

fact that the ancient stromatolite on Earth was found in 1977. The surface area of Mars and the land area of Earth are nearly equal, so it will not be an easy search! We will need intelligent roving vehicles, capable of gathering samples to be sent back to Earth for examination in our laboratories (Fig. 10.10).

These examinations will obviously span a much wider range of disciplines than biology. Even if life never developed on Mars, the ancient rocks that we should be able to ferret out from locales protected from that hostile, oxidizing surface environment will have a record of conditions during the first billion years on Mars. Unlike the ancient rocks from the Moon, the rocks from Mars will have a story to tell about an early atmosphere (was it full of hydrogen compounds?) on the only other planet we know that once had rivers of liquid water coursing over its cratered surface.

FIGURE 10.10 a) An artist's conception of a Mars rover mission. Future exploration of the planet might include vehicles like the one in the foreground that would roam over the surface, collecting samples to be returned to Earth. b) Valery Barsukov, Director of the Vernadsky Institute of Geochemistry and Analytical Chemistry in Moscow, is a strong advocate of an early sample return mission from Mars.

(a)

(b)

Summary

Today the atmosphere of Mars is a mere remnant of its former self. The surface pressure of 0.006 bar is so low that liquid water cannot exist, even on occasions when the temperature rises above the freezing point. Just as the pressure on Venus is about one hundred times greater than on Earth, the pressure on Mars is less than $1/100$ that of Earth. The composition of the martian atmosphere, however, is very similar to that of Venus in its three main gases (CO_2, N_2, Ar), but quite different in minor constituents.

How can we search for life on another planet? A reasonable first step is to ask how inhabitants of some other world might hope to find life on Earth. This approach led early investigators to look for evidence of seasonal changes and artificial modifications of the natural landscape on Mars, with apparently positive results. Subsequent studies, however, showed that these early observations were incorrect. More subtle efforts that looked for traces of atmospheric oxygen or other chemically unlikely gases were also negative, nor were any radio or television broadcasts from putative martian civilizations detected.

A closer look at all the manifestations of life on our own planet shows that microbes are the most numerous and diverse organisms that have evolved here. This awareness was deeply ingrained in the minds of scientists proposing experiments for the Viking mission, which included a variety of experiments to characterize the martian environment such as cameras and a molecular analysis instrument that could search for traces of organic compounds in the martian soil. The latter found no trace of carbon-containing molecules, with a sensitivity of less than one part in a billion for some substances.

This background made it all the more remarkable when the three Viking experiments specifically designed to look for martian microorganisms all gave initially positive results. The first of these provided a rich nutrient medium to a sample of soil. A gas chromatograph then tested gases in the chamber containing the soil and the nutrient to see if any changes occurred. In fact, oxygen was given off as soon as water vapor from the nutrient medium reached the soil, but a test showed that the same reaction occurred when the soil was heated (which would have killed any organisms that were present) before introducing the nutrient.

The two other life-detecting experiments — one involving a medium tagged with radioactive carbon, the other using an atmosphere with radioactive carbon — also produced positive results that were evidently not caused by biology. The dry, iron-containing soil of Mars, exposed to solar UV and fragments of oxygen-containing molecules, must be very reactive.

Although life seems quite unlikely on Mars today, conditions in the past were much more favorable. With the aid of the noble gases and isotopes of nitrogen, we can infer that Mars once had an atmosphere larger than that of the Earth today, together with enough water to cover the surface to a depth of more than ten meters. As a result of its low temperature and smaller surface gravity, however, Mars could not retain this atmosphere. It experienced a runaway refrigerator effect, just the opposite of the runaway greenhouse on Venus, that has led to its present dry, cold, and inhospitable climate.

Key Terms

biological contamination	labeled release (LR) experiment
gas exchange (GEX) experiment	pyrolytic release (PR) experiment
GCMS instrument	

PART FIVE
REVIEW QUESTIONS

1. What would an informed college student have thought about Mars shortly after the turn of the twentieth century? Many knowledgeable people believed then that Mars was populated by a race of beings more intelligent than humans. How does this attitude compare with current opinions on UFOs, and the serious efforts being made to search for radio signals from intelligent creatures on other planets orbiting other stars?

2. Compare what can be learned about Mars with terrestrial astronomical instruments with what hypothetical martian scientists, similarly equipped, could learn about the Earth.

3. Compare the capabilities of the Viking Mars landers with the exploration of the Moon carried out by Apollo astronauts. What did Apollo accomplish that Viking did not? Was the human presence critical? How important is it that we send humans to explore Mars?

4. Contrast the surface of Mars as seen from the Viking landers with that of Venus as seen from the Venera landers. Compare both of these with the Earth and the Moon.

5. Discuss how the age of the martian surface and of major geological features such as the Tharsis volcanoes or the Valles Marineris can be found. What assumptions must be made to apply the lunar chronology to Mars? How accurate do you believe the ages are in Table 9.3?

6. Compare and contrast impact craters on Mars, the Moon, and the Earth. What are the consequences of the different amounts of atmosphere on the three bodies? How much erosion takes place on each?

7. Compare the martian volcanic features with those on the Earth and the Moon. Explain why the shield volcanoes on Mars are so much larger than their terrestrial counterparts. Why do you think there are no large shield volcanoes on the Moon?

8. Describe the various kinds of martian landforms that seem to be the result of flowing water. Are there similar features on the Earth? How old are the martian features? What do they indicate about the past climate on that planet?

9. Think about the history of water on Mars. How much might have been present, where did it come from, and where has it gone? How much is present today in the atmosphere, in the polar caps, and in subsurface permafrost? What would be required to release some of this water and thereby perhaps create a more favorable environment on the planet?

10. What do the polar caps reveal about the climatic cycles on Mars? Describe what you would expect the polar caps to look like close up. What would you expect from a project to drill into the caps and extract a core? How would such a core compare with one obtained on Earth from the Greenland ice cap or from the Antarctic?

11. What causes the great martian dust storms? How can such a small atmosphere generate such powerful storms? Compare these storms with storms on Earth. Could there be similar large dust storms on Venus also?

12. Compare the compositions of the atmospheres of Venus and Mars, the two terrestrial planets whose atmospheres have not been modified by life. Both are composed primarily of CO_2, but otherwise they are surprisingly different, particularly with respect to the critical noble gases. What are the possible explanations of these differences, and what can they tell us about the origins of the atmospheres of the terrestrial planets?

13. Describe the Viking biology experiments. Why did some of them yield positive results? Do you believe that there is any chance that they actually succeeded in detecting martian life, and that we are misinterpreting the results?

14. What is the significance of the failure of the Viking GCMS to detect organic matter? Review the possible sources and sinks of organic matter on the martian surface. Can you imagine a life form that would eliminate organic matter in the soil, or do you agree that the absence of organic matter argues strongly against life on Mars?

15. Could there be oases on Mars, places where life might flourish even though apparently none did at the two Viking sites? What would be the nature of such an oasis? Can you suggest a strategy for locating oases as part of a future program of Mars exploration?

16. Much interest exists today in setting a long-term goal for the United States space program of landing humans on Mars, just as the United States goal in the 1960s was landing on the Moon. Discuss the pros and cons of setting such a goal. If the goal were adopted, how long would it take to fulfill it?

ADDITIONAL READING

Arvidson, R.E. *et al.* 1978. "The Surface of Mars." *Scientific American 238*:1, 76.

Carr, M.H. 1976. "The Volcanoes of Mars." *Scientific American 234*:1, 33.

*Carr, M.H. 1981. *The Surface of Mars*. New Haven: Yale University Press.

Chapman, C.R. 1982. *Planets of Rock and Ice* (Chapter 11). New York: Scribners.

Cooper, H.S.F. 1980. *The Search for Life on Mars*. New York: Harper and Row.

Ezell, E.C. and L.N. Ezell. 1984. *On Mars* (NASA SP-4212). Washington: U.S. Government Printing Office.

Goldsmith, D. and T. Owen. 1980. *The Search for Life in the Universe*. Menlo Park, CA: Benjamin-Cummings.

Hoyt, W.G. 1976. *Lowell and Mars*. Tucson, AZ: University of Arizona Press.

*Klein, H.P. 1979. "The Viking Mission and the Search for Life on Mars." *Reviews of Geophysics and Space Physics 17*, 1655.

Masursky, H. 1982. "Mars." In *The New Solar System*, 2nd ed., ed. J.K. Beatty, B.T. O'Leary, and A. Chaikin. Cambridge, MA: Sky Publishing Corp.

Matsunaga, S. 1986. *The Mars Project*. New York: Hill and Wang.

*Murray, B.C., M.C. Malin, and R. Greeley. 1981. *Earthlike Planets*. San Francisco: Freeman.

Pollack, J.B. 1982. "Atmospheres of the Terrestrial Planets." In *The New Solar System*, 2nd ed., ed. J.K. Beatty, B.T. O'Leary, and A. Chaikin. Cambridge, MA: Sky Publishing Corp.

Soffen, G.A. 1982. "Life on Mars?" In *The New Solar System*, 2nd ed., ed. J.K. Beatty, B.T. O'Leary, and A. Chaikin. Cambridge, MA: Sky Publishing Corp.

Viking: The Exploration of Mars (NASA EP-208). 1984. Washington: U.S. Government Printing Office.

Washburn, M.L. 1977. *Mars at Last*. New York: Putnam.

PART ◆ SIX
The Outer Planets

When ancient astronomers named the planet Jupiter for the king of the gods in the Greco-Roman pantheon, they had no idea of the planet's true dimensions. The name is entirely appropriate, since Jupiter is larger than all the other planets combined. It has a faint system of rings and sixteen known satellites. Jupiter itself has an internal source of heat; it is radiating about twice as much energy as it receives from the Sun. This giant also has the strongest magnetic field of any planet. If we could see Jupiter's magnetosphere with our eyes, it would appear larger than the apparent diameter of our Moon in the night sky, even though it is almost 400 times farther away.

Saturn is not far behind. Most famous for its beautiful, intricate system of rings, this planet also has an internal source of energy and twenty-one known satellites. Like Jupiter, Saturn has a strong magnetic field and belts of charged particles. The other two giants, Uranus and Neptune, are both smaller and less well known. Each is depleted in hydrogen and helium relative to Jupiter and Saturn, and each

◀ An artist's pre-encounter concept of the view from the Voyager 2 spacecraft as it leaves the Neptune system on its way to interstellar space. The large satellite Triton is in the foreground. Seas of liquid nitrogen on Triton's surface reflect the dim light of the distant Sun, while a thin haze in its atmosphere partially blocks the view of Neptune's night-side hemisphere.

has a correspondingly larger core of icy and rocky material. Uranus has a system of dark rings and fifteen regular satellites, while the two known satellites of Neptune are both irregular. Uranus is tipped on its side with its rotation axis slightly below the plane of the solar system, while Neptune rotates normally. Finally Pluto, the outermost planet, is smaller than our Moon, and its low density suggests that it is composed of half water ice and half rock. It therefore bears a closer resemblance to one of the icy satellites than to any of the other planets.

Why are Jupiter and Saturn, and to a lesser extent Uranus and Neptune, so enormous? Why also do these planets have a fundamentally different composition and internal structure, dominated by volatiles and the light gases hydrogen and helium, from that of the inner planets? The key lies in lower temperatures of formation and the ready availability of hydrogen. With hydrogen as the most abundant gas, we expect the atmospheres of the outer planets to consist of hydrogen-rich compounds, and that is indeed the case. Instead of carbon dioxide (CO_2) there is methane (CH_4), instead of molecular nitrogen (N_2) there is ammonia (NH_3). Free oxygen (O_2) is impossible; if it were produced, it would quickly combine with the plentiful hydrogen to form H_2O. This compositional difference applies also in Part VII when we look at the satellites and rings of the outer planets.

CHAPTER ELEVEN

Jupiter and Saturn: The Biggest Giants

11.1 The Pioneer and Voyager Missions

Reaching to the Outer Solar System

Knowledge of the Jupiter and Saturn systems has grown dramatically during the last decade. The new information has come in part from ever more sophisticated ground-based observations, but especially from two sets of spacecraft: Pioneers 10 and 11 launched in 1972 and Voyagers 1 and 2 launched in 1977 (Fig. 11.1).

It is considerably more difficult to travel to Jupiter and Saturn than to Mars or Venus. The greater distances the spacecraft must traverse mean longer trip times and thus a requirement for higher reliability of all of the instruments and the spacecraft systems that support them. In the inner solar system, electrical power can be obtained by means of solar panels that convert sunlight to electricity. Since the less intense sunlight at Jupiter's distance is inadequate to produce the required power, spacecraft destined for the outer solar system must carry small electric generators powered by the heat of radioactivity.

The large distances also create difficulties in communication. Just as sunlight decreases in intensity with the square of the distance, so do the radio waves carrying messages to and from the spacecraft. That means huge antennas and sensitive receivers must be used. During part of the

FIGURE 11.1 The launch of Voyager 1 on 5 September 1977, the first of the two Voyager spacecraft to make the journey to Jupiter and Saturn. It has crossed the orbit of Pluto and is heading toward the Oort comet cloud.

Voyager 2 encounter with Neptune, for example, the largest array of radio telescopes on Earth (the VLA near Socorro, New Mexico, Fig. 2.11) will be employed to record signals from the spacecraft. These signals traveling with the speed of light will take four hours to reach the Earth. Obviously at such large distances a spacecraft must be much more autonomous than if it were visiting Venus, where commands and responses can be transmitted in a matter of minutes.

The Pioneer Encounters

Despite all these difficulties, the missions to the outer solar system have been very successful. The first of these were Pioneers 10 and 11, which encountered Jupiter in 1974 and 1975. Before these pathfinder missions it was uncertain whether spacecraft could be sent safely to the outer planets. First, there was fear of destructive impacts with dust in the asteroid belt which must be crossed to reach Jupiter and beyond. Second, the Van Allen belts of the inner jovian magnetosphere also posed a hazard.

In addition to their role as scouts for the Voyagers, Pioneers 10 and 11 carried the first scientific instruments to the outer solar system. Well instrumented to map out the charged particle belts, the Pioneers made numerous discoveries about Jupiter's magnetosphere as they passed through the system. A few photographs and other measurements were made of the planet, but since this was a spinning spacecraft their quality did not approach that achieved by the Voyagers a few years later.

The trajectory of Pioneer 11 brought it close enough to Jupiter to allow a gravity-assisted deflection (Section 2.6) across the solar system for a close flyby of Saturn in 1979. This passage led to the discovery of Saturn's F Ring as well as the first description of the planet's magnetic field and belts of trapped charged particles.

Again, the Pioneer served as pathfinder, passing through the plane of the rings at the same distance from Saturn required by Voyager 2 for its gravity-assisted trajectory to Uranus and Neptune.

The Voyager Encounters

The Voyagers were among the most sophisticated spacecraft ever used for planetary exploration, rivaling in size and complexity the two Viking landers (Fig. 11.2). Launched in 1977, Voyagers 1 and 2 reached Jupiter in March and July of 1979. Each encounter actually stretched over many weeks, as the spacecraft approached the planet while making close flybys of the Galilean satellites. Gravity assists at Jupiter then sent both spacecraft on to Saturn where Voyager 1 arrived in November 1980, followed by Voyager 2 in August of 1981.

The Voyager 1 encounter with Saturn was targeted to provide a close passage of its large satellite Titan. Therefore, the spacecraft could not pass close enough to Saturn to get the necessary boost required to reach Uranus or Neptune. The trajectory of Voyager 2, however, was chosen with this latter objective in mind. Accordingly, this spacecraft flew through the Uranus system in January 1986, where a third gravity assist directed it to Neptune for arrival in August of 1989 (Fig. 2.15).

The Voyager spacecraft are operated from the Jet Propulsion Laboratory (JPL) in Pasadena, just as were the Vikings. A staff of several hundred is required to operate the spacecraft and the complex computers and communication systems at JPL. In addition, more than one hundred scientists associated with thirteen spacecraft instruments are members of the Voyager team, led by Project Scientist Edward Stone, a physicist and magnetosphere expert from the California Institute of Technology (Caltech) who heads the scientific operation and serves as project spokesperson (Fig. 11.3).

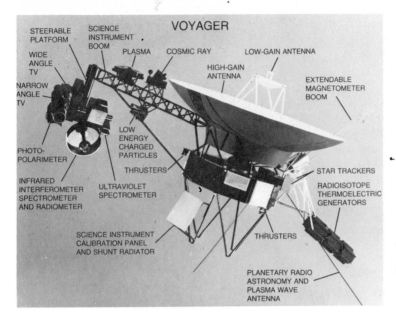

VOYAGER

STEERABLE PLATFORM
SCIENCE INSTRUMENT BOOM
PLASMA
COSMIC RAY
LOW-GAIN ANTENNA
WIDE ANGLE TV
HIGH-GAIN ANTENNA
NARROW ANGLE TV
EXTENDABLE MAGNETOMETER BOOM
LOW ENERGY CHARGED PARTICLES
PHOTO-POLARIMETER
THRUSTERS
STAR TRACKERS
INFRARED INTERFEROMETER SPECTROMETER AND RADIOMETER
ULTRAVIOLET SPECTROMETER
RADIOISOTOPE THERMOELECTRIC GENERATORS
SCIENCE INSTRUMENT CALIBRATION PANEL AND SHUNT RADIATOR
THRUSTERS
PLANETARY RADIO ASTRONOMY AND PLASMA WAVE ANTENNA

FIGURE 11.2 A diagram of the Voyager spacecraft showing the location of the various instruments. The big circular antenna at the top is about 4 m in diameter (cp. Fig. 2.18).

The two Voyager spacecraft have already returned such a large quantity of information from Jupiter and Saturn that the data are still being analyzed a decade after the initial encounters. As the description of their trajectories indicates,

FIGURE 11.3 Voyager Project Scientist Edward Stone of Caltech, who has supervised the Voyager science teams throughout the mission.

these are flyby missions — no instruments are sent into the planets' atmospheres. A complement of sophisticated equipment on the spacecraft makes measurements of the magnetic field, the properties of charged particles in the immediate environment, and uses remote sensing techniques to study the radio, light, and infrared radiation reflected and emitted by planets, satellites, and rings. In the following discussion, we rely almost exclusively on this wealth of new information.

Successful as they were, the Voyagers by no means represent the end of our efforts to understand the outer planets. A spacecraft named Galileo has been built to go into orbit around Jupiter after deploying a probe into the giant planet's atmosphere. This mission should occur in the mid-1990s. A Saturn orbiter with a probe into Titan's atmosphere is currently being studied as a joint mission called Cassini by NASA and the European Space Agency (ESA). If approved by both agencies, it will arrive in the Saturn system shortly after the year 2000.

11.2 Internal Structure: Journeys to the Centers of Gas Giants

Composition

We can see the basic difference between the gas giants and the inner planets by simply looking at their respective densities. The density of Jupiter is only 1.3 g/cm³, about the same as that of the Sun, while the density of Saturn is an even lower 0.7 g/cm³. For comparison, recall that the densities of the inner planets tend to increase with the planets' masses (except for anomalous Mercury), since gravitational compression forces the densities of material in the interiors to be higher than normal. Jupiter and Saturn would have densities much greater than 6 g/cm³ if they were made of the same materials as Earth. Thus these planets must be made of something far less dense than metal and rock.

The explanation, as we noted back in Section 1.6, involves the two most common elements in the universe, which are also the two lightest and the two most difficult to condense: hydrogen and helium. Early studies of Jupiter and Saturn quickly demonstrated that they could not be composed of pure hydrogen. If they were, their densities would be even lower. Adding helium to the models in about the same proportion that it is found in the Sun and other stars makes up the difference. There is also a need for a dense core in order to account properly for the planets' gravitational fields.

Constructing Theoretical Models

It is difficult to reach any further conclusions with great confidence. We have no seismic data to reveal properties of the interiors; nor can we expect to get any, since neither of these planets has a solid surface (Fig. 11.4). Instead we must use the data we obtain from remote investigations — mass, radius, rotation rate, heat balance, atmospheric composition, perturbations of satellite orbits — to construct models for the dis-

tribution and state of matter deep inside Jupiter and Saturn. This challenge is in many ways similar to that of modeling the interiors of the terrestrial planets (Section 6.4).

A basic problem in constructing such models of the giant planets is our lack of knowledge about the behavior of the planets' principal constituents, hydrogen and helium, at appropriately high pressures and temperatures. The central temperature of Jupiter must be about 25,000 K to be consistent with the emitted thermal radiation, while the pressure may be as much as 50 million bars, compared with just 4 million bars at the center of the Earth (Section 7.2).

Even before these extreme conditions are reached, we know that hydrogen will assume a metallic state — the pressure squeezes the hydrogen atoms so much that the electrons are no longer bound to the nuclei, giving the hydrogen the conductivity of a metal. This transition occurs at a depth of about 20,000 km, approximately 75% of the distance out from Jupiter's center; the exact range of pressures and temperatures for this transformation are not well known. Above this zone, hydrogen is in the molecular form of two atoms linked together (H_2), but both the molecular and metallic states are liquids. At still greater distances from the center, the hydrogen assumes the gaseous state that we observe in the outer layers of the atmosphere.

The Cores of the Giant Planets

Continuing our journey to the centers of these planets, we find that beneath the deep layer of liquid metallic hydrogen lies a core of rock and ice. Here the terms rock and ice are used very generally to denote compounds of silicon, oxygen, metals, and the heavy volatile elements. We have no idea exactly what materials are present. We only know that there is a strong concentration of mass at the center of each planet, and that this mass must consist of elements heavier than hydrogen and helium.

Both Saturn and Jupiter have cores of about

the same size: ten to fifteen times the mass of the Earth. Since Saturn is smaller than Jupiter, its core is relatively larger compared to the planet as a whole. In other words, Saturn is depleted in hydrogen and helium relative to Jupiter; if Saturn had its full quota of these two elements, it would be as large as Jupiter. We shall find that Uranus and Neptune also have cores of about ten to fifteen times the mass of Earth (Chapter 12). It appears that the formation of such a solid core may be a prerequisite for the formation of a giant planet (Chapter 15).

Calculations based on theories for the behavior of matter at the temperatures found in the giant planets indicate that Jupiter is just about as large in diameter as a planet can be. If we perform the thought experiment of adding more mass to Jupiter, we find that the self-compression caused by the increased mass would lead to a smaller diameter. Eventually, adding enough

FIGURE 11.4 Jupiter, as observed by Voyager 1, February 1979, at a distance of 30 million km. We are looking through an atmosphere of hydrogen and helium to layers of clouds made up of ammonia and other compounds. The white oval to the lower left is almost as big as Earth. Everything we see is in motion, currents moving along lines of latitude and eddies twisting clockwise and counterclockwise about their centers (cp. Plate 13).

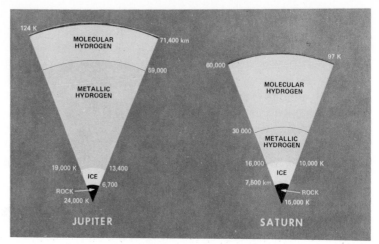

FIGURE 11.5 The interiors of Jupiter and Saturn probably consist of the same basic components, but they are distributed in different ways as these simplified diagrams indicate. The interior of Saturn is not as hot as that of Jupiter, allowing helium to condense. Both cores at their centers have masses approximately equivalent to 10–15 Earths.

mass would make the internal temperatures and pressures increase to the point where nuclear fusion reactions would begin. At this stage, corresponding to seventy to eighty times the current mass of Jupiter, a star is formed. Once internal energy is generated by nuclear reactions, a new adjustment of internal structure takes place allowing the object to grow in size. Meanwhile, it is interesting to note that a super-giant planet with ten or fifty times the mass of Jupiter would actually be smaller than Jupiter itself.

Fig. 11.5 illustrates the conditions one would encounter on voyages to the interiors of these two planets. There are no well-defined solid surfaces or even gas-liquid boundaries. Such transitions are only sharp under relatively low pressures. Instead we find immensely deep atmospheres in which the gases gradually become compressed to the point where they liquify.

Two Overheated Giants

Both of these giant planets radiate more energy than they receive from the Sun: Jupiter twice as much, Saturn nearly three times. We can think of each planet in terms of a giant light bulb; Jupiter radiates a total of 4×10^{17} watts, while Saturn glows with half this power at 2×10^{17} watts. Saturn is therefore equivalent to 2 million billion one-hundred-watt light bulbs, but its energy is obviously not emerging as visible light. The outer, visible layers of both planets are far too cold; instead, the energy is released as infrared radiation.

It is possible to calculate what the temperature of Jupiter should be from a knowledge of the local intensity of sunlight and the reflectivity of the planet. The answer is about 107 K (−166 C). An object at this temperature should radiate a predictable amount of infrared radiation. But observations of Jupiter at infrared wavelengths show that the amount of radiation it produces corresponds to a planetary temperature that is about 20 degrees warmer than this prediction. Since the flux of radiation increases as the fourth power of the temperature, this means that Jupiter radiates twice as much energy as it gets from the Sun, as stated above. What generates this extra energy?

Sources of Internal Energy

The answer does not lie in nuclear reactions, the ultimate source of sunshine and starlight. Hot as they are, the interiors of these planets are far too cold to permit nuclear reactions to begin. Higher temperatures require larger masses, and Jupiter is 70 to 80 times too small to become a star. Instead of invoking nuclear physics, let us return to Isaac Newton. If an apple falls from a tree, it acquires energy of motion from the Earth's gravitational field, which it gives up abruptly when it hits the ground. In the case of Jupiter, it is the contraction of the entire planet from an initial state when it was an extended cloud of gas and dust that produces the observed energy. This cloud was a condensation in the primordial solar nebula. Newton's apple may now be thought of as a piece of Jupiter itself, moving inward under the influence of the planet's gravitational field. The resulting energy of motion is transformed into faster motions of the gas molecules — an increase in the temperature of the gas. During the final stages of its formation process, when matter was streaming into the planet from considerable distances, Jupiter must have been very hot indeed, probably hot enough to produce a visible reddish glow. About 4.5 billion years ago, this hot young planet had a profound effect on its forming satellites, as we shall see in Chapter 13. Jupiter today is much quieter, slowly releasing primordial heat from that bygone era, possibly still contracting very slowly at a rate no greater than 1 mm/year.

Saturn: A Different Energy Source

Such a simple scheme will not account for all of the energy radiated by Saturn. Saturn, because of its smaller size, never reached the red-hot stage in its early youth. Hence there has been ample time in the ensuing 4.5 billion years for Saturn to cool much more than its giant neighbor. Why then is it still emitting so much energy?

The answer, ironically, is found in the lower *internal* temperature that Saturn exhibits today. At sufficiently high pressures and temperatures, liquid helium dissolves in liquid hydrogen in the same way that a cook can dissolve large amounts of sugar in hot water. But just as the cook has trouble stirring sugar into cold water, at the lower temperatures in Saturn's interior helium does not dissolve. Droplets of helium form in the liquid hydrogen, and being more dense move toward the center of the planet, like vinegar separating from oil in a salad dressing. This very slow helium rainfall, deep in Saturn's interior, takes us back to Newton's apple: once again gravitational energy is converted into energy of motion, which is transformed into heat and ultimately radiated into space.

This theoretical model for processes occurring deep inside Saturn can be tested by remote sensing. If helium is indeed raining out in the interior, it must be disappearing from the atmosphere. Therefore we expect to find the abundance of helium relative to that of hydrogen to be much smaller in Saturn's atmosphere than in Jupiter's. The Voyager Infrared Spectrometer measurements show exactly this result.

On Jupiter, the ratio of helium to hydrogen atoms, He/H, is close to the solar value of 7/100. But on Saturn, He/H is about 2/100. Even with the difficulties in the measurements, the amount of helium in Saturn's atmosphere is distinctly lower relative to hydrogen than the amount observed in the atmospheres of Jupiter or the Sun. The amount of missing helium is consistent with a rate of helium precipitation that would produce the extra energy that Saturn radiates. Hence there is good consistency between the model for Saturn's energy source and its two observable consequences: the amount of helium in the atmosphere and the amount of energy radiated by the planet.

11.3 Atmospheric Composition and Structure

Methane and Ammonia

The first gas to be identified in the atmospheres of Jupiter and Saturn was methane. Over fifty years ago, strong absorption bands in the spectra of both planets (Fig. 11.6) were found to coincide with absorptions in laboratory spectra of methane gas. Ammonia was discovered on Jupiter in the same way at about the same time, but whether or not·it was present on Saturn re-

mained a controversial problem until the late 1960s, when conclusive evidence for it was obtained.

At the pressures and temperatures in the visible regions of these atmospheres, ammonia can condense just as water vapor condenses in the lower part of our own atmosphere. Depending on the circulation of the atmosphere, we expect to find cloudy regions and clear regions when we look at these planets. The deeper we can see in the cloud-free region, the more gaseous ammonia vapor we find.

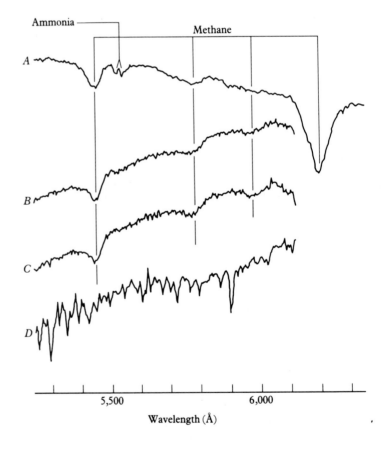

FIGURE 11.6 Spectra of Jupiter (*A*), Saturn (*B*), and Titan (*C*) showing methane and ammonia absorptions. These spectra have been divided by a spectrum of the Sun (*D*) to remove solar absorptions. If the planets had no atmospheres their divided spectra would simply be gently curving lines. Note that there is more ammonia on Jupiter than on the other two objects.

Variations in Composition with Altitude

Imagine looking down at the Earth from space. If you can only see the top of a thick white cirrus cloud, you are just looking through the stratosphere to the upper troposphere (Section 7.5). The air is very cold here, so only a little water vapor can be present. If your gaze wanders to a continent or an ocean peeking through the clouds, you are seeing down to the Earth's surface through a much warmer column of air, which can hold more water vapor.

The same thing happens when we view the giant planets, but we are less aware of it because these objects have no well-defined surfaces. Instead, we are looking at superimposed layers of clouds, the highest of which is mainly ammonia cirrus, clouds of ammonia crystals (Figs. 11.4 and 11.7). At deeper levels we expect clouds of liquid ammonia in combination with other substances as we shall see.

Since Saturn is farther from the Sun than Jupiter, it has a colder outer atmosphere so ammonia condenses more readily and less ammonia gas is present. Sunlight reflected by Saturn will therefore suffer less absorption by ammonia than sunlight reflected by Jupiter, exactly what we see in the spectra of Fig. 11.6. This explains the early controversy about the presence of ammonia on Saturn: it was difficult for the astronomers to detect these weaker absorptions.

To put this difference between the two planets on a scale, the temperature in the atmosphere at an altitude where the pressure is equal to the sea-level pressure on Earth (1 bar) is about 175 K (-98 C) on Jupiter but only 140 K (-133 C) on Saturn. This difference of 35 degrees means that a balloon floating in Jupiter's atmosphere at an altitude where the pressure equals 1 bar would be below the white ammonia cirrus, while a balloon on Saturn at the same pressure level would be surrounded by clouds of ammonia crystals. The lower gravity of Saturn leads to a more extended atmosphere, hence a thicker

cloud layer, which produces the much more uniform appearance of this planet as compared with Jupiter (Fig. 11.7; Plates 13 and 15).

Abundances of Atmospheric Gases

Although methane and ammonia dominate the spectra of these two planets (Fig. 11.6), they are only minor atmospheric constituents, occurring in smaller proportion on Jupiter than argon does in our own atmosphere. While it is 500 times more abundant than methane, hydrogen has a much weaker absorption spectrum because it is a molecule of two identical atoms that interacts only weakly with light. (Nitrogen and oxygen are identical-atom molecules so they are also nearly transparent to visible and infrared light, thus explaining their lack of importance for the terrestrial greenhouse effect.) Helium is even more difficult to detect, since its only absorptions at the temperature of Jupiter's atmosphere occur at very short wavelengths in the ultraviolet, which cannot be observed from the Earth's surface. Even from space, direct detection of helium is not easy because these short wavelengths are screened by hydrogen in Jupiter's atmosphere.

Despite these difficulties, we now have rather good estimates for the relative abundances of these two gases, as well as many others that have been detected by spectroscopy. The experimenters on the Voyager spacecraft succeeded in determining the helium abundances by measuring the effect of this gas on the observed spectrum of hydrogen. The hydrogen absorption bands are broader than they would be if only hydrogen were present. Some other undetected gas must be present to supply the pressure that produces the observed broadening, and helium is the only gas that would be sufficiently abundant to serve this function. The atmospheric abundances for both Jupiter and Saturn, based on

FIGURE 11.7 The atmospheres of both Saturn (a) and Jupiter (b) contain layers of clouds at different altitudes. Ammonia can condense over a greater altitude range on Saturn because of the planet's lower gravity. These diagrams show the approximate locations of major cloud systems on the two planets (After W.K. Hartmann.)

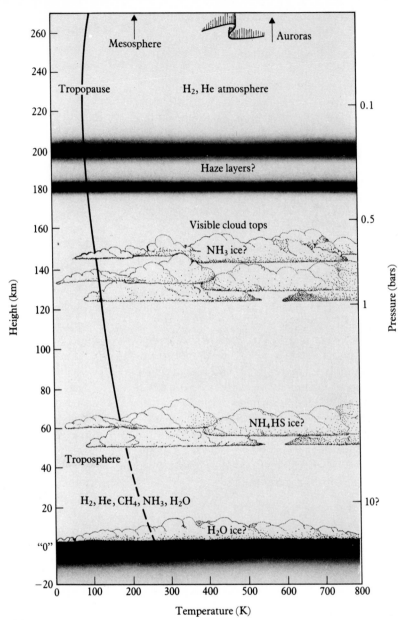

(a)

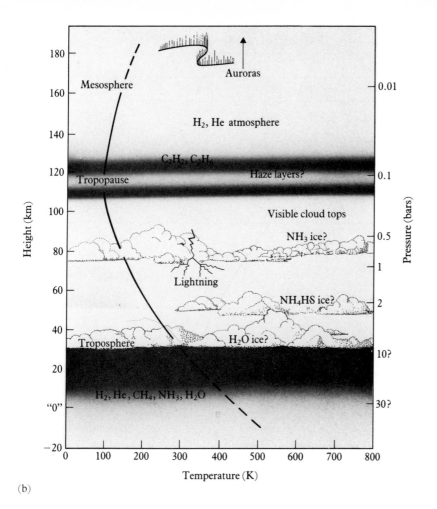

(b)

both Voyager and telescopic data, are summarized in Table 11.1.

Expected and Unexpected Gases

Table 11.1 is divided into two parts to distinguish the simple molecules that would be expected to form in these atmospheres from those whose presence is something of a surprise. As we said earlier, all the abundant elements should combine with hydrogen to make molecules. These should be the simplest molecules possible, given the huge excess of hydrogen and the relatively low pressures and temperatures in the upper atmospheres of these planets. The resulting compounds are indeed present and are shown above the dashed line.

In the regions probed by spectroscopy, the water vapor abundance is $1/10$ to $1/100$ of the

TABLE 11.1 The compositions of the atmospheres of Jupiter and Saturn

Main Constituents (percent)

Gas	Formula	Jupiter	Saturn
Hydrogen	H_2	86.1	92.4
Helium	He	13.8	7.4
Methane	CH_4	0.09	0.2
Ammonia	NH_3	0.02	0.02
Water vapor	H_2O	0.008(?)	—

Trace Constituents (parts per billion)

Gas	Formula	Jupiter	Saturn
Acetylene	C_2H_2	800	100
Ethane	C_2H_6	40000	8000
Carbon monoxide	CO	3	2
Hydrogen cyanide	HCN	2	7
Germane	GeH_4	0.6	0.4
Phosphine	PH_3	400	3000
Methyl acetylene	C_3H_4	?	trace
Propane	C_3H_8	trace	trace

amount that would be expected if oxygen were present on Jupiter in the same proportion (relative to carbon) as found on the Sun. Note that in the presence of hydrogen gas essentially all the oxygen will be in the form of H_2O. This missing water may be in clouds (Section 11.5) or in the planet's core (Chapter 15).

Table 11.1 contains a large number of other molecules as well, such as carbon monoxide (CO), hydrogen cyanide (HCN), acetylene (C_2H_2), and other hydrocarbons (carbon-hydrogen compounds). Evidently sources of energy other than local heat are acting to produce these substances. Photochemical reactions are taking place, in which solar ultraviolet breaks down methane, and its fragments form acetylene and other hydrocarbons. This is the same kind of photochemistry that produces ozone in the Earth's atmosphere and creates radicals like NH_2 and C_2 in the heads of comets.

In the convective region of the atmosphere where we see the clouds, lightning discharges (detected by the Voyager spacecraft, Fig. 11.8a) can contribute to these chemical processes and may be the main factor in the production of hydrogen cyanide. Still deeper, at temperatures around 1200 K, carbon monoxide is produced by a reaction between methane and water vapor. Since carbon monoxide is observable, vertical currents must be sufficiently strong to bring it up to the visible layers of the atmosphere where it can be detected (Fig. 11.8b).

Gases other than those listed in Table 11.1 are undoubtedly present in the atmospheres of both Jupiter and Saturn. In particular, neon should be about as abundant as ammonia, but like helium, it only absorbs radiation in the far ultraviolet making it difficult to detect. The apparent absence of sulfur is more of a puzzle. Recall that sulfur is far more abundant than phos-

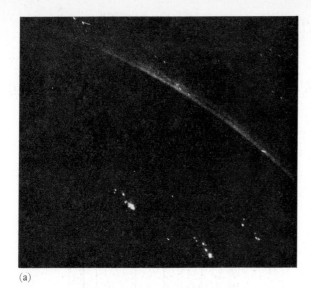

(a)

FIGURE 11.8 a) This Voyager picture of the dark side of Jupiter was a time exposure lasting 3 min. 12 sec. A double auroral arc can be seen at north polar latitudes to the right, with some lightning flashes illuminating clouds in the lower center of the frame. b) In addition to lightning and the bombardment of charged particles that cause the aurora, ultraviolet light from the Sun and internal heat from the planet are available to drive chemical reactions.

(b)

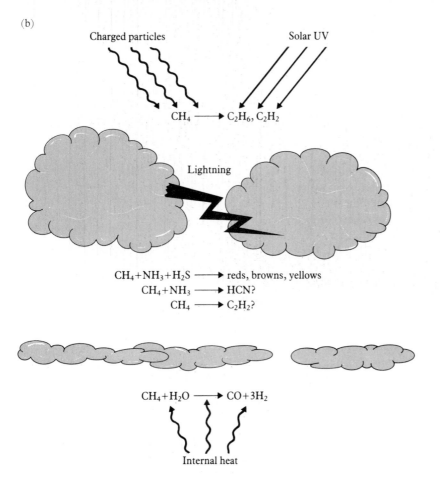

Charged particles

Solar UV

$CH_4 \longrightarrow C_2H_6, C_2H_2$

Lightning

$CH_4 + NH_3 + H_2S \longrightarrow$ reds, browns, yellows
$CH_4 + NH_3 \longrightarrow HCN?$
$CH_4 \longrightarrow C_2H_2?$

$CH_4 + H_2O \longrightarrow CO + 3H_2$

Internal heat

phorus or germanium in a cosmic or solar mixture of the elements (Table 2.1). One would expect hydrogen sulfide (H_2S) to form in such atmospheres. Yet careful efforts to detect it have failed, placing the ratio of sulfur to hydrogen in Jupiter's atmosphere at least 1000 times lower than the solar value. Where is it? Perhaps the missing sulfur is bound up in one of the layers of clouds.

The Change in Temperature with Altitude

All the abundances shown in Table 11.1 refer to the upper layers of the planets' atmospheres. We have already seen that the difference between the ammonia abundances above the clouds on Jupiter and Saturn is attributable to condensation. Observations made at radio wavelengths allow us to examine the atmospheres beneath the clouds (recall Venus, Chapter 8), and here we find the amount of ammonia to be similar on both bodies. Unfortunately, there is no comparable information yet about water, which condenses at even higher temperatures than ammonia. Therefore it remains unknown whether or not this important compound is actually deficient, or whether at deep enough levels in the atmospheres its abundance becomes normal.

To understand this problem of condensation and mixing, we need to know how the temperature varies with altitude (pressure) in these atmospheres. The best information about the change of temperature with altitude has been derived from the Voyager and Pioneer missions. These spacecraft passed behind each planet as seen from Earth. As they did, the radio signal they sent to the Earth had to pass through thicker and thicker layers of the atmospheres of the planets they were moving behind. This spacecraft maneuver is called an **occultation** (Fig. 11.9). Occultations also have provided a powerful means to study the rings of Saturn and Uranus, as we will see in Chapter 14.

In the Pioneer and Voyager occultations, the increasing attenuation of radio signals traveling through the planet's atmosphere provided a measure of the increase in atmospheric density, which is determined by the gases that are present and the local temperature and pressure (Fig. 11.7). Analyses of these data agree with independent determinations of temperature profiles based on studies of the thermal radiation escaping from the planets' interiors.

The temperature-pressure profiles shown in Fig. 11.7 represent averages of these various observations. As expected, Saturn is distinctly cooler than Jupiter at similar pressure levels, owing to its smaller size and greater distance from the Sun. Approximate locations of various cloud levels in the atmospheres are marked. It is interesting to note that on both planets, temperatures above the freezing point of water occur at pressures just a few times greater than the sea-level pressure on Earth. An astronaut in the gondola of a balloon could float quite happily here in shirtsleeves and scuba gear, if only a suitable gas could be put in the balloon! These balmy conditions are a consequence of the planets' internal energy, although some warming would occur at these levels simply from an atmospheric greenhouse effect.

Thermal Inversions

The increase in temperature just above the tropopause is caused by the absorption of short-wavelength (ultraviolet) sunlight by gases and aerosol particles. Since temperature normally decreases with height in a planetary atmosphere, a region like this in which the inverse occurs is known as a **thermal inversion.** All four of the giant outer planets except Uranus exhibit this effect; it is similar to the warming in the Earth's upper atmosphere produced by the ozone layer (Section 7.5). It is in this region of the atmospheres of Jupiter and Saturn that the hydrocarbons formed by photochemical reactions are found.

This solar ultraviolet radiation cannot penetrate to deeper layers, just as it can't on Earth. Despite the dramatic differences between our oxygen-rich environment and these hydrogen atmospheres, the physics of molecules dictates that most gases are excellent absorbers of ultraviolet radiation — hence the similarity in atmospheric structure between these very disparate planets. This same basic physics causes both Jupiter and Saturn to have ionospheres and exospheres. But the gravitational fields of these planets are so strong that even atomic hydrogen can escape only very slowly. So unlike the Earth, they have maintained hydrogen-rich environments throughout the lifetime of the solar system.

FIGURE 11.9 A schematic illustration of an occultation of a spacecraft by Jupiter. The way in which the radio signal is attenuated by the planet's atmosphere provides information about temperature, pressure, and composition as a function of altitude. Characteristics of the ionosphere are also obtained by this technique.

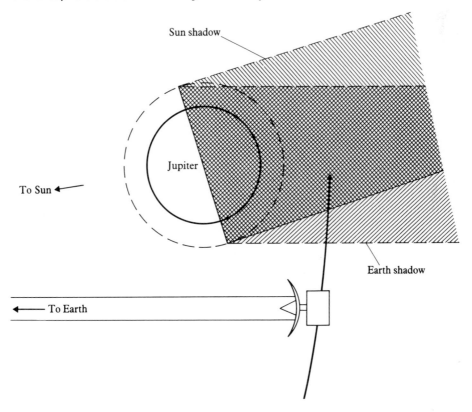

11.4 Weather and Climate

Appearance of Jupiter and Saturn

Even a modest telescope can show much detail on Jupiter. The region of the planet's atmosphere that we can see contains several different kinds of clouds. Individual pictures such as Figs. 11.10 and 11.11 and Plates 13 and 15 provide snapshot views of these clouds at particular instants in time. While such pictures may suggest that all the clouds we see are at the same level in the atmosphere of the planet, Fig. 11.7 reminds us that this may not be the case.

Even at telescopic resolution, approximately 2000 km for Jupiter, changes in the visible cloud systems can occur in a few hours. At Voyager resolution, the planet presents a constantly shifting pattern of clouds, but an underlying pattern of currents flowing parallel to lines of latitude has maintained its stability for decades. On Saturn, cloud changes are much more difficult to see, but again an underlying atmospheric circulation pattern exists. The appearance of these planets, especially Jupiter, has been described in terms of alternating bright zones and dark belts, but the currents seem to be more persistent than this cloud pattern (Fig. 11.12).

FIGURE 11.10 A high-resolution image of Jupiter taken by Voyager 1 at a range of 4 million km. The well-defined dusky band running diagonally through the white clouds defines the north temperate current with wind speeds of about 120 meters per second (270 mph). The wave-like pattern at the bottom of the frame is similar to that found in the "wake" of the Great Red Spot (cp. Plate 14).

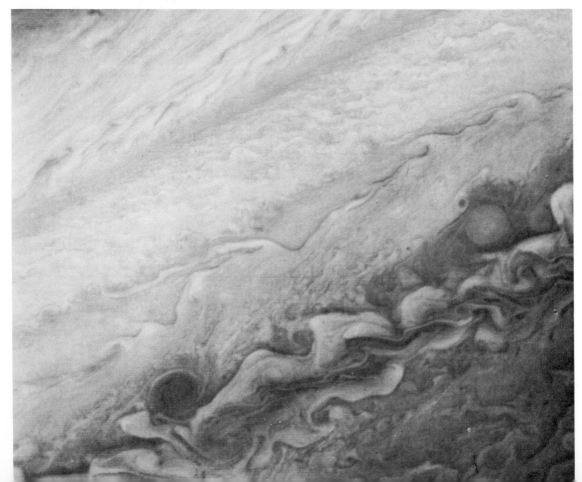

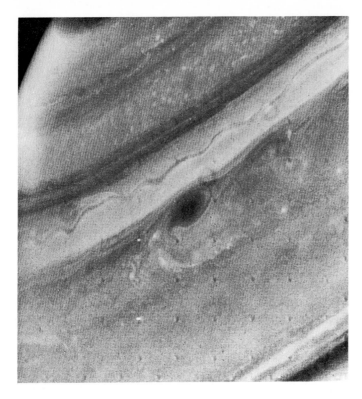

FIGURE 11.11 A close-up view of northern latitudes on Saturn. The contrasts have been greatly enhanced in this picture to reveal details that would otherwise be hard to see. The large dark oval, about 3000 km in diameter, is rotating in an anticyclonic direction, indicating that it is a high pressure disturbance.

Rotation Periods and Wind Speeds

Three rotational periods have been established on Jupiter. The two periods labeled Systems I and II (Table 11.2) are average values and refer to the apparent average speed of rotation at the equator and at higher latitudes, respectively. These periods are defined by observed motions of features in the planet's cloud layers, which are in fact tracking the local currents. Since no solid surface exists, we must look elsewhere for an absolute standard by which to judge these motions.

Studies of the jovian radio emissions made over the last twenty-five years have shown a definite, unchanging periodicity that must refer to the rotation of the planet's magnetic field. Since this field is generated deep within Jupiter's interior, the radio period, called System III, has become identified with the true rotation of the planet itself. Judged against this standard, we see that Jupiter exhibits an eastward-flowing equatorial jet stream (System I) with a relative velocity of about 300 km/h, comparable to the velocities of jet streams on Earth. At higher latitudes in each hemisphere, an alternating pattern of easterly and westerly winds occurs (Fig. 11.12).

TABLE 11.2 Rotation periods for Jupiter and Saturn

	Jupiter	Saturn
Deep Interior (System III)	$9^h55^m30^s$	$10^h39^m24^s$
Atmosphere (Averages):		
Equatorial (System I)	$9^h50^m30^s$	10^h14^m
High latitudes (System II)	$9^h55^m41^s$	10^h40^m

The visible clouds on Saturn are much more uniform in appearance than those on Jupiter. Only rarely do spots appear that can be observed by telescopes on Earth. While a belt-zone pattern is evident, it is much more subdued (Plates 13 and 15). The Voyager cameras were able to define enough discrete features to map out the circulation of Saturn, which is distinctly different from that of Jupiter (Fig. 11.12). Once again an equatorial jet is apparent. Now, however, it extends to 40° on either side of the equator, and its peak velocity is a remarkable 1300 km/h. At

FIGURE 11.12 a) A map of wind currents on Jupiter. There is a strong equatorial jet, with currents at higher latitudes alternating between easterlies and westerlies, roughly following the pattern of boundaries between dark and light belts and zones. b) The equatorial jet on Saturn is both stronger and wider than that on Jupiter. This planet does not have the same pattern of alternating currents as its larger neighbor, except at latitudes above 40°.

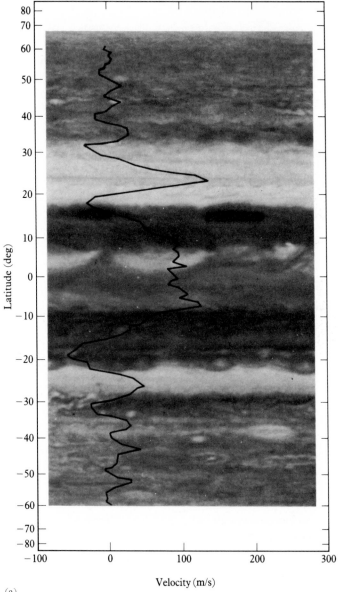

(a)

still higher latitudes, an alternating pattern of eastward and westward currents persists, but now it is not so closely tied to the faint belts and zones.

The most striking cloud features on Saturn recorded by the Voyager spacecraft were located near 40° N. This latitude marks a pronounced change in wind velocities: the great equatorial jet stream loses its strength and a high-latitude current takes over. Not surprisingly, we see a number of eddies and other atmospheric disturbances at this velocity boundary. One of the

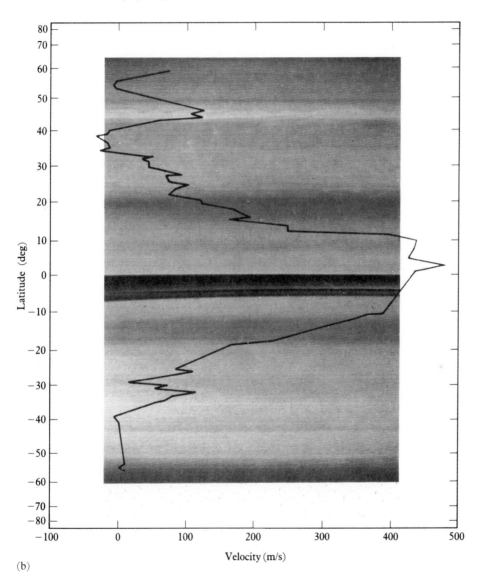

(b)

most striking of these is illustrated in Fig. 11.13. With tongue firmly in cheek, one of the Voyager scientists pointed out that the resemblance of this feature to the number 6 may have a deep, underlying significance, since Saturn is the sixth planet in order of distance from the Sun. (Unfortunately, no eddy in the shape of a 7 was found on Uranus!)

The General Circulations of the Atmospheres

The reasons for the basic differences in atmospheric circulation between Jupiter and Saturn are not yet clear. One interesting possibility suggested by planetary meteorologist Andrew Ingersoll of Caltech is that the circulation patterns actually extend to very deep layers. The currents that we see may represent the outermost edges of concentric cylinders of gas, rotating around the planet's rotational axis (Fig. 11.14). The sizes and shapes of these cylinders would then

be determined by the internal structures of the planets, so that the relative sizes of the solid/liquid cores of Jupiter and Saturn may play a role in determining the nature of the atmospheric currents. This is yet another illustration of just how different these gas giants are from the rocky inner planets.

We can compare the meteorology on these planets with the global circulation of the Earth's atmosphere (Section 7.7). On Earth, huge spiral cloud systems often stretch over many degrees of latitude and are associated with motion around high and low pressure regions. These cloud systems are much less confined to specific zones of latitude than the cloud systems on Jupiter and Saturn. Clouds on Earth can move in latitude as well as longitude, as the migration of hurricanes from the Caribbean along the east coast of the United States dramatically demonstrates each fall.

Planetary meteorologists like Ingersoll have simulated the general circulation of the giant

FIGURE 11.13 The plot of wind speeds in the northern hemisphere of Saturn superimposed on a picture of the appropriate area. The ribbon-shaped cloud at 47° N corresponds to a local maximum in wind speed. A transition between the two wind directions (E and W) occurs at 40° N, and an eddy shaped like the figure 6 can be seen just below this boundary.

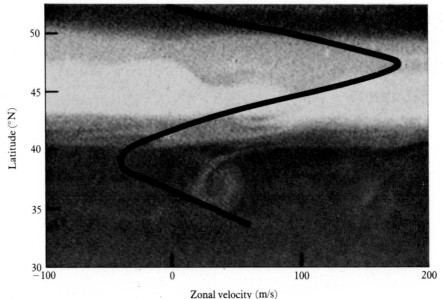

Latitude (°N)

Zonal velocity (m/s)

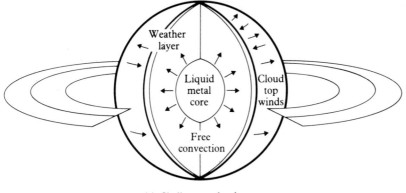

(a) Shallow weather layer

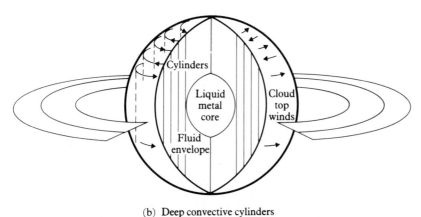

(b) Deep convective cylinders

FIGURE 11.14 Two models for the global circulation on Saturn apply to Jupiter as well: a) suggests that the circulation is driven in a rather shallow layer in the visible part of the atmosphere; b) invokes deep convection following a pattern of concentric cylinders (After M. Allison.)

planets by starting with a computer model of the Earth and simply increasing the rotation rate. The greatly enhanced Coriolis force (Section 7.7) appears to shrink disruptive eddies and produces alternating bands of strong winds that are extremely stable. Thus computer experiments play an important role in understanding the atmospheres of the outer planets, just as they do in tracing the thermal evolution of the inner planets and the operation of the greenhouse effect.

High rotation rate is certainly one factor that makes the circulation on the giant planets very different from that on Earth. Another is the lack

of a solid surface. Weather on Earth is often closely tied to the local environment, which in turn is determined by the varied nature of our planet's surface. On Jupiter and Saturn, the absence of physical boundaries contributes to the stability of the atmospheric circulation patterns.

The Great Red Spot

This permanence in the presence of constant change is most obvious on Jupiter, where one can see more features and variations in the cloud layers. The most famous of these features is the

Great Red Spot (GRS) (Fig. 11.15), which has existed for at least 100 and perhaps more than 300 years. Its present dimensions are about 26,000 km by 14,000 km, making it large enough to accommodate two planets the size of our Earth side by side. These huge dimensions are probably responsible for the feature's longevity.

The true nature of this giant cloud system is still unknown, despite extensive observations by the instruments on the Voyager spacecraft. The reddish clouds within the spot exhibit a counterclockwise rotation with a period of about 6 days. (Remember that counterclockwise motion in the southern hemisphere of a planet is anticyclonic, Section 7.7.) This cloud system thus appears to be an enormous anticyclone, a vortex or eddy whose center must be a region of locally high atmospheric pressure. Nevertheless, Voyager pictures failed to reveal any evidence of upwelling at the spot's center. The clouds here seem remarkably tranquil (Plate 14). Cyclones and an-

FIGURE 11.15 A close-up view of Jupiter's Great Red Spot. This giant storm system could swallow two planets the size of Earth, side by side. The white oval below it has existed for over forty years; the smaller "doughnuts" at still lower latitudes have shorter lifetimes (cp. Plate 14).

ticyclones are also found on Earth, but on our planet they are very short-lived. Furthermore, the clouds associated with them are white water clouds like all the others that form on Earth.

The huge lateral dimensions of the GRS are associated with a considerable vertical extent as well, allowing it to project well above and below the adjacent cloud deck. The upward projection has been verified by direct observation, but the lower reach of this enigma has not been established. The chemicals responsible for its color are also unknown.

Ovals, Dark Brown Clouds, and Hot Spots

On Jupiter cloud features much smaller than the GRS are also persistent and localized. The three white ovals found at a latitude just south of the Great Red Spot (Figs. 11.4 and 11.15) have existed for 40 years; white ovals of this size (about as large as the planet Mars) are found nowhere else on the planet. Dark brown clouds, which are evidently glimpsed through holes in the nearly ubiquitous tawny cloud layer, are found almost exclusively at latitudes near 18° N. The blue-grey or purple areas from which the strongest thermal emission is detected occur only in the equatorial region of the planet.

How does Jupiter "know" that it should have one type of cloud in only one location? These equatorial hot spots must be clear regions that allow us to look deep into the planet's atmosphere. We know that because of the intense thermal radiation we can detect there. Pictures of Jupiter made by recording this escaping energy reveal these equatorial windows as the brightest areas on the planet (Fig. 11.16). Yet there is no brown cloud layer here, as there is at 18° N. Why is there only one Great Red Spot? We do not yet have answers to these questions.

One piece of this network of riddles seems fairly clear: the larger the disturbance the longer it lasts. On Earth, our most powerful storms, hurricanes and typhoons, dissipate when they

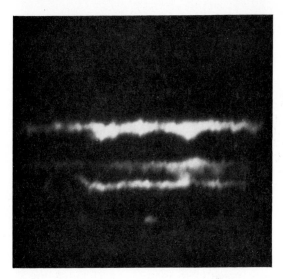

FIGURE 11.16 Images of the same hemisphere of Jupiter in visible light (top) and infrared light at a wavelength of 5 μm (bottom). The bright regions in the infrared picture are places where heat is escaping from the interior of the planet. These seem to be correlated with the darker regions in the cloud layers seen in visible light. Evidently these regions must include some holes in the clouds through which radiation from the warm interior can escape. The Great Red Spot appears dark in this image, indicating that it is opaque to this internal heat radiation.

cross land. Without the condensation of water vapor that comes from the oceans these storms lose their energy. (Evaporation of water requires energy — that's why perspiration cools us — while condensation liberates energy, hence the destructive power of thunderstorms and hurricanes.)

On Jupiter, there are no continents, and once these storms grow large enough to overcome the effects of encounters with smaller systems, they can evidently persist for long periods of time. Thus the largest such feature, the GRS, has existed for 300 years, while the smaller white ovals have been present for several decades. The still smaller white spots at more southerly latitudes have proportionately shorter lifetimes.

Seasons

Since both Jupiter and Saturn have major *internal* sources of energy, one might expect their weather to show no sign of seasons. This may be the case at low altitudes where the internal heat is the most important source of energy for atmospheric circulation, but the upper atmospheres (including most of the regions we see) are strongly influenced by the solar radiation they absorb. We have already mentioned one indication of this in the discussion of the temperature inversions. Thus the key property for establishing seasons on Jupiter or Saturn remains the inclination of the planet's axis of rotation, just as it is for Earth.

Jupiter, with an axial inclination of only 3°, exhibits very little seasonal modulation of its weather pattern, while Saturn, inclined at nearly 27°, shows strong seasonal effects. One manifestation of Saturn's seasonal change is the increased abundance of ethane during the summer as a result of the irradiation of methane by ultraviolet light from the Sun. Since Saturn's satellites share the planet's inclination, we can anticipate that the atmosphere of Titan will also be subject to seasonal effects.

♦

11.5 Clouds, Colors, and Chemistry

Photochemical Hazes: Smog on Other Planets

On Earth, we are accustomed to the formation of smog layers over our cities, since it is the cities that furnish the gases on which sunlight can act to produce the smog particles. On the outer planets, the hydrogen-rich atmospheres themselves contain gases that are easily converted to compounds that form hazes. Because of these organic gases, photochemistry plays a very important role in the outer solar system, in spite of the large distances of the planets from the Sun.

These thin hazes of photochemical smog are found in the planets' stratospheres, since the ultraviolet light from the Sun does not reach lower altitudes. It is absorbed by methane and ammonia in the process of splitting these molecules apart. Below the tropopause (Fig. 11.7), we begin to encounter the thick clouds that appear in the photographs of these planets. These clouds are formed by condensation of atmospheric gases.

Condensation Clouds

The highest, coldest clouds are the white ammonia cirrus. If Jupiter and Saturn were simply weird varieties of terrestrial planets, we could reasonably expect that the next cloud layer to be encountered in a descent through the atmosphere would be composed of liquid ammonia droplets. These would be white, puffy clouds analogous to terrestrial cumulus, or perhaps layers of ammonia stratus clouds. But in these enormously deep atmospheres, we must anticipate the presence of other gases and ask what effect they will have on cloud composition. In fact, the next clouds we do encounter are no longer white, but tawny in color. Evidently a change in chemistry has occurred.

There is no direct determination yet of the composition of these colored clouds. A table of cosmic abundances of the elements gives us a clue, however. As discussed in Section 11.4, no compounds of sulfur have been detected yet on Jupiter and Saturn even though the relatively high abundance of this element leads us to expect such compounds to be present. Chemistry tells us that ammonia and hydrogen sulfide will combine to form ammonium hydrosulfide (NH_4SH). Since there is more nitrogen than sulfur in a cosmic (or jovian) mixture of the elements, this compound can account for the absence of detectable sulfur above the tawny clouds. It is almost certainly the main constituent of this cloud layer. But ammonium hydrosulfide is white, so something else must be happening here. The answer may lie in the ability of sulfur to combine with itself to form compounds that are indeed yellow or brown in color. However, this attractive hypothesis has been challenged by recent evidence that the ultraviolet sunlight required to drive these chemical reactions does not penetrate deeply enough into the atmosphere to do the job.

At much lower levels in the atmosphere, water clouds should form. Once again, these will not be pure water. They will certainly contain ammonia and may well resemble a very dilute solution of household ammonia. But other soluble gases will also be involved. At this time, there are no observations of these deep clouds. In view of the uncertainty about the abundance of water vapor (Section 11.3), we cannot even be sure that any water clouds will form.

Chemical Clouds

With our present knowledge, it is already obvious that Jupiter and Saturn have remarkable atmospheres. Imagine floating on a world with colored clouds! The views would be magnificent, even if the aromas leaking into your spaceship might leave something to be desired. Ignoring their disagreeable odors, however, the presence of these nitrogen and sulfur compounds ensures a complex and interesting chemistry. Some of the reactions taking place on Jupiter today may resemble those that occurred on the primitive Earth, producing compounds essential for the origin of life.

We must admit that we know only part of this interesting story. Sulfur compounds may be responsible for the dark brown clouds on Jupiter, but they may not. Red phosphorus has been proposed as the substance responsible for the coloration of the GRS, but the theoretical calculations that led to this prediction have not been substantiated by laboratory experiment or by observations of the planet itself. On the other hand, yellow phosphorus is produced from phosphine (PH_3) in such experiments and may contribute to colors seen on both Jupiter and Saturn. The various colored products of reactions involving methane and ammonia provide additional possibilities.

These suggestions are based in part on theoretical studies by John Lewis, Ron Prinn, and their colleagues, as well as on a variety of careful laboratory experiments. Such experiments often produce colored materials; the problem is that there are usually no discrete absorption features in the spectra of these materials (the way there are for gases) that can lead to an identification on the planets. We may have to wait for a direct analysis by probes sent into these atmospheres to discover what the chemical agents are that produce the colors we observe.

11.6 Magnetospheres and Radio Broadcasts
Why Planets Transmit Radio Waves

Jupiter was the first planet found to be a source of radiation at radio wavelengths. This radiation was detected in 1955 at a frequency of about 20 megahertz, corresponding to a wavelength of 15

m. For comparison, a radio station on Earth that you tune in at 100 on your AM dial is broadcasting at 1 megahertz, or a wavelength of 300 meters, and a station at 100 on FM is broadcasting at 100 megahertz or 3 meters.

These signals from Jupiter are not news reports or rock music; they are simply radio noise produced by the interaction of charged subatomic particles — primarily electrons — with the planet's magnetic field and its ionosphere. These radio emissions, caused by electrons and ions moving at very high speeds, are called **non-thermal radiation,** because they are not caused by the normal emission of energy associated with the heat of the source. The intensities of the jovian non-thermal noise bursts are occasionally great enough to make Jupiter the brightest object in the sky at these long wavelengths, with the exception of the Sun during its most active periods.

Non-thermal radio bursts from Jupiter provided the first indication of a magnetic field on any planet other than Earth. Subsequent observations at wavelengths shorter than one meter revealed that Jupiter is also a source of steady radio emission. It has become customary to refer to these two types of emission in terms of their characteristic wavelengths: decameter radiation (the erratic bursts) and decimeter radiation (the continuous source). Except at the very shortest radio wavelengths, the decimeter radiation is generated non-thermally, primarily by electrons spiraling at very high speeds around magnetic lines of force. In addition to these trapped electrons, Jupiter is also surrounded by a doughnut-shaped distribution of protons and heavier ions (Fig. 11.17). In other words, this giant planet has belts of trapped charged particles analogous to the Van Allen belts surrounding Earth (Section 7.9).

At wavelengths of a few millimeters, Jupiter's radio spectrum is dominated by the thermal radiation from the planet, which can also be detected at slightly longer wavelengths once the ef-

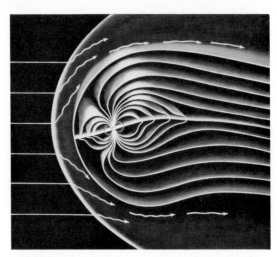

FIGURE 11.17 Jupiter's enormous magnetic field has trapped electrically charged particles, mostly electrons and protons, from the solar wind. When the electrons spiral around the magnetic field lines, they produce radio waves that can be detected from Earth or from spacecraft.

fects of the non-thermal emission produced by the trapped electrons have been removed. Such measurements provide the deepest available sounding of the planet's lower atmosphere.

The Influence of Io

The noise storms at decameter wavelengths are strongly influenced by the position of the satellite Io in its orbit, as viewed from Earth. An example of this is an enhancement that occurs when Io and Jupiter form a 90° angle with the Earth. Evidently a cluster of magnetic field lines (called a **magnetic flux tube**) links Io to the planet (Fig. 11.18). As Io moves in its orbit, the foot of this flux tube sweeps across Jupiter, like the shadow of a satellite that is causing a solar eclipse. Electrons spiraling around the field lines and interacting with the jovian ionosphere cause the observed bursts of radio noise.

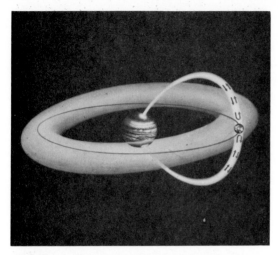

FIGURE 11.18 As Io moves through Jupiter's magnetic field, it generates an electrical potential of 400,000 volts across its surface. At certain positions in its orbit, an electrical current of 3 million amperes will flow along Jupiter's magnetic field lines to the planet, through Jupiter's electrically conductive ionosphere, where it triggers an auroral hot spot, and back to Io.

Another way to think about the situation is to consider the flux tube as a wire along which an electric current (the spiraling electrons) passes. This 5-million-ampere (!) current is generated by the motion of Io through the planet's magnetic field, just as a wire (armature) passing through a magnetic field generates a current in one of our power stations on Earth.

The flux tube and its influence on jovian radio bursts is not the only way Io interacts with the magnetosphere. As we will describe below, Io and its volcanoes are the primary source for heavy ions in the inner magnetosphere of Jupiter. These heavy, energetic particles are concentrated in a doughnut-shaped volume, the **Io plasma torus,** that surrounds Jupiter near the orbit of the satellite. This Io torus is the strongest part of what we might think of as the jovian Van Allen belts. It has been photographed from Earth in the glow from the various atoms and ions it contains, as shown for neutral sodium in Fig. 11.19.

FIGURE 11.19 The orbit of Io is surrounded by a doughnut, or torus, of atoms and ions, as shown in a picture obtained from Earth in the light from the sodium atoms it contains. Images of Io and Jupiter have been added at their appropriate positions relative to the sodium cloud.

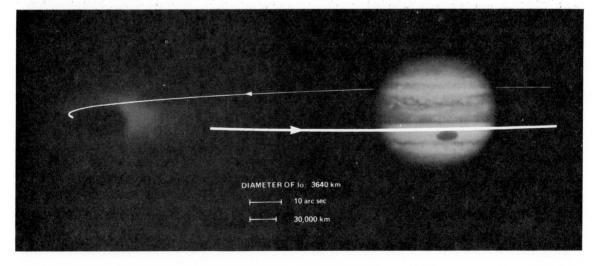

DIAMETER OF Io: 3640 km

10 arc sec

30,000 km

Magnetic Fields

The deductions about the magnetic field of Jupiter from these Earth-based observations were refined and extended by the spacecraft that penetrated the jovian magnetosphere. The magnetic field of Jupiter is dipolar (like a bar magnet, the same type as Earth's) and is generated by a natural dynamo driven by convection within the electrically conducting layers of the planet's interior. The intrinsic strength of this field is 19,000 times greater than that of Earth, leading to a field strength at the equator of 4.3 gauss, compared with 0.3 gauss at the Earth's surface. (The surface field strength is reduced because Jupiter's volume is so much greater than Earth's.) The orientation of the Jovian magnetic field is exactly opposite to the present orientation of the Earth's field, so a terrestrial compass taken to Jupiter would point south instead of north.

Saturn's magnetic field is somewhat weaker than Jupiter's but exhibits the same orientation at the present time. (We have no idea if these fields, like that of the Earth, reverse direction from time to time.) The field strength at the equator is only 0.21 gauss. The most unusual characteristic of Saturn's field is that it is not inclined to the planet's axis of rotation, whereas the fields of Jupiter and Earth both have inclinations of about 10°. This lack of inclination is surprising, since most theories for the generation of planetary magnetic fields require such a tilt. Perhaps there is a conducting region around the field-generating core in Saturn's interior which modifies the external appearance of the field.

Saturn's Radio Broadcasts

Another difference between these two planets is the lack of strong radio emission from Saturn. It was not until the first Voyager spacecraft approached the planet in 1980 that long-wavelength radiation was detected. This radio noise occurs at frequencies from 3 kilohertz to 1.2 megahertz, overlapping our commercial AM radio band, with a peak intensity at wavelengths near 2 km. Radio signals in this frequency range are reflected by the Earth's ionosphere. Thus the AM signals do not escape our atmosphere, and the Saturn radiation cannot enter — hence the need for a spacecraft to detect them.

The energy for this radio emission from Saturn is provided by electrons from the impinging solar wind. Therefore changes in solar wind pressure or speed produce large changes in the power radiated by Saturn. Since the interaction with the solar wind is tied to the configuration of the planet's magnetic field, it has been possible to determine the rotation period of the field, and hence of Saturn's interior, exactly as in the case of the System III period for Jupiter (Table 11.2). The resulting period of 10 hours 39.4 minutes is the standard against which the wind velocities can be measured.

The Van Allen Belts of Jupiter and Saturn

Both Jupiter and Saturn are surrounded by huge seas of plasma. The protons, ions, electrons, and neutral atoms in these plasmas come from three distinct sources: the solar wind, the atmospheres of the planets, and the surfaces of the satellites. The Earth's magnetosphere is not populated from this latter source (Section 7.9), which is one of the major differences between our Van Allen belts and the magnetospheres of Jupiter and Saturn. The satellite surfaces contribute the heavy ions that are present in addition to the electrons and protons already mentioned. These ions are atoms stripped of one or more electrons. Being charged, they are under the control of the planetary magnetic fields.

Oxygen ions have been found around both Jupiter and Saturn. Sulfur and oxygen ions around Jupiter presumably come from the volcanoes or surface of Io, which must also contribute the neutral sodium that is visible from Earth (Fig. 11.19). Some of these neutral atoms and ions escape directly from the volcanic eruptions on Io (Chapter 13), while others are released from the surface by a process called **sputtering.** Sputtering results when an impacting ion from the plasma torus has sufficient energy to eject additional ions from a solid surface, almost like an atomic version of impact cratering. The process is important at Io because the more heavy ions are injected into the plasma torus, the more "ammunition" there is to sputter other ions from the surface.

The heavy ions in the Io torus make the environment around this satellite lethal to human beings and very hostile to spacecraft. Pioneer 10 was nearly destroyed as it passed through this region, so considerable effort and expense have been lavished on the Galileo spacecraft to protect its electronic components from this threat.

In the Saturn system, sputtering from the surfaces of the icy satellites and the rings furnish oxygen ions, while nitrogen ions are supplied by the atmosphere of Titan. Saturn's huge ring system also acts as a sink for charged particles, absorbing those that strike it. The details of the interactions of the electrons, protons, and heavy ions with the ices in the main rings are still not understood. Some charging of small solid particles in both Saturn's and Jupiter's rings may take place, leading to observable effects (Chapter 14).

Having seen where the ions and electrons in the magnetospheres originate, we now ask how they are lost. There are three major sinks in the Jupiter and Saturn magnetospheres, compared with two in the simpler magnetosphere of the Earth (Section 7.9). Just as with the Earth, some escape from the outer part of the magnetosphere, and others are lost in collisions with the upper atmosphere of the planet. The third sink occurs in the surfaces of the satellites and rings, which act as both a source (through sputtering) and a sink (through absorption of particles that strike their surfaces). Hence each satellite in each sys-

tem that is actually within the respective Van Allen belts sweeps out a torus around its orbit that manifests itself as a decrease in the concentration of trapped protons and electrons. These absorptions are stronger in the Saturn system because of the large surface area of Saturn's rings and the closely spaced inner satellites. Therefore a major difference exists in the distribution of plasma in these two giant magnetospheres.

Dimensions of Magnetospheres

As we saw for the Earth in Section 7.9, a magnetosphere is bounded on the upstream side (facing the Sun) by the equilibrium between the internal pressure exerted by the planet's magnetic field and the external pressure from the solar wind. The magnetic fields of Jupiter and Saturn are much stronger than Earth's, while the solar wind pressure decreases at these greater distances from the Sun. The net result is that the two giant planets have magnetospheres that are so large they exceed the size of the Sun (Fig. 11.20).

Since the intensity of the solar wind is not constant but changes with the activity of the Sun, the upstream boundaries of these magnetospheres also change. A strong solar wind pushes the boundary closer to the planet. At such times, Saturn's satellite Titan actually moves outside the magnetosphere when its orbit brings it around to a position between the planet and the Sun. At other times, Titan remains within the magnetosphere throughout the course of its orbital journey.

In the downstream direction, the magnetosphere stretches out into a **magnetotail** (so named for its similarity to the plasma tail of a comet), coaxed along in the anti-Sun direction by the solar wind. Jupiter's magnetotail extends to such an extent that it appears to reach Saturn's orbit, a distance almost equal to the distance of Jupiter from the Sun.

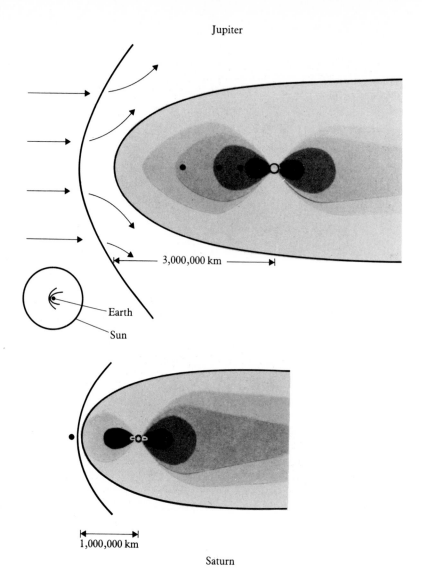

FIGURE 11.20 The small circle to the left in this diagram represents the Sun; inside it, to the same scale, is the Earth and its magnetosphere. The magnetospheres of Jupiter and Saturn are shown to scale for comparison. When the solar wind is strong, it can compress the magnetosphere of Saturn sufficiently that Titan can be outside it when this satellite is closest to the Sun in its orbit.

Energy to Power the Magnetospheres

Given the sources and sinks for the charged sub-atomic particles in the plasmas, what supplies their energy? Ultimately, it is the rotation of the parent planets that provides the energy to generate the planet's magnetic fields. Charged particles spiraling along magnetic field lines must revolve around the planet at the same rate as the field lines. This is the rotation period of the planet itself, as we discussed above.

As a result of this coupling of the plasma to the magnetic field, we have the curious situation that these spiraling electrons, protons, and ions are moving around the planet as if they were attached to it. They exhibit what is called rigid body rotation, rather than following Kepler's laws of planetary motion, because their motion is dominated by magnetic forces instead of gravity. Therefore a particle at Io's distance from Jupiter must follow the radio rotation period of System III: 9 hours 55.5 minutes, much faster than the period of revolution of Io, which is 42 hours 27.6 minutes. Hence the plasma flows past the satellite, leading to the unusual result that a wake *precedes* the satellite in its orbit about the planet (Fig. 11.21). The same thing happens with Titan in the magnetosphere of Saturn.

The charged particles become so energetic because of their rapid motion that when they strike the surface of a satellite they can sputter ions that then join the co-rotating plasma. Forcing this plasma to rotate with the field and to generate the radio noise that is equivalent to 10^{14} watts costs the planet some energy. It means that its rotation rate must be steadily decreasing, but at an unmeasurably slow rate.

Auroras

By analogy with the Earth, one would expect that charged particles from the radiation belts will sometimes follow the magnetic field lines into the planet's atmosphere. On Earth, this precipitation produces auroras in the north and south polar regions. Indeed, the same phenomenon has been detected on Jupiter both by direct photography and by ultraviolet and infrared spectroscopy (Fig. 11.8a). Not only is light produced by these interactions, new compounds are formed in the planet's polar atmosphere as molecules of methane are broken apart and the fragments recombine.

The same thing is presumably happening on Saturn as well, but the observations of this planet were not adequate to reveal these processes. The pictures of the dark side that might have been expected to show auroras and thun-

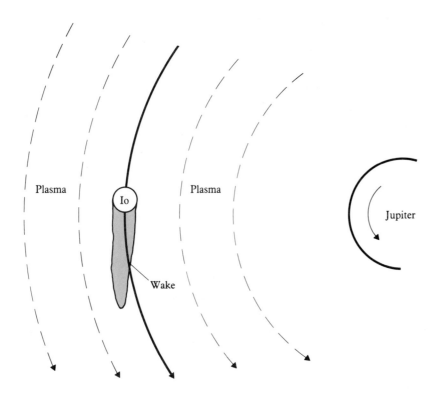

FIGURE 11.21 Jupiter's magnetic field is generated in the deep interior of the planet. Since it behaves as if it were rigidly connected to Jupiter, its period of rotation is constant with distance from the planet. Thus the magnetic field moves faster than the satellites move in their orbits, so they have magnetospheric wakes that *precede* them, instead of following them, as we might have expected.

derstorms were handicapped by the fact that the night sky on Saturn is never really dark. Light scattered by the rings always leaves the planet in a twilight glow. This phenomenon is illustrated in Fig. 11.22, which shows a view of Saturn we cannot capture from Earth. Looking back at the planet from Voyager 1, we see the shadow of Saturn crossing the magnificent system of rings. An observer floating in the atmosphere on the night side of the planet would see the bright rings

arching up from the horizon and disappearing in shadow overhead. Ring light would illuminate the surroundings. On the opposite side of the equator it would be darker, but still not dark enough (Fig. 2.16). The illustration reminds us that in leaving Saturn we have now visited all of the planets known to the ancients. As Voyager 2 ventures forward to Uranus and Neptune, it explores worlds unknown to Galileo, Shakespeare, and Newton.

FIGURE 11.22 A dramatic picture looking back at Saturn, taken by Voyager 1 after its spectacular encounter with the ringed planet. In a view we can never get from Earth, Saturn appears as a crescent with its dark shadow crossing the magnificent ring system. In fact, there is no total darkness on Saturn, because the rings scatter sunlight into the nighttime hemisphere (cp. Fig. 2.16).

———————◇———————

Summary

The two largest planets, Jupiter and Saturn, with their extensive retinues of satellites, resemble miniature solar systems. Because of their extraordinary masses, they have been able to retain thick atmospheres of hydrogen and helium throughout their 4.5 billion year histories. The atmospheres contain most of the mass of these planets, in sharp contrast to the inner planets whose atmospheres are relatively thin. Since no solid surfaces exist, there is no geology on Jupiter and Saturn.

Besides hydrogen and helium, these atmospheres contain other simple hydrogen-rich molecules. Some of these, such as methane and ammonia, are expected to form readily at the local temperatures and pressures. Others, like acetylene and hydrogen cyanide, require some additional source of energy such as solar ultraviolet, lightning, or heat coming from the planetary interiors.

Both Jupiter and Saturn radiate more energy than they receive from the Sun. Jupiter is simply radiating primordial heat, the result of its contraction from a giant proto-planetary phase. Smaller Saturn must have some other source of energy. Precipitation of helium deep in its interior is the best current explanation, supported by the observed deficiency of helium in the atmosphere.

Both planets have layers of metallic hydrogen surrounding solid cores composed of rock and ice. These cores appear to be similar in size, about ten to fifteen times the mass of the Earth. The difference in the relative size of Saturn's core may be responsible for a major difference in the general circulation of the two atmospheres: the much broader and faster equatorial wind system observed on Saturn.

Despite the continuous motion of currents and clouds, some features in these atmospheres have remained remarkably permanent. Foremost among these is Jupiter's Great Red Spot, which may be more than 300 years old. Its longevity is attributable to its size, which is large enough to swallow two planets the size of the Earth. The color of this giant high-pressure region, like the other colors found among the clouds in these atmospheres, is caused by traces of chemicals whose identities remain unknown. Suggestions for the GRS range from organic compounds to red phosphorus, with sulfur regarded as a possible component of some of the brown and tawny colored clouds.

Both Jupiter and Saturn are surrounded by giant magnetospheres. Each planet has a magnetic field thousands of times stronger than the Earth's, which traps energetic plasma. The atoms and fragments of atoms that make up the plasma originate from the solar wind, from the atmospheres of the planets, and from the satellites and rings. Jupiter is an intense source of non-thermal radio noise at both long and short wavelengths, while Saturn only radiates at very long wavelengths. Saturn's extensive system of rings serves as an effective absorber of charged atomic particles near the planet.

The plasmas surrounding these giants differ from the environment found in the Earth's Van Allen belts in that they contain large numbers of heavy ions. These ions are contributed by the surfaces of the satellites through a process known as sputtering and, in the case of Io, by volcanic eruptions.

———————◇———————

Key Terms

Io plasma torus

magnetic flux tube

magnetotail

non-thermal radiation

occultation

sputtering

thermal inversion

CHAPTER TWELVE

In Deep Freeze: Planets We Cannot See

12.1 Discoveries of the Outer Planets

Beyond Saturn only the unchanging stars were known until two centuries ago. In extending our perspective to Uranus, Neptune, and Pluto, we enter new territory. These planets were *discovered* in a literal sense: one by accident, one by calculation, and one as the result of a diligent search. In each case, the discovery generated at least as much public interest as we have seen in our time in association with a major spacecraft encounter.

The direct exploration of these distant worlds by spacecraft is just beginning. In January 1986 Voyager 2 passed through the Uranus system, yielding many discoveries; in 1989 this same spacecraft is scheduled to reach Neptune, and we again anticipate an enormous increase in our knowledge. Pluto, unfortunately, remains beyond our reach, since the current alignment of the planets does not allow either Voyager spacecraft to approach this planet.

Discovery of Uranus

The night of 13 March 1781 was clear and dark over the town of Bath in southwest England. A professional musician named William Herschel was pursuing his astronomical hobby that eve-ning, continuing a project to chart all stars down to the eighth magnitude. He was using a six-inch telescope he had built himself, and on this particular evening he was examining the stars in a portion of the constellation Gemini. As he recorded in his notebook, one of the stars that came into view as he moved the telescope from one place to the next seemed unusual. Herschel described it as "a curious either nebulous star or perhaps a comet." Instead of a point of light, in Herschel's telescope it appeared as a small disk.

Subsequent observations revealed that this object moved (unlike a star), but that the motion was too slow for a comet. Soon its path had been mapped well enough to establish that it was a new planet, orbiting the Sun at a distance of 19 AU (nearly three billion kilometers). A new world had been added to those known since antiquity, and the size of the solar system had suddenly been doubled. In recognition of his accomplishment, Herschel received a knighthood, a royal pension, and went on to become one of the most productive astronomers of his era.

These days we have become rather blasé about new astronomical discoveries — volcanoes on Io, black holes, colliding galaxies. Five years after the United States declared its independence from Britain, however, the world was a very different place. Adding a new planet to the solar system was an extraordinary event. Perhaps the most famous comment on the effect of this discovery is in John Keats' sonnet "On First Look-ing into Chapman's Homer," where the poet uses Herschel's discovery as an image for his own feelings on being introduced to Homer's great epic:

"Then felt I like some watcher of the skies
When a new planet swims into his ken"

Herschel named this new planet "Georgium Sidus" after his king and patron, King George III of England. We have been spared having a planet named George by the suggestion of Her-schel's contemporary, Johann Bode (the same person responsible for the Titius-Bode rule; Sec-tion 1.5) that the ancient tradition for planetary names be continued by using the name Uranus. In Greco-Roman mythology, Uranus was the fa-ther of Saturn who was in turn the father of Jupiter.

Discovery of Neptune

This new boundary to the solar system was soon transcended. Uranus is bright enough that even a small telescope can reveal it easily; in fact, it can just be glimpsed with the naked eye in a dark, clear sky. After Herschel's discovery, as-tronomers looked through records of previous observations and found some twenty cases where Uranus had been recorded. It had been mistaken for a star, since the earlier observers had not used enough magnification to show the disk. With these additional observations dating back to 1690, it was possible to develop a very accu-rate path for the new planet.

Astronomers used these measurements to calculate the orbit of Uranus, but even when all of the gravitational effects from the known planets were accounted for, it was impossible to fit the observations with an elliptical orbit. By 1830, the discrepancy was 15 seconds of arc, which meant that the predicted position of Ura-nus was more than four times the planet's di-

ameter from its observed place in the sky. As the validity of Newton's laws of planetary motion was universally accepted, it seemed likely to as-tronomers of the time that this discrepancy was the result of a **perturbation** caused by the grav-itational pull of some unknown body.

Since scientists had now accepted one new planet, it was reasonable to suggest that perhaps another might be present, orbiting the Sun at a still greater distance and perturbing the motion of Uranus by its gravitational attraction. How to find it among the millions of stars? The first clue astronomers used was the Titius-Bode rule (Sec-tion 1.5). Since Uranus very nearly obeyed this rule, perhaps the new planet would lie at the dis-tance from the Sun predicted by this example of numerology, about 30 AU. With this starting point, it was then a question of solving some for-midable mathematical equations that described various orbits for the hypothetical planet, whose mass was also one of the variables.

Two gifted mathematicians succeeded in solving this problem independently: John Couch Adams in England and Urbain Jean Joseph LeVerrier in France. Although Adams obtained the first solution in September of 1845, he was unable to persuade any of his countrymen to ini-tiate a telescopic search. After LeVerrier pub-lished the preliminary results of his calculations in June 1846, the British astronomers realized that these agreed with Adams' work. They then began to look for the object, but were unsuc-cessful.

Meanwhile, LeVerrier finished his analysis in August 1846 and sent a letter to the German astronomer Johann Galle in Berlin suggesting that he try to find the new planet. The Berlin observatory had a newly completed set of star charts covering the region of LeVerrier's predic-tion. Using these charts, Galle was able to find the planet on the evening of 23 September 1846, on his first attempt. It was just one degree (twice the apparent diameter of the Moon) from the po-sition predicted by LeVerrier and two and one-half degrees from Adams' prediction. Galle could not discern the planet's tiny disk, but he identified it by its appearance as a wanderer among the fixed stars. Neptune is only about one-tenth as bright as Uranus, much too faint to be seen without a telescope.

The discovery of Neptune represented a stunning triumph for Kepler, Newton, and gravitational theory. A search was made for pre-discovery observations, and once again several were identified, including one observation by the hapless English astronomer who first set out to find the planet two months earlier. Perhaps the most interesting early observation is one attributed to Galileo. In a sketch of observations of Jupiter made in December 1612 and January 1613, Galileo recorded a "star" that modern astronomers have been able to identify as Neptune by extrapolating the position of the planet backward in time along its orbit (Fig. 12.1). Even though Galileo's telescope was powerful enough to show Neptune, without charts he had no hope of identifying this tiny star-like object as a planet.

Understandably a dispute over the proper credit for the discovery of Neptune emerged. Matters were not helped by the intervention of the French astronomer François Arago, who suggested that the new planet be named Le-Verrier. Bowing to international disapproval, Arago later withdrew this idea and instead proposed Neptune, the Roman god of the sea. Happily, Adams and LeVerrier managed to disassociate themselves from these disputes, and the two have been credited jointly with the discovery. Note that in this case the discoverer is considered to be the person who made the mathematical calculations, not the one who actually first saw Neptune through the telescope.

Discovery of Pluto

Continued tracking of both Uranus and Neptune during the rest of the nineteenth century suggested that yet another planet was required to

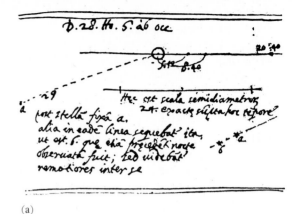

(a)

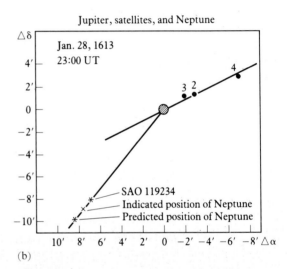

(b)

FIGURE 12.1 a) An entry from Galileo's notebook in which he recorded the relative positions of Jupiter, three of its moons, and two stars on 28 January 1613. b) A reconstruction of Galileo's observations by Charles Kowal. One of the "stars" was actually Neptune.

account satisfactorily for their motions. Although the evidence was only marginally convincing, a number of astronomers undertook calculations similar to those of Adams and LeVerrier. The most persistent of these individuals was Percival Lowell, already familiar to us from his work on Mars (Section 10.2). Lowell not only calculated an orbit for this hypothetical planet, he initiated some systematic searches for it at his observatory. Though unsuccessful, they did provide valuable experience. A special wide-field telescope was built at the Lowell Observatory in 1929, thirteen years after Lowell's death, and the search for a new planet was put in the hands of Clyde Tombaugh, a young assistant at the observatory. On 18 February 1930, Tombaugh discovered the trans-Neptunian planet on photographs he had taken with the new telescope the previous month (Fig. 12.2). The ninth planet was announced to the world on March 13, Percival Lowell's birthday.

Although this discovery might seem to be another triumph for gravitational theory, it was actually the product of meticulous astronomical observations. We now know that Pluto's mass is far too small to have caused the apparent perturbations in the motions of Uranus and Neptune. Recent studies of this problem suggest that the difficulty lies in the observations of these two planets, but interest in a possible tenth planet persists.

Pluto was the Roman god of the underworld, a fitting title for a planet at such a great distance from the Sun. The symbol for this planet, ♇ , has the additional attribute of commemorating the initials of Percival Lowell, the main catalyst behind its ultimate discovery. This is as close as one might properly come to naming a planet for an astronomer.

Pluto is so small and far from the Sun that it is 10,000 times too faint to see with the unaided eye. It is only within the last decade that we have learned anything more about it than Tombaugh knew shortly after he had discovered it.

FIGURE 12.2 The two photographs from the Lowell Observatory that led to the discovery of Pluto. In comparing the two, taken 6 days apart, Clyde Tombaugh noticed on 18 February 1930 that one of the "stars" (shown here with an arrow) had moved. It was the planet he had been seeking.

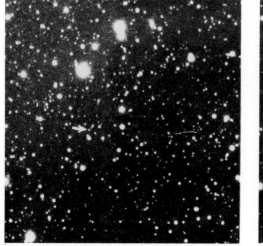

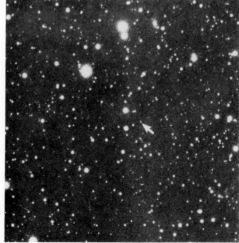

January 23, 1930   January 29, 1930

12.2 Three Distant Worlds

Outer Planet Overview

Before discussing Uranus, Neptune, and Pluto in detail, it is useful to put them in perspective by comparing them with planets already discussed. This comparison is summarized in Fig. 12.3 (see also Table 1.2), where we see that Uranus and Neptune are nearly twins. While they can be classed as giant planets, with masses roughly fifteen times that of the Earth, they are much smaller than either Jupiter or Saturn. Their densities (about 1.5 g/cm³) are also greater than those of their larger cousins, indicating that they must be made of different material.

In Section 11.2 we saw that Saturn was both less massive and less dense than Jupiter, which is consistent with their nearly identical bulk compositions. With a lower mass, Saturn suffers less internal compression, hence the overall density is lower, only 0.7 g/cm³. If Uranus and Neptune had the same composition as Jupiter and Saturn, we would expect these bodies to have densities even less than that of Saturn. Instead they are more dense than Jupiter. This leads us to the conclusion that Uranus and Neptune contain a higher proportion of heavier elements.

Ice and rock are the logical candidates for the dense material in Uranus and Neptune, as inspection of a table of cosmic abundances shows (Table 2.1). Oxygen is the most abundant reactive element after hydrogen, and these two combine readily to form ice (H_2O) at low temperatures. What we call rocks are materials composed primarily of the common elements silicon and oxygen (Section 2.1). Scientists who have constructed models for the interiors of these two objects find that they must contain massive cores composed of ice and rock, surrounded by envelopes of liquid and gaseous hydrogen. But Uranus and Neptune are too small to achieve the pressures and temperatures required for the formation of metallic hydrogen

FIGURE 12.3 The relative sizes of the outer planets. Uranus and Neptune are distinctly smaller than Jupiter and Saturn, while Pluto is about the size of our Moon.

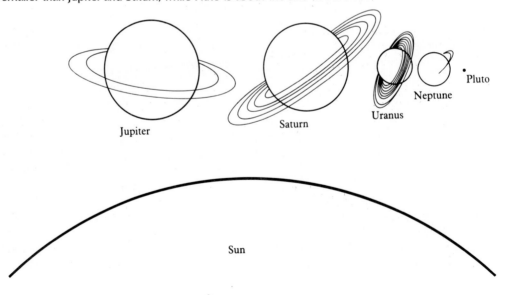

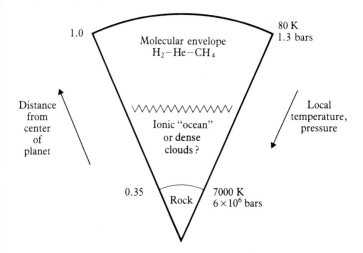

FIGURE 12.4 A schematic idea of what the interiors of Uranus and Neptune may be like. Pressures are not great enough for metallic hydrogen to occur (cp. Fig. 11.5). (After D. Stevenson.)

(Fig. 12.4). Thus the *cores* of these two planets may be very similar to the cores of Jupiter and Saturn, but the remainder of their internal structure and composition differs.

Some scientists currently suggest that both of these planets may have thick layers of water clouds in their lower atmospheres. There is no direct evidence for the existence of these clouds, since water vapor has not been detected on either planet, but there are some chemical arguments that support this idea. In particular, the apparent deficiency of ammonia gas in both atmospheres (Section 12.4) could be explained if the ammonia were dissolved in these water clouds below the observable layers.

Differences Between Uranus and Neptune

Closer inspection of Table 1.2 indicates that the initial appearance of twinship may be incorrect, just as we found for Venus and Earth. The densities of Uranus and Neptune are distinctly different, indicating some internal structural or compositional differences. Uranus also lacks a significant internal heat source, while Neptune (similar to Jupiter and Saturn) derives much of its heat from internal sources. Models for Neptune's interior suggest that like Jupiter, it could still be radiating *primordial* heat. Despite its small mass, the high proportion of rock and ice to total mass leads to very slow cooling. Uranus may have a similar source of heat, but it could be masked by the energy Uranus absorbs from sunlight by virtue of its smaller distance from the Sun.

Perhaps the most puzzling enigma posed by Uranus is the bizarre orientation of its rotational axis, which is inclined by 98°, placing it practically in the plane of the planet's orbit (Fig. 12.5). The orbits of Uranus' satellites and rings are in the planet's equatorial plane, just as they are for Jupiter and Saturn, so the whole uranian system is tipped on its side. It is often suggested that a glancing impact by another planet-sized object, which then became added to the mass of Uranus, could explain this strange orientation, but there is no detailed theory yet that satisfactorily describes such an encounter and its effects.

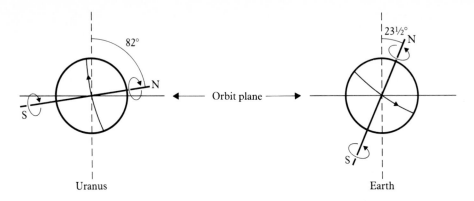

FIGURE 12.5 The axis of rotation of Uranus lies nearly in the plane of the planet's orbit. At the time of the Voyager 2 encounter, the axis pointed almost directly at the Sun (cp. Fig. 12.9).

Pluto: A Special Case

The above discussion has focused on Uranus and Neptune because these two planets are similar in many respects to Jupiter and Saturn. When we turn to Pluto, we are dealing with a different kind of object altogether. This planet is slightly smaller than our own Moon and has a density of 1.7 g/cm³. This combination of small size and low density suggests that Pluto is composed partly of water ice and probably resembles one of the large icy satellites of Saturn (Chapter 14). In no way can it be considered one of the jovian planets; nor is it a displaced inner planet, roaming the cold and dark domains of space beyond Neptune. At best, it might be representative of the kinds of objects that collided with one another to form the icy cores of the giant planets, a fate it somehow avoided.

Pluto's orbit is both the most eccentric and the most highly inclined of any known planet (although comets beat it by a wide margin in both categories). This means that its orbit crosses that of Neptune, but the relative motions of these two bodies are such that they never come closer than 18 AU. On the other hand, there are periods such as the present one (1979–99) when Pluto is actually closer to the Sun than Neptune. Such a large change in solar distance will obviously have a big effect on Pluto's climate. As we shall see, the total mass of the atmosphere may change substantially as the planet moves around its orbit.

There have been several proposals that Pluto is actually an escaped satellite of Neptune and not a planet in its own right. The discovery of a plutonian satellite in 1978 tends to weaken this hypothesis. Another relevant discovery was the asteroidal object Chiron, orbiting the Sun between Saturn and Uranus. Other such bodies may yet be discovered, and Pluto may simply be the biggest of these — a remnant of the early solar system that was neither accreted into a planetary core nor captured as a satellite because it happened to form in the stable orbit in which we find it today.

12.3 Atmospheres of Uranus and Neptune

Atmospheric Composition

Through a telescope, the tiny disks of Uranus and Neptune appear greenish in color. This color results from the fact that the atmospheres of both planets are unusually free of clouds. Hence the pictures of Uranus obtained by Voyager 2, although beautiful in their own right, are singularly free of detail (Plate 16). Sunlight penetrates to great depths before it is scattered back into space, and all along its path it is absorbed by methane gas (CH_4). The same absorption bands that are present in spectra of Jupiter and

Saturn appear in the spectra of Uranus and Neptune, but they are very much stronger. Many new absorptions are also evident (Fig. 12.6).

The hydrogen (H_2) absorption lines on Uranus and Neptune are also stronger than in spectra of Jupiter and Saturn, but the increase is not as great as for methane. In other words, the proportion of methane in the atmospheres of Uranus and Neptune is much higher than in the atmospheres of Jupiter and Saturn (Table 12.1). This is consistent with the higher proportion of ices in the total masses of these smaller planets. Uranus and Neptune are clearly enriched in heavier elements compared with Jupiter and Saturn, and this difference shows up in the composition of their atmospheres as well as in their average densities.

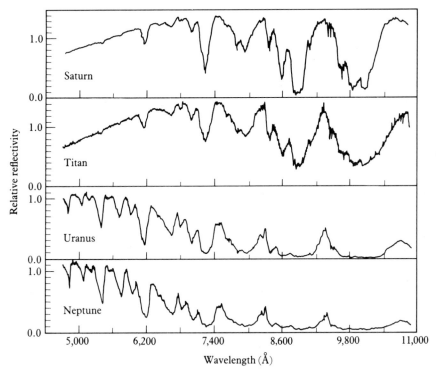

FIGURE 12.6 Spectra of Saturn, Titan, Uranus, and Neptune, each divided by the spectrum of the Sun. If these objects had no atmospheres, the spectra would be smooth, nearly flat curves. The deep absorptions are caused by methane, which is producing a much larger effect on Uranus and Neptune than on Titan and Saturn. (After R. Danhy.)

TABLE 12.1 Atmospheric compositions of Uranus and Neptune

A. Main Constituents (percent)

Gas	Formula	Uranus	Neptune
Hydrogen	H_2	84	84 ?
Helium	He	14	?
Methane	CH_4	2	2–3

B. Trace Constituents (parts per billion)

Acetylene	C_2H_2	200	present
Ethane	C_2H_6	?	present

ammonia, although this gas may be present at lower, warmer levels; at still greater depths water vapor should exist as well.

The Voyager results indicate that there is a tenuous haze in the upper atmosphere of Uranus centered over the rotational pole (Fig. 12.7; Plate 16). Evidently this haze is formed by the steady irradiation of the planet's upper atmosphere by solar ultraviolet light, so the haze is concentrated in the sunlit hemisphere. Hence even at this immense distance from the Sun, some atmospheric chemistry is taking place, forming a very thin smog layer that preferentially absorbs short wavelength light.

The Voyager infrared spectrometer was able to determine that the relative abundance of helium in the atmosphere of Uranus is even more similar to that in the Sun than is the case for Jupiter. This result is consistent with a two-step model for the formation of the giant planets. First, a huge core of ice and rock, ten to fifteen times the mass of the Earth, forms, and second, an envelope of gas (mostly hydrogen and helium) from the solar nebula collapses around the

At a level in their atmospheres where the pressure is the same as the sea-level pressure on Earth, the temperatures on both Uranus and Neptune are about −200 C, about the temperature of liquid nitrogen. This is much colder than conditions on Jupiter or Saturn at the same pressure level (Section 11.3). It is far too cold for

FIGURE 12.7 Voyager pictures of Uranus (left to right) taken through orange-, violet-, and methane-sensitive filters. The haze shown in Plate 16 is easily seen in the violet image. The orange and methane images show two narrow cloud systems at temperate latitudes on Uranus. The bright edge of the image in the methane picture is caused by scattering from a high altitude haze.

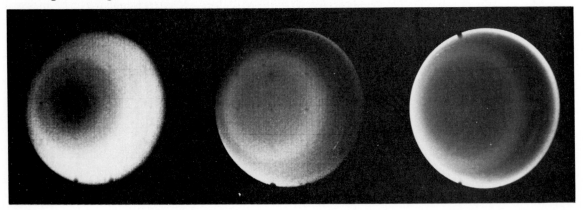

core. The core contributes the excess methane, while the bulk of the atmosphere is representative of the solar nebula. Evidently Uranus and Neptune acquired less massive envelopes of solar nebula gas than Jupiter and Saturn, and therefore they have a relatively higher proportion of ice and rock.

We know much less about Neptune than about the other three giant planets, pending the Voyager encounter in 1989. Methane is abundant, but helium has not yet been detected. Further, Neptune has extensive high-altitude clouds, while Uranus does not. Neptune is definitely not an identical twin of Uranus.

Atmospheric Temperatures

At infrared wavelengths, where both Jupiter and Saturn exhibit strong emissions from methane and from products of methane photochemistry, the spectrum of Uranus is essentially blank. In contrast, Neptune has emissions from both methane and ethane (Table 12.1). Why this difference? Evidently there is a temperature inversion in the atmosphere of Neptune but not on Uranus.

When temperature increases with altitude in such an inversion, absorption bands produced by gases in a normal atmosphere become emission bands, as seen from the outside, because the local gas temperature is higher than that of the planet beneath. Emission bands produced in inversion regions can be detected with spectrometers tuned to the infrared, typically at wavelengths near 10 μm, where little reflected sunlight can interfere.

Once again we have a fundamental difference between these two bodies for which there is no explanation. Uranus is unique in the outer solar system in having no thermal inversion in its upper atmosphere; even Saturn's satellite Titan exhibits this phenomenon (Chapter 13).

Temperatures Measured at Radio Wavelengths

Going to still longer wavelengths, we find more puzzles in observations made with radio telescopes. No indications of non-thermal emissions were detected from either of these planets until Voyager reached Uranus. Telescopes on Earth have not noted noise bursts at decameter wavelengths or steady signals from spiraling electrons in the decimeter region. What the radio observations do show, however, is very interesting indeed.

At these wavelengths, we detect thermal radiation from deep within the atmospheres, so the corresponding temperatures are much higher than those detected elsewhere in the spectrum. In fact, one expects the measured temperatures to be higher as the wavelength of the observations increases. Since the atmospheric gases become increasingly transparent to radiation of longer wavelengths, energy from lower, warmer levels of the atmosphere can escape more freely into space. This is how the high surface temperature of Venus was discovered (Chapter 8). The expected increase of temperature with increasing wavelength indeed occurs on Jupiter, Saturn, and Neptune, but once again, Uranus differs. Evidently, the temperature in this planet's atmosphere does not increase steadily with depth, as it does in the atmospheres of the other giants.

Unlike some of the peculiarities of Uranus already discussed, this one can be explained. Recall that Uranus has no detectable source of internal energy. In other words, it does not have an excessively hot interior. That means that only a small amount of thermal energy will attempt to escape from deep within the planet. In that case, there will not be enough energy to maintain convection in the lower atmosphere, so the increase in temperature with depth at these deep levels will be very gradual. At higher altitudes, sunlight provides the energy to maintain convection. The situation is rather like that in our oceans, where only the upper layers participate in the seasonal cycle driven by sunlight. The murky depths below remain at essentially the same temperature throughout the year. This combination of external sunlight and a weak internal heat source leads to the unusual temperature profile shown in Fig. 12.8.

The radio spectra of these planets reveal another unusual characteristic. Radio brightness measurements for Jupiter and Saturn agree with calculations for the thermal emission expected from these planets, but the Uranus and Neptune data do not. The discrepancy is particularly marked in the 3 to 10 cm region of the spectrum. The planets are warmer here than the calcula-

tions would suggest. This means that the real atmospheres are more transparent at these wavelengths than the model atmospheres used in the calculations. This higher transparency indicates a lower amount of ammonia, since ammonia absorbs at these wavelengths. Why then is ammonia depleted? Perhaps the answer lies in those hypothetical water clouds described in the previous section. Since ammonia dissolves readily in water, it might be kept out of the atmosphere above the clouds. This is a promising idea, but by no means an established fact. We need still more data on the structure and composition of the deep atmospheres of these planets.

FIGURE 12.8 The variation of temperature with height on Uranus. The solid line shows the Voyager results; the dashed line is based on radio telescope studies from Earth. Temperature increases more slowly with depth on Uranus than it would if the planet had a large internal heat source.

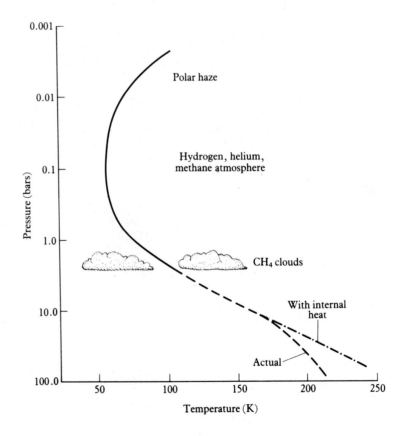

12.4 Climate, Clouds, and Weather

Seasons on Uranus and Neptune

The seasons on Earth occur because the Earth's axis is inclined to the plane of its orbit by 23.5°, and the direction of the axis is fixed in space as the planet revolves around the Sun. The resulting changes in the intensity and the duration of the illumination of a given region of the planet's surface have a profound effect on the weather. Now imagine the seasons on Uranus, whose axis is tipped nearly into the plane of its own orbit. This planet requires 84 years to circle the Sun. That means that for 42 years, one pole receives some illumination from the Sun while the other pole is in total darkness. The pole-on configurations represent the winter and summer sol-

stices, while the equinoxes occur when the equator faces the Sun (Fig. 12.9).

One might think that this orientation of the axis of rotation should effect the circulation of the atmosphere. When one of the poles is facing the Sun, it receives continuous illumination that is independent of the planet's rotation. In this configuration, one would expect the circulation to consist of warm air rising at the pole that points toward the Sun and then moving past the equator to the opposite pole where it would sink. This is the simplest type of Hadley cell (Section 8.3).

However, Voyager found that the general circulation on Uranus is dominated by the planet's rotation, and the clouds, therefore, exhibit a banded pattern lying along latitude lines similar to that of Jupiter and Saturn. This conclusion

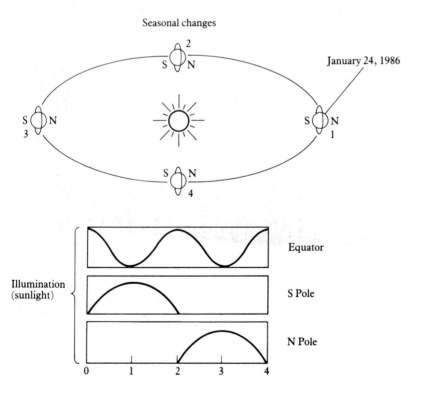

FIGURE 12.9 Seasons on Uranus last for 21 years; note that the equatorial region experiences two summers and two winters each year.

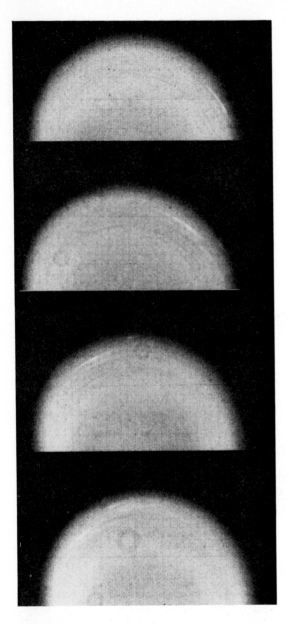

FIGURE 12.10 A sequence of four Voyager 2 images of Uranus taken through an orange filter indicates the motions of two narrow, bright clouds over 4.6 hours. The larger of the two clouds is at a latitude of 33° S; the smaller cloud, seen most clearly in the lower two images, is closer to the edge of the visible disk, at a latitude of 26° S.

was not easily reached, since very few clouds were detected on Uranus. Fig. 12.7 shows two. The apparent rotation of the planet was revealed in sequences of pictures such as those shown in Fig. 12.10.

These clouds are probably condensed methane, rather than the ammonia or ammonium hydrosulfide clouds encountered on Jupiter. They appear to be found at an altitude where the temperature is 80 K and the pressure is 1.3 bar. The fact that we can only see them at rather low latitudes in the Voyager pictures indicates that there must be some additional thin haze above them at higher latitudes.

Weather on Neptune

Although we have no spacecraft pictures of Neptune yet, ground-based observations obtained with special filters under excellent conditions do reveal clouds in the planet's atmosphere (Fig. 12.11). The presence of these high-contrast clouds on Neptune is yet another significant difference between these two planets. The composition of these clouds remains unknown. Neptune's atmosphere is cold enough for ethane, methane, argon, and carbon monoxide to condense. As we have frequently noted, it is difficult to determine the composition of a cloud from remote sensing. Remember that it was not until the last decade that the cloud composition on Venus, the nearest and brightest planet, was determined.

Changes in the cloud cover on Neptune on an almost global scale can occur within a few days, as if a clear atmosphere suddenly turned hazy over the entire planet. These changes have been detected by their effect on the reflection of infrared radiation by the planet, as well as in pictures such as those reproduced here. Weather, then, exists on Neptune, even though the nature of the clouds and the general circulation are not yet known.

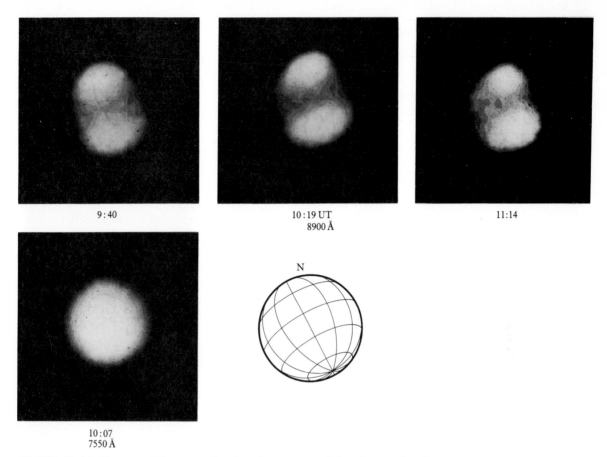

9:40

10:19 UT
8900 Å

11:14

10:07
7550 Å

FIGURE 12.11 Pictures of Neptune, showing the rotation of the planet, taken by B.A. Smith and R. Terrile through a filter centered on a strong methane absorption band. Bright clouds are clearly revealed since there is less absorbing gas above them.

Rotation Periods and Wind Speeds

The accurate determination of the rotation period of Uranus was not possible before the Voyager encounter in 1986. As shown in Fig. 12.10, the spacecraft pictures revealed clouds that could be tracked as rotation carried them around the planet, indicating a period near 16 hours. As in the case of Jupiter or Saturn, however, what is actually measured in such observations is the sum of the underlying rotation and of the wind speeds at the location of the clouds. To disentan-

gle these two effects, one must have an independent measurement of the rotation period of the deep interior, determined from observations of the magnetic field of Uranus to be 17 hours 14 minutes. Then the true wind speeds can be determined. As shown in Table 12.2, the speed varies with latitude on Uranus, as it does on Jupiter and Saturn, but in the opposite sense.

On Neptune it has been easier to determine a rotation period from remote observations. Even though this planet is farther from us than

TABLE 12.2 Rotation periods for Uranus and Neptune

Uranus	Neptune
Atmosphere:	Atmosphere:
30° S: 16.5 hours	Middle latitudes:
40° S: 16 hours	17.3 hours
70° S: 14.25 hours	
Interior:	Interior:
17.24 hours	no data

Uranus, the contrast of clouds in its atmosphere is much greater, so it is actually possible to determine the period from ground-based pictures of the planet. The result is 17 hours 18 minutes (a value, of course, that includes the average wind speed in the atmosphere). Thus the rotation periods of Uranus and Neptune are nearly identical.

A curious paradox is implicit in the rotation periods of Uranus listed in Table 12.2. The basic physics that governs moving air tells us that since the atmosphere of Uranus gets most of its heat at the pole, winds should increase in speed toward the equator. However, the reverse is true: the fastest winds are near the pole. Furthermore, the Voyager measurements showed that the temperature of the atmosphere is the same at the north pole, which has just experienced 21 years of total darkness, as it is at the south pole, which has been heated for 21 years. Evidently some dynamical processes, not yet understood, are redistributing the heat deposited from the Sun in ways that can account for both the unusual temperature distribution and the wind pattern.

The last peculiar characteristic of the rotation of Uranus is that it is retrograde. Like Venus and Pluto, this planet rotates backwards compared to its direction of motion around the Sun (Fig. 12.5).

Magnetospheres

In the absence of any detectable non-thermal radio emission from Neptune, we have no evidence yet that this planet has a magnetic field. Voyager did give us a detailed picture of the unusual magnetic field and magnetosphere of Uranus. Perhaps the most surprising fact is that the magnetic field is inclined at a large angle (60°) to the rotation axis, unlike the situations on Earth, Jupiter, and Saturn, where the magnetic axis and the rotational axis are roughly parallel.

The close alignment of these two axes is what leads inhabitants of Earth to say that a compass on our planet points north, since the magnetic pole is not too far from true north, as defined by the axis of rotation. Imagine the plight of a traveler on Uranus, trying to establish directions on a cloudy day. The magnetic axis is not only inclined by 60°; it is also offset from the rotational axis by one third of the planetary radius (Fig. 12.12). Hence the direction that a compass needle would point on Uranus (relative to the orientation of the rotational axis) would vary greatly with position on the planet. The strength of the field would also vary, from 0.1 to 1.1 gauss (compared with a nearly constant 0.3 gauss on the Earth).

The question is still open as to how this magnetic field is generated. There is no metallic hydrogen in the interior of Uranus, so the conducting medium required for a dynamo must be something else. Perhaps there is a layer of water that is sufficiently electrically conducting to serve this purpose.

Uranus also has a magnetosphere, generally similar in size to that of Saturn. However, the composition of the atomic particles in the magnetosphere is simpler, consisting almost entirely of protons and electrons derived primarily from hydrogen escaping from the planet's atmosphere. The magnetosphere also shares the high inclination of the planetary magnetic field to which it is bound. A magnetotail stretches out

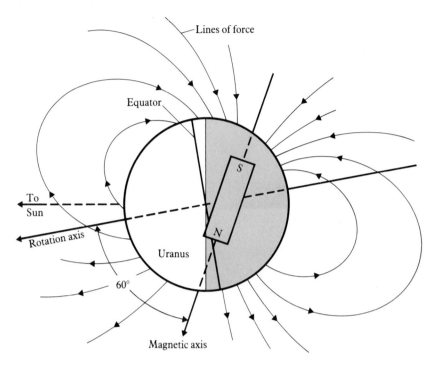

Lines of force

Equator

To
Sun

Rotation axis

Uranus

60°

Magnetic axis

S

N

FIGURE 12.12 The magnetic field of Uranus behaves as if there were a bar magnet inside the planet, tipped at an angle of 60° to the axis of rotation and offset as shown in the diagram.

tens of planetary radii behind the planet, but because of this field inclination, it rotates like a corkscrew as the planet turns on its axis.

The entire sunlit side of Uranus is bathed in a glow of ultraviolet light emitted by escaping hydrogen atoms. This "electroglow," as it is termed, was also seen on Jupiter, Saturn, and Titan. Auroral activity, produced by the collision of magnetospheric ions and electrons with the planet's upper atmosphere, also occurs. On Uranus the aurora is not near the north and south poles; instead it takes place near the equator, since the magnetic poles are there. The Uranians (if there were any) would speak of the beauties of the "tropical lights" instead of the "northern lights" that we extol. But would they even consider the region around their equator to be the tropics? From our familiar terrestrial perspective, Uranus is a confusing place.

12.5 Pluto and Its Moon
The Pluto-Charon System

Is Pluto really a planet? This is a question that scientists have been asking ever since the discovery of this unusual object. You might think that since it moves about the Sun in a roughly circular orbit it must be a planet by definition. However, comets and asteroids also orbit the Sun. Pluto's orbit is the most eccentric and highly inclined of any of the planets. Pluto is also small, smaller than our Moon. Its rotation period is unusually long; at 6 days 9 hours and 17 minutes, it is exceeded only by Venus and Mercury. Perhaps Pluto should be considered the largest asteroid rather than the smallest planet.

As discovered in 1978, however, Pluto does have a satellite, named Charon for the mythical boatman who carried the dead into the realm of

Pluto. Although Charon is so small and close to Pluto that ground-based photographs lack the resolution to distinguish it clearly from the image of the planet (Fig. 12.13a), astronomers have used several ingenious techniques to investigate the Pluto-Charon system. The most useful tools have been provided by a series of natural occultations taking place between 1985 and 1991, in which the satellite passes alternately in front of and behind Pluto (Fig. 12.13b). Accurate timing of these events, as well as measurements of the changing brightness of the system as one object passes behind the other, have yielded a great deal of new information about both the planet and its satellite. Much of the data presented here has been derived from occultation measurements, primarily by David Tholen and Marc Buie of the University of Hawaii.

Charon orbits Pluto at a distance of just under 20,000 km, closer than any other natural sat-

FIGURE 12.13 a) The discovery photograph of Charon, Pluto's only known satellite. Charon appears as a small bulge at the top of this highly magnified image of Pluto obtained by James Christy and Robert Harrington on 2 July 1978 with the 1.54-m reflector of the U.S. Naval Observatory. b) As Pluto moves along its orbit around the Sun, the aspect of Charon's orbit changes as viewed from Earth. We are now in a period of time when eclipses and occultations can occur in this system. Observations of these phenomena are providing new information about both Pluto and Charon.

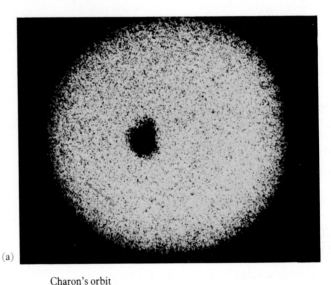

(a)

Charon's orbit

Pluto's orbit

1994

Last eclipses
1991

1988

First eclipses
1985

1982

ellite except Mars' Phobos. The period is equal to the period of the planet's rotation. Here is another case of tidal resonance in the solar system, and one that is different from all the others we know. Presumably Charon was able to pull Pluto into a tidal lock because of its large relative mass, which is only about one-tenth that of Pluto. The Pluto-Charon system has the largest satellite/planet mass ratio in the solar system, with that of the Earth-Moon being second at a value of 1/80.

Mass and Density of Pluto

Until the discovery of Charon, the mass of Pluto could only be estimated indirectly by the apparent effect of Pluto on the motions of Uranus and Neptune (Section 12.1). These estimates were very uncertain, with values ranging all the way from 0.2 to 7 Earth masses. The corresponding uncertainties in the density of Pluto made it impossible to guess at the composition of this planet.

From Charon's orbit, it is possible to use a simple application of Newton's laws to determine Pluto's mass. The result is a mass of $\frac{1}{400}$ that of Earth, much smaller than the previous estimates. When this mass is combined with the newly determined occultation diameter (2400 km), a density of 1.7 g/cm³ is derived, consistent with the idea that this planet may be a remnant of the kind of rock-and-ice planetesimal that accreted to form the cores of the giant planets.

Although Pluto's density of 1.7 g/cm³ is low by the standards of the inner planets, it is significantly higher than the densities of many of the icy satellites we shall encounter in Chapters 13 and 14. Thus Pluto appears to contain more rock (or less ice) than some other objects that are closer to the Sun, an unexpected result that we will return to when we discuss the origin of the solar system (Chapter 15).

The orbit of Charon defines the inclination of Pluto's axis of rotation, since it is reasonable to assume that a system tidally locked in this fashion will have the satellite's orbit in the equatorial plane of the planet. This was another surprise: the inclination of Charon's orbit to the orbit of Pluto around the Sun is 118°. If the satellite is indeed moving in the planet's equatorial plane, this means that Pluto, like Uranus, has its rotational axis very nearly in the plane of its orbit. Three planets then rotate backwards: Venus, Uranus, and Pluto (Fig. 12.14).

FIGURE 12.14 To an observer located above the plane of the solar system, both Uranus and Pluto would seem to rotate in a retrograde direction. (Sizes not to scale.)

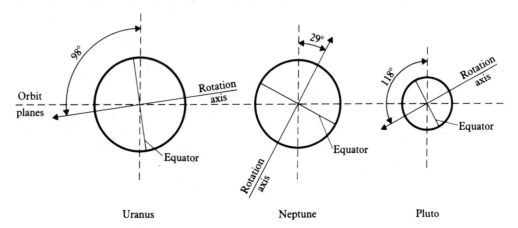

The Surface and Atmosphere of Pluto

The surfaces of both Pluto and Charon are fairly bright, reflecting between 35% and 50% of the weak sunlight that falls on them. The distribution of bright and dark material is not uniform, however, and Pluto is lighter toward its poles than it is near the equator (Fig. 12.15). A few years before the discovery of Charon, spectroscopic observations of Pluto revealed the presence of solid methane on the planet's surface. Methane ice might indeed be expected at Pluto's distance from the Sun, where average surface temperatures should range from about 47 K when it is farthest from the Sun to 60 K when it is closest. Somewhat surprisingly, observations obtained during the 1987 Pluto-Charon occultations show that the satellite surface lacks methane, but is composed instead of ordinary water ice.

At these low temperatures, most gases except hydrogen, helium, and neon will form solids or liquids. Yet as we know from our experience with water on Earth, a small amount of gas always exists in equilibrium with the solid or liquid. In the case of the methane on Pluto, there should be some gas in equilibrium with methane ice on the surface. Hydrogen and helium can easily escape from such a small planet, even one that is this cold, but other gases may well be present and simply not yet detected.

Because of the 13 degree temperature difference expected between the aphelion and perihelion points of Pluto's orbit, we can anticipate that the amount of gas in the atmosphere changes dramatically with time. The vapor pressure of methane, for example, would be about 10 times greater at perihelion than at aphelion. Similar changes would occur for other gases in equilibrium with solids or liquids. We are presently observing Pluto near the time its atmosphere is at its maximum pressure, because perihelion occurs in 1989.

Pluto is the only outer planet that we could stand on rather than float in the gondola of a balloon. The surface is covered with ice or snow, and the sky overhead would probably be black, even at midday, since the atmosphere is very thin and the Sun very far away. Seen from this vast distance, our star would be some 16,000 times fainter than it appears to us, too small even to show a disk without optical magnification. Charon would be visible only from one hemisphere of Pluto, as a consequence of the synchronous rotation of the planet.

Unfortunately, we cannot expect the same rush of new information about Pluto that is occurring for the other outer planets as a result of the Voyager encounters. No way exists to deflect the Voyager 2 spacecraft at Neptune in order to allow it to reach Pluto, and visits to Pluto by other spacecraft will probably have to wait until the next century; they just aren't in the current budgets. Unless the space exploration program is accelerated, most of us will never see a close-up photo of Pluto in our lifetimes.

FIGURE 12.15 A crude computer "image" of Pluto reconstructed from photometric data by M. Buie of the University of Hawaii. The main features are lighter polar regions and a single prominent dark spot near the equator.

———————————◇———————————

Summary

The three outermost planets were not known in antiquity; they had to be discovered. The stories of their discoveries illustrate an interesting variety of methods. Uranus was found accidentally, at a time when no one expected any planets beyond the five visible to the unaided eye. The search for Neptune involved theoretical predictions based on gravitational theory and the most sophisticated mathematical techniques of the nineteenth century; its discovery was a triumph for Newtonian celestial mechanics. The search for Pluto was similarly inspired by theoretical work, but ultimately the planet was found as the result of a very careful observational search. We now know Pluto is far too small to have perturbed the motions of other planets in a measurable way.

Uranus and Neptune form a pair of planets, substantially smaller than Jupiter or Saturn but much larger than the terrestrial planets. Their composition is also intermediate between that of the giants, with their nearly cosmic proportions of the elements, and that of the rocky, volatile-depleted terrestrial bodies. Uranus and Neptune are composed primarily of the common ices probably mixed with silicates; they may have been built up largely from primordial icy planetesimals which were probably similar (in composition, if not in size) to the comet nuclei we see today.

Uranus is notable for its rotation, with its axis tipped on one side, for its magnetic field inclined at 60° to the rotation axis, and for its lack of a major internal heat source. Neptune has no regular satellites, but otherwise seems more normal. Both planets have atmospheres composed of hydrogen, helium, and methane. The atmosphere of Uranus is largely free of clouds, but that of Neptune exhibits rapidly changing high-altitude clouds. Surprisingly, the circulation and temperatures in the Uranus atmosphere do not seem to be influenced by the peculiar orientation of its rotation axis.

Pluto is an anomalous object, resembling an icy satellite more closely than the other planets. Smaller than our Moon, Pluto has a frigid surface of frozen methane and a density of 1.7 g/cm^3, the lowest of any solid planet. Its orbit is quite eccentric, bringing it closer to the Sun than Neptune during part of its 249-year period of revolution. Pluto has a satellite, Charon, which is the largest satellite in the solar system relative to its primary, and like Uranus, Pluto is tipped on its side with a retrograde rotation.

———————————◇———————————

Key Terms

perturbation

PART SIX
REVIEW QUESTIONS

1. Summarize the main distinctions between the giant planets (Jupiter, Saturn, Uranus, Neptune) and the terrestrial bodies studied in Parts III, IV, and V.

2. Summarize the main distinctions between the two largest giants (Jupiter and Saturn) and their smaller, denser cousins (Uranus and Neptune).

3. Compare and contrast Jupiter and Saturn. Indicate how their differing masses and distances from the Sun can explain many of the observed differences in the atmospheres and interiors of these two planets.

4. Compare and contrast Uranus and Neptune. Indicate how their differing masses and distances from the Sun can explain many of the observed differences in the atmospheres and interiors of these two planets.

5. Compare the compositions of the four giant planets with those of the Sun, the comets (which may have formed in the same part of the solar system), and with what we know of the primordial solar nebula.

6. One of the ways the giant planets differ from the terrestrial planets is in the amount of hydrogen available in their atmospheres. Compare the chemistry of atmospheres dominated by hydrogen with those that are not. Specifically, consider the behavior of the abundant elements oxygen, carbon, and nitrogen in these two chemical environments.

7. Characterize the interiors of Jupiter and Saturn. How are the magnetic fields generated in each? How complete is our knowledge, compared with that of the interiors of the terrestrial planets? What factors determine which planet we are better able to understand?

8. Fundamental properties of planets include the magnitude of the internal heat source and the rate of heat escaping to the surface. Discuss these factors for the giant planets. How is the energy generated for each? Why is Uranus so different from the other three?

9. Characterize the atmospheric structure for each of the giant planets. What are the roles of internal heat and of heating from sunlight? What determines which clouds form and where? Why are the clouds of Jupiter so much more colorful and dynamic than those on Saturn? What determines whether there is a temperature inversion?

10. Discuss atmospheric circulation for Jupiter and Saturn, and compare these planets with Earth and Venus. What are the basic factors that cause different circulation regimes for each of these four planets?

11. Describe the magnetospheres of Jupiter, Saturn, and Uranus. Make a table showing the sources and sinks for charged particles in the magnetospheres of Earth, Jupiter, Saturn, and Uranus. What causes the main differences between the magnetospheres of Jupiter and Saturn?

12. Compare the radio emissions, both thermal and non-thermal, from Jupiter, Saturn, and Uranus. What do studies in this part of the spectrum tell us about the magnetospheres, the rotation rates, and the atmospheric structures of each of these three planets?

13. The axis of rotation of Uranus has a remarkable tilt. How might this situation have come about? What consequences does this axial orientation have for the properties of Uranus? Specifically, discuss how it influences the weather.

14. Compare and contrast Pluto and our Moon. What initial differences are the most fundamental?

15. Explain how the discovery of Charon has helped us to understand Pluto itself. What things can be measured now that were unknown before this discovery?

16. Is Pluto really a planet? Explain.

17. Review the conditions of discovery for Uranus, Neptune, and Pluto. How do the methods used in each case reflect the state of science and of the societies in which the discoveries were made? If you were to undertake a search for a trans-plutonian planet today, how would you go about it?

ADDITIONAL READING

Alexander, A.F.O.'D. c1962, 1980. *The Planet Saturn*. New York: Dover Publications.

Alexander, A.F.O'D. 1965. *The Planet Uranus*. New York: Dover Publications.

Cooper, H.S. 1982. *Imaging Saturn*. New York: Holt, Rinehart and Winston.

*Dessler, A.J., ed. 1983. *Physics of the Jovian Magnetosphere*. New York: Cambridge University Press.

Fimmel, R.O., Van Allen, J., and Burgess, E. 1980. *Pioneer — First to Jupiter, Saturn, and Beyond* (NASA SP-446). Washington: U.S. Government Printing Office.

*Gehrels, T. ed. 1976. *Jupiter*. Tucson: University of Arizona Press.

*Gehrels, T. and M.S. Matthews, eds. 1984. *Saturn*. Tucson: University of Arizona Press.

Hoyt, W.G. 1980. *Planets X and Pluto*. Tucson: University of Arizona Press.

*Hubbard, W.B. 1984. *Planetary Interiors*. New York: Van Nostrand, Reinhold.

Ingersoll, A. 1982. "Jupiter and Saturn." In *The New Solar System*, 2nd ed., ed. J.K. Beatty, B. O'Leary, and A. Chaikin. Cambridge, MA: Sky Publishing Corp.

Ingersoll, A. C. 1987. "Uranus." *Scientific American* 256:1, 38.

Morrison, D. 1982. *Voyage to Saturn* (NASA SP-451). Washington: U.S. Government Printing Office.

Morrison, D. and J. Samz. 1980. *Voyage to Jupiter* (NASA SP-439). Washington: U.S. Government Printing Office.

Poynter, M. and A.L. Lane. 1981. *Voyager: Story of a Space Mission*. New York: Atheneum.

Smith, B.A. 1982. "The Voyager Encounters." In *The New Solar System*, 2nd ed., ed. J.K. Beatty, B. O'Leary, and A. Chaikin. Cambridge, MA: Sky Publishing Corp.

Washburn, M.L. 1983. *Distant Encounters*. San Diego: Harcourt Brace Jovanovich.

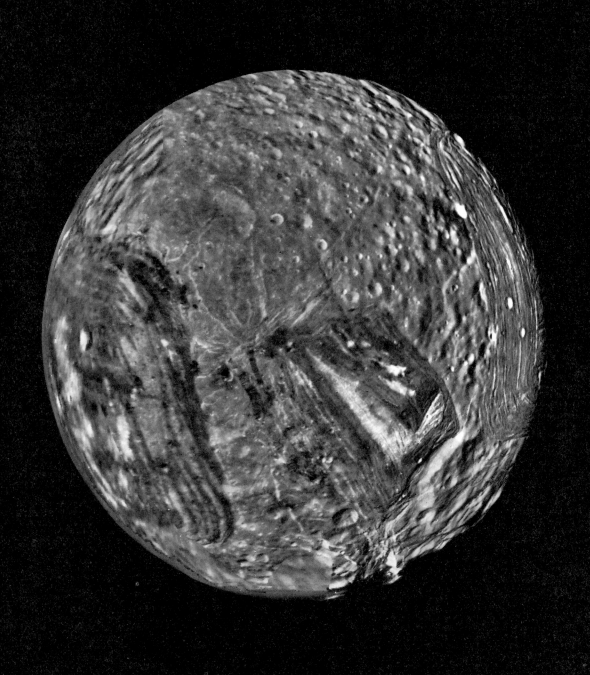

PART ◆ SEVEN
Satellites and Rings

The outer solar system consists of much more than the five planets discussed in Part VI; here are also found more than fifty satellites and four ring systems, with a diversity of phenomena worthy of half a dozen chapters in this text. As would be expected so far from the Sun, the solid objects that formed in the outer solar system are different in composition from the terrestrial planets. Instead of being worlds of rock and metal, the satellites of the outer solar system were formed, for the most part, with substantial quantities of water ice and other volatile materials. The typical solid object beyond Jupiter is, like a comet nucleus, made of a mixture of about half rock and half ice. In the larger of these outer solar system objects, differentiation has resulted in a rocky core with an icy mantle.

The largest of the satellites of the outer solar system are substantial worlds, comparable in size to the smaller terrestrial planets. The biggest — Ganymede, Titan, and Callisto — are almost as big as the planet Mars. Many of the smaller satellites are also interesting objects, while the rings of Jupiter, Saturn, Uranus, and Neptune have turned out to be remark-

ably varied, with quite different compositions, origins, and appearances. Recently many planetary scientists have become interested in the complex structures of ring systems, and understanding the dynamics of planetary rings may help us in the study of many other astrophysical phenomena, such as the origin of the solar system and the spiral structure of galaxies.

When we turn our attention to the discoveries made by the Voyager spacecraft in the satellite systems of the outer planets many questions come to mind. Why does Titan have an atmosphere, while Ganymede and Callisto, which are equally large, do not? What caused Ganymede to have a much more active geological history than its near-twin, Callisto? In studying the different ring systems, we wonder: Why are the Saturn rings bright and broad with a few narrow gaps, while the Uranus rings are dark and narrow with wide gaps? When we compare the large satellites with the terrestrial bodies in the inner solar system we also ask: How do the atmospheres of Titan and Triton compare with those of Earth, Venus, and Mars? Does the impact cratering in the outer solar system indicate that these objects were bombarded by the same meteoroidal projectiles that cratered the inner planets? We cannot promise answers to all of these questions, but perhaps the questions themselves will help us to pull together our knowledge of the inner and the outer solar system so that we may better understand the entire planetary family.

◀ Miranda, the daughter of Prospero in Shakespeare's *The Tempest*, speaks the famous lines "Oh brave new world . . ." Voyager 2 discovered that the satellite bearing this character's name is itself a new world beyond the imaginings of the scientists who designed the mission. This picture is a mosaic of the best images obtained by the flyby in January 1986. Note the rich variety of geological features on this small (500 km) icy satellite.

CHAPTER THIRTEEN

Worlds of Fire and Ice: The Large Satellites

13.1 Satellite Systems

Satellite Discoveries

In the inner solar system satellites are rare. Neither Mercury nor Venus has one, and our own Moon is a peculiar object whose very existence is difficult to understand (Section 6.6). Only Mars has more than one satellite, and these are both tiny objects presumably captured from the early population of asteroids (Section 4.5). In contrast, satellites are the norm for the giant planets, and it is quite reasonable to think of these systems as miniatures of the solar system itself.

The first satellites were discovered by Galileo in 1610, on the first night he turned his newly built telescope toward the sky. These are known as the Galilean satellites of Jupiter. Titan, the largest satellite of Saturn, was found in 1655 by Christian Huyghens of the Netherlands. As summarized in Table 13.1, discoveries continued over the next two centuries at a rate of nearly one satellite per decade, with Cassini and Herschel each equalling Galileo's record of four.

In the twentieth century, the most successful discoverer was Seth Nicholson of Mt. Wilson Observatory, who spotted four faint outer satellites of Jupiter between 1914 and 1951. Another important contributor of the pre-space era was Gerard P. Kuiper, the Dutch-born astronomer who did most to advance planetary studies in the

difficult decades from 1940 to 1960. Kuiper, who was with the University of Chicago before founding the first university department of planetary science at the University of Arizona, discovered one satellite each for Uranus and Neptune (Fig. 13.1).

Recent satellite discoveries have been made primarily with the Voyager cameras, although ground-based work continues to be productive. During the last decade, three new satellites of

FIGURE 13.1 Gerard P. Kuiper at his desk in the Lunar and Planetary Laboratory of the University of Arizona. Kuiper was almost the only U.S. scientist fully committed to planetary studies during the decade after the Second World War.

TABLE 13.1 The satellites

Planet	Satellite Name	Discovery	Semi-major Axis (km × 1000)	Period (days)
Earth	Moon	—	384	27.32
Mars	Phobos	Hall (1877)	9.4	0.32
	Deimos	Hall (1877)	23.5	1.26
Jupiter	Metis	Voyager (1979)	128	0.29
	Adrastea	Voyager (1979)	129	0.30
	Amalthea	Barnard (1892)	181	0.50
	Thebe	Voyager (1979)	222	0.67
	Io	Galileo (1610)	422	1.77
	Europa	Galileo (1610)	671	3.55
	Ganymede	Galileo (1610)	1070	7.16
	Callisto	Galileo (1610)	1883	16.69
	Leda	Kowal (1974)	11090	239
	Himalia	Perrine (1904)	11480	251
	Lysithea	Nicholson (1938)	11720	259
	Elara	Perrine (1905)	11740	260
	Ananke	Nicholson (1951)	21200	631 (R)
	Carme	Nicholson (1938)	22600	692 (R)
	Pasiphae	Melotte (1908)	23500	735 (R)
	Sinope	Nicholson (1914)	23700	758 (R)
Saturn	Unnamed	Voyager (1985)	118.2	0.48
	Unnamed	Voyager (1985)	133.6	0.58
	Atlas	Voyager (1980)	137.7	0.60
	Prometheus	Voyager (1980)	139.4	0.61
	Pandora	Voyager (1980)	141.7	0.63
	Janus	Dollfus (1966)	151.4	0.69
	Epimetheus	Fountain, Larson (1980)	151.4	0.69
	Mimas	Herschel (1789)	186	0.94
	Enceladus	Herschel (1789)	238	1.37
	Tethys	Cassini (1684)	295	1.89
	Telesto	Reitsema et al. (1980)	295	1.89
	Calypso	Pascu et al. (1980)	295	1.89
	Dione	Cassini (1684)	377	2.74
	Helene	Lecacheux, Laques (1980)	377	2.74
	Rhea	Cassini (1672)	527	4.52
	Titan	Huyghens (1655)	1222	15.95
	Hyperion	Bond, Lassell (1848)	1481	21.3

Diameter (km)	Mass (10^{20} kg)	Density (g/cm³)	Reflectivity	Surface Material
3476	735	3.3	0.12	silicates
23	1×10^{-4}	2.2	0.05	carbonaceous
13	2×10^{-5}	1.7	0.06	carbonaceous
20	—	—	0.05	rock ?
40	—	—	0.05	rock ?
200	—	—	0.06	rock, sulfur
90	—	—	0.05	rock ?
3630	894	3.6	0.6	sulfur, SO_2
3138	480	3.0	0.6	ice
5262	1482	1.9	0.4	dirty ice
4800	1077	1.9	0.2	dirty ice
15 ?	—	—	—	?
180	—	—	0.03	carbonaceous
40	—	—	0.05	carbonaceous
80	—	—	0.03	carbonaceous
30 ?	—	—	—	?
40 ?	—	—	—	?
40 ?	—	—	—	?
40 ?	—	—	—	?
15 ?	3×10^{-5}	—	—	?
15 ?	3×10^{-5}	—	—	?
40	—	—	0.5	ice ?
80	—	—	0.5	ice ?
100	—	—	0.5	ice ?
190	—	—	0.5	ice ?
120	—	—	0.5	ice ?
394	0.4	1.2	0.8	ice
502	0.8	1.2	1.0	pure ice
1048	7.5	1.3	0.8	ice
25	—	—	0.6	ice ?
25	—	—	0.9	ice ?
1120	11	1.4	0.6	ice
30	—	—	0.6	ice ?
1530	25	1.3	0.6	ice
5150	1346	1.9	0.2	cloudy atmosphere
270	—	—	0.3	dirty ice

TABLE 13.1 The satellites (*continued*)

Planet	Satellite Name	Discovery	Semi-major Axis (km × 1000)	Period (days)
	Iapetus	Cassini (1671)	3561	79.3
	Phoebe	Pickering (1898)	12950	550 (R)
Uranus	Cordelia	Voyager (1986)	49.7	0.34
	Ophelia	Voyager (1986)	53.8	0.38
	Bianca	Voyager (1986)	59.2	0.44
	Cressida	Voyager (1986)	61.8	0.46
	Desdemona	Voyager (1986)	62.7	0.48
	Juliet	Voyager (1986)	64.6	0.50
	Portia	Voyager (1986)	66.1	0.51
	Rosalind	Voyager (1986)	69.9	0.56
	Belinda	Voyager (1986)	75.3	0.63
	Puck	Voyager (1985)	86.0	0.76
	Miranda	Kuiper (1948)	130	1.41
	Ariel	Lassell (1851)	191	2.52
	Umbriel	Lassell (1851)	266	4.14
	Titania	Herschel (1787)	436	8.71
	Oberon	Herschel (1787)	583	13.5
Neptune	Triton	Lassell (1846)	354	5.88 (R)
	Nereid	Kuiper (1949)	552	360
Pluto	Charon	Christy (1978)	19.7	6.39

Jupiter, ten of Saturn, and ten of Uranus have been added to our catalog of outer solar system objects.

Table 13.1 summarizes the orbital and physical data for all of the satellites in the solar system, including the Moon and the martian satellites for comparison. This table provides the background for discussions of individual objects in this and the next chapter.

Regular and Irregular Satellites

The satellites of the outer planets are conveniently divided into two groups on the basis of their orbits, following a classification suggested by Kuiper. All but one of the large moons and many of the small ones are **regular satellites,** which revolve in orbits of low eccentricity located near the equatorial plane of their planet. The regular satellites are the analogs of the planets in the solar system, which revolve in nearly circular orbits close to the Sun's equatorial plane.

The second group of moons are called the **irregular satellites.** Their orbits are peculiar, having either large eccentricity or high inclination or both. Several of the irregular satellites even revolve in a retrograde direction. The irreg-

Diameter (km)	Mass (10²⁰ kg)	Density (g/cm³)	Reflectivity	Surface Material
1435	19	1.2	0.5	ice/carbonaceous
220	—	—	0.06	carbonaceous ?
40	—	—	—	?
50	—	—	—	?
50	—	—	—	?
60	—	—	—	?
60	—	—	—	?
80	—	—	—	?
80	—	—	—	?
60	—	—	—	?
60	—	—	—	?
170	—	—	0.07	carbonaceous ?
485	0.8	1.3	0.3	dirty ice
1160	13	1.6	0.4	dirty ice
1190	13	1.4	0.2	dirty ice
1610	35	1.6	0.3	dirty ice
1550	29	1.5	0.2	dirty ice
3500	—	—	0.4	methane ice
500 ?	—	—	—	?
1200	—	—	0.4	ice

ular satellites have orbits more like those of comets or asteroids around the Sun.

Because they are so closely bound to their planets, the regular satellites probably formed with the planets from a primordial planetary nebula, just as the planets formed from the primordial solar nebula. In contrast, the irregular satellites may be interlopers captured from elsewhere. In the case of Neptune, however, an original regular satellite system may have been disrupted to produce the peculiar orbits we see today. As noted in Section 12.2, it is even possible that the planet Pluto was once a part of the Neptune system.

The Jupiter System

The four most important members of the Jupiter satellite system are the Galilean satellites: Callisto, Ganymede, Europa, and Io. These are all large objects, ranging in size from a little smaller than our Moon to larger than the planet Mercury. The orbits of the inner three are evenly spaced, so that the period of revolution for Europa is twice that for Io, and Ganymede revolves in twice the period of Europa.

The even spacing of the Galilean satellites is more than a numerical coincidence, like the Titius-Bode rule for planetary distances. It represents a gravitational coupling or resonance. If one of these orbits should change, the other two would change with it to restore the resonant relationship. The orbital positions of the three satellites are also coupled, so that all three, for example, can never be on the same side of Jupiter. This relationship is called the Laplace resonance, named for the eighteenth century French mathematician who first explained it.

The other twelve known members of the jovian system can be classified into three groups of four satellites each. Close to the planet, orbiting inside the Galilean satellites, are four small objects, three of which were discovered by Voyager. Far beyond the Galilean satellites are eight small irregular satellites, half in direct orbits and half in retrograde.

The Saturn System

Saturn has the largest regular satellite system, just as it has the largest system of rings (perhaps for the same reasons). Fig. 13.2 compares the jovian and saturnian satellite systems. Of the sixteen satellites that are clearly regular, seven are co-orbital or nearly so; that is, they represent situations in which two or more satellites occupy nearly the same orbits. These cases, which are unique to the Saturn system, are discussed in Section 14.4.

Saturn's largest satellite is Titan, which is about the same size as the two largest Galilean objects, Ganymede and Callisto. Next in size are six satellites with diameters of 400 to 1500 km, corresponding to surface areas ranging from that of France to the whole of Europe. Jupiter has no satellites of this intermediate size; in fact, the only objects of this size we have encountered previously are the largest asteroids, such as Ceres and Vesta.

The three outer satellites of the Saturn system are more irregular. Hyperion, although its orbit fits the definition of a regular satellite, has a resonance with Titan that introduces an irregular, or chaotic, rotation. Iapetus has a moderate inclination and eccentricity, making it the Pluto

FIGURE 13.2 The satellite systems of Jupiter, Saturn, and Uranus are dramatically different, even in outline. Saturn has only one large satellite and only one distant irregular satellite. Ice is the dominant material composing all of Saturn's satellites, while it appears as a major component in only two of Jupiter's. Uranus has no satellites as big as our own Moon.

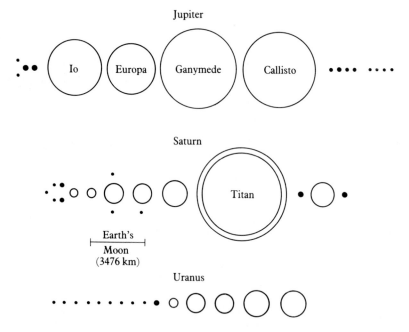

of the Saturn system. Only outermost Phoebe, which is in a retrograde orbit at a large distance from the planet, is clearly an irregular satellite.

The Uranus and Neptune Systems

The fifteen satellites of Uranus are all regular. Since they orbit in the equatorial plane of the planet, they share its high inclination. At the time of the Voyager 2 encounter, the polar axis of Uranus was directed nearly toward the Earth, so that the satellite orbits made a bull's-eye pattern on the sky. The five larger Uranus satellites are all in the same size range as the medium-sized Saturn satellites, while ten more newly discovered objects are less than 200 km in diameter.

Neptune is unique in that it is the only giant planet without a regular satellite system. Triton, its one large satellite (similar in many ways to the planet Pluto), is in a retrograde circular orbit. Triton is almost as close to Neptune as Io is to Jupiter. The second and much smaller satellite, Nereid, has the highest eccentricity (0.75) of any satellite; it is the comet of the Neptune system.

13.2 Impact Craters and Surface Ages

Should We Expect Craters in the Outer Solar System?

Before we look at the satellites individually, we should ask ourselves (as did the Voyager scientists before the Jupiter and Saturn encounters) what cratering we expect to find on solid bodies in the outer solar system. Will numerous impacts have taken place, and if so will the resulting craters have been preserved for us to see?

Let us address the second question first. The surfaces of almost all of the outer satellites are composed of ice or of ice-rock mixtures, like the comets. We know on Earth that ice is a plastic material, and that in the form of glaciers ice

can flow like a highly viscous fluid. At the distance of Saturn and beyond, however, the low prevailing temperatures cause ice to become as strong as rock, and craters should be preserved indefinitely unless there is local heating or erosion. The jovian satellites have slightly warmer temperatures, so the ice is not strong enough to preserve large craters for billions of years. Thus the craters of Rhea (Fig. 13.3) are nearly indistinguishable from those on the Moon, while the craters illustrated on Callisto have been considerably modified (Fig. 13.5).

FIGURE 13.3 The surface of Rhea bears a superficial resemblance to the highland regions of our Moon (cp. Fig. 5.2), but there are no indications of mare-like features on this mosaic of pictures of the north polar region obtained by Voyager 1 in November 1980. We are looking here at an icy surface, much more reflective than our Moon's dark rock. The smallest visible detail is only 1 km across.

The first question is more complicated. We clearly know less about the meteoroid population in the outer solar system than we do about potential impacting bodies close to the Earth. If, for example, most of the impacts on the Earth and Moon were from asteroids, we might expect a much smaller flux near the outer planets. So to answer the question of impacts in the outer solar system, we must first refine our understanding of the meteoroids closer to home.

The most complete census of meteoroids in the solar system has been carried out by Eugene Shoemaker and his colleagues. They have concluded that both asteroids and comets in the size range of about 1 km or larger can produce visible impact craters in the inner solar system. Shoemaker calculates that for the Earth and the Moon the two sources are of comparable importance; that is, about half of our craters are due to comets (both long-period and short-period) and about half to asteroids. Of course, these conclusions refer to the present. Different sources may have been important in the past, and certainly a different and much larger population of impacting bodies was present at the time of the terminal bombardment, 3.9 billion years ago.

Extending these results to the outer solar system, we see that the meteoroid population is cut in half without the asteroidal contribution. Impacts with short-period comets will also be less important at Jupiter and negligible at Saturn and beyond. Consequently, the impacting flux for the satellites should be less than half as great as we see in the inner solar system. Very roughly, the calculations of Shoemaker and others indicate a cratering rate about one-quarter that of the inner planets, resulting primarily from impacts with long-period comets.

Variations in Cratering Rates

Once the meteoroid flux has been calculated, it should be possible to use crater densities on the satellites to calculate approximate crater retention ages, as described in Section 5.4. The uncertainties in the flux are substantial, probably amounting to a factor of two or three, but crater densities should at least tell us whether a surface age is measured in billions, or hundreds of millions, or perhaps tens of millions of years. First, however, an important correction must be made to the calculated impact rates.

We saw when studying cratering on the Earth and the Moon that the size of the crater excavated by a projectile depends on the impact speed. The Earth, being larger, attracts meteoroids more strongly than the Moon, and the resulting craters are larger. In comparing crater densities on the Earth and the Moon, we must allow for the gravity field of the Earth.

In the outer solar system, both the gravity of the satellite and the gravity of its central planet are important, especially the latter. Imagine each giant planet at the bottom of a gravity slope. Meteoroids, mostly comets, are attracted toward the planet. The closer they come, the faster they move as they fall down this slope. Therefore a satellite far down the slope receives more impacts occurring at higher speeds. This effect can be very large: Mimas, the innermost large satellite of Saturn, can expect twenty times the cratering rate at Iapetus, which orbits far from the planet. Therefore, the inner satellites of Jupiter and Saturn should actually have higher cratering rates than the Earth or the Moon, despite the smaller flux of meteoroids.

Crater Retention Ages on the Satellites

Consequently, crater densities on the satellites are related to age in a manner not too different from that derived for the inner solar system. For example, we can conclude that a satellite crater density similar to that of the lunar maria indicates an age of several billions of years. As we will see when we look at individual objects in this and the next chapter, many satellites include terrains (called lightly cratered plains) with crater densities of 100 to 200 10-km craters per million square kilometers, and these areas are probably of similar age to the lunar maria and the northern volcanic plains of Mars.

Like the Moon, the satellites of the outer solar system also include surfaces with high crater densities. Such high densities cannot easily be explained from estimates of current cratering rates, which predict a maximum of a few hundred for the crater density even on a surface more than 4 billion years old. Thus we are drawn to the same conclusion described in Section 5.4 for the Moon: there was an early period in the outer solar system when cratering rates were much higher than they are today.

This is an important conclusion, as has been emphasized by Laurence Soderblom of the U.S. Geological Survey. Soderblom is the Deputy Team Leader and the senior geologist on the Voyager Imaging Team (Fig. 13.4). He played an important role in defining the standard cratering history for the inner solar system, with its late heavy bombardment ending about 3.8 billion years ago. Soderblom and his colleagues conclude that this unifying paradigm developed for the inner planets appears to apply to the outer solar system as well. This bombardment is therefore associated with the entire solar system, rather than being limited to the terrestrial planets. With this perspective, we now turn to the individual satellites.

FIGURE 13.4 Larry Soderblom (right) shares a humorous moment with Bradford Smith (left) during the Voyager encounters with the Saturn system.

composed about half of H_2O ice, and it was with considerable anticipation that Voyager geologists awaited the first high-resolution photos of these objects in March of 1979.

13.3 Callisto and Ganymede: The Large Galileans

Callisto and Ganymede are twins, in much the same way as are Venus and the Earth. They have nearly identical sizes and densities (Table 13.1), and they occupy similar regions of space. They are the first solid objects to be studied that are

Callisto: Basic Facts

Callisto, the outer Galilean satellite of Jupiter, has had the simplest geologic history of the large satellites, one dominated by impact cratering. Although its surface looks rather like the lunar highlands, we must remember that this satellite, like most of the others studied in this chapter, is composed in large part of ice (Fig. 13.5).

Callisto has a diameter of 4840 km, almost identical to the planet Mercury; yet its mass is less than one-third as great. From this simple comparison, we are immediately aware of a fun-

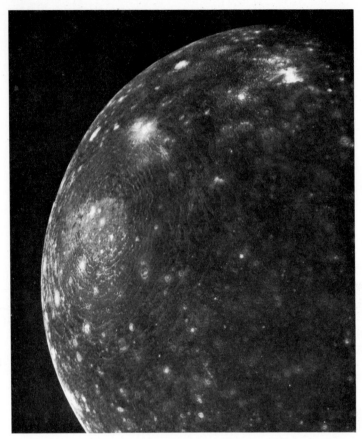

FIGURE 13.5 Callisto revealed a heavily cratered surface to the Voyager 1 cameras in March 1979. This mosaic shows the huge ringed feature called Valhalla (center left), with a central bright spot about 600 km across. The concentric bright rings can be traced to a distance of 1500 km from the center, making this feature bigger than the somewhat similar ringed basin, Mare Orientale, on the Earth's Moon (cp. Fig. 5.5).

damental difference in composition. Detailed calculations suggest that for a body the size of Callisto, its density of about 1.9 g/cm^3 implies that water ice is about equally as abundant as rocky materials in its interior.

The surface temperature of Callisto varies from 150 K near noon to about 100 K at night. Under such frigid conditions, less than a meter of ice is expected to have evaporated from the surface during the past 4.5 billion years. A bit nearer the Sun, however, in the main asteroid belt, an icy object could not survive. (Recall that comets made of water ice begin to turn on in the region of the asteroids, when their surface temperatures rise above 200 K.) Thus abundant

water ice is expected in the outer solar system, while ice and water can only survive in the inner solar system on the surfaces and in the atmospheres of planets with sufficiently high gravity, such as the Earth and Mars.

Ice is also observed directly on the surface of Callisto, where its presence is revealed by prominent infrared absorption bands. This ice is by no means pure, however, and the reflectivity of Callisto is actually rather low, only 18%. This dirty crust may be primitive, but more likely it is simply old. If there has been no resurfacing with fresh ice for billions of years, the accumulation of meteoritic dust mixed with the original ice crust would be expected to be fairly dark.

Fig. 13.6 illustrates infrared spectra of all four Galilean satellites, showing the strength of the ice bands on each.

Geology of Callisto

We begin our detailed study of outer planet satellites with Callisto because it is relatively simple to understand. The Voyager pictures reveal a heavily cratered world, with a landscape that is almost exclusively the product of impacts (Figs. 13.5 and 13.7). Over most of the surface, these impact craters are nearly as densely packed as those in the lunar highlands. Expressed in the same units that we have used before, the crater density on Callisto is 250 10-km craters per million square kilometers. Barring some improbable recent blizzard of impacts, it appears that the surface of Callisto is at least as old as the lunar maria, and that during this vast span of time lit-

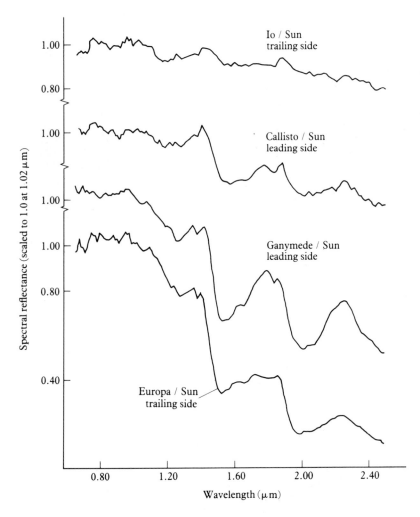

FIGURE 13.6 Long before the Pioneer and Voyager spacecraft reached the Jupiter system, we knew that Callisto, Ganymede, and Europa were covered with ice. This deduction is easily made from infrared spectra such as these, obtained with the Kitt Peak Solar Telescope.

tle or no internally produced geological activity has been present to destroy the accumulating impact scars.

It also seems probable that many of these craters were formed at a time of higher cratering rates, the outer solar system equivalent of the late heavy bombardment on the Moon. It is tempting to assert that this late heavy bombardment took place on Callisto about 4.0 billion years ago as it did on the Moon. Unfortunately, this attractive hypothesis cannot be tested at present.

A closer inspection (Fig. 13.7) reveals some important differences between the cratered surface of Callisto and the lunar highlands, however. The craters are not the same shape. Instead

of having bowl-shaped profiles, the Callisto craters are remarkably subdued in topography, as though they had been flattened or the crustal material had undergone **plastic deformation** since the time the craters were formed. Apparently, we see here one of the differences between a rocky and an icy object; at the temperature of Callisto's surface, ice, being less rigid, cannot preserve the sharp contours of an impact crater over hundreds of millions of years. Instead, it acts like a highly viscous fluid, which flows slowly, evening out topographic features.

Callisto also lacks the very large craters or basins that we have encountered on the Moon, Mercury, and Mars. Perhaps the population of impacting debris did not include the asteroidal-

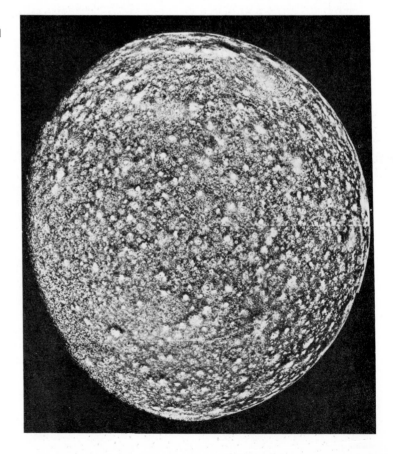

FIGURE 13.7 If you like craters, you'll love Callisto! This computer-enhanced mosaic of Voyager 2 pictures reveals that the satellite's surface is virtually covered with craters up to diameters of 100 km. Yet larger craters are strangely absent, suggesting that the icy surface cannot support such large structures over geologic time.

sized projectiles that struck the inner planets during the era of heavy bombardment. Such a deficiency of large impacts might be expected if the major source of projectiles in the outer solar system has been the smaller comets rather than asteroidal fragments. More likely the absence of large craters is another result of the icy composition of this satellite. Large topographic features are harder to preserve in a plastic material than small ones, and probably any large craters would simply have subsided into the surface.

This suggestion is supported by the presence of a few large circular white spots, surrounded by low concentric ridges spaced about 10 km apart to make a bull's-eye pattern (Fig. 13.5). These features resemble the pattern of waves formed by tossing a stone into a pool of water, and this analogy may be instructive. Apparently the bull's-eye basins are the ghostly remnants of great impacts, nearly swallowed up by the icy surface of the satellite. Nearly all topography has been destroyed by plastic deformation, and all that remains to mark the basin is the lighter surface and the concentric rings.

While we cannot probe the interior of Callisto and the other Galilean satellites, good reasons exist to believe that these are differentiated objects. Forming in the outer solar system where temperatures were low, they began their existence with a good deal of water ice mixed with rocky and metallic materials. Presumably the rocky component contained its fair share of radioactive elements as a heat source, while the water ice has a much lower melting temperature than rock. Inevitably, then, the interior temperature would soon rise above the melting point of the ice, even if the satellite formed cold. The heavier materials then sank to form a rocky or muddy core, and the liquid water rose and eventually refroze to form a crust and mantle of ice. The resulting internal structure, illustrated in Fig. 13.8, should apply to both Callisto and Ganymede, and probably to Saturn's Titan as well.

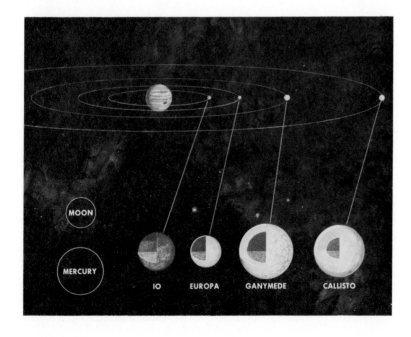

FIGURE 13.8 Conditions in the interiors of the satellites of Jupiter must be inferred from observations of their surfaces, sizes, and densities. Here we see that Io, with the highest density, must have a molten interior to account for its active volcanism. Callisto and Ganymede must be about 50% ice and 50% rock, while rocky Europa simply has a low-density crust consisting of ice or water and ice. The diameters of Mercury and the Moon are shown at left for comparison.

Geology of Ganymede

Ganymede is the largest satellite, and one that has experienced a unique geologic history. In the Galilean satellite system, it occupies the next orbit inward from Callisto. Ganymede has a nearly identical density to that of Callisto, and we therefore believe its composition and interior structure are similar. Both objects also have similar low surface temperatures, and both show the spectral signature of surface ice. Based on such overall comparisons, it would be reasonable to expect the two satellites to have experienced similar geologic histories; yet even a glance at the images of Ganymede and Callisto reveals fundamental differences (Figs. 13.5 and 13.9).

About half the surface of Ganymede is fairly dark and closely resembles the ancient cratered surface of Callisto, including such features as crater ghosts and concentric bull's-eye ridge patterns. These old terrains, however, are interspersed with lighter, less cratered areas that have been modified by internal activity. The crater densities on these modified areas are typically 100 to 200 10-km craters per million square kilometers, significantly lower than those on the heavily cratered areas of either Callisto or Ganymede, although still indicating crater retention ages measured in the billions of years.

Indications of Internal Activity

Many of the younger, less cratered areas of Ganymede also have systems of parallel moun-

FIGURE 13.9 A distant view of Ganymede obtained by Voyager 2 shows that the surface has both dark and light terrains, reminiscent of the appearance of our own Moon to the naked eye. Bright ray craters are visible everywhere.

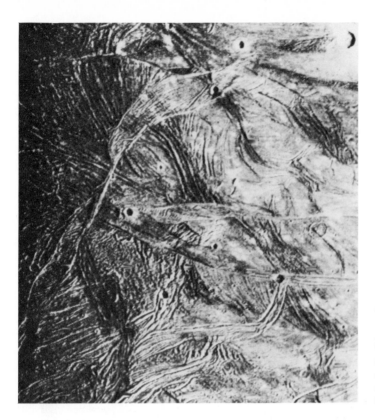

FIGURE 13.10 Surface features as small as 1 km across are visible in this high-resolution picture of Ganymede's grooved terrain. This kind of landscape is found only on the lighter regions of the satellite's surface. It is reminiscent of the Appalachian Mountains on Earth or the Maxwell Mountains on Venus. We will see it again on Miranda.

tains and valleys (Fig. 13.10). In scale these mountain systems on Ganymede resemble the low Appalachian Mountains of the eastern United States and the parallel mountain chains of the Ishtar upland of Venus, with ridges separated by 10 to 15 km. However, the mechanism of their formation is different.

While the terrestrial mountains, and possibly those of Venus, were created by wrinkling of crust caught between colliding continents, the mountains of Ganymede appear to result from long cracks or faults separating strips of land that have been alternatively lifted up or depressed (Fig. 13.11). They are the product of tension rather than compression in the crust. Many of the depressed areas in turn look as if they had been flooded with liquid — presumably liquid

water — from the interior. We do not know exactly how long ago this activity took place, but it happened after the era of heaviest meteoroidal bombardment and appears to have stretched over a span of hundreds of millions of years.

A compositional difference between the older and younger terrains on Ganymede can also be inferred. Both contain water ice mixed with dirty contaminants, but the geologically younger areas are lighter in color, suggesting less contamination of the ice. (Note that this is just the reverse of the case on our Moon, where the darker areas are younger.) The reflectivity of these younger terrains is about 40%, as compared with 25% for the heavily cratered areas. Ganymede is further distinguished by a few very bright craters and crater rays, which appear to

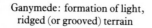

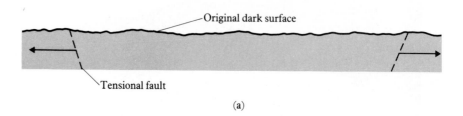

(a)

(b)

(c)

FIGURE 13.11 While we do not completely understand the processes that formed the grooved terrain on Ganymede, this diagram describes one possibility. a) The original primitive crust is cracked by tensional stresses. b) Subsidence, flooding, and freezing produce a new surface. c) Additional faulting and subsidence produce the ridges and valleys seen today.

have splashed nearly pure water across the surface (Fig. 13.12). No similar bright craters occur on Callisto, suggesting the presence of fairly clean ice below the dirty crust of Ganymede but not of Callisto.

The evidence from the surface of Ganymede, although ambiguous, suggests that this satellite experienced a series of internal upheavals sometime during the first billion years of its existence. One suggestion is that these events were triggered by phase changes as the ice in the interior altered its crystalline form with the slow cooling of the core. The resulting expansions and contractions could have cracked the surface and resulted in the submersion of low lying areas by a lava-like flow of liquid water.

At one time the crust of ice may even have floated on a viscous mantle of slushy ice, resulting in an episode of icy plate tectonics analogous to those of the Earth. The coupled subjects of internal evolution and surface geology on an icy world such as Ganymede are fascinating fields just beginning to be addressed by planetary geologists.

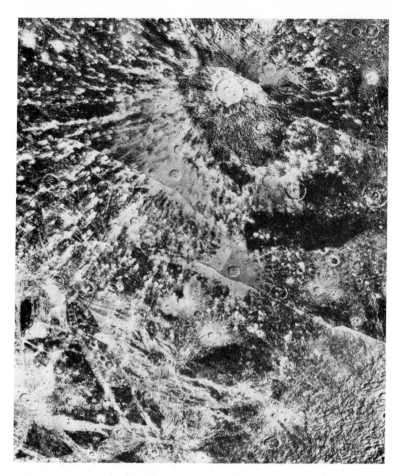

FIGURE 13.12 Impacts on Ganymede's icy surface have left a record of bright craters like these, surrounded by systems of rays. No dark craters have been found, indicating that the material that darkens some regions of Ganymede is probably confined to a thin surface layer.

13.4 Europa and Io: The Active Galileans

Geology of Europa

Of all of the Galilean satellites of Jupiter, Europa remains the most enigmatic. It is a smaller satellite, similar in size to Io and our Moon. Its density of 3 g/cm³ indicates a primarily rocky composition with only 10% water ice, much less than is present in Callisto or Ganymede, which are each about half composed of ice. Yet the surface of Europa is the brightest of any of the Galilean satellites (70% reflectivity), and its spectrum indicates a surface composition of relatively pure water ice. In this respect Europa is like the Earth, which is made of dense materials on the inside but has a coating of water and ice over most of its surface. What is special about the icy surface of Europa relative to those of Callisto and Ganymede is its comparative purity, suggesting that some kind of continuing process periodically resurfaces the planet with fresh material from below.

The visual appearance of Europa is also different from that of any other planetary body we

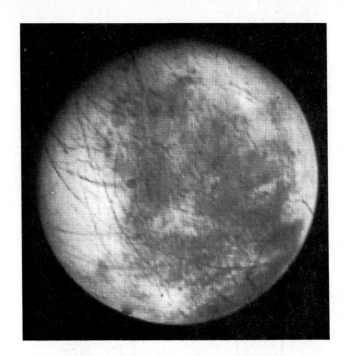

FIGURE 13.13 Europa is the smoothest satellite known. The absence of easily visible impact craters means that we are looking at a relatively young surface.

have studied (Fig. 13.13). Almost no impact craters are visible at the Voyager resolution of a few kilometers, indicating that the record of early heavy bombardment by comets and asteroids has been erased. The crater densities on Europa, significantly less than on the lunar maria, are roughly similar to those on the continents of the Earth.

Linear Markings

Little topographic relief of any sort occurs on Europa; in fact, this satellite is the smoothest planetary object in the solar system. The markings that are visible on its surface (Plate 17; Fig. 13.14) are mysterious. Most take the form of long, narrow, light or dark lines stretching for hundreds or even thousands of kilometers across the landscape. Some of these lines are double or multiple, with one common form consisting of one light line running between two dark lines. If this all sounds familiar, it is because Europa ac-

FIGURE 13.14 Viewed at high resolution and under oblique lighting, the surface of Europa reveals a pattern of remarkable curving, white ridges, some 5 to 10 km wide and just a few hundred meters high. Together with a network of dark streaks, evocative of Lowell's martian "canals," they give the satellite the appearance of a ball of string.

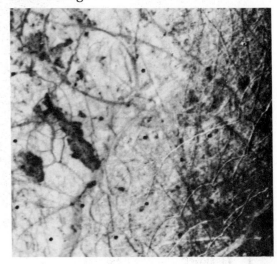

tually has much of the appearance of the fanciful Mars globes drawn by Percival Lowell at the beginning of the twentieth century. The famous canals of Mars have found their true home on a satellite of Jupiter!

Topographically, the canals of Europa are not depressions but shallow ridges a few kilometers in width and a few hundred meters high. Perhaps these ridges are made of viscous material squeezed up through cracks in the icy surface. But what lies below? Many people believe that beneath the ice crust of this satellite is a global ocean of liquid water as much as a hundred kilometers deep. They suggest that the relatively thin crust occasionally cracks, resulting in a release of liquid from below. A few speculative individuals have even gone so far as to suggest that the oceans of Europa might conceivably support some exotic life forms, maintaining themselves on the weak sunlight that could filter down through the thinner ice where recent cracking of the crust had taken place.

Surface Composition of Io

The innermost of the large jovian satellites is by far the most spectacular, with its active volcanoes and constantly changing surface. Following Voyager, the name Io has almost become a household word (except that no one can seem to agree on how to pronounce it: EE-O or EYE-O). Before we look at the remarkable volcanoes of Io, let us discuss how it compares in other respects with the three other Galilean satellites.

The density of Io is 3.6 g/cm^3, slightly higher than that of Europa and indicative of a rocky object with little or no water. In fact, Io has nearly the same size and density, and therefore probably a very similar bulk composition, to our Moon. Like Europa, Io has a high surface reflectivity, but here the similarities end. Instead of a bland white ice surface like that of Europa, or the grey dirty-ice surfaces of Ganymede and Callisto, nature has given Io a multi-hued face of yellow, red, brown, black, and white materials (Plate 19; Fig. 13.15). No water in any form can

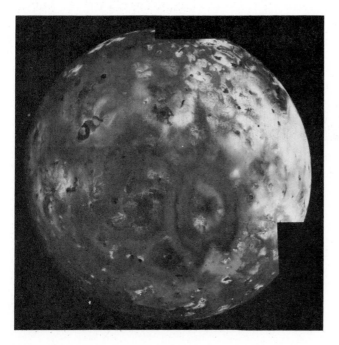

FIGURE 13.15 A partial mosaic of Io compiled from Voyager 1 pictures shows the astonishing surface features caused by this satellite's active volcanism. Again there are no impact craters, but now the surface we see is very young indeed, being continually reformed by volcanic eruptions. The "lava lake" with a central "island" in the upper left is called Loki; it produces so much heat it can easily be detected from Earth (cp. Fig. 13.20).

be identified; probably Io is the driest place in the outer solar system.

The brightly colored materials on the surface of Io ought to be identifiable, but unfortunately the evidence from spectra is not entirely clear on this subject. The one material positively detected is sulfur dioxide (SO_2), which is an acrid volcanic gas on Earth but freezes on Io as a white frost (Fig. 13.16). A thin atmosphere of sulfur dioxide gas has also been detected, while the breakdown products of this material, sulfur and oxygen, are major contributors to the charged particle population of the jovian magnetosphere, as described in Section 11.6.

It seems probable that sulfur dioxide is the primary component of the white areas on Io, but what of the more colorful yellows, oranges, browns, and blacks? The most likely candidate is elemental sulfur, which has the property of taking on many colored forms depending on its temperature and the way it is cooled from the liquid state. Another possibility, however, is provided by compounds of sulfur and sodium, which also display an appropriate diversity of hues. Even with this evidence for abundant sulfur and sulfur compounds, we do not know if a thick crust of these materials exists or if they constitute only a thin coating, perhaps disguising more ordinary rocks below.

Geology of Io

The underlying geology of Io is also unique. There is a complete lack of impact craters, indicating a geologically young surface. Note that this is true in spite of two factors that should increase the crater density relative to the other Galilean satellites: the greater cratering rate so close to Jupiter, and the absence of ice in Io's crust. Just like the Earth, Io is capable of erasing its craters as fast as they are formed. Calculations of the resurfacing rate required to destroy the cra-

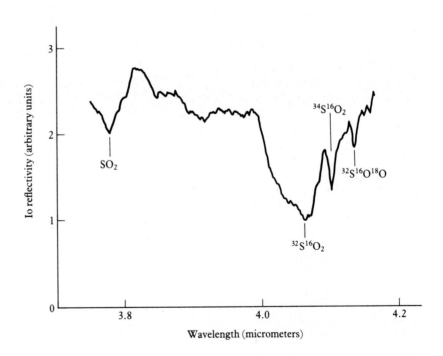

FIGURE 13.16 A spectrograph attached to an Earth-based telescope recorded this evidence of sulfur dioxide absorption on Io. Volcanic eruptions produce the gas. (After D.P. Cruikshank and R. Howell.)

ters on Io indicate that hundreds of meters of new crust must be formed each million years, a value similar to that of areas of high sedimentation on the Earth.

Instead of impact craters, Io displays numerous volcanic features, including layered lava plains and flow fronts, volcanic mountains, calderas, and vents (Fig. 13.17). Long twisting flows issue from some vents, while others are surrounded by white aprons, apparently deposits of sulfur dioxide frost. There are a few low shield volcanoes, and near the south pole there are also some isolated high mountains of un-

known origin. With altitudes as great as 9 km, the mountains of Io are the highest of any of the Galilean satellites (Fig. 13.18).

All of this evidence of volcanic activity is dramatic enough to the geologist, but even more exciting was the Voyager discovery that ongoing eruptions were visible as the spacecraft flew past. For the first time outside the Earth, it was possible to see geological changes as they took place, rather than being forced to infer them from evidence collected after the fact. The volcanoes of Io therefore became the focus for study of this remarkable world (Fig. 2.18).

FIGURE 13.17 Seen in a close-up view by Voyager 1, the surface of Io reveals a rich variety of volcanic features. Here we see several large flows of "lava" (sulfur-dominated silicates?) emanating from volcanic calderas.

FIGURE 13.18 Near the north pole of Io, this mountain (Haemus Mons) soars to an altitude of 9000 m (30,000 ft). To reach such heights, it must be composed in large part of silicates, since simple sulfur compounds would not have sufficient strength to support such tall structures.

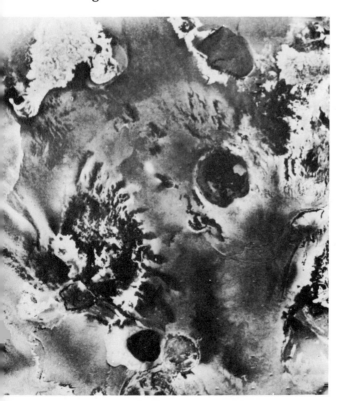

13.5 The Volcanoes of Io

Plume Eruptions

The most spectacular volcanic events on Io are the **plume eruptions** discovered by the Voyager cameras (Fig. 13.19). In March 1979, Voyager 1 photographed nine plume eruptions fountaining up to 300 km above the surface. Four months later, Voyager 2 found that one of these had ceased activity and one was unobservable, but the others were still going strong. The plumes arise from rifts or caldera-like vents, and associated with them are characteristic patterns of white or dark red material deposited around the vents. By decision of the International Astronomical Union, the volcanoes of Io are named for volcano or fire gods from ancient cultures, such as Pele, the Hawaiian volcano goddess, and Loki, the Norse fire god.

Plume eruptions on Io apparently occur in two varieties. The first kind is violent and short-lived, producing a dark red surface deposit. The best example is Pele (Plate 18), the largest plume

photographed by Voyager 1, which had ceased activity four months later when the Voyager 2 flyby took place. These plumes also apparently have the hottest vents, with temperatures up to 700 K, and the highest ejection velocities, 1000 m/s. Probably the primary material ejected is elemental sulfur. The second type consists of long-lived eruptions of a white material, very likely sulfur dioxide. These eruptions have lower temperatures and lower ejection velocities. Most of the plumes photographed by Voyager are of this sulfur-dioxide class.

Both types of plume eruptions are produced by the high-pressure release of hot underground liquid, which turns to gas as it rushes upward and then condenses again as snow when released into the cold of space. Plume eruptions resemble steam **geysers** on Earth, such as the famous Old Faithful in Yellowstone National Park. However, since there is no water left on Io, the hot fluid that drives the eruptions is liquid sulfur or sulfur dioxide. Analysis of the Voyager images indicates that about 100,000 tons of material are erupted each second in these plumes, enough to

FIGURE 13.19 A sequence of volcanic eruptions on Io. (a) Prometheus as it appears from straight above. The vent is surrounded by a ring of bright material deposited by the plume. As it apparently moves toward the horizon (b) we get a clearer view of the material raining down on the surface, while at (c) we see the plume in projection, with its characteristic mushroom shape.

(a)

(b)

(c)

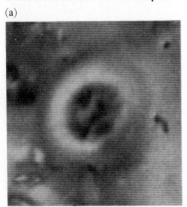
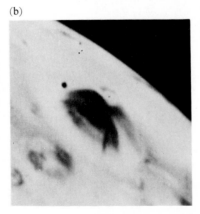
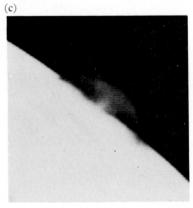

cover the entire surface of Io to a depth of tens of meters in a million years, or to alter the color of an area of thousands of square kilometers in a few weeks.

Most of the sulfur and sulfur dioxide in the eruption plumes rains back onto the surface, but a small fraction, probably amounting to ten tons per second, escapes from Io. Quickly broken down by sunlight and the impact of energetic magnetospheric particles, this material contributes most of the charged ions of sulfur and oxygen that are the primary constituent of the inner magnetosphere. Thus the volcanoes of Io contribute directly to the population of charged particles trapped in the huge magnetosphere of Jupiter.

The Hot Spots of Io

Io's remarkable level of volcanic activity produces a net flow of heat from the satellite that can be measured with Earth-based telescopes as well as by the Voyager infrared detectors. Most of this heat is radiated from a few **volcanic hot spots** not necessarily associated with currently active plumes. In 1979, the largest hot spot was a sort of lava lake, 200 km in diameter, near the Loki eruption site (Fig. 13.20). The "lava" may be liquid sulfur, which has a melting temperature of 385 K.

Heat radiation from the Io hot spots can be monitored from Earth by measuring the infrared emission from the satellite when it is in eclipse. Once in each orbit, Io passes through Jupiter's shadow. With the reflected sunlight removed, the glow of the hot spots becomes more apparent, like city lights after sunset. In addition, the rest of the surface of Io cools during the two hours it is in the shadow, further increasing the contrast of the volcanic radiation. Eclipse observations extending back to eight years before Voyager show these effects, although they were not recognized at the time. However, these tele-

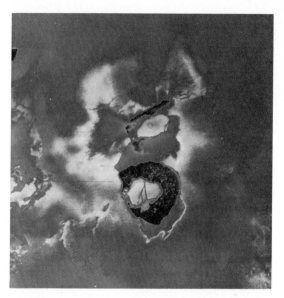

FIGURE 13.20 A close-up of the "lava lake" Loki (cp. Fig. 13.15). The lake itself is about 200 km across. At the time of the Voyager 1 encounter, this lake was about 150 C hotter than the surrounding plains.

scopic studies cannot pinpoint the sources: all they reveal is the total emission and the fact that it is emanating from 1–2% of the surface where the average temperature is about 300 K, similar to the surface temperature of the Earth.

Beginning in 1983, ground-based infrared observations began to provide additional information on the nature of the hot spots. Observations carried out with sensitive infrared detectors at Mauna Kea Observatory confirmed that the Loki lava lake was still the single largest source of radiated heat. Thus we have learned two more important properties of the hot spots since Voyager: most of the energy is coming from only one or two major sources, and these sources have lifetimes of many years. Other telescopic observations indicate that volcanic plume eruptions are also still taking place, but from the ground it has been impossible to locate specific eruption sites.

Other Volcanic Activity

The large hot spots and the two kinds of geyser-like plume eruptions do not exhaust the range of volcanism taking place on Io. The Voyager pictures show many surface flows similar to the long twisting lava flows on the slopes of terrestrial volcanoes (Fig. 13.17); these are probably sulfur, but some of them could be basaltic rock not so different from that produced by other volcanoes in the solar system. These quieter flows probably do more to create the surface landscape of Io than the more spectacular plume eruptions.

No astronaut is ever likely to stand on the surface of Io; the intense bombardment by energetic particles from the magnetosphere would kill a human in a matter of minutes. But perhaps someday we will be able to view the scene remotely, through the cameras of a lander spacecraft. From one hemisphere of Io, Jupiter itself would dominate the view, with fifty times the apparent diameter of the Moon in the skies of Earth. Even at night, the colored landscape of Io would be illuminated brightly by Jupiter. In the darkness of the hemisphere facing away from Jupiter, however, we might see the black sky glowing faintly with electrical discharges in the tenuous atmosphere of Io. The most spectacular sight would surely be one of the huge plume eruptions, sending a fountain high above the surface. Anywhere within hundreds of kilometers of such a vent we would find ourselves in a gentle snowfall of crystals of sulfur or sulfur dioxide. After a few weeks, the surrounding landscape and the spacecraft would acquire a fresh coating of this falling material. Meanwhile, we should look over our shoulders, for who knows when a tongue of molten sulfur from some other vent might be headed in our direction?

Energy to Power the Volcanoes

Io's level of volcanic activity is unique. Yet Io is a small body, similar in size and composition to our Moon. Why is the face of Io constantly changing due to volcanic activity, while the Moon's volcanic fires cooled more than 3 billion years ago? What is supplying the heat that maintains such a high level of activity on just this one satellite? The answer is found in tidal heating, which we first discussed in Section 6.2 on Mercury.

Io owes its special character primarily to its unique location. It is about the same distance from Jupiter as is our Moon from the Earth, yet Jupiter is more than 300 times more massive than Earth, thereby causing tremendous tides on Io. These tides pull the satellite into an elongated shape, with a several-kilometer-high bulge extending toward Jupiter. This tidal bulge would not contribute to heating the interior if Io always kept exactly the same face turned toward Jupiter, the state toward which it would naturally evolve. However, the gravitational pull of Europa and Ganymede will not permit Io to settle into an exactly circular orbit. Instead, the Laplace resonance forces Io's orbit to be slightly eccentric, with the result that it twists back and forth with respect to Jupiter on each orbital circuit, while at the same time moving nearer and farther from the planet. The twisting and flexing of the tidal bulge heats Io, much as repeated flexing of a wire coat hanger heats the wire. In this way, the complex interaction of orbit and tides pumps tremendous energy into Io, melting its interior and providing power to drive its volcanic eruptions.

The size of the tidal energy source of Io can be estimated from the heat radiated by the hot spots, which is much greater than the energy of the plume eruptions. The Io power plant generates about 100 million megawatts, more than ten times the total energy consumption of humans on Earth. Ultimately, the source of this energy is the spin of Jupiter itself, which is slowed very slightly by the interaction with Io. At the same time, the coupled orbits of Io, Europa, and

Ganymede are being slowly pushed outward, just as tidal forces are pushing the Moon away from the Earth.

After billions of years, this tidal heating has taken its toll on Io. Any water or other ices, as well as most carbon and nitrogen compounds, have long since been eliminated. Sulfur and sulfur compounds are the most volatile materials remaining, accounting for their role as the working fluids for the volcanoes of Io, while a similar role on Earth is played by water and carbon dioxide. The inside is entirely liquid, with a solid crust probably no more than 25 km thick. The crust is constantly recycled by volcanic activity, and geologic features are probably short-lived. Although Io has been well mapped from the Voyager images, it may present a partly unfamiliar face when the Galileo spacecraft gets another close look fifteen years after the Voyager encounters.

◆

13.6 Titan: The Cloudy Satellite
A Satellite with an Atmosphere

Titan, the only satellite with a major atmosphere, is one of the special places in the planetary system. Saturn's largest satellite, it has essentially the same size and the same density (1.9 g/cm³) as Ganymede and Callisto. Thus one might expect it to look similar to these two giant moons of Jupiter, but sunlight reflected from Titan carries a message suggesting that this expectation will not be met.

As viewed with a telescope from Earth, Titan appears as a tiny, barely resolvable disk with a distinct reddish color. Ganymede, Callisto, and all of Saturn's other satellites are essentially neutral in color. In the middle decades of this century, when few astronomers were interested in the planets, Gerard Kuiper carried out a systematic spectroscopic survey of the solar system. In 1944, when he examined Titan with a spectrograph attached to the eighty-two-inch telescope of the McDonald Observatory, Kuiper found that Titan's spectrum showed absorption bands of methane gas in addition to the expected solar lines (Fig. 12.6). Unlike Ganymede and Callisto, Titan has an atmosphere.

It may seem surprising that a satellite could have a substantial atmosphere; our Moon certainly doesn't. Recall that the ability of a planet or satellite to retain an atmosphere is determined primarily by that body's mass and temperature. Titan is both more massive and colder than the Moon. Thus it has a stronger gravitational field, and molecules must move faster to have sufficient energy to escape into space. Yet the lower temperature means that molecular velocities will be slower. Thus both size and temperature favor retention of an atmosphere. But why Titan and not Ganymede or Callisto? We shall return to this question after we learn more about Titan's atmosphere.

Temperature Measurements

After Kuiper's discovery, knowledge about Titan grew slowly, and it was not until nearly twenty-five years later that the next major discoveries were made. The first of these was the infrared measurement of the temperature of the satellite. It was found to be surprisingly warm, although still a frigid 100 C below zero. The second was the discovery that the atmosphere was cloudy, with the surface probably hidden below a haze of unknown composition.

The measured temperature was twice as high as the expected value of 85 K, leading to speculation that Titan might have a major greenhouse effect. Recall from Section 8.2 that the first measured radio temperatures of Venus were also about twice as high as expected for its distance from the Sun, and that the greenhouse effect provided an explanation for the heating of the surface. However, there was no assurance that the infrared radiation from Titan was coming from the surface rather than from some warmer level in the atmosphere. Only in 1979, with the completion of the Very Large Array (VLA) radio telescope (Fig. 2.11), could the surface temperature of Titan be determined as had been done for Venus twenty years earlier. By using many telescopes linked by computer, this radio telescope provides high resolution as well as good sensitivity. The resolution permits a clear distinction between Saturn and Titan, which had not been possible with earlier radio telescopes. The answer from the VLA was that the surface temperature was between 78 K and 96 K, indicating that Titan was not supporting a Venus-like greenhouse effect.

By the eve of the 1980 Voyager encounter with Titan, many questions concerning its atmosphere remained unanswered. Only methane and some other hydrocarbons had been detected spectroscopically, but the presence of other gases such as nitrogen or argon was suspected. The composition of the clouds was unknown. A fundamental problem also remained in estimating the total amount of the atmosphere and its surface pressure. Various scientific opinions favored different possible models, in which the surface pressure ranged from a value almost as low as that on Mars up to several times greater than the surface pressure on the Earth. The information available was insufficient before the Voyager flyby to distinguish among these possibilities.

Voyager Results

The pictures of Titan sent back by the Voyager spacecraft were disappointing, since the surface was completely obscured by a ubiquitous smog layer (Fig. 13.21). Most of the exciting results came from the other Voyager instruments.

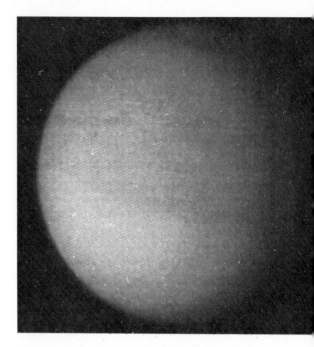

FIGURE 13.21 From a distance, Titan shows a remarkably well-defined contrast boundary between its northern and southern hemispheres, even though this equatorial division is simply set in a haze layer rather than on a solid surface. The north polar hood is also slightly darker than the adjacent haze.

As the Voyager 1 spacecraft passed behind Titan on November 11, 1980, its radio signal was attenuated by the atmosphere and ultimately blocked entirely by the satellite's solid surface. This occultation permitted Voyager scientists on Earth to determine how the density of the satellite's atmosphere varied with altitude above its surface. Assuming a composition based on results from the spectrometers on board the spacecraft, they deduced a surface pressure of 1.5 bars. As is often the case, the models had bracketed the true value.

With its surface pressure of 1.5 bars, Titan has a larger atmosphere than either Mars or

Earth. Since its gravity is much less than that of the Earth, more gas is required on Titan to exert the same pressure as it would on Earth. Thus the amount of gas above each part of Titan is approximately ten times greater than that above an equal area on Earth, and the atmosphere of Titan also stretches nearly ten times farther into space than does our own.

Voyager made important discoveries about the composition of Titan's atmosphere as well as its structure. The methane detected by Kuiper turned out to be a minor constituent. Titan's at-

mosphere is mostly nitrogen, as is ours. Impacting electrons from Saturn's magnetosphere and ultraviolet light from the Sun have sufficient energy to break molecules of methane and nitrogen apart. As the fragments recombine new species are created, and a number of these have been detected in Titan's atmosphere by the Voyager spectrometers (Fig. 13.22). Table 13.2 shows us that some very interesting chemistry is taking place on Titan, especially when considered in the context of the prebiological chemistry required for the origin of life on Earth. Hydrogen cyanide

FIGURE 13.22 A section of the infrared spectrum of Titan recorded by Voyager 1 shows emission bands from several gases that are present on Titan in trace amounts (top). The two laboratory spectra of individual gases show how these substances can be identified in Titan's atmosphere.

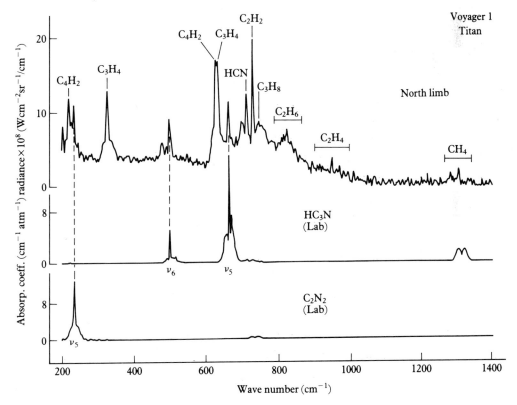

TABLE 13.2 Composition of the atmosphere of Titan

A. Main Constituents (percent)

Gas	Formula	Amount
Nitrogen	N_2	82–99%
Argon	Ar	0–12[a]
Methane	CH_4	1–6[a]

B. Trace Constituents (parts per million)

Hydrogen	H_2	2000
Hydrocarbons		
Ethane	C_2H_6	20
Propane	C_3H_8	20
Ethylene	C_2H_4	0.4
Diacetylene	C_4H_2	0.1–0.01
Methylacetylene	C_3H_4	0.03
Nitrogen compounds		
Hydrogen cyanide	HCN	0.2
Cyanogen	C_2N_2	0.1–0.01
Cyanoacetylene	HC_3N	0.1–0.01
Oxygen compounds		
Carbon monoxide	CO	50–150
Carbon dioxide	CO_2	0.015

[a]The presence of argon can only be deduced indirectly. There may be none at all, in which case the abundance of nitrogen would increase. The amount of methane varies with altitude and is still poorly determined.

(HCN) is a starting point for the formation of some of the components of DNA. The presence of both carbon monoxide (CO) and carbon dioxide (CO_2) in this mixture opens the possibility of the formation of amino acids. We still have no idea whether anything this interesting is happening on Titan; all we know is that the possibilities exist.

13.7 Organic Chemistry on Titan
Photochemical Smog

In Titan we have discovered a world nearly frozen in time, where we can examine a primitive environment in which chemical reactions taking place today resemble those that preceded the evolution of life on our own planet. It would be fascinating to study the end products of this chemistry, to use Titan as a natural laboratory for testing our ideas about chemical evolution. Are certain pathways toward complexity preferred? We know that there are more complex substances on this satellite than those we have already found because of the existence of the ubiquitous haze layer. But what are these compounds?

The haze is all that we can see in the detailed pictures sent back by Voyager (Fig. 13.21). The cameras on the spacecraft actually obtained pictures with much higher resolution than the one reproduced here, close-up views that show a small area of the satellite at high magnification. These pictures show no detail: no gaps in the haze, no structure that could be used to map winds. The haze of Titan is even more uniform than the clouds of Venus. Only when viewed from the side does it reveal structure in the form of distinct layers extending to hundreds of kilometers above the surface (Fig. 13.23).

Observations of a stellar occultation by the ultraviolet spectrometer revealed additional absorbing layers between 300 and 500 km (Fig. 13.24). Some organic molecules formed by the recombination of fragments of nitrogen and methane form long chains called polymers. The reddish-brown haze that pervades Titan's atmosphere is presumed to be a combination of these polymers and condensed organic compounds.

As the haze particles grow in size, they will gradually drop out of the atmosphere onto the satellite's surface forming a layer of solids and liquids that by now may be several hundred meters thick. Since the lower atmosphere and surface of Titan are colder than the upper atmosphere, most of the products of this precipitation and photochemistry cannot evaporate and mix with the atmosphere again, the way water does on Earth. The only exceptions will be gases like methane and ethane, which may exhibit something similar to our familiar water cycle.

Evolution of the Atmosphere

Although the surface of Titan is preserved in a deep freeze, the atmosphere continues to evolve. As electrons and ultraviolet photons break apart the methane, the resulting hydrogen will escape, since hydrogen atoms and molecules move fast enough even at Titan's low temperatures to reach escape velocity. A small amount of hydrogen is always present in Titan's atmosphere, since it is continuously being produced, but over the lifetime of the solar system a large amount of hydrogen must have escaped into space. This suggests that deuterium should be enriched on Titan as it is on Venus, although not to the same extent. There is evidence that such enrichment has occurred, in the form of a high measured abundance of CH_3D — methane in which one of the hydrogen atoms is replaced by an atom of deuterium.

The loss of hydrogen must be balanced by a loss of carbon; this loss is represented by the precipitation of aerosol particles and hydrocarbons onto Titan's surface. Ethane is the most abundant end product of the reactions occurring in the atmosphere, and Titan's surface is sufficiently cold that this gas would condense on it. Thus we may imagine lakes, seas, even oceans of liquid ethane. Methane and propane would also form liquids, but the other compounds in Table 13.2 will solidify at these temperatures and sink to the bottom of an ethane ocean.

FIGURE 13.23 A detached haze layer can be seen in projection at the edge of Titan's visible disk. It lies above the thick haze that obscures the surface of the satellite from view. The image was recorded after the spacecraft had passed the satellite, hence the crescent phase.

Titan offers us the apparent paradox of an atmosphere that has undergone extensive evolution without losing its primitive characteristics. The chemistry occurring on Titan today must be very similar, if not identical, to that of 4 billion years ago. This situation, which is totally different from the more rapidly evolving conditions found among the inner planets, is due to Titan's very low temperature.

The average surface temperature on Titan is 94 K, a frigid 177 C below zero. Titan is so cold that liquid water — so vital to life as we know it — is an impossibility. Even water vapor will be almost totally absent from this atmosphere, thereby allowing methane and other hydrogen-

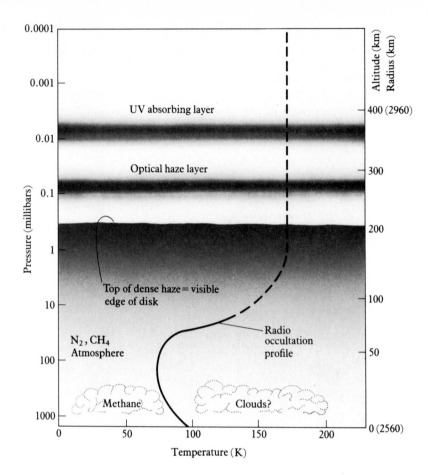

FIGURE 13.24 Unlike the atmospheres of Earth, Mars, and Venus, Titan's atmosphere is much warmer in the stratosphere than it is near the satellite's surface. There are several high-altitude haze layers.

rich compounds to remain abundant. If water vapor were plentiful, Titan's atmosphere would resemble that of Mars. The water molecules would be broken apart just like the molecules of methane and nitrogen, and the liberated oxygen would convert all the hydrocarbons to carbon dioxide. The only source of oxygen on Titan is offered by the ice particles the satellite sweeps up as it orbits Saturn. This source is sufficient to form the small amounts of CO and CO_2 detected but is hopelessly inadequate to oxidize all of the methane.

We can explain the existence of this atmosphere by a process of trapping gases from the solar nebula in water ice as the satellite formed.

This same process is probably responsible for the gases in comets and may have contributed gases to the atmospheres of the giant planets. The chemistry we now find on Titan resembles the processes proposed to explain the origin of life on Earth. We are eager to know what compounds are being produced and in what proportions.

The Surface of Titan

Now we have the components of a most unusual landscape! The basic structural element should be water ice, covered by layers of aerosol particles and crossed by rivers of liquid hydrocarbons emptying into lakes or seas. Some residual topography from the cratering events that have marked the surfaces of other solar system bodies may be present, modified by these two processes. As on Earth, the existence of an atmosphere will screen out the smaller impacting bodies. Finally, there is the distinct possibility that the crust of Titan is subject to the icy equivalents of volcanism and plate tectonics caused by mantle convection, leading to mountain chains, ridges, scarps, volcanic peaks, and blocks of ice strewn over the surface.

What would it be like sitting in a boat on an ethane ocean on Titan (Fig. 13.25)? It would be cold certainly, and probably gloomy. The level of illumination at Titan's surface is estimated to be similar to that from a full moon at midnight on Earth. The entire sky would appear to be glowing faintly, as sunlight filtered through the orange smog layer. The Sun itself would be invisible, as when heavy smog covers one of our cities. Yet the horizontal visibility would be good (unless, of course, we were caught in an ethane fog bank), since the thick part of the haze is many miles above the satellite's surface.

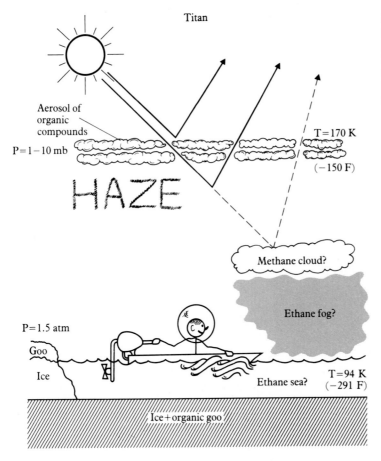

FIGURE 13.25 The surface of Titan may be covered by a global ocean of ethane, or it might be predominantly ice, with a covering of organic matter and lakes and seas of hydrocarbons. In either case, it may be an interesting place to explore by boat!

We don't expect to be rocked by big waves, because global temperature differences on Titan are small, and thus there should not be strong surface winds. Local weather may exist if the ocean is not global in extent, but the surface temperature is expected to be very uniform as a result of the moderating effect of the large atmosphere. It is possible that enough ethane is produced by the atmospheric photochemistry to form a global ocean of this compound, in which case no topography at all would be visible. We don't yet know whether or not this is the case, since the Voyager spacecraft could not give us any direct information about surface conditions. What we really need is a spacecraft equipped with radar and an entry probe; such a mission is currently under consideration as a joint enterprise between NASA and the European Space Agency.

13.8 Triton: The Satellite with Nitrogen Seas

The Largest Irregular Satellite

Triton, the largest satellite of Neptune, is the one remaining big moon in the planetary system. Since no spacecraft has yet reached the Neptune system, we know much less about it than the other bodies discussed in this chapter. Even the size of Triton, now estimated to be about the same as the planet Pluto, was unknown before 1982 when heat radiation from this distant satellite was first measured with ground-based telescopes.

The satellites of Neptune are irregular: Triton orbits the planet in a retrograde direction, while the much smaller Nereid has the most elongated orbit of any satellite. This system is distinctly peculiar, although there is no consensus among scientists as to how it might have originated.

Methane Ice and Nitrogen Oceans

Because it is much farther from the Sun than the satellites of Jupiter and Saturn, Triton is capable of maintaining deposits of frozen methane on its surface. A small amount of methane gas is also present, but the total amount of atmosphere on Triton remains unknown. Since the density of Triton has not been measured, we have no evidence concerning its bulk composition.

Perhaps the most interesting recent discovery about this frigid little world is the probable existence of ponds or seas of liquid nitrogen. Astronomer Dale Cruikshank of the University of Hawaii has made extensive use of the high infrared transparency of the skies above Mauna Kea to search for unknown constituents on objects in the outer solar system. It was Cruikshank and colleagues who first identified methane ice on the surface of Pluto and sulfur dioxide ice on Io. In 1983 Cruikshank began a study of the spectrum of Triton.

He discovered that in addition to absorptions from methane ice and methane gas, Triton has a feature in its spectrum that might be caused by a normally invisible band of molecular nitrogen. Cruikshank and his colleagues noted that this band does occur when nitrogen is in the liquid, rather than gaseous, state. In order to produce the feature measured, the nitrogen must be at least several meters deep and must cover a substantial fraction of the surface. If this interpretation is correct, it is the first direct spectroscopic identification of a liquid on the surface of any extraterrestrial object (page 320).

We see reflected sunlight only from the warmer, summer hemisphere of Triton, which is turned toward the Earth during the 1980s, and it is here that the liquid nitrogen has been detected. It is possible that the liquid nitrogen may freeze during the winter on Triton. Alternatively, if truly deep seas of this material do exist, they may form a global nitrogen ocean that never freezes solid.

There must also be nitrogen in the atmosphere of Triton just as there is water vapor in ours. Calculations suggest that the surface pressure of this atmosphere is about $\frac{1}{10}$ bar, substantially smaller than the surface temperature on Titan but much greater than that on Mars.

If Triton's oceans of nitrogen are real, we can see that this satellite has much in common with Titan. If Triton were moved from 30 AU, its present distance from the Sun, to Saturn's distance of 9 AU, its temperature would rise enough to boil away the nitrogen seas and sublimate the methane ice. In that case it would have a nitrogen atmosphere perhaps as large as that of Titan. It would not stay long in a Titan-like condition, however, for Triton is too small to retain a massive nitrogen atmosphere, which would escape into space.

◆

13.9 Comparing the Large Satellites

The outer solar system contains six large satellites, which are described as individuals in the preceding sections of this chapter. In this context, large means about the size of our Moon or bigger. A comparison of the properties and history of these six satellites is as instructive as similar comparisons of the four terrestrial planets and the Moon.

The three largest satellites, Ganymede, Callisto, and Titan, are an especially enigmatic group. They all have nearly the same sizes, densities, and bulk compositions. Yet their histories have been very different, ranging from geological inactivity (Callisto) to activity (Ganymede), and from the naked surfaces of Callisto and Ganymede to the deep atmosphere and remarkable photochemistry of Titan. The puzzle of understanding these cases of divergent evolution is comparable to that of deciphering the different histories of Earth and Venus.

Ganymede and Callisto

The first challenge lies in determining why Ganymede maintained a substantial level of geological activity for hundreds of millions of years while Callisto did not. The distinctions between these two objects are found in their distances from Jupiter, and in small differences in their sizes and densities. Could any of these factors be responsible for their different histories?

Because Ganymede is closer to Jupiter, it is subject to larger tidal stress, and it also experiences more meteoroidal impacts because of the attraction of Jupiter's gravity. Neither of these effects is large, but both could have resulted in slightly greater heating for Ganymede than Callisto early in their respective histories.

Ganymede is larger than Callisto and also slightly more dense, indicating that it contains more radioactive material and therefore had, and still has, a greater internal heat source. In addition, Ganymede's larger size means that it can hold heat slightly more easily than Callisto. These effects also point toward a higher internal temperature and slower cooling for Ganymede.

Each effect discussed above could contribute to a more active geology on Ganymede than on Callisto. The problem is that each individual factor is small, and even when added together they do not suggest that Ganymede should have been a great deal warmer than its twin moon. What must have happened is that these differences, even though small, managed to trigger a major change in the internal processes on Ganymede that was out of all proportion to the temperature differences. Such a major change can occur with ice, because it is a substance that assumes many different forms, each with its own density, depending on the temperature and pressure.

Calculations show that just a small difference in temperature early in the history of Ganymede may have led to convection in the upper mantle, perhaps sufficient to drive some kind of plate tectonic activity for a period of the satellite's history. Subsequently, about a billion years after the formation of Ganymede, a change of phase in the interior may have caused a small expansion, sufficient to crack the crust and initiate the period of mountain building. According to these calculations, Callisto, by being just a little cooler, escaped these internal events.

The Atmosphere of Titan

In the case of Titan, we do not know whether it has experienced stages of major geological activity like Ganymede or has led a quiet existence like Callisto. We do know that Titan, unlike either of its cousins in the jovian system, has developed a major atmosphere. The mystery of the origin of Titan's atmosphere is even greater than that of the differences between Callisto and Ganymede. Why does Titan have an atmosphere? Where did it come from?

There is an easy test we can make to determine whether the mixture of gases we find on Titan today is a relic of a dense, primary atmosphere captured by the satellite as it formed from the solar nebula. If that were the case, we would expect neon to be as plentiful as nitrogen in the present atmosphere, since these elements are nearly equally abundant in the Sun and hence in the solar nebula (Table 2.1). In fact, the ultraviolet spectrometer on the Voyager 1 spacecraft detected no evidence of neon at all, leading to an upper limit of 1% for the abundance of this gas in Titan's atmosphere. Thus we conclude that the satellite did not capture its gases directly from the nebula, but instead the atmosphere was formed by the release of gases from the interior, in much the same way as the inner planets developed their atmospheres. The big difference in the case of Titan is that approximately 50% of its mass was water ice, which now forms a crust and mantle around the satellite's rocky core.

The Role of Ice

Ice formed at low temperature (less than 135 K) is an excellent carrier of gases. There is room in the irregular lattice structure of this low-temperature ice for molecules of gases to be accommodated. When the ice is warmed above 135 K, the lattice structure changes, becoming regular, and most of the trapped molecules are released. The ability of ice to capture gases in this way depends on the size and electrical properties of the gas molecules. Hydrogen, helium, and neon are not trapped except at temperatures much lower than those anticipated in the vicinity of Saturn during the time the planet and its satellites were forming, but argon, methane, nitrogen, and carbon monoxide can all be captured in this way.

These facts probably explain why Titan has the atmosphere it does and why Ganymede and Callisto have no atmospheres. The jovian satellites were simply too warm for ice to trap large amounts of gas, warmer even than their present temperatures, since they were heated during formation by Jupiter itself. At the distance from the Sun where Titan formed, however, gas-rich ice was able to form and serve as the carrier of nitrogen and methane, which in turn were the building blocks of the other exotic gases now found in Titan's atmosphere.

Triton presumably formed at still lower temperatures, in a part of the solar nebula where possibly even methane ice was available. Undoubtedly when spacecraft finally reach this distant satellite, they will find it to be different in fundamental ways from any of the satellites of Jupiter and Saturn.

Europa and Io

When we turn our attention to the jovian satellites Europa and Io, we encounter other mysteries. Why are these two satellites so depleted in water relative to Callisto and Ganymede? Much of this difference is probably the result of the high temperature of Jupiter early in its history, when it was contracting and forming its satellite system. Calculations of temperatures within the nebula surrounding this proto-Jupiter suggest that it may have been cool enough for ice to condense at the distances of Ganymede and Callisto, but not closer to the planet where Europa and Io formed. In that case, we should expect both Europa and Io to be dry, whereas Europa today probably contains about 10% water.

The failure of this simple explanation in terms of temperatures in the Jupiter nebula comes from its neglect of the contribution of impacting bodies from elsewhere. We know that the solar system was full of debris of various kinds early in its history, and that a substantial part of that debris was cometary. The impacts of many icy comets must have supplied a great deal of water to all of the Galilean satellites, probably sufficient to account for Europa's present icy surface. This is the same way we believe Venus, Earth, and Mars acquired most of their water. In the case of Io, however, this early infusion of water was doomed by the resonant lock that developed between Io, Europa, and Ganymede. Held in the gravitational grip of the other two satellites, Io could not circularize its orbit, with the result that tidal heating drove off all of the water and other volatile substances.

---◇---

Summary

The large satellites of the outer solar system are planet-sized worlds of surprising variety and complexity. The six largest are Ganymede, Titan, Callisto, Io, Triton, and Europa. These include two objects with substantial atmospheres and one, Io, that is the most volcanically active body in the solar system.

Ganymede, Titan, and Callisto have nearly the same size and density, inviting comparison among them. With densities less than 2.0 g/cm^3, they are of obviously different composition from the inner planets, which they resemble in size. Models indicate that they are nearly half composed of H_2O, and that their interiors are differentiated.

Callisto's heavily cratered icy surface shows little sign of internal activity. Ganymede, in contrast, once experienced extensive geologic activity, with resurfacing and the widespread formation of long mountain ridges and valleys. We don't know what the geology of Titan is like, because it is shrouded by a cloudy atmosphere, probably the result of contributions to this satellite of ices containing methane and ammonia that were absent in the Jupiter system.

Titan is a remarkable satellite. Its atmosphere, composed primarily of nitrogen, is larger than that of the Earth. Complex photochemical reactions in its atmosphere give rise to extensive hazes of organic smog. Because of the low temperature, this organic material can accumulate, and today the surface of Titan may be covered with material similar to that formed in the early history of our own planet. Titan has appropriately been described as a terrestrial planet preserved in deep freeze.

Triton, like Titan, has an atmosphere that again contains methane. Although we know much less about this moon of Neptune than about the other large satellites discussed in this chapter, recent observations indicate that it too is a strange world, possibly with a methane ice surface and an extensive ocean of liquid nitrogen.

Io and Europa in the Jupiter system differ from these other satellites in that they are not composed in equal parts of ice and rock. Both of these satellites are about the size and density of our Moon and presumably have a similar composition. Europa has a surface of ice, possibly floating on a global ocean of water as much as 100 km deep. There are few craters, indicating that the surface we see was formed fairly recently, as well as mysterious patterns of linear markings. Much more spectacular, however, is brightly colored Io, which has been extensively modified by internal heating and volcanic activity.

Io's energy source is tidal heating, a result of tides raised by Jupiter acting in conjunction with a non-circular orbit. This heating is sufficient to power widespread surface volcanism. Over its history, Io has outgassed and lost highly volatile materials, such as water and carbon dioxide, and today its surface is covered by sulfur and sulfur compounds, which are recycled through its volcanic eruptions.

Most of the internal heat of Io is radiated into space through hot spots, which appear to be large lakes of liquid sulfur heated from below. In addition, there are two kinds of plume eruptions, in which fountains of sulfur or sulfur dioxide rise tens or hundreds of kilometers into space and fall back to recoat the surface. Some of the sulfur dioxide from these eruptions also escapes, where it becomes ionized and contributes significantly to the inner magnetosphere of Jupiter.

Key Terms

geyser

irregular satellite

plastic deformation (of ice)

plume eruption

regular satellite

volcanic hot spot

CHAPTER FOURTEEN

Small Satellites and Rings

14.1 Inner Satellites of Saturn

The Satellite System of Saturn

The five medium-sized inner satellites of Saturn, together with the rings and about a dozen small satellites, form a regular and tightly knit group. This section focuses on four of these medium-sized satellites: Rhea, Dione, Tethys, and Mimas. The special cases of two unusual Saturn satellites, Enceladus and Iapetus, are discussed in Section 14.2, and the small satellites of Saturn are the subject of Section 14.4. We turn to the rings of Saturn in 14.5 and following sections.

In looking at the Saturn system, we are naturally inclined to make comparisons with Jupiter. In Section 13.1 we noted that Saturn has only a single giant satellite, while Jupiter has the four Galilean moons. It is also apparent that the satellites of Saturn (excluding Titan) show less variety than those of Jupiter. In particular, they do not exhibit the range of density, and hence of bulk composition, of the Galilean satellites (Table 13.1). These differences, as well as the existence of Saturn's extensive rings, clearly set this system apart from that of Jupiter. The reasons may have to do with the smaller mass of Saturn, the lower temperatures in the nebula that surrounded it as the system formed, or perhaps other factors which have not yet been identified.

Rhea

Rhea, as the largest Saturn satellite after Titan, provides a convenient point of reference for studying the medium-sized icy satellites of the outer solar system. Its diameter is 1530 km, just half as big as Europa, but it is 60% larger than the largest asteroid, Ceres. Several other satellites of Saturn and Uranus have about the same size and density. Fig. 14.1 compares the size of

FIGURE 14.1 Three of Saturn's medium-size satellites superimposed on a map of the United States. From the left, they are Dione, Rhea, and Mimas. Dashed circle is size of Earth's Moon.

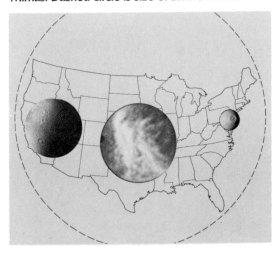

Rhea and two other medium-sized Saturn satellites with a map of the United States.

The density of Rhea is only 1.3 g/cm³, substantially lower than that measured for any solid body previously discussed (although we presume comet nuclei have similarly low densities). This does not mean, however, that the composition of Rhea is much different from that of Titan or the large icy satellites of Jupiter; rather, the density of Rhea is less primarily because of its smaller size. Ice is a fairly compressible material, resulting in a higher density for large satellites. Rhea, being smaller, has a less compressed interior. Taking this effect into account, we can estimate that Rhea, like Titan, has a composition that is roughly half water ice and half silicate minerals.

Rhea is highly reflective (60%), and its spectrum is dominated by the absorptions of water ice. The fact that this satellite is nearly as bright as Europa might be evidence that it too has occasionally resurfaced itself with fresh ice, but it is equally probable that the environment at Saturn is simply less dirty than at Jupiter. If most of the impacting debris is icy, a satellite surface can remain relatively clean even after billions of years of exposure.

The surface of Rhea is heavily cratered, as shown in Fig. 14.2a and 13.3. Unlike the subdued craters of Callisto and Ganymede, the craters of Rhea look remarkably lunar-like. In fact, most geologists, looking at this picture, would be hard put to distinguish Rhea from Mercury or the Moon, unless they were told that they were viewing a brilliant white surface of ice instead of a dark grey surface of rock. At the low temperatures (about 100 K) prevailing at the distance of Saturn, ice behaves very much like rock when a

FIGURE 14.2 a) Craters are shoulder-to-shoulder in this close-up view of Saturn's satellite Rhea. The smallest features are 2.5 km across (cp. Fig. 13.3). b) A distant view of Rhea, showing the bright streaks that came to be known as wispy terrain.

(a)

(b)

crater-forming impact takes place. The colder ice is, the less plastic and the more brittle it becomes.

The crater density on Rhea is about 1000 10-km craters per million square kilometers, as great as the value for the lunar highlands. Further, there is little if any indication of internal geologic activity to erase or distort craters. Lack of geological activity should not surprise us; after all, what is there to heat a small icy world out at the distance of Saturn?

In the midst of all this evidence of a dull history, however, Rhea does display one peculiarity. While its leading side — the hemisphere that faces forward in its orbit as it moves around Saturn — is unmarked except by craters, its trailing side shows prominent streaks (Fig. 14.2). Although the Voyager pictures are not good enough to reveal the exact nature of these markings, it seems probable that they are the remnants of some long-ago episode of activity in which water was released from the interior and condensed on the surface. Similar light streaks on the trailing hemisphere are observed on the next inner satellite, Dione.

Dione, Tethys, and Mimas

Although Dione is smaller than Rhea (1120 km in diameter) and slightly denser (1.4 g/cm³), in most respects it is a similar icy world. Its surface is heavily cratered, and on its trailing hemisphere the bright streaks (what geologists call wispy terrain) are even more prominent, reflecting up to 70% of the incident sunlight and contrasting strongly with the underlying surface, which reflects only half as much sunlight (Fig. 14.3).

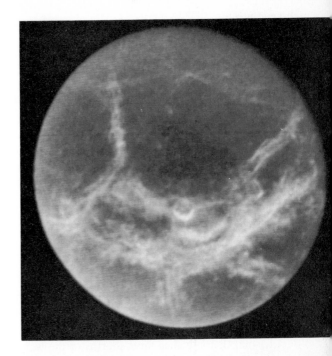

FIGURE 14.3 (top) A distant view of Dione shows the same type of wispy terrain seen on Rhea (Fig. 14.2). Could this terrain be an assemblage of the bright deposits along the edges of cracks in the satellite's surface? (bottom) In addition to the expected craters, the surface of Dione shows sinuous valleys (lower right) sometimes bordered with bright deposits (upper left). ▶

The most significant new characteristic of Dione, however, is its evidence of internal geologic activity. Over substantial parts of its surface, flooding or some other type of resurfacing has obliterated old craters and resulted in lower crater densities, reminiscent of the lunar maria or the old lava plains of Mercury. Also visible on some Voyager pictures are valleys of unknown origin (Fig. 14.3), apparently associated with the bright streaks. Perhaps the bright material represents deposits from gases released from the interior that froze along the rims of these cracks in the crust.

Tethys, the next satellite in toward Saturn, is a close twin of Dione in almost every way. It too has a mixture of surface terrains, some very heavily cratered, others modified by geologic activity. The reflectivity of the icy surface is a very high 80%. Instead of several small valleys, however, Tethys has one giant valley system (Ithaca Chasma) that stretches three fourths of the way around the satellite (Fig. 14.4). The surface area of Ithaca Chasma, which is about 100 km wide over most of its length, is comparable to that of Valles Marineris on Mars. Such a feature could have been formed if the interior of Tethys expanded enough to increase its surface area by 10%, not an impossibly large figure for some kinds of water ice.

The innermost of the large satellites of Saturn, Mimas and Enceladus, form a pair as do Tethys and Dione (Table 13.1). Both have di-

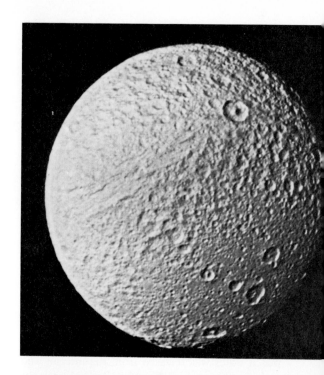

FIGURE 14.4 (top) Here, Tethys looks rather like a dried orange. The large crater in the upper right is almost on the huge valley system called Ithaca Chasma that can be seen extending toward the lower left. (bottom) Mimas has one large crater, called Herschel, in honor of the satellite's discoverer. If the object that made this crater had been slightly larger, it would have shattered Mimas entirely. ▶

ameters of about 500 km, and both have experienced the same environmental effects. Yet they are very different; so much so that we defer discussion of Enceladus to the next section.

Mimas is heavily cratered (1000 or more 10-km craters per million square kilometers) and shows little evidence of internal activity. Its most notable feature is a single very large crater on the forward-facing or leading hemisphere that has a diameter about one-third that of the satellite itself (Fig. 14.4). This is one of the largest craters in the solar system in relation to the size of the body struck (not in absolute dimensions!). The energy released by such an impact does not fall far short of that required to shatter and disrupt Mimas itself.

Comparing These Four Satellites

Rhea, Dione, Tethys, and Mimas have many things in common. All have surfaces of relatively pure water ice, and from their densities we can infer a bulk composition that is about one-half water ice as well. All are heavily cratered, testifying to a heavy meteoroidal bombardment. However, both Dione and Tethys show evidence of a surprising amount of past geological activity, although this activity may have been confined to the earliest period of planetary history.

Relative to the Galilean satellites of Jupiter, these four inner satellites of Saturn make up a compact group. It is interesting to ponder the effects of being so close to Saturn. Probably the most important influence of the giant planet is gravitational, as noted in Section 13.1. Passing meteoroids are pulled inward, converging toward the planet and increasing both the impact rate and the impact speeds for the inner satellites. The closer one is to Saturn, the larger these effects will be. Thus the same flux of cometary impacts that will just build up a heavily cratered surface on Rhea will result in a much more severe pounding of Mimas, where Saturn's gravity produces a convergence of impacting bodies.

14.2 Enceladus and Iapetus: Two Puzzling Satellites

Two of the medium-sized Saturn satellites stand out sharply from their rather bland, icy, cratered siblings. Enceladus appears, in spite of its small size, to have remained highly active geologically. It also has a ring, the E Ring of Saturn, associated with it. Iapetus, in the outer part of the Saturn system, is a unique two-faced moon, with its leading hemisphere covered by a mysterious deposit of black material.

Volcanic Enceladus

Among the inner Saturn satellites, Mimas and Enceladus would be expected to resemble each other. Both are icy bodies with diameters of about 500 km, and both are close enough to Saturn to have experienced similar heavy bombardment by cometary projectiles, but they are not alike at all. In fact, Enceladus offers some major challenges to understanding the Saturn system.

Even from a distance, Enceladus looks strange. Its surface is blindingly white, reflecting nearly 100% of the incident sunlight. As far as is known, Enceladus has the highest reflectivity of any naturally occurring surface in the solar system. In addition, Enceladus seems to be the origin of Saturn's E Ring (Fig. 14.5a). This faint, tenuous cloud of very small, spherical particles fills much of the space between the orbits of Mimas, Enceladus, and Tethys, with its maximum brightness at the orbit of Enceladus. Since the E-Ring particles are so small, they cannot survive long in their present orbits; instead, radiation pressure would be expected to disperse them like the dust in the tail of a comet. Therefore we conclude either that there is a continuing source of particles or that the E Ring is young, having formed in some recent event. In either case, Enceladus seems implicated as the most likely source of the E-Ring particles.

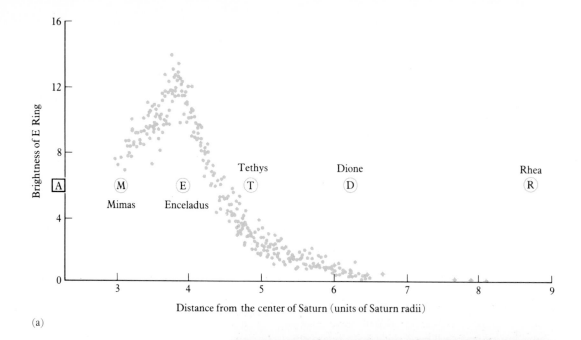

(a)

(b)

FIGURE 14.5 a) Saturn's E Ring is closely associated with the satellite Enceladus, as revealed in this plot of ring brightness versus distance from the planet. Positions of the other satellites and the A Ring (box on the y-axis) are shown for scale. b) Enceladus is one of the most mysterious moons yet discovered. A large portion of the surface seems melted, with no evidence of impact craters, but the cause of this smoothing has not yet been established. The reflectivity of the surface is close to 100%.

Seen at closer range, Enceladus lives up to its billing as a strange place (Fig. 14.5b). Over much of its surface all impact craters have been erased, a sure sign of high levels of geological activity. Although the total cratering flux is not well known in the Saturn system, it appears that these smooth plains are no more than a few hundred million years old, indicating major activity within the epoch we call Phanerozoic on Earth. Some of these smooth plains also show ridges and flow marks. Here, surely, we are seeing evidence of water volcanism. Even on the cratered terrain of Enceladus many individual craters have been deformed by plastic flow in the crust, the result of higher temperatures below the surface.

It is tempting to compare Enceladus with Io, the one other outer planet satellite known to be geologically active. Granted, the current activity rate on Enceladus is much lower, but then Enceladus is a much smaller body than Io. Both objects present essentially the same challenge: to find a relatively large source of internal heating that is capable of maintaining geologic activity in spite of the rapid escape of heat from the interior.

In the case of Io that mechanism has been identified in the form of tidal heating. Could the same thing be happening to Enceladus? The difficulty with this theory is that nearby satellites do not force Enceladus to revolve in a non-circular orbit the way Europa and Ganymede constrain Io. Efforts have been made to construct scenarios in which Enceladus is occasionally forced into an eccentric orbit, resulting in episodic heating of the satellite, but to date these ideas have failed to convince skeptics. The problem of how Enceladus maintains its internal activity remains an open one.

The origin of the E Ring is also uncertain. Calculations of the lifetime of the particles suggest that the event or events that formed the ring took place within the past few tens of thousands of years, extremely recently on any astronomical

or geological time scale. One suggestion is that a cratering event punched through to the liquid interior of Enceladus (if there really is a liquid interior), producing a fountain of fine water droplets that froze to produce the E Ring. The same mechanism might be responsible for the high reflectivity of the surface of Enceladus, which is coated with this fresh layer of tiny ice spheres. If this reasoning is correct, then the E Ring and the high reflectivity are only indirectly connected with the geological evidence of major internal activity.

Another hypothesis associates the expulsion of E-Ring particles from Enceladus with the activity sometimes observed on distant comets. Comet Schwassman-Wachmann I, in an orbit just outside that of Jupiter, is a remarkable object that exhibits explosive outbursts in which a cloud of particles is expelled from the nucleus. If this comet were in orbit around Saturn, such an outburst would produce a tenuous ring like the E Ring. Unfortunately, no one knows the origin of these explosive eruptions of the comet either. Although there may be a link, there is not yet any real understanding.

Two-Faced Iapetus

Iapetus presents us with a different set of mysteries. Ever since its discovery in 1671 by the Director of the Paris Observatory, Jean Dominique Cassini, the strangeness of Iapetus has been recognized by astronomers. It is a two-faced satellite, with a dark leading hemisphere and a bright trailing hemisphere. Since Iapetus, like most satellites, always keeps the same face toward its planet, we see the brightness of the satellite vary dramatically as it moves around its orbit, presenting us first its dark side, then its bright. The bright side is water ice with the typical reflectivity of about 50%. The dark side, however, is covered with a reddish-black material, probably organic in composition, that reflects only 3% of the incident sunlight.

Efforts to identify this dark coating, carried out by Dale P. Cruikshank and his colleagues in Hawaii, have revealed that organic material extracted from the Murchison carbonaceous meteorite (Section 3.4) matches the spectrum of sunlight reflected from the dark side of Iapetus. Is it really the same material? Our abilities to identify solids by remote measurements are not yet good enough to give us a definitive answer. Similar dark, reddish material seems to be common in the outer solar system, on some asteroids, and on comet nuclei. Other dark material in the outer solar system, such as that making up the rings of Uranus, has equally low reflectivity but lacks the reddish color (which would indicate strong absorption of blue light) of Iapetus and the other objects just listed. Perhaps all of these surfaces represent blends of primitive, carbon-rich substances from the solar nebula.

Although Voyager unfortunately did not come any closer to Iapetus than 1.1 million km, its cameras confirmed the two-faced nature of the satellite and provided additional information on the distribution of the dark material (Fig. 14.6). Voyager also provided the first measurement of the mass of Iapetus. In combination with the diameter of 1460 km, this mass yields a density similar to that of the other icy satellites of Saturn. Thus Iapetus appears to be similar to the others in bulk properties, and the dark deposit seems to be of external origin. The zebra is revealed as a white horse with dark stripes, rather than the other way around.

Voyager pictures of the icy side of Iapetus show a heavily cratered surface with at least as great a crater density as Rhea, its twin in size. The surface of this side of Iapetus is therefore very old. Unfortunately, there is no way to establish the age of the dark hemisphere, since the surface was too dark for the Voyager cameras to detect any features.

Since the dark spot is symmetric with respect to the orbital motion of Iapetus, it must surely be the result of some external material striking the surface. What might this material be? The next outer satellite, Phoebe, is dark and

FIGURE 14.6 a) Iapetus exhibits a leading hemisphere that is roughly ten times darker than the trailing side. Here special processing brings out the boundary of the dark side (lower left) but provides no detail. b) A map of Iapetus based on the Voyager photographs. The large dark patch is the leading hemisphere, still totally unexplored.

(a)

(b)

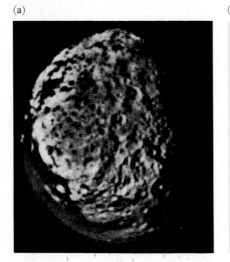
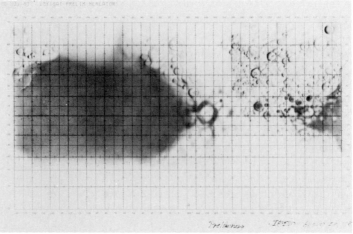

has been suggested as a source for the dark material, but the color of Phoebe does not match that of the dark side of Iapetus. Perhaps it is Phoebe-dust that is hitting Iapetus, and the dust interacts with the surface material on Iapetus to change its color. Indeed, the dark material may well be indigenous to Iapetus, concentrated on the leading side surface by impacts that selectively evaporate away the water ice. Such a process would be similar to that by which a comet develops a dark surface by concentrating carbonaceous material as the ice evaporates.

These are intriguing ideas, but unfortunately they beg the question of the uniqueness of Iapetus. Why should this satellite alone have a hemispheric deposit of dark material on an underlying icy base? We do not know, and the Voyager data are probably not sufficient to provide a satisfactory answer. Unfortunately, it may be decades before we get a closer spacecraft look at Iapetus.

Enceladus and Iapetus are strange places, much stranger than had been anticipated before the Voyager missions. They are examples of the repeated experience that when we actually send spacecraft to previously unexplored regions in the solar system, we always seem to find more variety and more intriguing mysteries than we could have imagined in advance.

◆

14.3 Satellites of Uranus
The Satellite System of Uranus

Uranus has five medium-sized satellites, with diameters and masses similar to the members of Saturn's regular satellite system. In order of their distance from Uranus, these five satellites are Miranda, Ariel, Umbriel, Titania, and Oberon — four named after Shakespearean figures, and Umbriel deriving from Pope's poem "The Rape of the Lock." An additional ten small satellites, discovered by Voyager during its 1986

encounter, have orbits between the larger satellites and the rings.

All fifteen satellites of Uranus are regular in their orbits. They move in the equatorial plane of the planet, which is tilted nearly perpendicular to the orbital plane of the planetary system (Section 12.2). Also orbiting in this same plane are a dozen narrow rings. In the 1980s, when the planet is nearly pole-on to the Sun, the ring and satellite orbits form a bull's-eye pattern as seen from the Earth.

Because the five main uranian satellites are so distant, they are difficult objects for astronomical investigation. Before the Voyager encounter, our meager knowledge of these frigid worlds had to be obtained from the light collected by large telescopes on the surface of the Earth. As a measure of the challenge of studying this system, we note that it was not until 1977 that the rings were discovered, 1982 that the satellite sizes were first measured even approximately, and 1983 that the masses of two of the satellites were first determined by their mutual gravitational influence. Now that Voyager has left Uranus and headed for Neptune, future progress in studying this system will again depend on Earth-based astronomical observations, challenging the technology and ingenuity of planetary astronomers.

Titania, Oberon, Umbriel, and Ariel

In size, the satellites of Uranus resemble the inner satellites of Saturn (Table 13.1). Titania and Oberon are largest, each about the size of Rhea; Miranda is smallest, about the size of Mimas. The densities of the uranian satellites are also fairly similar to those in the saturnian system, although at 1.3–1.6 g/cm³ they suggest a slightly larger proportion of silicate materials and correspondingly less ice. In general terms, however, it is fair to say that the satellites of both planets are composed about half of rock and half of ice, with only minor variations from one object to another.

The surface compositions of the satellites of Uranus also resemble those of Saturn. Water ice is detected spectroscopically, but reflectivities are only 20–30%, suggesting that the ice is relatively dirty. Perhaps these objects are darker because their surfaces have been exposed for longer to infalling debris, or perhaps there is more dark carbonaceous material present near Uranus to contaminate their icy surfaces.

In their appearance the outer two uranian satellites, which have diameters of about 1600 km, are similar to Rhea, Dione, and Tethys. The main difference, aside from a lower surface reflectivity, is the absence of the wispy light streaks seen on the trailing hemispheres of the Saturn satellites. Both Titania and Oberon are heavily cratered at the Voyager resolution (about 10 km), with little indication of internal geologic activity. This cratering does show, however, that some sort of early heavy bombardment of satellite surfaces extended even out to the uranian system, more than two billion kilometers from the Sun.

Umbriel, the middle satellite of the five, is only about the size of Dione or Tethys (diameter 1190 km). It is the darkest of these uranian satellites and displays the least indication of any internal activity. Only one item of any special interest shows up on the Voyager photos: a bright ring about 30 km in diameter near the edge of the illuminated hemisphere. (Since the satellites share the peculiar orientation of the planet, only one hemisphere of each can be seen; the other is experiencing a decades-long winter of darkness.) The nature and origin of this bright feature are unknown, and it will be very many years before we have another look at it.

Ariel is about the same size as Umbriel but is much more interesting. Voyager was able to photograph it at 2 km resolution (Fig. 14.7), revealing a variety of geological structures. In addition to cratering there are long valleys similar to Ithaca Chasma on Tethys, and smoother areas that appear to have been resurfaced since the period of heaviest bombardment. Some of the val-

FIGURE 14.7 This mosaic of the four highest-resolution images of Ariel represents our most detailed view of this satellite of Uranus. Numerous valleys and fault scarps criss-cross the cratered terrain.

ley floors have been filled in, and there is even evidence of flow patterns reminiscent of some of the valleys of Mars. Evidently this satellite, in spite of its small size, experienced substantial geological activity early in its history, probably including a global expansion that stretched and cracked its crust.

Bizarre Miranda

Miranda proved to be the big surprise of the 1986 Voyager encounter. The spacecraft passed only 36,000 km from this satellite, permitting excellent photography. The resolution, better than 1 km, equalled the best that had been obtained at Io or Rhea. However, the choice of Miranda as the best-studied uranian satellite had not been a happy one, since most geologists expected it to be a dull cratered little world, perhaps resembling Mimas. The only reason Miranda had been selected for close viewing was the requirement that Voyager pass through a point near Miranda's orbit in order to receive the gravitational boost necessary to reach Neptune in 1989. Hence Miranda was the only satellite for which a close approach was possible.

By good fortune, however, Miranda did not resemble Mimas at all, as is apparent in the mosaic on page 378. Instead of being a dull cratered satellite with little geological history of its own, Miranda turns out to have a surface extensively modified by internal forces. There are great valley systems up to 50 km across and 10 km deep, apparently produced by large-scale stresses in the crust. Peculiar oval or trapezoidal mountain ranges cover about half of the surface (Fig. 14.8a). In some areas the craters are apparently

mantled and softened by overlying material, while elsewhere the craters are sharp and fresh looking. One huge cliff seen at the border of the illuminated area is more than 10 km high (Fig. 14.8b).

Interpreting the images of Miranda is not easy. Many of its features look similar to those seen previously, such as the grooves of Phobos (Section 4.5), the mountain ranges of Ganymede (Section 13.3), or the valleys of Tethys (Section 14.1). Why are these phenomena all found together on this one small satellite? What caused Miranda, which is less than 500 km in diameter, to be so much more active than its larger neighbors? At this writing, the answers to these fundamental questions remain unknown. One suggestion is that Miranda was shattered after it differentiated, with different pieces falling back together randomly like a jumbled jigsaw puzzle.

FIGURE 14.8 a) A close-up of part of the white "chevron" near the pole of Miranda (left), showing the white and dark grooved terrain with its sharp boundaries. To the upper right, a uniformly dark grooved terrain is visible. An older heavily cratered section of the surface lies between these two features. b) Looking toward the edge of Miranda below the bright chevron, an approximately 15-km-high scarp is visible. This would be a dramatic sight from the satellite's surface, or during the close-up view afforded during the 9 minutes (!) it would take for a falling object to reach the bottom.

(a)

(b)

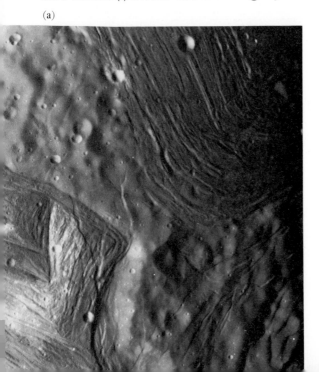
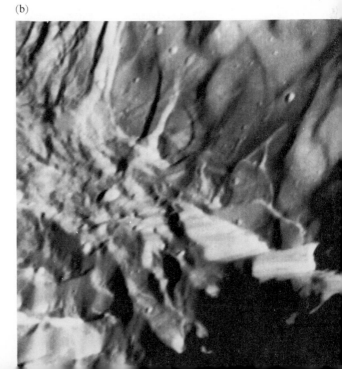

14.4 The Smallest Satellites

Saturn has ten small satellites, most of them in the inner part of the system. Several of these inner satellites occupy unusual orbits, sharing nearly the same path with each other or with the larger satellites. Some also interact strongly with the nearby rings, a topic we return to in Section 14.8. Jupiter has twelve small satellites, while Voyager discovered ten small uranian satellites in 1986.

Primitive Satellites

Three of the small satellites of Saturn and Uranus — Phoebe, Puck, and Hyperion — are examples of objects that probably retain a chemically primitive surface. Each may actually be a captured icy planetesimal. We begin with two objects of 200 km diameter, Phoebe (the outermost Saturn satellite), and the similar-sized inner satellite of Uranus called Puck (Fig. 14.9a). Unlike other satellites we have discussed, both are dark objects, with 5–7% reflectivity. Their surface material is presumably composed of various types of carbonaceous matter remaining from the formation of the outer solar system. Phoebe is in a retrograde orbit, and as noted in Section 14.2, debris from this satellite may contribute to producing the dark deposit on the leading hemisphere of Iapetus.

Hyperion, which occupies a moderately eccentric orbit outside that of Titan, is another apparently primitive object. Voyager images (Fig. 14.9b) show it to be irregular in shape, and spectroscopy has revealed that Hyperion has a surface of dirty ice perhaps similar to those of the satellites of Uranus (Section 14.3). Because of its small size, many scientists believe that Hyperion

FIGURE 14.9 a) Puck, discovered on 30 December 1985. Several craters are visible on the dark surface of this object, which is only 150 km in diameter. b) Hyperion is unique for its irregular shape and apparently chaotic rotation.

(a)

(b)

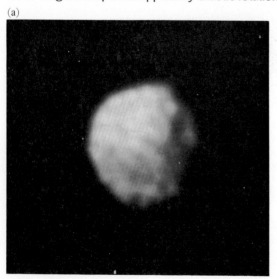
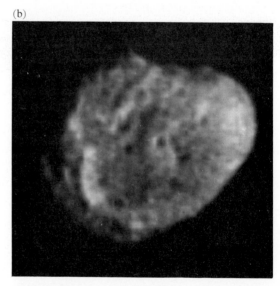

never was heated or differentiated. Like a fresh comet, this satellite may provide a glimpse of the original, primitive dirty-ice material of the outer solar system. This is an interesting idea that should stimulate additional study.

One of the most remarkable aspects of Hyperion does not concern its composition, however, but its rotation. As we have noted repeatedly, nearly all satellites are in synchronous rotation, always keeping the same side toward their planet. In Section 6.2 we saw how tidal forces lead to synchronous rotation, but Hyperion is different. The problem first arose when members of the Voyager imaging team found they could not align their series of pictures of Hyperion on the assumption that the satellite rotated synchronously. Different rotation periods were suggested, but none seemed satisfactory.

Then several theorists calculated that it was possible for Hyperion not to have a fixed rotation period at all! This strange state, termed **chaotic rotation,** is possible because of the satellite's eccentric orbit and its close coupling with the much more massive Titan. In response to the many gravitational forces tugging it, Hyperion exchanges energy between orbital motion and rotation, causing it to tumble in a seemingly irregular way like a top that is running down. Careful telescopic observations made in 1983 and 1984 confirmed this unique and previously unknown rotational state for Hyperion.

Small Inner Satellites of Saturn and Uranus

The small inner satellites of Saturn are all icy objects with bright surfaces. Fig. 14.10 is a composite group photo. Several were discovered telescopically in 1966 and 1980, when the rings were turned edge-on to Earth, reducing the glare

FIGURE 14.10 A group portrait of Saturn's tiny satellites. At left is Atlas, guarding the outer edge of the A Ring, then Prometheus (top) and Pandora, which shepherd the F Ring, Janus (top) and Epimetheus, which occupy nearly the same orbit, the two Lagrangian satellites of Dione and the single Lagrangian of Tethys. Note the crater in the latter — again a slightly larger object would have shattered it.

and allowing faint satellites in orbits close to the rings to be photographed. Others were discovered by the Voyager cameras in 1980 and 1981 (Table 13.1). As a matter of fact, Voyager photographed several more moons that are not listed in our tables of known satellites. There is little doubt as to the reality of these objects, but without a series of pictures their orbits cannot be determined.

The most surprising aspect of these small inner satellites of Saturn is their orbits, each of which is special in some way. The two innermost satellites have orbits that are embedded within the rings, and the next three skirt the very edge of the rings and interact with them, as discussed in Section 14.7. Others share orbits with the larger satellites Dione and Tethys, providing the only known examples of Lagrangian orbits within a satellite system. Perhaps most special are the co-orbital satellites, Janus and Epimetheus.

Co-Orbital Satellites

The **co-orbital satellites** were first spotted in 1966, when interference from the rings was at a minimum, but observers then failed to recognize that two objects might occupy essentially the same orbit. By treating the two objects as if they were one, astronomers initially calculated the wrong orbit, and for more than a decade textbooks listed a nonexistent tenth satellite of Saturn. A later re-analysis of the observations suggested that two satellites might be present, and this possibility was quickly verified when the rings were again edge-on in 1980. Meanwhile, the Voyager 1 spacecraft was approaching Saturn, and its cameras soon confirmed the presence of two satellites in nearly the same orbit, about 13,000 km beyond the rings. The larger is about 200 km in its longest dimension, and the smaller is about 150 km.

If the two satellites had exactly the same orbital period, they might be able to avoid interference, just as Lagrangian satellites stay clear of each other, but the so called co-orbitals actually have orbits differing by about 50 km in radius. The orbital period of the inner satellite is 16.664 hours, and that of the outer one 16.672 hours. This small difference causes the inner one to catch up with the outer at a relative speed of 9 m/s. About once every four years the inner satellite laps its slower-moving sibling, but since the space between orbits is smaller than the objects themselves there is no room to pass. How is a collision to be avoided? What happens is that short of a collision, the two satellites attract each other gravitationally and exchange orbits. They then slowly move apart, and the four-year cycle starts all over again. This strange orbital dance is unique, as far as we know, in the planetary system.

The co-orbital satellites are elongated and irregular in appearance, looking very much like fragments from an ancient catastrophe. Many scientists think that these satellites were once joined together, and that some ancient impact fractured their parent body, leaving these two pieces in nearly the same orbit. All of the small inner satellites of Saturn similarly look as if they might be remnants of past collisions. Recall the argument at the end of Section 14.1 indicating that the inner part of the Saturn system was subject to very heavy impact cratering early in its history. Extending that argument to these smaller bodies, perhaps we should not be surprised to see a few remnants of objects that were destroyed by such impacts. Indeed, the rings themselves might have been formed in this way, but we will return to the rings later.

The Inner Satellites of Uranus

In its 1986 flyby, Voyager discovered ten new satellites of Uranus, all but one a little outside the rings (orbital radii from 1.9 to 3.3 times the radius of the planet). We have already mentioned Puck, the only one of these objects photographed in any detail. Like it, the others appear to have dark surfaces, making them similar to the dark rings but sharply distinguished from the larger satellites discussed in Section 14.3.

Irregular Shapes of Small Satellites

The small satellites of Saturn illustrate very well the point raised in Section 4.5 concerning the shapes of asteroids and other members of the planetary system. Small bodies can have irregular shapes, while larger bodies are more nearly spherical. If an object is large enough, its self-gravity overcomes the inherent resistance of its rock (or ice) and forces it into a spherical shape. In a similar way, the maximum height that a mountain can assume is determined by the point at which gravity overcomes the strength of rock (Section 9.5).

For the icy satellites of the Saturn system, the limiting diameter below which objects still have the strength to resist gravity and remain non-spherical is about 400 km. Look at Mimas (Fig. 14.4b) and Enceladus (Fig. 15.5b). Both have diameters of about 500 km, and both are clearly spherical. In contrast, Hyperion (Fig. 14.9b), with a diameter of nearly 400 km, and the larger co-orbital (Fig. 14.10), with a diameter of about 200 km, are quite irregular. Consider the small satellites of Jupiter, which are made of rock. Do you expect the same transition from irregular to spherical objects to apply there?

Jupiter's Small Satellites

The twelve small satellites of Jupiter are not at all like those of Saturn. The outer eight were probably captured early in solar system history, possibly from the same parentage as some of the primitive asteroids. Today they are found in two groups: the outer four in retrograde orbits at about 22 million km from Jupiter (nearly half the distance from Earth to Mars at closest approach); the inner four in direct but highly inclined orbits at about 12 million km distance. Physically, these satellites are all smaller than 200 km diameter. They have dark surfaces with a reflectivity of only 4%, similar to the dark, primitive asteroids. Like the Trojan asteroids, they may be a remnant of the original population of planetesimals near the orbit of Jupiter.

Three of the four known inner satellites of Jupiter were discovered by Voyager. All three are less than 50 km in diameter, and the innermost two orbit very close to the planet, just at the outer edge of its tenuous ring. More substantial is Amalthea, which was discovered in 1895. It is an irregular cratered object about 250 km in its longest dimension, sometimes described as looking like a healthy potato.

♦

14.5 Rings of Saturn
The Nature of the Rings

Many people believe that the rings of Saturn are the most beautiful sight that can be seen through a telescope (Fig. 14.11). Even after the discovery of rings around Jupiter and Uranus, the Saturn rings remain *the* planetary ring system. As both the most extensive and the best studied, the rings of Saturn are an appropriate place to begin a survey of planetary rings.

The rings of Saturn were first seen by Galileo in 1610, but recognition of the nature of this new phenomenon did not come for another fifty years. Galileo thought he was glimpsing bumps on Saturn, or perhaps a triple planet. Some other seventeenth century observers drew the rings as handles extending on either side of the planetary disk. Not until 1659 did the Dutch astronomer

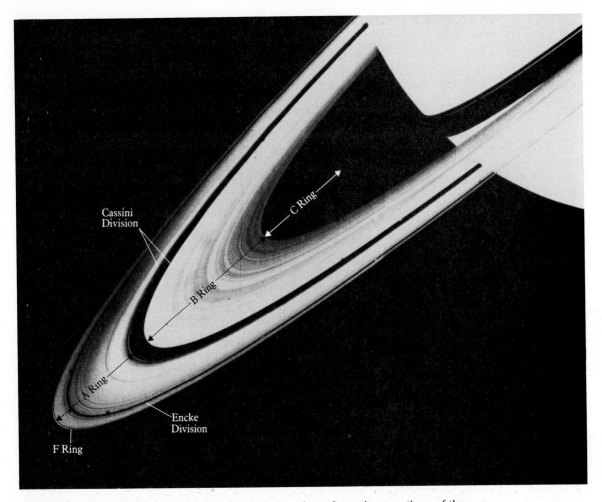

FIGURE 14.11 A Voyager picture of Saturn labeled to show the various sections of the ring system. (The C Ring is not visible in this high-contrast picture.) Note the shadow of the rings on the planet, appearing as a dark band with a white line in it. The white line is caused by sunlight passing through the Cassini Division (cp. Fig. 14.14).

Christian Huyghens recognize that Saturn was surrounded by a "thin flat ring, nowhere touching," lying in the equatorial plane of the planet. In 1675 Cassini, who also discovered the two-faced nature of Iapetus, found that there were two rings, not one. The dark division that separates the two parts is still called the Cassini Division.

The fact that the rings are not solid but are composed of billions of tiny moons orbiting the planet was not demonstrated until the second half of the nineteenth century, although many astronomers had suspected this for some time. According to the laws of planetary motion (Section 1.3), the inner particles travel faster than the more distant ones. At the inner edge of the main rings, one circuit of Saturn requires only 5.6 hours, while the period at the outer edge is 14.2 hours. This was demonstrated by James Keeler's study of the spectrum of the rings. Keeler used the Doppler effect to show the difference in speed of the inner and outer edges of the rings.

Characterizing the Ring Particles

Further understanding of the nature of the ring particles awaited the development of modern astronomical instrumentation. Infrared spectroscopy revealed in 1970 that the particles were composed primarily of water ice. Interpretation of the first radar signals bounced from the rings in 1973 led to information on the sizes of the particles, which were found to be typically in the tens of centimeters size range. The ability of the rings to serve as strong reflectors of radar further indicated that the particles must be rather loosely distributed, not collapsed into a single layer just one particle in thickness.

Some additional information was provided by the 1979 flyby of the spacecraft Pioneer Saturn, but most of what we now know about the rings is the direct result of the Voyager encounters in 1980 and 1981. At that time Saturn and its rings appeared on the covers of both *Time* and *Newsweek*, something planetary scientists applauded as well-deserved recognition for both a great planet and a great mission.

The rings of Saturn consist of a thin sheet of small icy particles, spanning sizes from grains of sand up to house-sized boulders. Just as with the asteroids (Section 4.1) and other populations of debris, there are many more small particles than large ones. The individual particles are bright, reflecting 50–60% of the incident sunlight and exhibiting the strong spectral signature of water ice. An insider's view of the rings would probably resemble a bright cloud of floating hailstones with just a few snowballs and larger objects, many of them loose aggregates of smaller particles.

Each ring particle follows an almost perfectly circular orbit around Saturn in the equatorial plane of the planet. It is easy to see why this must be. Imagine a particle with an eccentric orbit. During each circuit of Saturn it would move in and out, crossing the orbits of other particles. Low-speed collisions would result, and the particle would lose energy, almost as if it were rubbing against its neighbors. The same sort of fender-benders would be encountered by a particle in an inclined orbit, swinging back and forth across the plane of the rings. Either way, the effect of the friction is to circularize the orbit and bring it into the same plane with the rest of the ring particles. This is generally what we see at Saturn, and yet there remain many surprises when we look closely: narrow gaps, waves, eccentric rings, kinky rings, and even braided rings.

Overview of the Rings of Saturn

The rings of Saturn are very broad and very thin. Their dimensions are summarized in Table 14.1. The main rings, those visible from Earth, stretch from about 7000 km above the atmosphere of the planet outward for a total span of more than 70,000 km. The distance from one edge of the rings through the planet to the opposite edge is almost as great as the distance from

TABLE 14.1 Dimensions of the main Saturn rings

Name	Outer Edge (R_s)	Outer Edge (km)
D	1.235	74510
C	1.525	92000
B	1.949	117580
Cassini Division	2.025	122170
A	2.267	136780
F	2.324	140180

the Earth to the Moon. Yet the thickness of this vast expanse is only about 20 m, the width of a typical house lot. If we made a scale model of the rings out of paper the thickness of the sheets in this book, we would have to make the rings more than one kilometer across — about eight city blocks. On this scale, the planet would loom as high as a hundred-story skyscraper.

Three distinct rings can be seen from the Earth, called the A, B, and C Rings; an additional narrow F Ring was discovered in 1979 by Pioneer. Voyager, however, revealed much greater complexity. The ring material is organized into tens of thousands of ringlets visible in Voyager photographs. These ringlets are not generally separated from each other by gaps, but rather represent local enhancements or thinning of the concentration of ring particles. Only a few empty gaps exist, providing natural boundaries between parts of the rings. The origin of these gaps is discussed in Section 14.6. First let us take a tour of the rings, noting the major features as we proceed outward from the planet (Fig. 14.12).

The Inner Rings

Between the upper atmosphere of the planet and the inner edge of the C Ring, Voyager discovered several thin rings invisible from Earth, collectively called the D Ring. The substantial part of the rings — the C Ring — starts at a distance of 14,000 km, where the particles are packed densely enough to reflect a fair amount of sunlight. The planet is easily visible through the C Ring. There are two major gaps in this ring, each several hundred kilometers wide.

Inside one of the C Ring gaps is a remarkable narrow, **eccentric ring.** This entire ribbon behaves as if each of its particles orbited Saturn with the same eccentricity, and with all of the individual orbits exactly aligned. The color of this ring is subtly but definitely different from that of its neighbors, suggesting that the particles that compose the eccentric ring have a different composition. There are only three of these eccentric ringlets in the Saturn system, each very narrow and each lying in an empty gap. As we will see later, the rings of Uranus are also narrow and eccentric.

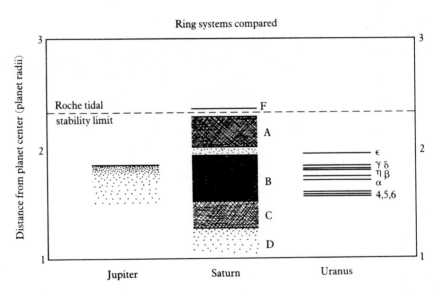

FIGURE 14.12 The three well-observed ring systems in the solar system are compared here in a diagram that presents each one in terms of its distance from its respective planet.

The A and B Rings

At a distance of 32,000 km from Saturn the ring particles suddenly bunch together, and the structure of the concentric ringlets becomes more complex. This is the edge of the B Ring, the brightest part of the ring system and the part that contains most of the particles. Throughout much of the B Ring, which stretches out to 57,000 km, the particles are so closely spaced that the ring is nearly opaque. Particles range typically from tens of centimeters to meters in diameter. There are no empty gaps in the entire 25,000-km span of this ring.

Perhaps the most enigmatic features of the B Ring seen by Voyager are the dark radial spokes that turn as the ring rotates like the spokes on a wheel (Fig. 14.13). These transient features consist of very fine particles hovering in a cloud above the plane of the rings. Each long cloud is formed within a few tens of minutes and then survives for as long as one rotation of the rings, about ten hours. No one knows how these mysterious clouds are formed, but the mechanism probably involves electrically charging fine particles in the ring so that they can rise out of the ring plane.

At the outer edge of the B Ring lies the 3500-km-wide Cassini Division, the one break in the rings that can clearly be seen from the Earth. As shown in Fig. 14.14, however, the Cassini Division is by no means an empty gap. Within the division are several true gaps, including one that contains an eccentric ring, and a great deal of fine structure visible in the spacecraft photos. At least one small satellite also orbits inside this division. At a time when it was still thought that the Cassini Division was empty, a proposal was considered to target the Pioneer Saturn spacecraft, which reconnoitered the planet in 1979, to pass through the rings within this division. Had this been attempted, the Pioneer Saturn mission might have come to a very sudden end!

FIGURE 14.13 a) A picture centered on Saturn's B Ring. The mottled appearance of the ring to the left center is caused by the superposition of so-called "spokes." b) Here we are looking back at the rings after the spacecraft passed them. The spokes appear bright (the diagonal streak to the left), indicating that we are looking at small particles that are preferentially scattering the light forward, along the same direction it was traveling.

(a)

(b)

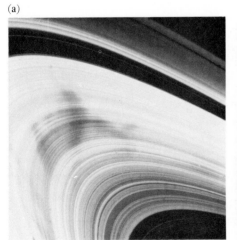

FIGURE 14.14 Seen in this greatly magnified view, the Cassini Division in Saturn's rings, which looks dark and empty when viewed from Earth, is actually found to contain several ring structures of its own. The classical Cassini Division is just to the right of the center, containing five bands of material bounded by dark gaps.

Outside the Cassini Division, beginning at 62,000 km from the planet, lies the last major ring, the A Ring. The A Ring is intermediate in brightness and transparency between the opaque B Ring and the translucent C Ring. Its most outstanding feature is in one of its gaps, the 360-km-wide Encke Division, which contains two discontinuous, kinky ringlets (Fig. 14.15a) and one known small satellite. These peculiar ribbons of material are only about 20 km wide and were observable from the Voyagers only when they passed very close to the rings. The problem of the origin of kinky rings is discussed in Sec-

tion 14.8, but don't expect any answers. Kinky rings remain a bit beyond our current understanding.

Beyond the Main Rings

The A Ring ends abruptly 77,000 km from the planet. There is no tapering off of ring particles, no stray ringlets drifting off into space. There is, however, the fascinating F Ring 4,000 km farther out. Unlike the rings discussed so far, the F Ring is an isolated bright ribbon less than a hundred kilometers wide. It is the third eccentric ring in the Saturn system. It also has the most complex and peculiar structure of any Saturn ring, including areas where it divides into multiple strands that appear to be intertwined or braided (Fig. 14.15b).

The discovery of braiding in the F Ring was one of the sensations of the Voyager encounters, leading to newspaper headlines asserting that these rings disobeyed the laws of physics. They don't, but they do point out that scientists are sometimes unable to interpret the laws of nature to explain very complex situations. What makes the F Ring both complicated and fascinating is the presence of two small inner satellites of Saturn that orbit the planet on either side of it. The boundaries of the ring are clearly defined by the presence of these satellites, and it is their gravitational influence that generates its fine structure. We will return to this subject when we discuss satellite-ring interactions in Section 14.7.

Probably the two most fundamental questions raised by the rings of Saturn are those of origin and of structure. The problem of the origin of ring systems is the subject of Section 14.8. To address properly the structure of these rings, we need to look at the fine-scale organization of the ringlets revealed by Voyager, and also to compare the structure of the rings of Saturn with the very different rings of Uranus.

14.6 Occultations and the Fine Structure of Rings

One of the most powerful tools of planetary science for probing the fine structure of planetary atmospheres is provided by occultations. Occultations produced when a star or a spacecraft passes behind a ring system are equally useful. By measuring the changes in the starlight or radio transmission as it passes through rings, the astronomer can map out details that could never

FIGURE 14.15 a) A ring that appears to defy Kepler's laws, this discontinuous kinky segment lies in the Encke Division in the A Ring. b) Saturn's F Ring consists of at least five separate strands, of which two are easily seen here. They exhibit a braided appearance with knots of greater density (moonlets?) embedded within them.

(a)

(b)

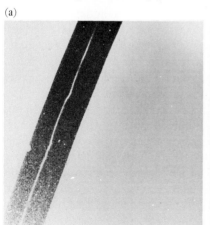
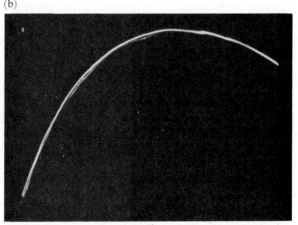

be seen directly. In this section, we discuss the use of occultations of stars to reveal the high-resolution structure of rings.

Discovery of the Rings of Uranus

Occultations of stars by planets first attracted the attention of most planetary scientists in 1977, when they yielded the unexpected discovery of the rings of Uranus. This discovery was the by-product of another experiment aimed at studying the atmosphere of this planet.

Only about once per decade does the apparent path of Uranus across the sky take it in front of a star bright enough to be useful for occultation studies of the atmosphere. On March 10, 1977, a particularly favorable event occurred, except for one thing: the occultation of the planet could not be seen from any of the more populated parts of the Earth. Only from Antarctica and the Indian Ocean was it sure to be seen, with marginal conditions in Australia, southern India, and South Africa. Nevertheless, observers set up equipment at all of the observatories that might be in the path, to await this exceptional event.

To supplement the ground-based observations, James Elliot and his colleagues from MIT proposed to use NASA's Kuiper Airborne Observatory (Fig. 14.16) to fly above the southern Indian Ocean, well within the predicted zone for the occultation. To accommodate this project, the airplane had to fly to the southern hemisphere and operate out of an Australian airbase.

As expected, Elliot's team succeeded in measuring the occultation of the star by Uranus, while the ground-based observers saw the planet skim by the star without blocking its light. The most important results that night came not from the planet at all, however, as Elliot has recounted in detail in his recent book on rings (see the additional reading list). As measured from both the airplane and the ground, the occulted star began to wink out about 40 minutes before it should have first been affected by the upper atmosphere

of the planet. Several times the star dimmed dramatically for intervals that lasted from 2 to 8 seconds, then returned just as suddenly to full brightness (Fig. 14.16).

Clearly something was briefly blocking the starlight, and at first Elliot and his colleagues on the plane thought they might be seeing a swarm of small uranian satellites. The occultation events were not randomly spaced, however. As Uranus moved beyond the star, the exact same sequence of occultations occurred, but in reverse order. Further, the occultation pattern was the same at the different observing sites. What had been discovered was a series of narrow, opaque rings, invisible in reflected sunlight, but detected by their ability to block the light from a distant star. The situation resembles counting the number of cars in a train on a dark night by recording the visibility of a streetlight viewed through the passing cars.

The uranian rings had not been detected before because they are narrow and composed of dark particles. Even if they had been seen in reflected sunlight, however, the occultation technique would have revealed much more than could be photographed with any telescope. The reason is that the occultation resolution is not limited by the size of the telescope or the shimmering of the Earth's atmosphere. The level of detail probed by an occultation depends on how rapidly the rings appear to move across the sky and how often we can sample the changing brightness of the star. In the case of the Uranus occultation, structure as small as a few kilometers could be measured.

Since their discovery, the rings of Uranus have been repeatedly probed using occultations, with the results described in Section 14.8. This technique is very well suited to the task for three reasons: (1) Since the rings are dark, they produce almost no reflected light to interfere; (2) At near-infrared wavelengths the planet itself is also very dark, further reducing interference; and (3) The high tilt of the planet permits the rings to be seen nearly face-on at the present time, pre-

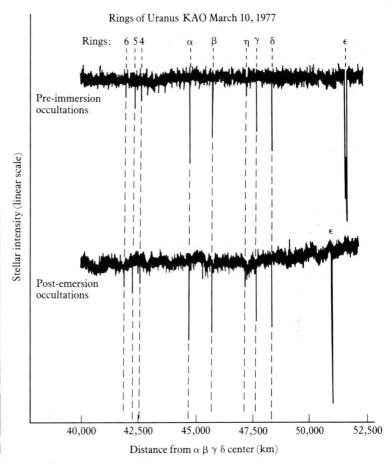

FIGURE 14.16 (left) James Elliot making an adjustment at the 0.9-m telescope of the Kuiper Airborne Observatory (KAO), with which he discovered the rings of Uranus. (right) The two intensity tracings that revealed the presence of the rings. Top: As Uranus passed in front of a star, the intensity of starlight recorded by a photometer on the KAO telescope suddenly dimmed, then recovered, and repeated this behavior nine times. Bottom: The same thing happened on the other side of the planet, except that the eccentric Epsilon Ring was closer to Uranus and narrower on this side of the planet.

senting a wide target. In contrast, stellar occultations cannot be used from Earth to study the saturnian rings, which are very bright and are viewed obliquely. Ironically, therefore, the fine structure of the faint and distant uranian rings can be studied better from the Earth than that of the much brighter saturnian rings.

The Voyager Occultation of the Rings of Saturn

One way to increase the resolution of the occultation technique is to get closer to the ring system being studied. A logical extension of the Earth-based observations, therefore, involves measuring an occultation from a spacecraft. Voyager carried a small telescope and detector capable of making occultation measurements, and the trajectory of Voyager 2 at Saturn permitted a two-hour occultation observation as the spacecraft sped below the rings. By observing the star from the dark side of the rings, the spacecraft experiment was also able to eliminate the major interference problem that makes Earth-based occultations by the saturnian rings unobservable. Two ring occultations were also observed by Voyager 2 at Uranus in 1986.

In the Voyager experiments, the brightness of the star was recorded 100 times per second as it winked on and off through the wide expanse of the rings. At Saturn, the result was approximately a million separate measures of the transparency of the ring, along a line 82,000 km long. Thus features could be resolved down to 100 m in size, compared to the best Voyager imaging resolution of a few kilometers.

A stunning complexity of structure in the rings of Saturn was revealed by the Voyager oc-

cultation data and by the high-resolution photographs obtained while the spacecraft were close to the rings. The narrow F Ring, for instance, was resolved by the spacecraft cameras into a series of ringlets each a few kilometers across, while in the occultation data the brightest ringlet itself showed even more structure at the hundred-meter level. The many regular patterns revealed by both imaging and the occultation data in the A and B Rings were equally interesting.

Remarkably, the fine structure of the rings varies with time and location. Fig. 14.17 illustrates the structure in the outer part of the B Ring, at the border of the Cassini Division. Four different photos, taken at different times, are shown side by side. While the major features line up well, it is clear that the smaller ringlets do not. These ring structures are not fixed, but shift from hour to hour. We should think of them as transient waves, flowing back and forth over the rings' spinning surfaces like waves on the ocean.

FIGURE 14.17 These four radial segments of the outer edge of the B Ring of Saturn show that the fine structure in the ring is not uniform around the planet. The outer edge of the B Ring itself is eccentric so it occurs at different distances from the planet at different places along its circumference.

Waves in the Rings

Tens of thousands of ringlets can be identified, many of them grouped into patterns of regularly spaced bright and dark lines. Many of these patterns are examples of **spiral density waves** (Fig. 14.18). As the name implies, these are waves that follow a spiral pattern, like the grooves in a phonograph record. Density waves had been predicted theoretically but never observed before Voyager arrived at Saturn. They are a phenomenon peculiar to a flat spinning disk in which individual particles can interact gravitationally.

Density wave theory had been worked out to explain the spiral structure of galaxies, but it applied equally well to the saturnian rings. The presence of these waves demonstrates that the ring particles interact with each other as they circle the planet. Even though the individual particles rarely touch, the ensemble of billions of particles behaves in many ways like a thin sheet of rubber because of mutual gravitational attractions. This is especially true in the B Ring, where the particles are most closely packed together.

Another similar spiral wave pattern can occur when the thin sheet of spinning particles is bent upward or downward by the gravitational pull of one of the inner satellites. These **bending waves** (also shown in Fig. 14.18) are wrinkles in the rings, and they help to give it some thickness instead of letting the particles collapse into a single very thin layer, as they would do without the influence of the satellites.

Dozens of individual spiral wave patterns have been identified primarily in the A Ring, but a great deal of the structure, particularly that seen in the B Ring, cannot be accounted for by density waves or bending waves. Some of this structure is probably the result of the superposition of unidentified wave patterns, but much of it may have quite different causes. We could understand the behavior of the rings better if we had moving pictures of the changing ring pat-

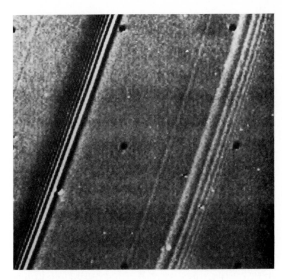

FIGURE 14.18 The same 5 : 3 gravitational resonance with the satellite Mimas generates both these systems of waves. The one on the left is a bending wave with ripples 1–2 km high, sufficient to cast shadows. On the right is a spiral density wave, consisting of local concentrations of material within the ring plane.

terns. Unfortunately, however, there was only a single Voyager occultation, and the high-resolution images are only snapshots. A Saturn orbiter spacecraft such as the one proposed for the Cassini mission would be required to obtain the data we need.

Voyager Observations of the Rings of Uranus

As we have already noted, the rings of Uranus are just the opposite of the Saturn rings in several ways. At Saturn we have broad rings interrupted by a few narrow gaps; at Uranus we have very narrow rings separated by broad gaps. The Saturn ring particles are bright and composed of ice; those at Uranus are dark and apparently made of some sort of carbonaceous material.

The dimensions of the rings of Uranus are shown in Table 14.2. There are five main rings, named the Alpha, Beta, Gamma, Delta, and Epsilon Rings in order outward from the planet (Fig. 14.19). Their distances from the upper atmosphere of Uranus range from 19,000 to 26,000 km. An additional complex ring called the Eta Ring lies between the Beta and Gamma rings at 22,000 km from Uranus. All six of these rings, as well as three less prominent rings inside the Alpha Ring, were known from occultation observations before the first spacecraft reached Uranus.

In 1986, the Voyager cameras found one more narrow ring, as listed in Table 14.2, bringing the total to ten. Two more occultations of stars by the rings were also measured, yielding

TABLE 14.2 Rings of Uranus[a]

Ring Name	Distance (km)	Width (km)	Eccentricity
6 Ring	41,850	1–3	0.0010
5 Ring	42,240	2–3	0.0019
4 Ring	42,580	2	0.0011
Alpha	44,730	8–11	0.0008
Beta	45,670	7–11	0.0004
Eta	47,180	2	0
Gamma	47,630	1–4	0
Delta	48,310	3–9	0
1986U1R	50,040	1–2	?
Epsilon	51,160	22–93	0.0079

[a]Many additional rings were seen in forward scattered light.

FIGURE 14.19 The Uranus rings as revealed by the Voyager 2 cameras corroborated and extended the occultation observations. a) A distant view of the planet shows only the Epsilon Ring, revealing its variation in thickness. (Compare upper right with lower left.) b) Here we see all of the pre-Voyager rings. Starting from the bottom, they are 6, 5, 4, α, β, η, γ, δ, ε. Note that η is distinctly thicker than all the others except ε.

(a)

(b)

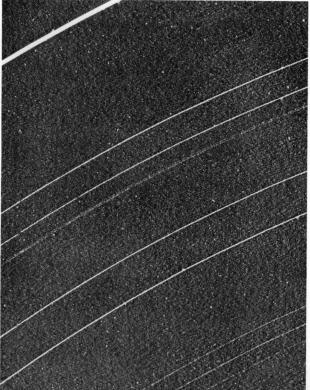

ring profiles of substantially higher resolution than those obtained either by the Voyager cameras or Earth-based occultation observations.

Most of these dark rings of Uranus are nearly circular and exceedingly narrow, no more than 10 km in width. In other words, their lengths are nearly 100,000 times their breadth, like a piece of spaghetti several city blocks long. Yet the density of particles within each ring is high, sufficient to block most of the starlight when the ring passes between us and an occulted star. These particle densities are similar to those in Saturn's A and B Rings.

Two of the rings have a special structure, nearly as peculiar as that of the F Ring of Saturn. The Epsilon Ring, which probably contains as much mass as all of the other rings combined, is both eccentric and variable in width (Figs. 14.16 and 14.19). Where it comes closest to Uranus, its width is 20 km; at the opposite side, where the ring is 800 km farther from the planet, it has a much greater width of 100 km. Like the eccentric rings of Saturn, the Epsilon Ring must be held together by some gravitational force in order to maintain its shape. The Eta Ring is equally misshapen, consisting of a relatively broad, shallow ring 60 km wide with a narrow, denser component at its inner edge.

Because the rings are so dark and difficult to photograph, even the best Voyager images are rather disappointing. However, the non-imaging data from the spacecraft are more exciting. The best stellar occultation profiles resolve features as small as 10 m in size, revealing intricate structure and demonstrating that the thickness of these rings, like those of Saturn, is no more than a few tens of meters (Fig. 14.20). In addition, the results of the occultation of the spacecraft by the rings, during which the effects of the rings on the transmitted radio signal were measured, show that at least the Epsilon Ring is composed primarily of larger particles than those characteristic of Saturn's rings. Very little of the mass

FIGURE 14.20 A profile of the Epsilon Ring of Uranus obtained by the Voyager 2 spacecraft as the ring occulted a star (cp. Fig. 14.16).

at Uranus is to be found in particles smaller than a few centimeters in diameter. Apparently such fine dust is either not formed, or else it has long since been swept out of the rings, leaving only the larger particles behind. This is another, unexpected difference between the rings of Uranus and those of Saturn.

14.7 Ring Dynamics and Satellite-Ring Interactions

In the preceding sections we have described ring structure but have not addressed in any detail the question of its cause. Why are the edges of the rings so sharply defined? What causes the Cassini Division and the smaller gaps in the rings of Saturn? Why are three saturnian ringlets as well as most of the uranian rings eccentric? Why are the rings of Uranus so narrow? What generates the spiral wave patterns? It turns out that the answers to all of these questions probably involve interactions between the rings and the satellites.

Sharp Edges of Rings

Left to itself, a planetary ring will slowly spread. Over hundreds of millions of years, the interactions between closely spaced ring particles will force them apart, with some spiraling inward to disintegrate like meteors in the atmosphere and others moving out to ever greater distances from the planet. To be stable a ring must be bounded.

The binding force that limits the outward spreading of the rings of Saturn is gravitational. The outer edge of the A Ring is exactly in a ⅔ resonant orbit with Mimas, and the gravitation of Mimas holds the ring particles firmly trapped at this particular distance from Saturn.

Shepherd Satellites

A different kind of gravitational influence holds the thin F Ring of Saturn in place and even explains its slightly eccentric shape. In this case there are two satellites, Prometheus and Pandora, one on each side of the ring (Fig. 14.21). These are called the **shepherd satellites** for their role in keeping the particles of the F Ring narrowly confined. Probably it is the influence of

FIGURE 14.21 The F Ring of Saturn appears to be held in place by the two small shepherding satellites Prometheus and Pandora. This view shows a segment of the F Ring with Prometheus in the same field of view.

the shepherd satellites that also generates the braids in the F Ring, but the theory for such a process has not been worked out.

The rings of Uranus were discovered at a time when the Saturn rings were still thought to be broad and relatively unstructured, and at first the existence of such narrow rings utterly confounded astronomers. It was in response to the challenge posed by the narrow rings of Uranus that theorists first worked out the idea that a ring could be confined by shepherd satellites.

These shepherds, predicted for Uranus, were confirmed in concept by the discovery of Pandora and Prometheus on either side of the F Ring of Saturn. It was, therefore, with great interest that the Voyager cameras searched for similar shepherds at Uranus. Unfortunately, however, the search was only partially successful.

Two Uranus shepherds (Cordelia and Ophelia) were discovered in association with the Epsilon Ring, each a small dark satellite less than 50 km in diameter. These shepherds orbit about 2000 km on either side of the ring, very much like the F Ring shepherds at Saturn. However, no other shepherd satellites were seen near the other rings. Since the cameras were only sensitive to dark objects 10 km or more in diameter, such objects may be present, but if so, they are very small. Thus we still have no confirmation of the postulated additional shepherds needed to explain all of the other known rings.

Embedded Satellites

If there are satellites inside the rings, they can have gravitational effects also. Within the broad main rings of Saturn, the presence of unseen satellites is thought to produce the few genuine gaps. By its gravitational herding effect, a small satellite could sweep clear a lane much wider than its own diameter. In order to confirm this explanation, very careful searches were made of the Voyager images for these **embedded satellites,** but to the great disappointment of the Voyager scientists none was found. In the Cassini Division gaps, where the most thorough search was made, it seems clear that no object as large as 10 km in diameter could have escaped detection.

Normally when a theory fails such a test it is discarded, but in this case no alternative explanation for ring gaps has been proposed. Many scientists therefore still believe that some version of the embedded satellite theory will ultimately prove correct, but will involve satellites smaller than 10 km in diameter or too dark to have been imaged.

Embedded satellites have also been suggested as an important agent for forming narrow, eccentric rings. Contradictory though it may seem, calculations have shown that under some circumstances — generally involving closely packed ring particles — an embedded satellite will produce a narrow ring rather than clearing a gap. If the orbit of the embedded satellite is eccentric, so will be the path of the thin ring it controls. This is an especially attractive theory for trying to understand the kinky rings in the Encke Division of the A Ring, and it has also been suggested as an alternative way of forming some of the rings of Uranus.

Although no embedded satellites were imaged directly by Voyager, later analysis of the fine structure of the rings of Saturn provided fairly convincing evidence for the existence of two of these objects, one in the Cassini Division and one in the Encke Division. Both satellites revealed themselves by the waves generated in the ring from their passage, much like the wake that stretches far behind a ship moving across the oceans of Earth. From the properties of their wakes, the positions and masses of these embedded objects can be calculated, and both turn out to be less than 15 km in diameter, perhaps small enough to have been missed by the Voyager cameras.

Resonant Effects of the Larger Satellites

The larger external satellites also influence the ring structure by their gravitational effects. We have already noted how Mimas, the closest of the larger satellites of Saturn, limits the outer extent of the A Ring. The Cassini Division also is caused by Mimas. At the inner edge of the division a ring particle orbits Saturn exactly twice for each orbit of Mimas, and this strong resonance is apparently responsible for the resonant gap located here, just as the resonant gaps in the asteroid belt are caused by Jupiter (Section 4.2).

Weaker resonances with other, smaller satellites do not produce gaps, but they do disturb the ring particles having orbital periods that are simple fractions of those of the satellites. The two co-orbital satellites generate a large number of these resonances. The disturbances in these cases produce the spiral density waves. Each spiral represents a wave pattern that originates at the resonant position and then winds its way outward.

Mimas is the main cause of the spiral bending waves, which result from a gravitational tug out of the ring plane. Since Mimas does not orbit Saturn in exactly the same plane as the ring, it exerts such forces each time it circles the planet. These bending waves are seen to extend inward, rather than outward from their place of origin.

By invoking the presence of satellites both seen and unseen, we can explain a great deal of complex structure found in the rings of Saturn. Not everything makes sense, of course, but most of the prominent features seem understandable. The main problem with this approach is that it depends in part on the existence of a variety of unseen satellites embedded within the rings. These satellites have been sought but not found. If they are present, we wonder how they have eluded detection; if they are absent, then our understanding of ring structure is much poorer than we would like to think. Similarly, the presence of two shepherds for the Epsilon Ring of Uranus seems to provide an adequate explanation for the existence of this ring, but the satellites to control all of the other rings of Uranus remain at present only speculation.

14.8 Other Ring Systems

Rings of Jupiter

The rings of Jupiter were discovered by Voyager in 1979 (Fig. 14.22) and confirmed within a few days by ground-based observations (it helps if an astronomer knows just what to look for!). Owen led the Voyager effort to search for this ring, which began with a single long-exposure image

that was recorded as Voyager 1 passed through Jupiter's equatorial plane. The camera was aimed at a point in space that would correspond to Saturn's B Ring, halfway between the orbit of Amalthea and the jovian cloud tops. This single frame succeeded in catching the rings. Using the results from Voyager 1, the Voyager 2 cameras were programmed to take a variety of ring pictures, and most of what we know of the rings is derived from the Voyager 2 data.

The primary ring is 54,000 km from the planet and about 5000 km wide. It is much more tenuous than any of the other known rings, blocking only about one millionth of the light from a source seen through it. For comparison, a sheet of clear glass absorbs several percent of incident light, so one could say that the ring of Jupiter is ten thousand times more transparent than window glass! A still more insubstantial doughnut or torus of material surrounds the main ring.

Most of the particles in the Jupiter ring are very small, no larger than grains of dust. They are in fact made of dark silicate dust, since grains of water ice would evaporate at the temperatures found in the jovian system. As a consequence of their size, they easily acquire small electrical charge from interactions with the jovian magnetosphere, and the mutual repulsive force of these charged particles lifts them up out of the ring plane. The processes involved in the formation of this extended cloud may be similar to those that generate the spokes in the B Ring of Saturn (Section 14.5), but neither phenomenon is very well understood.

Rings of Neptune?

With rings known to circle the other three giant planets, we naturally ask whether Neptune shouldn't also have rings. Of course, such rings would be more difficult to see than those of the

(a)

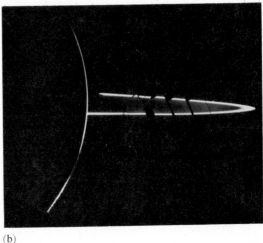
(b)

FIGURE 14.22 a) Owen (left) discussing the discovery picture of Jupiter's ring with Candy Hansen (right) who designed the Voyager camera sequence that acquired it, while Jeff Plescia looks on. b) Jupiter's ring, revealed in a mosaic of pictures obtained with the spacecraft in the planet's shadow so small particles in the ring produce strong forward scattering of light.

nearer three giant planets. Neptune is more than four billion kilometers away, and it has never been visited by a spacecraft. It is no surprise, therefore, that no rings have been observed directly; certainly, rings at Neptune such as those at Uranus and Jupiter would be invisible to Earth-based observers.

Occultations provide the one tool with which to search for rings at Neptune. As we discussed above, the measurement of occultations is nearly independent of distance to the planet. All that is required is that the ring system pass in front of a distant star, and that the ring material be packed densely enough to block the starlight. A tenuous dust ring like that of Jupiter would not be detectable, but any of the major rings of Saturn or Uranus would be readily discovered by occultations if they were present at Neptune.

At this writing the occultation results for Neptune are both contradictory and tantalizing.

Several events have been detected — cases where starlight was briefly occulted by something near Neptune. However, most of these occultation events have not repeated at the symmetric position on the other side of the planet. Recall that it was the symmetry of the occultation pattern that demonstrated that the original Uranus occultations were caused by rings and not by isolated satellites.

In the case of Neptune, occultations by something orbiting close to the planet have been observed, but the occultations are not symmetric. Thus no continuous rings that stretch all the way around the planet have been seen. It is difficult to imagine various short arcs of rings, present on only one side of the planet at a given time, yet that is apparently what is being found. Clearly, the continuing study of the partial rings of Neptune is likely to provide more surprises and to stimulate still more theories for the formation and dynamics of planetary rings.

14.9 Origin of Ring Systems

The planetary ring systems have several features in common: they are composed of small particles of a wide variety of sizes; they are close to their planets; and much of their structure appears to be controlled by a few larger objects, including the external satellites. The interactions with satellites in particular seem to be necessary to bound a ring and keep it from gradually spreading out and fading away. What could have produced the original swarm of billions of particles in a disk surrounding the planets?

There are two basic theories of ring origin. First is the breakup theory, which suggests that the rings are the remains of a shattered satellite. The second theory, which takes the reverse perspective, suggests that the rings are made of particles that never were able to come together to form a satellite in the first place.

The Tidal Stability Limit

In either theory of formation, an important role is played by tidal forces. When tides were discussed in Section 6.2, we noted that they are very sensitive to distance, varying as the inverse cube of the separation between two bodies. The effect of tides is to distort a satellite, raising bulges toward and away from a planet. The closer a satellite comes to its primary, the larger the tidal distortion. Ultimately the satellite can be torn apart, if its internal strength is not great enough to withstand the tidal stress.

Around each planet there exists a **tidal stability limit,** the distance within which tidal forces can destroy an intruding satellite. This limit was first calculated by the nineteenth century French mathematician Edouard Roche, and

it is often called the Roche limit. Its exact value depends on the density of the satellite, but for most solar system objects the tidal stability limit is at about 2.5 planetary radii from the center of the planet. As shown in Fig. 14.12, the three known planetary rings all lie within the stability limit.

It should be emphasized that the tidal stability limit is calculated for an intruding satellite with no intrinsic strength of its own. It tells us where a liquid satellite, or one made of disconnected bits of gravel, would be pulled apart by tidal forces. The stronger a satellite, the less it is affected by tides. As an extreme example, consider the Space Shuttle or any other low-orbit artificial Earth satellite. These operate well inside the stability limit, yet they are not pulled apart by tides. On the other hand, a loose tool left floating in the Shuttle bay will drift away under tidal forces and be lost in space. This is what it means to be inside the Roche stability limit.

Another way of looking at the stability limit is to think of it as the distance within which individual particles will not come together under their own gravity to form a larger body. If a very large disk of particles once surrounded Saturn, for instance, it is easy to imagine them coalescing to form individual satellites everywhere except inside the stability limit, where they might remain to this day as the ring particles. In such a case, the ring particles would probably all be very small, and they would retain their primitive composition.

Ring Formation by Satellite Breakup

In the breakup theory of ring formation, we might imagine a satellite or even a passing comet coming too close to a planet (well inside the Roche limit) and being torn apart by tidal forces.

There is an even more likely scenario, however, in view of the very heavy meteoroidal bombardment that took place early in the history of the Saturn system. We have already suggested that a number of small inner satellites might have shattered during this early bombardment, leaving fragments such as the co-orbital and shepherd satellites. If a satellite were disrupted near the tidal stability limit, it would be unable to reform itself, and the fragments would spread into a ring. In this case we would expect a wide range of particle sizes, including a few large fragments several kilometers across. The two embedded satellites found in the rings of Saturn may be examples of such fragments. If the presence of bodies in the size range from 1 to 10 km could be confirmed, they would also argue for the breakup theory.

How much material is actually present in the three ring systems we are discussing? This is difficult to determine, since what we see is primarily the smaller particles, while most of the mass may be contained in just a few particles (or moonlets) near the upper end of the size distribution. This caution about unseen large particles aside, however, it has proved possible to measure the mass of the main B Ring of Saturn from Voyager data. Since the spiral density waves depend on the gravitational interactions between ring particles, their behavior is sensitive to the mass of material present. Within the B Ring, there is nearly 1 ton of material per square meter, spread through the 20-meter thickness of the ring. Summing the entire ring system, we obtain a total mass of about 10^{15} tons. The mass of the Uranus rings is guessed to be perhaps a thousand times less, and that of the Jupiter rings a million times less, than the Saturn rings.

The total mass of the saturnian rings of 10^{15} tons is equal to that of an icy satellite about 250 km in diameter — about the same size as the larger co-orbital satellite, Janus. This mass is therefore entirely consistent with the idea that the rings might have resulted from the breakup of a typical inner satellite near the tidal stability limit. Alternatively, it is easy to imagine there being this much leftover material that never formed a satellite in the first place.

We are left with the mystery of the different chemical compositions of the three known ring systems. Saturn seems to make the most sense, since both its rings and its satellites are made of the same icy material. Jupiter can be understood also when we remember that icy bodies, either satellites or ring particles, could not have survived near the planet early in its history, when it radiated more heat than at present. The dusty rings of Jupiter represent erosion from rocky satellites.

Uranus remains perhaps the most strange. Its large satellites are icy, but both the rings and the small inner satellites are made of very dark, presumably carbonaceous material. The association of the dark satellites with the dark rings suggests that the rings might have been formed by the breakup of one or more inner satellites. But we cannot say why the ice and the carbonaceous material are separated in the uranian system, with an apparent demarcation line at about 100,000 km from the planet.

Summary

There are a great many bodies in the outer solar system that have no counterparts nearer the Sun. For the most part, these bodies contain substantial quantities of water ice, a material that could not have condensed in warmer parts of the solar nebula. Dark, apparently carbonaceous material is also relatively abundant. Thus we tend to identify both the satellites and the ring systems with primitive material, although somewhat to our surprise we have also found cases where even relatively small bodies appear to have experienced considerable thermal evolution.

The best-studied icy satellites in the size range of 400 to 1600 km are those that form the regular satellite system of Saturn. Four of these — Rhea, Dione, Tethys, and Mimas — are basically similar objects with low density, high reflectivity, and heavily cratered surfaces, although there are some interesting variations in geologic history. The odd object in this set is Enceladus, which in spite of its small size has experienced widespread, recent geologic activity. Enceladus also seems to be the source for the tenuous E Ring of Saturn, and it has a possibly related surface coating of highly reflective small ice grains. The other equally mysterious satellite of Saturn is Iapetus, which ought to be the twin of Rhea but isn't. Iapetus has been coated on its leading hemisphere by dark, probably carbonaceous material of unknown origin, giving it a strangely two-faced appearance.

The five major satellites of Uranus have also been studied in some detail, thanks to the 1986 Voyager flyby. These objects have about the same bulk composition as the Saturn satellites, and their surfaces are also icy, although with a larger proportion of dark (possibly carbonaceous) material present. Geologically they are also similar, except for Miranda, which displays a wide range of bizarre features that planetary geologists are only beginning to interpret.

The smaller satellites of Saturn are also primarily icy bodies. We saw how their irregular shapes can be understood in terms of small size and an environment of intense meteoric bombardment early in their history. Many of these satellites have remarkable dynamical properties, ranging from the chaotic rotation of Hyperion to the ring-related shepherd satellites to the coorbitals, which periodically exchange orbits.

The small satellites of Jupiter are primarily rocky and may be the shattered remnant of one or more asteroids.

Each of the giant planets has a ring system made of small bodies orbiting within the tidal stability limit. The rings of Saturn are composed of icy objects in the size range from ten meters to dust; the Uranus rings are made of very dark material and the particles span a more limited size range from a few centimeters up to tens of meters; and the tenuous ring of Jupiter is little more than a smoke ring of dust eroded from the inner satellites.

Structurally, the rings display a remarkable variety of forms. Within the Saturn rings, which are the best studied, we see examples of eccentric rings, kinky rings, and braided rings. In some places the rings are limited by shepherd moons or embedded objects and in some by resonances with more distant satellites. Other resonances stimulate spiral density waves and bending waves that propagate across the rings, generating a complex structure of thousands of ringlets. The rings of Uranus are still more mysterious, since the particles are confined to narrow ribbons. The outer Epsilon Ring has two shepherds like the F Ring of Saturn, but for the others no embedded or shepherding satellites have been discovered.

In discussing the rings of Saturn and Uranus we noted how occultation techniques could be applied to reach much higher resolution than can be obtained by imaging. The Uranus rings were discovered by occultation, and subsequent observations by this technique have probed them to resolutions of a few tens of meters. Voyager occultation measurements of the rings of Saturn revealed detail down to the hundred-meter level.

We still do not know the origin of the rings of Saturn and Uranus, but tidal forces, which become strong enough to disrupt large objects, must play a part. Three possibilities exist. Perhaps the rings represent material near the planet that was prevented by tidal forces from ever forming larger satellites, or perhaps a satellite that formed elsewhere came too close and was tidally torn apart. Finally, and some think most likely, is the scenario in which a satellite near the tidal stability limit was broken apart by impacts during the period of heavy bombardment and could not reform itself.

Key Terms

bending wave

chaotic rotation

co-orbital satellite

eccentric ring

embedded satellite

shepherd satellite

spiral density wave

tidal stability limit

PART SEVEN
REVIEW QUESTIONS

1. Compare and contrast the satellite systems of the outer planets. Is there any relationship between the nature of the satellite systems and the rings?

2. Compare the satellite systems of the outer planets with the solar system itself. Make a list of the basic properties of the solar system and see how many of these are found in each satellite system. Do you believe that processes that formed satellite systems are essentially the same as those that gave rise to the solar system?

3. What are the probable sources of cratering impacts in the outer solar system? How can our knowledge of these sources lead us to a scheme for dating the surfaces of the satellites? Assess the uncertainties in satellite dating from crater counts, and judge how these uncertainties affect our understanding of the evolution of the icy satellites.

4. Compare the three largest satellites. Trace their evolution from formation to their current state. Can the Voyager observations of diversity among them be understood in terms of known processes of planetary evolution?

5. Compare the evolution of icy satellites such as Ganymede and Callisto with that of such rocky objects as the Moon and Mercury. How do they differ? What would you expect for an object that was almost all water and other volatiles, with very little rocky material?

6. Compare the geology of Ganymede with that of the Earth and Mars. What are the major differences? Are the differences understandable in terms of differences in composition? Do you believe that Ganymede ever experienced anything like the terrestrial plate tectonics?

7. Compare the evolution of Europa and Io. Why are these satellites so different from each other? Why are they different from the Moon, which is about the same size?

8. Describe the volcanic activity of Io. Which kinds of features are due to which materials? Where is most of the tidal energy released? How do both the level of activity and the nature of the volcanoes compare with the Earth?

9. Compare the atmosphere of Titan with that of the Earth today and with that of the Earth 4 billion years ago. Which differences are due to composition of the planet? Which are due to size? Which are due to the distance from the Sun?

10. Compare the geology of Rhea with that of Callisto. Which differences are due to the size of the objects? Which are due to distance from the Sun?

11. Compare Iapetus with Rhea and Enceladus with Mimas. In each case, why are these near-twins so different?

12. In what ways are the small satellites of Jupiter and Saturn like Phobos and Deimos or like the asteroids? Do you believe the irregular satellites were captured from the asteroid belt? Why? How could we test this hypothesis?

13. Compare the four outer planet ring systems. What do they have in common? Why are they different in so many ways?

14. Describe the individual particles in the Saturn rings. How do these compare with the particles in the other rings? Why are they different?

15. Explain how the occultation technique is used to probe ring structure. What exactly is measured? What determines the resolution? How do ground-based and spacecraft data compare? Why have telescopic measurements been more productive for the rings of Uranus than for the rings of Saturn?

16. Ring structure is determined in part by satellite interactions. Explain the role of shepherd satellites, of embedded satellites, and of satellite resonances. Which of these produces what structure for the known ring systems?

ADDITIONAL READING

*Burns, J.A. and M.S. Matthews, eds. 1986. *Satellites.* Tucson: University of Arizona Press.

Chapman, C.R. 1982. *Planets of Rock and Ice* (Chapter 9). New York: Scribner.

Davis, J. 1987. *Flyby: The Interplanetary Odyssey of Voyager 2.* New York: Atheneum.

Elliot, J. and R. Kerr. 1984. *Rings: Discoveries from Galileo to Voyager.* Cambridge, MA: MIT Press.

*Greenberg, R. and A. Brahic, eds. 1984. *Planetary Rings.* Tucson: University of Arizona Press.

*Hunten, D.M. 1984. "Titan." In *Saturn,* ed. T. Gehrels and M. Matthews. Tucson: University of Arizona Press.

Johnson, T.V. and L.A. Soderblom. 1983. "Io." *Scientific American 246*:6, 100.

*Morrison, D., ed. 1982. *The Satellites of Jupiter.* Tucson: University of Arizona Press.

Morrison, D. 1985. "The Enigma Called Io." *Sky and Telescope 69,* 198.

Owen, T. 1982. "Titan." *Scientific American 246*:2, 98.

Pollack, J.B. 1982. "Titan." In *The New Solar System,* 2nd ed., ed. J.K. Beatty, B. O'Leary, and A. Chaikin. Cambridge, MA: Sky Publishing Corp.

Soderblom, L.A. and T.V. Johnson. 1982. "The Moons of Saturn." *Scientific American 246*:1, 100.

Soderblom, L.A., R.H. Brown, and T.V. Johnson. 1987. "The Moons of Uranus." *Scientific American 256*:4, 48.

EPILOGUE
Back to the Beginning

Now that we have surveyed the individual components of the planetary system, we want to try to understand how the system itself came to be the way we find it today. This basic question has been addressed in bits and pieces throughout the book. In this final part, we will pull these individual discussions together and address some additional issues that have not yet been confronted.

In spite of the wealth of new information we now have on the members of the planetary system, an understanding of origins remains difficult. Direct evidence of the events of 4.5 billion years ago is, of course, very limited. The problem is made more difficult by our lack of information on other planetary systems with which we might compare our own.

This final chapter also addresses ways in which we might remedy this situation, by searching for direct evidence of other planets circling other stars.

In writing about the origin of the planetary system, we find ourselves reviewing much of the material presented in previous chapters. Especially relevant are the data on the meteorites, comets, and asteroids presented in Part II, on impacts and dynamical evolution from Parts III through VI, and on the chemistry of the outer solar system in Parts II, VI, and VII. Thus we hope that this epilogue, in addition to summarizing current ideas on the origin of the planetary system, will serve as a review of many important concepts from earlier chapters.

◀ Shiva, the Hindu god of destruction and rebirth, is here shown in his form as the cosmic dancer, Nataraja. The Hindu concept of the recurring cycle of life, death, and rebirth is a good analogy for the formation of the planetary system from the gas and dust produced in the interiors of multiple generations of stars.

CHAPTER FIFTEEN

The Origin of Planets

15.1 Basic Properties of the Planetary System

Context of the Planetary System

In this book we have discussed more than fifty worlds, some in considerable detail. These planets, satellites, asteroids, and comets display an incredible diversity of composition and history. Yet they were all presumably formed at about the same time, condensing from the primordial solar nebula that also gave birth to the Sun. In spite of their individual diversity, these bodies carry many clues concerning their origins.

Lurking in the background is the question of the likelihood that there are other, perhaps similar, planetary systems around other stars. If we can understand the processes that formed the planets we know, we can then try to predict how probable it is that these same processes produced other planets, including some like ours. Similarly, if we could find planets orbiting other stars, their existence might help us to understand the origins and evolution of our own system.

Having stated what we would like to do, we must admit right away that it is not yet possible to do it. We are unable to work backward from the wealth of data on the present state of the solar system to derive a unique, detailed picture of how the system began. Neither can we work for- ward from a theory of star formation to the production of a solar system with all the properties we find today. Even the first and most important step, the formation of the Sun itself, is only poorly understood. Instead of a unique and all encompassing theory, we must work with a collection of reasonable explanations for those properties of the solar system that seem especially basic.

This somewhat unsatisfactory state of affairs may change dramatically within the next decade or two. A number of techniques are being developed that might enable a planet as large as Jupiter to be detected if it orbited one of the nearer stars. Space-based telescopes launched before the end of this century should have the capability to survey hundreds of nearby stars for planets substantially smaller than Jupiter. At the same time, infrared and radio astronomy are constantly revealing more about the formation and early evolution of stars. The direct detection of other planetary systems, together with a deeper understanding of the star-formation process, may provide a much sharper perspective on these problems than is possible with our present limited information.

Fundamentals

Much of this book has been devoted to reviewing the known properties of the individual planets, satellites, rings, and smaller bodies that make up the planetary system. No theory of planetary formation can deal with this wealth of detail. To begin, therefore, we should try to identify the really basic properties of the system that may be understandable in terms of its origin. We have made a list of 12 of these properties in Table 15.1. In the subsequent discussion, we shall refer to them as Fact 1, Fact 2, etc.

These facts will be used in the following sections to constrain theories of the origin and early evolution of the planetary system. Of course, the concept that emerges must be consistent with other data as well, particularly the detailed chemical and isotopic composition of the planets

and the primitive bodies of the system, such as the comets and meteorites. In addition, any theory of origins will make use of the increasingly detailed astronomical information on star formation that astronomers are developing.

◆

15.2 The Life of a Star
The Ages of the Sun and Stars

At the beginning of the twentieth century, many astronomers thought that the planets were formed as the result of a remarkable accident,

---

TABLE 15.1 Facts that any theory of origins should explain

1. The most ancient age recorded in the solar system is just over 4.5 billion years, even though the galaxy that contains the solar system is much older than this.

2. All planets move around the Sun in the same direction that the Sun rotates and nearly in the plane that passes through the Sun's equator.

3. Although the Sun has 99.9% of the mass in the solar system, the planets have 99.7% of the system's angular momentum.

4. The inner planets are both smaller and more dense than the outer planets, being composed primarily of the cosmically rare silicates and metals. The giant planets, in contrast, have a more nearly cosmic (or solar) composition, and their satellites are rich in water ice and other volatiles.

5. The asteroids, which represent a composition intermediate between the metal-rich inner planets and the volatile-rich outer solar system, are located primarily between the orbits of Mars and Jupiter.

6. The primitive meteorites are composed of compounds representative of solid grains that are expected to have formed in a cooling gas cloud of cosmic (solar) abundance at temperatures of a few hundred K.

7. Comets, like the surfaces of some outer-planet satellites, appear to be composed primarily of water ice, with significant quantities of trapped or frozen gases like carbon dioxide, plus silicate dust and dark carbonaceous material.

8. Volatile compounds (such as water) must have reached the inner planets in spite of the fact that the bulk composition of these bodies suggests formation at temperatures too high for volatiles to form solid grains.

9. Despite the general regularity of planetary revolution and rotation, Venus, Uranus, and Pluto all rotate in a retrograde direction.

10. All of the giant planets except Neptune have systems of regular satellites orbiting in their equatorial planes, rather like miniature versions of the solar system itself.

11. The giant planets Jupiter, Saturn, and Neptune have one or more irregular satellites — either in retrograde orbits or with very high eccentricities.

12. The Galilean satellites of Jupiter exhibit a decrease in density with increasing distance from Jupiter that mimics the decrease in planetary densities with increasing distance from the Sun.

such as the near-collision of the Sun with another star. Since then, we have come to realize that stars naturally form in a cloud of dust and gas, what we have called the solar nebula in the case of our own system. Further, a variety of kinds of matter have been located orbiting nearby stars, and while none of these has yet turned out to be another planetary system like our own, their existence strengthens the idea that stars do not form alone. Finally, we now know that the ages of the Sun and of the planetary system are approximately the same. As we have seen, the Moon, the meteorites, and the Earth all formed 4.5 billion years ago. Astrophysicists who study stellar evolution give this same value for the age of the Sun. From all of these arguments, we conclude that the Sun and the planets probably formed together from a common source of material (Fact 1).

Of course, not all stars have the same age. The galaxy is at least 12 billion years old, and many of the stars have been present since its formation. Stars that formed early in the life of the galaxy contain much smaller quantities of the heavier elements, and for this reason they may be less likely to have formed planets. The very existence of the building blocks of our planetary system depends on the presence of heavy elements formed in previous generations of stars and ejected back into interstellar space before our own system formed.

On a galactic time scale, the Sun is a relative newcomer. We are not among the latest arrivals, however. Stars that are more massive than the Sun have much shorter lifetimes. Their internal temperatures are hotter and their nuclear fires burn with greater intensity. The bright blue-white stars that dazzle us at night are all younger than the Sun. Some of them have ages of only a few millions of years instead of billions. The most brilliant and massive of these stars are destined to explode as supernovas, generating a very special group of elements that can only be created in the unique conditions that briefly occur during these cosmic cataclysms.

Stellar Nurseries

Even younger stars exist, since they are being formed today. Star formation takes place in clouds of interstellar gas and dust such as those in the constellation of Orion (Fig. 15.1). A typical interstellar cloud in which star formation is occurring has a mass hundreds of times greater than that of the Sun. It is composed primarily of hydrogen and helium, the predominant elements in the stars that it will spawn. The other elements that can be studied appear to be present in roughly the same proportions as they are found in the Sun and other young stars, exactly as one would expect (Table 2.1).

What may seem surprising, however, is the richness of the molecular chemistry that takes place in these clouds. Instead of just simple compounds like methane and ammonia, a large array of molecular species is being formed. A list of those known at the time of this writing would include more than sixty entries. Among the more interesting interstellar compounds, we call attention to ethyl alcohol (CH_3OH), the essential ingredient of liquor, beer, and wine; formaldehyde (CH_2O), used for preserving corpses; and hydrogen cyanide (HCN), a deadly poison that is also a vital compound in experiments designed to simulate chemical evolution on the primitive Earth. New molecules are constantly being discovered as astronomers use more sensitive radio telescopes and study new segments of the radio spectrum.

It is interesting to compare a list of interstellar molecules with the molecules found in comets (Table 4.2). There are many species in common: H_2O, CO_2, HCN, OH, etc. As we noted in Chapter 4, some scientists think that the icy nuclei of comets contain unaltered interstellar material, trapped during the earliest stages of the formation of the solar system. Alternatively, it may be that the similarities simply reflect the universality of the processes that produce these compounds, whether they take place in the solar nebula or in interstellar clouds.

A Star Is Born

One of the first things to notice about stars is that most of them are members of multiple systems. Doubles and triples are more common than singles, but it seems unlikely on dynamical grounds that such multiple star systems will have many planets. Depending on the masses and distances of the stars, however, there may be regions around one or more of the stellar components where planetary orbits could be stable. Since we have neither the observations nor a theory for such systems, we will concentrate our attention on the much simpler case of single stars, like the Sun.

We start with a slowly rotating cloud of interstellar gas and dust that may itself be part of a much larger complex such as one of the giant clouds in Orion. At some point the cloud begins to collapse. Perhaps some gravitational instability has been created in its interior, by a random coming together of some of the material, or a nearby star has exploded as a supernova, perhaps seeding the cloud with short-lived radioactive elements and sending out shock waves that begin to compress the cloud. The collapse is possible as long as the energy of motion (internal pressure) of the gas in the cloud is less than the gravitational energy represented by the mass of the cloud and the distance through which it collapses.

As the cloud becomes smaller, three things happen: (1) its rate of rotation increases; (2) it flattens into a disk; and (3) it heats up, especially near the center. The heating is simply the conversion of gravitational energy to thermal energy (like the source of internal energy for Jupiter). The increase in rotation rate results from conservation of angular momentum: as the mass of the cloud comes closer to the center, the angular velocity must increase to keep the momentum con-

FIGURE 15.1 The Horsehead Nebula in Orion, as photographed with the 4-m telescope of the Kitt Peak National Observatory. Deep within the clouds of gas and dust that lie behind this obscuring veil, new stars are forming now, some perhaps with planets (cp. Fig. 1.1).

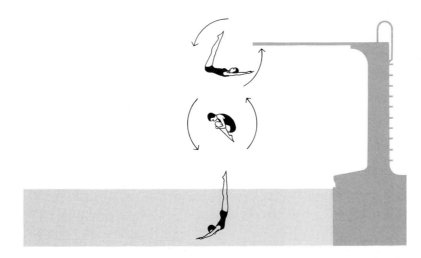

FIGURE 15.2 A high-diver who contracts her body in the direction perpendicular to her spin axis will increase her rate of spin. Her angular momentum remains constant as she falls, and the angular momentum depends on the product of her rate of spin and the square of her size in the direction perpendicular to her axis of spin.

stant (Fig. 15.2). The more rapid spin in turn causes the material to flatten into a disk.

In the disk, the gas and dust can radiate energy to space much more easily than in the center, where a spherical condensation develops. Here the temperature continues to rise until it finally reaches the point where nuclear fusion of hydrogen to helium can occur. At this stage, the Sun turns on and this spherical assemblage of matter begins its life as a star.

The Mass and Dimensions of the Disk

There is still a great deal of dispute about the mass and the dimensions of the original solar nebula and of the disk itself. The central condensation must have had a mass approximately equal to that of the present Sun, but what about the disk?

We can gain an idea of the minimum amount of mass that must have been present if we simply ask how much material of cosmic composition would be required to make all of the present planets. The idea behind this calculation is that the solar nebula started with cosmic composition and then the individual planets formed

from it with compositions reflecting the local temperature. If it had been cooler close to the Sun, massive planets like Jupiter and Saturn might have formed there.

To discover what these hypothetical planets would have been like, we can perform the thought experiment of adding hydrogen and helium to the existing planets until the ratio of these light elements to a key heavy element like silicon or iron is the same as it is in the Sun. The masses of the resulting planets are indeed similar to those of Jupiter and Saturn. Thus we might conclude that the initial disk must have had a mass roughly equal to at least ten times the present mass of Jupiter, or about 1% of the solar mass. In this case, the minimum mass for the entire nebula is about 1.01 solar masses.

More likely, planet formation will not be efficient, and there was probably much more material available that has since been lost by the blowing away of the nebula or by gravitational ejection of larger bodies after the planets formed. Hence these days scientists generally adopt a value for the entire nebula of about 1.1 to 1.2 times the mass of the Sun. Recent observations of similar disks of matter around young stellar objects indicate masses of this magnitude. However, some theories hold out for much higher masses, on the order of twice the solar mass, which simply demonstrates how much more work needs to be done in this field to achieve some real certainty.

15.3 The Solar Nebula Revisited

Dynamics of the Disk

We now have a newly born star at the center of a disk of gas and dust. This configuration can still be thought of as the primordial solar nebula, despite the fact that deep in the interior of the central condensation, nuclear reactions are beginning to convert hydrogen to helium.

The disk is revolving around this central condensation in the same sense that the condensation itself is rotating on its axis. This provides a ready explanation for Fact 2: all the planets revolve around the Sun in the same sense in which the Sun rotates, and they move approximately in the Sun's equatorial plane. Planets forming out of this disk will have exactly that configuration.

It was considerations such as these that led Immanuel Kant and Pierre Simon Laplace to propose (independently) nebular theories for the origin of the solar system as early as the eighteenth century. The main reason these theories were challenged was the attention given to Fact 3: how could the Sun slow down by transferring angular momentum to the disk and hence the planets?

It is enlightening to put this problem in context. It turns out that all stars with masses 15% or more greater than that of the Sun rotate much more rapidly than our star, whereas stars with comparable or smaller masses exhibit low rotation rates. Furthermore, if one calculates the rotation rate the Sun should have, given conservation of angular momentum as the original cloud collapsed, this rate turns out to be similar to that of the more massive stars. It is also the rate that would result if the present angular mo-

mentum of the planets were put back into the Sun. We conclude that the planets have the correct angular momentum for the system as it formed; some later process has acted to slow down the Sun.

Magnetic Braking

One solution to this problem invokes **magnetic braking.** Here one imagines the magnetic field of the Sun moving through the disk of material around it. The material in the disk is following orbits defined by Kepler's laws. Thus it is orbiting the Sun more slowly than the Sun is rotating. The situation is analogous to the interaction of the material in the Io plasma torus with the rapidly spinning magnetic field of Jupiter. The material in the inner part of the disk is ionized. The Sun's magnetic field encounters stiff resistance from this plasma as the Sun rotates, slowing down the spin. When the material in the disk is later dissipated, it takes the excess angular momentum with it.

This idea has several problems. It does not explain why only stars with low masses exhibit slow rotation, since it is well known that some stars with large masses have strong magnetic fields, far stronger than the Sun's. A solution to this dilemma would be to postulate that only low-mass stars form with disks, but we now know that disks are not uncommon about young stars with masses greater than the Sun's. Finally, magnetic braking would transfer angular momentum to the disk and hence to the planets. Yet as we have seen, the problem is not that the planets have too much momentum, but only that the Sun has too little.

The Solar Wind: Blowing the Problem Away

A second solution involves the solar wind. Recall that the gases in the outer fringes of the solar atmosphere have enough energy to escape into space, flowing steadily outward through the solar system at speeds of about 400 km/s. The amount of matter lost in this way is a tiny fraction of the Sun's total mass. Yet it is carrying angular momentum with it, which is lost by the Sun. Furthermore, observations of very young stars with small masses like the Sun's indicate that they generate particularly intense stellar winds shortly after they form. While it is difficult to make an accurate estimate of this effect, it appears sufficient to account for the present slow rotation of the Sun.

An especially appealing aspect of this theory of solar wind braking is its natural explanation of the mass-dependence of stellar rotation. It turns out that only stars with masses comparable to or less than that of the Sun have the proper atmospheric structures to produce steady stellar winds. Hence the same strong wind that ultimately clears residual gas and dust from the disk can slow down the rapidly rotating star that generates it. More massive stars will not produce such winds, so they will continue to exhibit rapid rotation.

15.4 Evolution of the Disk: Condensation, Aggregation, and Accretion

Early History

Temperatures in the early disk were not uniform. The initial condensation would generate heat as the gravitational energy of the extended cloud was converted to thermal energy. Local sources of heat from short-lived radioactive elements and electromagnetic discharges were also available. The collapsing cloud contained so much mass that the energy was sufficient to raise the internal temperature of the central condensation above the ignition point for thermonuclear reactions. While this was occurring, the surrounding nebula was heating up as well.

Temperatures above 2000 K were reached in the disk near the Sun, according to present estimates. This means that the dust in this region was vaporized and the nebula was totally gaseous here. Subsequent cooling of this hot gas allowed the condensation of molecules and the formation of grains again, but with the minerals sorted according to distance from the Sun (Fig. 15.3). Calculations indicate that the collapse of the cloud to a disk took 10 million years. At this point, the young Sun provided a source of energy to keep the temperatures high in the inner parts of the disk, inhibiting the condensation of most volatiles in the region of the inner planets.

This scenario provides an easy explanation for Facts 4 through 7. The inner planets are deficient in light elements compared with the cosmic distribution. Our model suggests that the inner planets formed at relatively high temperatures, at which abundant compounds of light elements (H_2O, CO_2, etc.) could not condense. The increasing fraction of volatiles in the solid objects we find today as we move outward through the asteroids to the outer solar system, and the comets, reflects the temperature gradient in the original solar nebula. Note that the gradient in composition extends only as far as Saturn, however. Judging by their densities, the large satellites of Uranus and the planet Pluto all

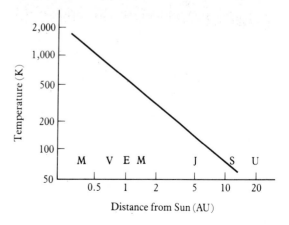

FIGURE 15.3 This graph illustrates current ideas about the decrease in temperature with increasing distance from the Sun. The positions at which the various planets formed are marked.

contain a smaller fraction of volatiles than the Saturn satellites. This modest reversal of the trend toward more volatiles at greater distances was only discovered in 1986 and thus far remains unexplained.

Formation of solid bodies in the disk did not proceed as uniformly as this picture would suggest, however. There must have been some radial mixing, in which material from the outer part of the disk (large orbital radius) reached the inner part (small orbital radius) because it was moving in eccentric orbits. This allowed material formed under one set of pressure and temperature conditions to be mixed with material from another environment. A prime example is the acquisition of water by Venus, even though this planet formed in a part of the nebula too hot for the condensation of water or ice from the vapor state. This radial mixing explains Fact 8: despite the temperature gradient in the solar nebula, some volatile constituents found their way to the surfaces of the inner planets.

The early vaporization of the interstellar grains in the inner part of the nebula helps us to understand the age limit in Fact 1: despite the comparatively young age of the Sun, nothing older than 4.5 billion years has been found in the solar system. In other words, we have not yet detected interstellar material in sufficient quantity to determine its age, although there are indications of the presence of interstellar carbon in some meteorites. As mentioned above, interstellar grains may also be present in comets, which formed and remained at low temperatures in the outer nebula. If we could obtain large enough samples from comets for study in the laboratory, we might find ages older than the 4.5-billion-year limit.

Condensation in the Solar Nebula

It is possible to calculate what compounds and minerals would form from this cooling cloud by investigating various possible chemical reactions among the elements that were present. Such calculations indicate that there is a definite temperature sequence in which various compounds form.

As the hot gas cools, one of the first things to condense is spinel, a magnesium-aluminum compound with the formula $MgAl_2O_3$. This is the same substance found in the white inclusions of the Allende meteorite. Further cooling leads to the condensation of a wide variety of silicates, which must have appeared as dust grains in the inner nebula. Since silicon and oxygen are both abundant elements, silicates should have become the dominant minerals in the inner disk. This is exactly what the composition of the inner planets suggests.

The inner boundary for the condensation of ice in the solar nebula was located at a distance of about 4 AU from the Sun. This is a major threshold, since in a cosmic mixture ice is potentially far more abundant than rock, even though the reverse is true on Earth and the inner planets. Recall that oxygen is approximately eighteen times as abundant as silicon in the Sun and stars. This means that in the solar nebula, H_2O should be approximately nine times as abundant as SiO_2. After forming all possible oxides and silicates, there is still plenty of oxygen left to make ice. As we saw in Chapter 11, crossing the threshold for the formation of ice enables the growth of giant planets. But how are planets formed from the tiny grains of ice and dust that condense in the cooling nebula?

Aggregation of Grains

The next step toward making planets after condensing grains from the gases in the solar nebula is for these grains to accumulate into larger aggregates. This process is aided by the fact that at any given point in the nebula, most of the material should be in nearly identical orbits. There may be some mavericks moving along highly eccentric ellipses, but in general, the relative velocities between nearby grains will be small. This enhances the chances that two colliding grains will be able to stick together, although it tends to reduce the total number of impacts. Sticking is especially likely if the grains are fluffy.

The evidence we have for the appearance of grains in the solar nebula suggests that they indeed had this desirable fluffy characteristic. Fig. 4.25 shows some of the grains, probably of cometary origin, collected in high-altitude flights by a project designed to detect interplanetary particles. These should be the most unmodified grains remaining in the solar system, since they have never been heated.

These fluffy particles form clumps that will gradually grow through additional collisions. While this is occurring, the solid material continues to settle toward the mid-plane of the nebula, the plane that cuts through the middle of the entire assemblage.

Formation of Planetesimals

Gravitational instabilities in the disk foster close encounters among the fluffy clumps of grains that are steadily growing through collisions. Some of these become large enough to attract other grains and other clumps gravitationally. At this stage, bodies as large as a few kilometers are forming, and we can speak of them as **planetesimals.** When diameters exceed values near 100 kilometers, random collisions again dominate over these instabilities, and the probability of a destructive impact increases.

The idea that the planets formed from intermediate-sized bodies called planetesimals was originally the work of Soviet theorist V.A. Safronov, who worked on this problem in the 1950s and 1960s. Previously, many scientists had thought that the formation of planets probably had taken place directly from the solar nebula and its tiny dust grains. We now recognize in the asteroids and comets surviving examples of planetesimals, and increasingly sophisticated calculations support Safronov's hypothesis.

It is possible to estimate the length of time required for the various stages in the formation of planetesimals. The theoretical models suggest that it may have taken as long as 100,000 years for 1-μm particles to reach the mid-plane of the disk, whereas 1-cm particles would settle out in only ten years. The initial loose condensations of dust should evolve into solid bodies with diameters of about 10 km in roughly 1000 years.

Confirmation of these amazingly short time scales is provided by investigations of the abundances of isotopes formed from radioactive parents with very short half-lives. Such studies indicate that solid material formed from the solar nebula within a few million years. Most of the meteorites that scientists have examined were formed from these grains within the first 20 million years after the nebula collapsed to a disk. The Earth and the other solid planets were essentially complete in 100 million years.

From Planetesimals to Planets

The inner planets and the cores of the giant outer planets formed from the continued accretion of planetesimals. As these proto-planets grew, their gravitational reach extended farther and farther, and the violence of the impacts increased as planetesimals struck them with higher velocities. In view of the rapidity with which the proto-planets grew, it seems inevitable that they were heated by the energy of these collisions once again above the melting point of silicates. Thus the planets probably differentiated as they formed, and no primitive material survived the violence of the accretionary process.

A number of calculations have recently been made of possible scenarios for the buildup of the inner planets. Some of these models suggest that a great many objects of lunar or larger size were formed, in addition to the growing proto-planets themselves. According to this picture, the late stages of the accretionary process were characterized by impacts of incredible violence as these lunar-sized objects impacted the planets (or, in some cases, were gravitationally ejected from the solar system). The possibility of a few discrete, random impacts of this magnitude can explain some of the peculiarities of the planets: in particular, such catastrophic events are probably implicated in the retrograde rotation of Venus (Fact 9) and the formation of Earth's Moon.

Dissipation of the Nebula

At some point after the planets formed, a relatively short-lived but highly intense blast from the solar wind cleaned out the remaining gas and small dust. Observations of spectra of young stars elsewhere in the galaxy have revealed a similar early active phase. The gases that comprise the major mass in the inner nebula are driven out, leaving behind the large cinders we know as the terrestrial planets (Fig. 15.4).

◆

15.5 Small Bodies and the Early History of the Earth

Debris of the Formation Process

As we stressed in Chapters 3 and 4, the small bodies in the solar system represent some of the most primitive material still in existence. They are remnants of the events we have just described, not having been accreted by forming planets. The asteroids evidently represent a case of planetesimal growth in the region between Mars and Jupiter that was halted at an early stage, when planetesimals had only grown to a maximum size of 100–1000 km. The gravitational effect of nearby Jupiter removed most of the planet-forming material from the region of

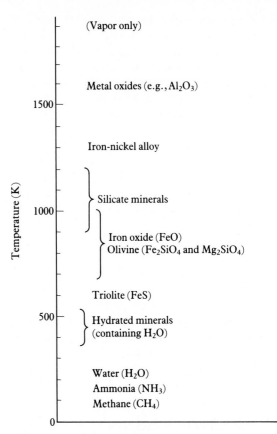

FIGURE 15.4 The chemical condensation sequence in the solar nebula, showing the primary chemical species that would be expected to form in a cooling gas cloud of cosmic (solar) composition. As the temperature drops, each material forms and condenses into liquid droplets or solid grains. The calculations upon which this diagram is based, originally carried out by John Lewis of the University of Arizona, presume that the grains remain in chemical equilibrium with the remaining gas. A slightly different condensation sequence results if the grains are immediately swept up into planetesimals and taken out of contact with the gas, but the general results are the same. These solid grains then become the building blocks of the planetesimals, and ultimately of the planets themselves.

the asteroid belt before it could accumulate into a full-sized planet (Fact 5). Perhaps the gravitational influence of Jupiter also was responsible for the circumstance that Mars is so much smaller than Earth and Venus.

We know from our detailed study of the impact history of the lunar surface that all of the inner planets must have continued to accumulate mass for the next 600 million years. Although this late bombardment effectively created the surfaces we see today, it only contributed a few percent to the total mass of an individual planet.

The First Billion Years

Approximately one billion years passed between the formation of the Earth and the production of the ancient fossil stromatolites depicted in Fig. 7.16b. It was during this interval that the chemical evolution on the Earth led to the origin of life, and life evolved to the level of sophistication that allowed such fossils to be produced. Similar processes could have begun on Mars.

The lunar cratering record suggests to many geologists that there was a discrete burst of impacts — the late heavy bombardment — about 600 million years after the formation of the Moon. Where did these impacting bodies come from, at a time when the inner solar system had been largely cleared of debris left over from the accretionary process? Many scientists think they may have originated near the orbits of Uranus and Neptune.

Because the distances between the outer planets are so great, it required a much longer time to clear the debris from this region. Quite possibly, the late heavy bombardment consisted largely of icy bodies gravitationally scattered inward by Uranus and Neptune. Perhaps this was the time when most of the volatiles that make up the terrestrial planet atmospheres arrived (Fact 8). At the same time, an even larger number of such objects should have been scattered outward, where they may have become the comets of the Oort comet cloud.

In addition to contributing volatiles to the atmospheres of the terrestrial planets, the cometary ices would also have been added to the pri-

mordial soup of chemicals accumulating in the oceans of Earth, ultimately to become part of organisms as life began. You may be pleased to realize that some of the carbon, nitrogen, and oxygen atoms in your cells were formed in stellar interiors, shot into interstellar space by supernova explosions, trapped in the ices of comets, and delivered to the surface of the primitive Earth. Some of the water you drank today may also have come from comets, but the alcohol in your beer was produced by local yeast, not by an interstellar brewery!

In the outer solar system, encounters of planetesimals with the giant planets led to capture for some of them (such as Saturn's Phoebe), presumably as a result of frictional drag when these bodies penetrated the extended atmospheres (or nebular disks) of the forming planets. This viscous drag would allow free-roaming planetesimals to lose energy and enter orbits about the planets, providing a natural explanation for Fact 10: all the outer planets except Uranus have one or more satellites in retrograde orbits. Satellites forming with the planets would not assume such orbits, but captured satellites could.

Formation of the Comets

A large percentage of the icy bodies that were not incorporated into the giant planets were forced into orbits with very high eccentricities that took them out to 50,000 AU from the Sun to form the Oort cloud of comets that we discussed in Chapter 4. Alternatively, some of the comets could have formed directly in the outer reaches of the solar nebula, but most scientists do not think this is likely.

The "new" comets we see are coming from this distant region, where the gravitational perturbations of passing stars occasionally disturb their orbits and bring them close enough to the Sun for us to see. A much larger number of icy bodies probably exists undiscovered between the orbit of Pluto and the Oort cloud at 50,000 AU, but these objects are bound to the Sun more tightly, and stellar perturbations are thus unable to deflect them inward and make them detectable. This unobserved region is called the inner Oort cloud.

As we write this, there is heady speculation by some scientists that clusters of comets may be broken loose from this inner comet cloud periodically, perhaps every 25–30 million years. Collisions of these comets with the Earth are held responsible for the apparently periodic mass extinctions of flora and fauna on Earth. The cause of the apparent periodicity is thought by some to be an undiscovered tiny star that is gravitationally bound to the Sun, making ours a double system after all. Alternatively, the agent that releases these comet showers might be the passage of the solar system through the plane of our galaxy, something that happens about every 30 million years. See Goldsmith's book *Nemesis* (in the additional reading list) for a fascinating discussion of these ideas.

Arguments about the reality of the periodicity claimed for the mass extinctions on Earth still persist. If the impacts are not clustered in time, but rather come at random intervals, there is no need to seek an exotic process to disturb the comets of the inner comet cloud. At this writing, the jury is out on this interesting question.

15.6 The Giant Planets and Their Satellite and Ring Systems

Growth of the Giant Planets

From this discussion of the formation of the solar system, it seems natural to expect that the giant planets with their systems of satellites and rings must have formed in a similar manner to the Sun and its retinue of planets. In the outer reaches of the solar nebula, enough matter was present to form giant proto-planets surrounded by disks from which the rings and satellites developed.

As we saw in Chapters 11 and 12, the giant planets consist of cores of rocky and icy material surrounded by huge envelopes of gas. All four of these planets have cores amounting to 10 to 15 Earth masses. These massive cores are apparently the key to the further attraction of gas from the solar nebula that built up the hydrogen and helium atmospheres observed today.

When the rock and ice cores of the outer planets achieved their maximum masses, they caused the surrounding gas in the nebula to collapse, simply from the strength of their gravitational attractions. Thus the cores are now surrounded by atmospheres that are a mixture from two sources: the outgassed atmosphere produced by the forming core and a primordial mix of gases collected from the solar nebula.

It is this mixture that explains the inference that all four giant planets started with atmospheres with a solar ratio of hydrogen to helium, while methane is enriched to various degrees. The hydrogen and helium were contributed by the solar nebula, while the extra methane was produced by the growing core as one of the components of its outgassed atmosphere. On Saturn, we find a slightly greater enrichment of methane than on Jupiter, while on Uranus and Neptune a much larger enrichment exists. Presumably this very large enrichment (by factors of 10 to 30 over the solar abundances) reflects the fact that these planets were able to capture much less of the primordial hydrogen and helium from the nebula. The increase in the ratio of methane to hydrogen results from less hydrogen, not more methane.

Formation of Satellite Systems

Just as a disk of gas was left around the forming Sun, the matter that formed the giant planets produced both central condensations and disks. The disks shared the sense of rotation of the forming planets, so satellites forming in these disks orbited their planets in the same direction that the planets rotated. These became the regular satellites that we find today (Fact 10).

Within the proto-planetary disk, one could expect to find a similar temperature and pressure gradient to the one we just discussed for the solar nebula. Near the planet, the density of gas would be greater and the temperature higher than at large distances. Obviously these proto-planetary disks never got as hot as the solar nebula near the Sun. Nevertheless, this change in physical properties with distance from the planet should leave some signature on the forming satellites.

In the case of Jupiter, the hottest of these planets because it was the largest, we can see the effects of this heat on the forming satellites when we examine them today (Fact 12). Recall that at Jupiter's distance from the Sun, H_2O ice is stable, even over the lifetime of the solar system. Thus we expect to find satellites that reflect the cosmic proportions of the elements, roughly half ice and half rock by mass. This is indeed the case with the two outer satellites, Callisto and Ganymede.

Io and Europa, close to Jupiter, have higher densities (3.0 to 3.6 g/cm^3), exactly as this scenario would predict. The corresponding proportion of water is no more than 10% on Europa, while Io has been baked dry by tidal heating and volcanic activity. This low proportion of water on Europa can be understood in terms of higher nebular temperatures than those in the regions where Callisto and Ganymede formed.

The satellite systems of Saturn and Uranus do not show a similar variation of satellite density with distance from the planet. All of their satellites apparently have approximately the half ice, half rock composition of Callisto and Ganymede. Thus it appears that these planets did not generate enough heat to inhibit the condensation of ice in their miniature solar nebulas. Nevertheless, there may be variations in the other volatiles trapped in these ices. For example, such variations provide the most probable explanation for the absences of methane-nitrogen atmospheres on Ganymede and Callisto, and the presence of such an atmosphere on Titan, which formed in a colder region.

Formation of Rings

Who do some planets have rings while others don't? What determines the kinds of rings a given planet may have? These are questions that are under active investigation at the present time. In Chapter 14 we discussed several scenarios for the origin and evolution of ring systems.

The general picture of rings as debris that did not accrete into large bodies provides a possible explanation for the existence of the rings of Saturn and Uranus. However, the close association of these two ring systems with a number of inner satellites has led Eugene Shoemaker and others to argue in favor of the breakup of one or more satellites as a more likely cause. Since it is clear that the bombardment of inner satellites by infalling debris must have been very heavy in the first few hundred million years of planetary history, it seems inevitable that satellites close to the planet, if they existed, were disrupted by large impacts. The debris of such events, if it were inside the tidal stability limit, could not have pulled itself together gravitationally to reform the satellite.

If these ideas are correct, the rings did not form directly from a nebula surrounding the giant planets. Instead, the ring systems are the natural result of the formation of satellites, combined with the high intensity of impacts produced close to the planet by its massive gravitational field.

15.7 The Search for Other Solar Systems

The Comparative Approach

One of the most powerful methods available to scientists is that of comparison. By examining different manifestations of the same kind of phenomenon, it is possible to derive the general laws and principles that are at work. For example, finding that craters on the Moon obeyed the same diameter/depth ratio as bomb craters on the Earth, scientists were able to deduce that explosive impacts rather than volcanism were responsible for these features of the lunar landscape. Newton discovered the law of gravity by noticing the effects of this force on the motions of apples and the Moon. The comparative approach to planetary science has been a central theme of this book.

It would be wonderful if we could apply this same approach in our efforts to understand the origin of the solar system. Many scientists have concluded that we will never be able to gain a full understanding of our own system without having some additional examples. We would begin by examining several different solar systems, to see what common properties they exhibited. These properties would help us determine which aspects of our own system are truly fundamental, requiring a common theory for their explanation, and which are random, the result of chance events. Do all systems consist of small planets close to their star and massive planets farther out? Do the planets always revolve around the star in the same direction? Are comets and asteroids always left behind when solar systems form? There are many questions we would love to ask.

Unfortunately, we can't expect any answers in the immediate future. At the present time, ours is the only planetary system we know, despite the existence of billions upon billions of other stars in our galaxy. We do find a higher probability for stars to form in pairs or multiplets than alone. So we might speculate that when a single star is observed, there was not enough material in the collapsing cloud to form a stellar companion, and a planetary system formed instead. Nature's preference for forming systems of small bodies in orbit around more massive ones is well illustrated by our own system's giant planets and their satellites. On the other hand, the excess material may have been lost and the star may truly be alone.

These ideas are now only speculation. We need real data on other planetary systems if we are to make much progress with such ideas. Fortunately, just such data are beginning to become available, and the prospects for learning more about other planetary systems during the next few years are exceedingly bright.

Stellar Nebulas and Disks

Because individual planets are so small and faint, it turns out that the easiest observational searches for other solar systems are those that look for stellar nebulas, equivalent to the primordial solar nebula that gave birth to our own planetary system. The temperature of such a nebula — a few hundred K — makes it a strong source of infrared radiation. For the past decade, astronomers have been regularly measuring the cocoons of dust and gas that appear always to attend the formation of stars. Recently Steven Strom of the University of Massachusetts has identified disks of gas and dust around HL Tau and several other very young stars that appear to have the dimensions (100-AU radius) expected for a disk like the solar nebula.

The identification of dust disks analogous to the solar nebula is important, but it does not provide direct evidence of the processes of aggregation and accretion that we think led to planetesimals and eventually to the planets themselves in our own system. Hence the significance of the 1983 discovery by the IRAS infrared satellite of excess thermal radiation from a number of bright, relatively nearby stars, including the first-magnitude stars Vega and Fomalhaut. Analyses of these objects soon showed that the radiation was coming from disks or rings of orbiting solid particles. While the sizes of the emitting particles are not known, they are definitely larger than the dust associated with forming stars. The dimensions of these disks are a few hundred AU, and estimates of the amount of material present have ranged up to several times the mass of the Earth.

The next step was to image some of this newly discovered material. Bradford Smith of the University of Arizona (who is also the Voyager Imaging Team Leader; Fig. 13.4) and Richard Terrile of the Jet Propulsion Laboratory had previously designed a special viewing system to permit the faint inner satellites of Saturn to be imaged in spite of the scattered light from the much brighter planet and rings. They modified this camera and with it obtained the image shown in Fig. 15.5 of the star Beta Pic, one of those that IRAS had discovered to be emitting excess infrared radiation. What is clearly revealed is a disk of solid material, in orbit around the star, seen nearly edge-on.

The Beta Pic disk has a diameter of several hundred AU. Smith and Terrile have suggested from their images that the disk does not extend down to the star, however, but that the inner part is empty. Thus the feature is more properly a ring rather than a disk, although the location of the inner boundary of the ring is in dispute. Essentially, this disk occupies the position in the Beta Pic system of the inner comet cloud for our own system, possibly extending inward to the equivalent of the orbit of Jupiter.

FIGURE 15.5 A composite of a positive image of the star Beta Pictoris divided by a negative image of the star Alpha Pictoris. The stray light from the two stars is nearly cancelled out, leaving dark rings around the resulting composite image. Dark and bright star images from the two original pictures appear in the surrounding field. The disk of material surrounding Beta Pic is visible as a bright diagonal band, which actually stretches some 200 AU on either side of the star.

The Beta Pic disk and other similar features discovered by IRAS are not planetary systems. They may be analogous to our inner comet cloud, or to some other form of solid material left over from the formation of the star. Certainly there is no way to tell from these data whether there might also be planets present, but these discoveries have shown us that other single stars have formed with solid material in orbit around them. In this sense, at least, our Sun is not alone. The next step is to look for evidence of planets, a much more challenging problem.

Astrometric Searches for Planets

The most powerful technique attempts detection by looking for evidence of the gravitational effect of a planet on its parent star. This is commonly called the **astrometric method,** since it involves very precise measurements of stellar positions. (Astrometry is the term astronomers use for the measurement of positions of stars.)

The astrometric technique has been in use for many years, as a by-product of the measurement of stellar distances and motions. In this approach, the astronomer observes the motion of a star through space, watching for any deviation from a simple orbital path. Stars move around the center of the galaxy in elliptical orbits, just the way planets move around the Sun. If there is also a companion moving around the star, the star's path through space is wavy rather than straight, since the companion actually orbits the **center of mass** of the system (star plus companion) rather than the center of the star itself. The center of the star also moves around this center of mass with the same period as the companion. If the mass of the companion is a large enough fraction of the mass of the star, and if the orbit of the companion is not too small, a distant observer will be able to notice the star moving around the center of mass. It is this orbital motion that causes the star to exhibit a wavy path as it follows its apparent track through the sky (Fig. 15.6).

By examining the motions of nearby stars with very high precision over many years, it has been possible for astronomers to identify a few that do indeed follow slightly wavy paths. Unfortunately, none of the sub-stellar companions have masses as small as Jupiter. Rather, they appear to be more like small, faint stars.

The minimum mass for a star to sustain itself by nuclear reactions in its core is 0.07 times

FIGURE 15.6 If a star has one or more sufficiently massive planets around it, the star itself will move in a small orbit around the center of mass of this planetary system. The center of mass of the system will move along an elliptical orbit about the center of the galaxy, while the star (which is what we observe) will execute a periodic wiggle across this orbit as it pursues its path.

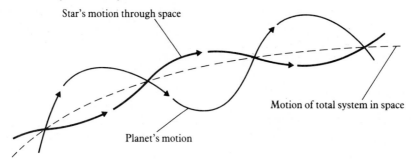

Star's motion through space

Motion of total system in space

Planet's motion

the mass of the Sun, or approximately 70 times greater than the mass of Jupiter. Thus there is a substantial range of masses — up to 70 times the mass of Jupiter — that corresponds to neither stars nor planets, as we are familiar with the terms. Such objects are called **infrared dwarfs** or **brown dwarfs.** Invisible companions in this mass range are the objects most likely to be discovered by ground-based astrometric searches, if such infrared dwarfs actually exist in substantial numbers in the galaxy. So far the evidence is mixed; a few such objects may be present, but they are not as abundant as many astronomers had expected.

Improved Searches

An object with the mass of Jupiter will not produce deviations in the motions of stars that are detectable with present equipment. This is a solvable problem, however, if a specially designed telescope could be built and operated on an ideal site. Designs for such an astrometric instrument indicate that it should be able to distinguish planets at least down to the mass of Jupiter if they orbit nearby stars. The search requires about a decade, however, since such planets are likely to have periods of at least several years.

A much more powerful search, but also proportionately more expensive, could be carried out with an astrometric instrument in Earth orbit. Above the blurring effects of the atmosphere, with its optical alignment continuously monitored and held constant, a rather small telescope with an aperture of just 1.5 m would be able to detect changes in star positions as small as the apparent width of a dime seen at the distance of the Moon!

With such a space instrument, it should be possible to detect planets the size of Uranus or Neptune in orbits as small as Jupiter's out to a distance of about 30 light years. In other words, any star within this limit would suffer a sufficient

perturbation by a Uranus-sized planet that we could detect the effect on the star's motion provided the planet were as close to the star as Jupiter is to the Sun. We choose a planet with the mass of Uranus because all four of the outer planets in our system have cores of about this mass. Since Jupiter and Saturn are actually much more massive than this, we are making a conservative assumption about the sizes of the planets we hope to detect. If they are actually the size of Jupiter or Saturn, it will be that much easier to discover them.

We have very little chance of finding planets like our own with this method, even if they were as close to their star as Earth is to the Sun. But the discovery of 10-Earth-mass planets (or larger) would be a great step forward. It would at least let us know how common planetary systems are among the hundreds of stars that are within the reach of this instrument.

Signals from Extraterrestrial Civilizations

An obvious way to accelerate the process is to search for a very special kind of planetary system: one in which *intelligent* life has evolved. If such a place exists, and if its inhabitants are transmitting high-powered radio signals in our general direction, then it is well within the capabilities of current radio telescopes to detect these signals. The discovery of such signals would not only demonstrate the existence of other planets, but would of course tell us a great deal more. If we could ever manage to decipher the signal, or even to establish a two-way communication, we could learn all we wanted about their system simply by asking.

The search for extraterrestrial intelligence (abbreviated **SETI**) is a matter of great interest that extends far beyond the realm of astronomy or planetary science. The first serious SETI program, aimed at surveying the entire sky with a radio receiver sensitive simultaneously to 8 million different frequency bands, has begun in the mid-80s, supported by a public-interest group called The Planetary Society. The president of The Planetary Society is Carl Sagan, long a proponent of the search for such signals, and major funding for the 8-million-channel receiver was provided by Steven Spielberg, maker of such science-fiction films as *E.T.*

The next step planned for SETI is a program to be funded by NASA that would use a 10- to 100-million-channel spectrum analyzer for both an all-sky survey and a targeted search. The survey will cover the entire northern and southern sky, with a sensitivity about 300 times greater than that of any current surveys. The targeted search will examine in greater detail more than 700 nearby stars that are similar to the Sun, those considered most likely to have planets. It would also respond immediately to any hints of signals discovered by the all-sky survey. Both

modes will cover thousands of new frequencies that have not been previously explored.

This program is estimated to require about ten years to complete. The search for signals will be about 100,000 times more extensive in its combination of sensitivity, number of targets surveyed, and frequencies covered than previous efforts. Nevertheless, the detection of extraterrestrial intelligence cannot be guaranteed. It may be that a much larger collecting area than any currently available antennas is needed for the detection of extraterrestrial signals (Fig. 15.7). We will not know until we try.

At the end of this program, we will have a new perspective on our place in the galaxy. How probable is it that there are other technically advanced civilizations living on planets that orbit other stars? In the year 2000 we will be able to give a much better answer than the one we could provide today.

FIGURE 15.7 If civilization on Earth decides to get really serious about SETI, it might consider B. Oliver's proposal, called Project Cyclops, to erect a giant array of 100-m telescopes similar to this one. Connected to one another and operated by a computer, this array would function as a single telescope with a sensitivity far greater than any existing instrument.

Summary

This final chapter reviews many ideas presented throughout the book, considering them here in the context of the formation and early evolution of the planetary system. Planetary scientists would like to consider the solar system as but one example of a more general phenomenon, but unfortunately at this writing there is no way to discover — or clearly to exclude — the existence of planetary systems accompanying other stars. Thus we must first look to our own system to see if we can generalize from the knowledge we have gained concerning its members. Table 15.1 summarizes some basic properties of our planetary system that are related to its origin.

There is a consensus among astronomers that the solar system was born about 4.5 billion years ago from a condensation in the interstellar medium called the solar nebula. Similar dense interstellar clouds of gas and dust are observed widely in the galaxy as the birthplaces of stars. One approach to understanding the origin of the solar system is to study these other, similar processes that are taking place around us today.

The solar nebula was the nursery of both the Sun and the planetary system. The Sun, forming in the center, attracted most of the mass and evolved into a self-sustaining star. The rest of the gas and dust collapsed into a disk, where it was heated by the proto-Sun. As part of this process, the proto-Sun lost most of its angular momentum, leaving most of the rotation in the spinning nebular disk.

As temperatures dropped in the disk, solid grains condensed, sorted compositionally by the temperature structure in the disk. These grains rather quickly aggregated into larger bodies, building up what are called planetesimals, according to the scenario originally proposed by Safronov. These planetesimals, tens to hundreds of kilometers in diameter, became the building blocks of the planets. The planets, growing by accretion and heated by impacts, differentiated at about the same time they formed.

For the first few hundred million years of solar system history, a great deal of solid debris remained between the planets, and impacts were much more frequent than today. The heavily cratered surfaces of the Moon and other solid bodies throughout the system date from this time. A small fraction of the leftover icy planetesimals from the outer solar system contributed volatiles to the atmospheres of the terrestrial planets, while others were gravitationally ejected to form the Oort comet cloud, which still remains a major source of the much rarer large impacts that occasionally take place on the planets today.

In the outer solar system, solid planetary cores formed that were large enough (10–15 Earth masses) to attract and hold part of the remaining gases of the solar nebula, thus creating the giant planets. These planets also formed their own miniature "solar nebulas," which gave rise to their regular satellite systems. The rings may have formed in the same way, but they could equally be the product of later impact fragmentation of inner satellites. Ultimately the remnants of the original nebula were blown away by strong solar winds from the young Sun.

This picture for the origin of our system is consistent with the basic properties in Table 15.1. It also suggests that the formation of other systems like our own should be a common aspect of the origin of single stars. In this case, the search for other planetary systems is especially compelling. Unfortunately, existing techniques fall just short of being able to detect other planets, although they have already revealed such interesting phenomena as disks of solid material orbiting some other stars. Probably a space-based astrometric telescope will be required for a definitive search for other planetary systems.

Finally, there is a question that extends far beyond the existence of other planetary systems.

Might there not also be other planets with life, even intelligent life, orbiting nearby stars? We do not know, but the search — SETI — has begun for radio signals from such other technical civilizations. The discovery of such signals, opening the possibility of establishing communication with extraterrestrial intelligent creatures, would be one of the most significant events in human history.

Key Terms

astrometric method

brown dwarf

center of mass

infrared dwarf

magnetic braking

planetesimals

SETI

EPILOGUE
REVIEW QUESTIONS

1. What is meant by the age of the solar system? Are all of the objects in the system the same age? How can we tell?

2. Review the steps in the formation and collapse of the solar nebula. Why did the nebula heat up? What is the evidence for a wide range of temperatures in the nebula? What events and processes caused parts of the gaseous nebula to condense into liquid droplets and solid grains?

3. Contrast the conditions in the solar nebula that led, on the one hand, to the formation of the inner rocky planets, while farther from the Sun the giant planets with their icy satellites formed. According to this scheme, where did the asteroids and comets form originally?

4. The concept of planetesimals was developed by theorists such as Safronov as an intermediate step between nebular grains and full-fledged planets. Is there any direct evidence for the existence of planetesimals? Do any of these objects survive in the planetary system today?

5. Summarize the evidence, pro and con, for the existence of other planetary systems around other stars. Should we be surprised that no such systems have yet been conclusively identified? What would it take to convince you one way or the other that other planetary systems do or do not exist?

6. Some people have argued that the Earth is being visited by aliens from other planetary systems. Others say that no such visits are taking place and argue that the absence of alien space-travelers provides strong evidence against the presence of other intelligent beings in our galaxy. How do you evaluate the logic of either of these positions?

7. There are proposals today to build special astrometric telescopes in Earth orbit to search for other planetary systems. There are also proposals to build sensitive multi-channel radio receivers with which to carry out a careful search (SETI) for radio signals from other civilizations. Both of these enterprises are clearly within our technological reach, and each one would cost less than the purchase of a single modern military aircraft. Should they be undertaken? Which is more likely to be successful? How would you assign priorities to the two projects?

ADDITIONAL READING

*Black, D.C. and M.S. Matthews, eds. 1985. *Protostars and Planets II*. Tucson: University of Arizona Press.

*Brahic, A., ed. 1982. *Formation of Planetary Systems*. Paris: Editions Cepadues.

*Dermott, S.F., ed. 1978. *The Origin of the Solar System*. New York: John Wiley and Sons.

Goldsmith, D., ed. 1980. *The Quest for Extraterrestrial Life: A Book of Readings*. Mill Valley, CA: University Science Books.

Goldsmith, D. and T. Owen. 1980. *The Search for Life in the Universe*. Menlo Park, CA: Benjamin/Cummings.

Lewis, J.S. 1982. "Putting It All Together." In *The New Solar System*, 2nd ed., ed. J.K. Beatty, B. O'Leary, and A. Chaikin. Cambridge, MA: Sky Publishing Corp.

McDonough, T.R. 1987. *The Search for Extraterrestrial Intelligence: Listening for Life in the Cosmos*. New York: John Wiley and Sons.

Quantitative Examples and Applications

---◇---

Introduction

The main text of this book concentrates on the most important ideas and concepts of planetary science without introducing mathematical formulations or exercises. For most students this non-quantitative approach is appropriate. But those who are comfortable with algebra and enjoy numbers may appreciate a brief description of some more mathematical aspects of this material. Indeed, these students will find many concepts easier to understand when they are expressed in algebraic formulas than when we are limited only to words. For their benefit, we include a number of simple mathematical examples and applications in this supplement, keyed to individual chapters and sections in the main text.

A.1 Kepler's Laws of Planetary Motion (Section 1.3)

Kepler's laws provide a fundamental description of planetary motion that can be generalized to any system of bodies of small mass in orbit about a large body (such as the satellite systems of the giant planets). While Newton and his successors developed many ways of improving the accuracy of orbital calculations by taking into account the gravitational influence of each planet upon the others (perturbations), the basic utility of Kepler's three laws remains.

Kepler's first law states that planetary orbits are ellipses with the Sun at one focus. The formula for an ellipse is

$$r = \frac{a(1 - e^2)}{1 + e \cos \theta}$$

where r is the distance from the Sun, a is the semi-major axis, e is the eccentricity, and θ is the angle that specifies the position of the planet along the ellipse, as seen from the Sun. Perihelion, the smallest value of r, corresponds to $\cos \theta = 1$ and is given by

$$r_{min}/a = (1 - e^2)/(1 + e) = 1 - e$$

The maximum value of r (aphelion) is similarly

$$r_{max}/a = 1 + e$$

Kepler's third law can be expressed as

$$a^3 = Kp^2$$

when a is in AU and the period, p, is in years. The constant K equals 1 AU^3/yr^2. Newton later showed that the masses of both the Sun and the planet should be taken into account, in which case the third law is written

$$a^3 = K(M + m)p^2$$

where M is the mass of the Sun and m is the mass

of the planet, both expressed in units of the combined mass of Sun and Earth, so that the term in parentheses still stays equal to 1.0 for the case of the Earth orbiting the Sun.

One application of the third law allows us to calculate the mass of Jupiter. Ganymede, Jupiter's largest satellite, can be observed to orbit the planet with a period of 7.16 days (1.96×10^{-2} years) with a semi-major axis of 7.15×10^{-3} AU. We thus have

$$M + m = a^3/p^2 = 9.5 \times 10^{-4}$$

Since the mass of Ganymede is small compared to that of Jupiter, we have that the mass of Jupiter is about $1/1000$ the mass of the Sun.

Now consider the orbit of Comet Halley, which has a period of 76 years and a mass that is negligible in comparison to the Sun (Fig. 1.20). The semi-major axis of its orbit is

$$a^3 = (76)^2 = 5776$$
$$a = 18 \text{ AU}$$

To find the perihelion distance of the comet from the Sun, we note that its orbital eccentricity is 0.97. Substituting for a and e in the formula for the perihelion distance, we have

$$r_{min} = 18 \times (1 - 0.97) = 0.54 \text{ AU}$$

This is near the orbit of Venus. The actual orbit of Comet Halley varies due to perturbations from Jupiter and the other planets.

EXERCISES

1. If the Sun were a star with only one-tenth its present mass, calculate the periods of revolution of the Earth and Jupiter, assuming they were in their present orbits.

2. A spacecraft on a trajectory to Saturn follows an ellipse with perihelion at the Earth's orbit (1 AU) and aphelion at Saturn's orbit (9 AU). If the semi-major axis of this transfer ellipse is halfway from Earth to Saturn (5 AU), what is the time required for the trip from the Earth to Saturn? Using similar reasoning, find the trip times to Mars and to Neptune.

A.2 The Escape of Planetary Atmospheres (Section 2.3)

A molecule or atom can escape from a planetary atmosphere when its energy is sufficient to overcome the gravitational attraction of the planet. Consider a molecule (or atom) of mass μ in the atmosphere of a planet with radius R, located sufficiently high that if it begins moving outward with sufficient speed, it will not encounter any other molecules. Its kinetic energy is given by the formula

$$E = \frac{1}{2}\mu v^2$$

where v is the speed of the molecule. The condition necessary for escape is that this energy be greater than the molecule's potential (or gravitational) energy, given by

$$E = GM\mu/R$$

where M is the mass of the planet and G is the universal constant of gravitation ($G = 6.7 \times 10^{-11}$ newton m^2 kg^{-2}).

If we equate the kinetic and gravitational energies, we can solve for the speed at which the molecule is just able to leave the planet, called the *escape velocity*, v_e:

$$v_e = \sqrt{2GM/R}$$

Note that the mass μ of the molecule is not a part of this equation, since it appears in the expressions for both kinetic and gravitational energy. Thus the escape velocity for a rocket, a baseball, or a molecule will be the same on the same planet. On the Earth, this escape velocity is 11 km/s.

In a planetary atmosphere, the speeds achieved by atoms and molecules depend on their masses and temperatures. From thermodynamics, we know that in a mixture of gases each species will have the same kinetic energy. If the energy is the same, then the less massive molecules in a mixture must be moving faster while the more massive ones must be moving more slowly. The average thermal velocity (v_t) of a molecule of mass μ is given, according to an expression derived by the nineteenth century Scottish physicist James Clerk Maxwell, by

$$v_t = \sqrt{3R^*T/\mu}$$

where T is the local temperature and R^* is the universal gas constant ($R^* = 8.31$ joule deg^{-1} mole^{-1}).

If a planet is to retain its atmosphere over the lifetime of the solar system, we require that the average thermal velocity of the molecules be considerably less than the escape velocity, or

$$v_t \ll v_e$$

Massive molecules at low temperatures have the best prospect of remaining bound to their planet, since their thermal velocities will be lowest. We can express this using the expressions already given for the thermal and escape velocities:

$$\sqrt{3R^*T/\mu} \ll \sqrt{2GM/R}$$

Since there is a distribution of molecular velocities about this average value, at any given temperature some of the faster molecules will still escape. Calculations that allow for the range of speeds present indicate that a gas will be retained over billions of years if it meets the condition

$$\sqrt{3R^*T/\mu} < 0.2\sqrt{2GM/R}$$

EXERCISES

1. If the average temperature in the atmosphere of the Earth is 260 K, find the mass μ of the lightest atom that can be retained over the age of the solar system. Repeat the calculation if the actual temperature of the upper atmosphere where escape takes place is 600 K.

2. Calculate the smallest molecular mass μ that would remain on Neptune's satellite Triton for 4.5 billion years, if its atmosphere has the same temperature as the surface. What gases could you suggest that would satisfy this constraint?

A.3 Radiation Laws (Section 2.4)

Everything in the universe that has a temperature above absolute zero (-273 C) emits radiation across the entire electromagnetic spectrum. This is called thermal radiation, to distinguish it from other more exotic processes that can produce non-thermal radiation, such as the output of a terrestrial radio transmitter or the radiation from the magnetosphere of Jupiter. One of the characteristics of pure thermal radiation is that its spectrum and intensity depend only on the temperature of the source, and are independent of the chemical composition or other properties of the emitting material. Thus physicists in the nineteenth century were able to deduce quite general formulas, called *radiation laws*, to describe this ideal thermal radiation, which is sometimes called *black-body radiation*. Real objects such as planets and stars are not perfect thermal radiators, but the basic radiation laws often describe their emission of energy reasonably well.

One radiation law, called the *Stefan-Boltzmann law* after its discoverers, describes the total emitted power from a perfect thermal radiator. As noted above, this total energy depends on temperature only. This temperature must be expressed relative to absolute zero, and thus we use the Kelvin scale for our calculations. The formula is

$$P = \sigma T^4$$

When P is expressed in units of power (watts), the constant of proportionality, called the Stefan-Boltzmann constant, is given by

$$\sigma = 5.670 \times 10^{-8} \text{ W m}^{-2} \text{ deg}^{-4}$$

As an example, we can calculate the relative powers emitted by one square meter on the Sun and by an equal area on the Earth. The solar surface has a temperature of 5700 K, while the so-called effective temperature of the Earth is about 260 K. Thus the ratios of emitted power are

$$\frac{P_{\text{Sun}}}{P_{\text{Earth}}} = \frac{(5700)^4}{(260)^4} = 2.31 \times 10^5$$

The higher the temperature of the source, the shorter the wavelength at which the maximum power is emitted. The *Wien radiation law* expresses this as

$$\lambda_{\text{max}} = 0.29/T$$

when λ_{max} is in centimeters and T is in degrees Kelvin. The solar spectrum has a maximum at 0.29/5700, or about 5×10^{-5} cm (5000 Å). This peak occurs right in the spectral interval to which our eyes are sensitive — visible light — a fact which is surely no coincidence. In contrast, the peak of the terrestrial thermal emission spectrum is at 0.29/260 = 11 µm, in the infrared. The peak for Pluto, with a temperature of about 55 K, is at about 50 µm.

Both of these radiation laws are incorporated in the much more general expression derived in 1899 by Max Planck, which describes the thermal emission spectrum — the power emitted at each wavelength — for a perfect thermal radiator (or black body). The *Planck radiation law* gives the flux F_λ (in watts/m²) per centimeter of wavelength:

$$F_\lambda = \frac{(3.74 \times 10^{-8})\lambda^{-5}}{e^{c_2/\lambda T} - 1}$$

If the wavelength λ is measured in centimeters, then the constant c_2 has the value $c_2 = 1.439$.

Note that all perfect radiators emit at all wavelengths, although of course this emission is strongest near the peak given by the Wien law. Also, at any given wavelength, a hotter source emits more energy; in other words, the Planck curves corresponding to the emission spectra of sources of different temperatures never cross each other.

EXERCISES

1. Calculate the wavelengths of peak emission and the total emitted radiation from each square meter of the surfaces of Mars (250 K), Callisto (125 K), and Titania (75 K).

2. Use the Planck law to calculate the emitted power from the surfaces of each of the objects in Exercise 1 at wavelengths of 10 µm, 100 µm, and 1 mm. Do your results support the statement made that the emission curves for objects of different temperature never cross?

◆

A.4 Radioactive Age Dating (Section 3.3)

Rocks are dated — that is, their solidification ages are determined — by measuring the amount of a radioactive parent isotope that has changed into a daughter isotope since a particular crystal or grain of material formed. This process is relatively straightforward if the only source of the daughter isotope is decay of a single parent. In this case, we can assume that the rock solidified with little of this daughter present, and the current measured ratio of daughter to parent in a mineral grain will establish the age. In practice, however, the situation is a little more complicated.

One common method of dating the formation of the Earth involves the decay of uranium and thorium to produce the lead isotopes Pb-206 and Pb-208, respectively (Table 3.3). But not all of the Pb-206 and Pb-208 present on the Earth today resulted from decay of thorium and uranium. To determine the original quantity of these two isotopes, we look at a third lead isotope, Pb-204, which is not produced by radioactive decay. One can calculate what the primordial ratio of Pb-206 and Pb-208 must have been to Pb-204, and use the measured amount of Pb-

204 to subtract the original quantity of the other two isotopes before calculating the age.

One of the simplest ways to use the presence of other, non-radiogenic isotopes to ensure that correct ages are being calculated is to graph the results, as we now describe. Consider the decay of Rb-87 to Sr-87, which is one of the most useful reactions for determining the ages of meteorites. After a time t the amount of Sr-87 is the sum of the initial quantity present (Sr-87*) plus that produced by decay of Rb-87:

$$\text{Sr-87} = \text{Sr-87*} + \text{Rb-87*}(e^{kt} - 1)$$

where k is a constant related to the half-life of the reaction. How do we determine Sr-87*? We note that since all isotopes of the same element have the same chemical properties, the amount of Sr-87 initially bound into the minerals of the rock should be exactly proportional to the amount of other strontium isotopes, such as Sr-86.

For each individual crystal or grain in a rock, make a plot of the measured ratio of the two strontium isotopes Sr-87/Sr-86 against the ratio Rb-87/Sr-86. Such a plot is illustrated in Fig. A.1, which shows data from the Guareña meteorite. Grains with a great deal of initial Rb-87 will plot toward the upper right, while those of different chemical composition with little Rb-87 will be toward the lower left. But consider what happens as the rock ages. Rb-87 is converted to Sr-87, causing the Rb-rich grains to shift down and to the right on the plot. The older the rock, the more the line drawn through the data points rotates in a clockwise direction, until eventually it will be nearly horizontal. Such a line is called an *isochron*, and its slope is a measure of the age of the rock, independent of any assumptions about the original quantity of Sr-87 present. In the example illustrated in Fig. A.1, the Rb-Sr isochron indicates an age of (4.46 ± 0.08) × 10⁹ years.

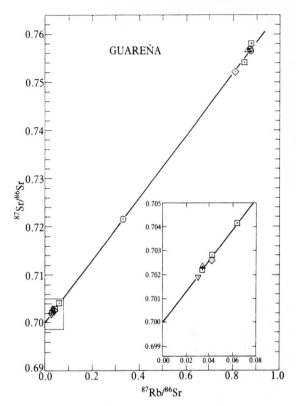

FIGURE A.1 The Rb-Sr isochron for the Guareña H6 chondrite. The derived age is 4.46 ± 0.08 billion years. (From G. J. Wasserburg *et al.*, *Earth Planet. Sci. Lett.* 7:33, 1969.)

EXERCISE

Look up some real data for meteorites or lunar samples as measured in modern laboratories to see how well the isochron ages agree for different dating methods. Possible sources include Dodd's *Meteoritics* and Wasson's *Meteorites*.

A.5 Asteroid Collisions (Section 4.2)

Although there are about 10^5 asteroids larger than 1 km in diameter in the main belt, they are widely spaced and collisions are rare. It is quite easy to make some approximate calculations to estimate collision frequencies.

First let us estimate the average spacing between asteroids. The asteroid belt can be considered a volume stretching around the Sun from 2.2 to 3.3 AU and about 0.2 AU thick. Its volume (circumference times width times height) is thus about 4 AU^3, or about 10^{25} km^3. Dividing this total volume by the number of asteroids gives 10^{20} km^3 per asteroid. The average spacing is approximately the cube root of this, or 5 × 10^6 km. Thus it should come as no surprise that spacecraft crossing the asteroid belt rarely come within a million kilometers of any asteroids unless they are specifically targeted to do so.

We can now calculate the frequency of collisions. Observations have shown that the typical asteroid has a speed relative to its neighbors of about 4 km/s, resulting from their different orbital eccentricities and inclinations. Take an asteroid with diameter 10 km as average. Its cross-section (given by πR^2) is about 10^2 km^2, and it thus sweeps out a volume in space relative to its neighbors equal to its cross-section times its speed, or 400 km^3/s. Since there are 3 × 10^7 seconds in a year, the volume swept out each year is equal to about 10^{10} km^3.

To estimate the frequency of collisions, we compare the volume swept out by each asteroid in a year with the volume of 10^{20} km^3 associated with each asteroid, which we calculated above. The ratio is 10 billion years, older than the age of the solar system. This tells us that most asteroids do not collide with others, even over billions of years, although if there were ten or one hundred times more asteroids in the past, you can see that collisions would have been more fre-

quent and nearly every asteroid would suffer several such catastrophes.

Finally, we can estimate how often a collision of the sort described above takes place in the asteroid belt today. This is simply the probability of an impact for one asteroid in one year (10^{-10}) times the number of main belt asteroids (10^5). The answer is one collision every 100,000 years. This frequency is sufficient to generate a great deal of dust and meteoritic fragments, and such collisions are probably responsible for the prominent asteroid dust bands discovered by the IRAS satellite (Fig. 4.26).

EXERCISE

Use reasoning similar to that presented above to estimate the probability that a spacecraft crossing the asteroid belt will collide with an asteroid of 1 km or larger diameter. Assume that the spacecraft has an area (cross-section) of 10 m^2 and that it spends six months crossing the belt at an average speed of 8 km/s.

A.6 Albedos and Temperatures (Section 5.1)

In this text we use the simple term "reflectivity" to indicate the fraction of incident sunlight that is reflected from a planetary surface. Generally, what we mean is the "normal reflectivity," that is, the fraction of the light reflected straight back from a beam incident normal (perpendicular) to the surface. But there are several other ways that reflectivity can be defined, depending on the application.

The normal reflectivity is a property of a particular area of the surface. Unless we have close-up photos from a spacecraft, however, we may not be able to distinguish individual surface elements. Suppose, instead, that we are interested in the global properties of a planet or asteroid. Planetary astronomers define a quantity called the *Bond albedo* as the ratio of total incident light to total reflected light. The Bond albedo, which is generally numerically smaller than the normal reflectivity of the surface, is a measure of the global reflection and absorption of sunlight.

Since the Bond albedo (called A) tells us how much solar energy is absorbed, we can use it to calculate an *effective temperature* for a planet. Remember that the Stefan-Boltzmann law expresses the radiation emitted by a perfect radiator at temperature T. Now a planet is not a perfect radiator, nor is its surface all at the same temperature, but we can define effective temperature as the value of T that satisfies the energy balance when the absorbed sunlight is equated to the emitted thermal power:

$$A \times (P_{Earth}/r^2)(\pi R^2) = \sigma T^4 \times 4\pi R^2$$

Here P is the sunlight per unit area at the distance of the Earth from the Sun (the solar constant), r is the distance of the planet from the Sun (in AU), and R is the radius of the planet. Note that the area (πR^2) on the left side is the cross-section of the planet, while that on the right side $(4\pi R^2)$ is the total area of a sphere, since we assume that the planet radiates from its entire surface, not just the part facing the Sun. Simplifying this equation, we have

$$T^4 = A(P_{Earth}/4\sigma r^2)$$

Another kind of albedo is defined to describe the apparent brightness of an object. This is the *geometric albedo* (called p), and it is the ratio of the global brightness, viewed from the direction of the Sun, to that of a hypothetical, white, diffusely reflecting sphere of the same size at the same distance. Numerically, the geometric albedo is approximately the same as the reflectivity quoted in the main text. Note that some objects can be more reflective than the hypothetical white reference sphere, so that geometric albedos greater than 1.0 are possible; the geometric albedo of Saturn's satellite Enceladus is about 1.1. The Bond albedo can never be greater than 1.0, however.

Finally, there is a relationship between the two albedos for any object, called the *phase integral, q*. The phase integral, which depends on the scattering properties of the surface, is defined by the relation

$$q = A/p$$

For the Moon and many other objects in the solar system, the value of the phase integral has been measured to be near 2/3; in other words, for typical surfaces of planets, the Bond albedo is equal to about two thirds of the geometric albedo.

EXERCISES

1. Saturn's satellite Enceladus has a geometric albedo of 1.1 and a Bond albedo of 0.8. Find its phase integral and effective temperature.

2. The lunar highlands have a geometric albedo of about 0.16 and the maria have a geometric albedo of about 0.08. If the phase integral is 0.6 for both types of terrain, calculate their Bond albedos and their equilibrium temperatures when illuminated from overhead by the Sun. For comparison, note that a black body ($A = 0$) at the Moon's distance would have an equilibrium temperature of about 380 K.

◆

A.7 Impact Energies (Section 5.3)

In order to determine the size of a crater formed from the impact of an asteroid or comet, we need to calculate the energy released by such an impact. This energy depends on the mass (m) and speed (v) of the projectile, as expressed by the standard equation for kinetic energy:

$$E = \frac{1}{2}mv^2$$

The speed is made up of two components: the speed of the projectile in space before it approaches the planet, and the additional speed it gains as it falls to the surface accelerated by the planet's gravitation. The minimum impact speed is the escape velocity of the planet, to which the initial relative speed must be added. For large planets, the speed gained during the fall generally exceeds the initial speed by a substantial margin.

The escape velocity of the Earth is 11 km/s or 1.1×10^4 m/s. Thus we calculate the minimum impact energy per kilogram of the projectile to be 6×10^7 joules. This is about ten times larger than the chemical energy of TNT, which is 4×10^6 joules/kg, which is why we have emphasized that the crater-forming event is in many ways similar to a bomb explosion.

Since the escape velocity of the Moon is only 2.4 km/s, impacts of much lower energy are possible. Typically, a projectile strikes the Moon with only about one-third the speed with which it would strike the Earth, resulting in an energy only about one-ninth as great. The mass (and volume) of material ejected in the impact is roughly proportional to the energy, and of course the linear dimensions (diameter and depth) of the crater are proportional to the cube root of the volume. Thus a projectile striking the Moon would produce a crater with diameter only $^3\sqrt{(1/9)} = 0.5$ times the diameter of the crater it would make on the Earth. This calculation is the basis for the assertion in the text that a given impacting body makes a lunar crater only half as big as a terrestrial crater (about ten and twenty times the projectile diameter, respectively).

Just how large are these impacting energies? A 10-km asteroid with a density of 2.5 g/cm³ has a mass of about 1.3×10^{12} kg. Striking the Earth with a speed of 15 km/s, it would have an energy of about 1.4×10^{21} joules. We have already noted that one kilogram of TNT is equivalent to 4×10^6 joules, so the energy of impact can also be expressed as 4×10^{14} kg = 4×10^{11} tons = 400,000 megatons of TNT.

EXERCISES

1. Suppose a meteoroid is capable of generating a 1-km-diameter crater on the Earth. What size crater would it make on Ceres, which has an escape velocity given by $\sqrt{(2GM/R)}$?

2. It can be estimated that the gravitational binding energy of the Moon is given approximately by GM^2/R, where G is the universal gravitational constant ($G = 6.67 \times 10^{-11}$ newtons m² kg⁻²) and M and R are the Moon's mass and radius, respectively. If an impact releases more energy than this amount, it is capable of breaking the Moon apart. Calculate the diameter of an impacting asteroid that would generate this amount of energy. Also calculate the size of the resulting crater, using the simple arguments for crater size scaling given here, and compare this size with the diameter of the Moon itself. This will give you an idea of the largest possible crater or basin that could be made on the Moon without disrupting the object entirely.

◆

A.8 Radar and the Doppler Effect (Section 6.1)

Radar provides one way to determine the rotation rate of a planet, using the Doppler effect. As an illustration, let us calculate the effect of the rotation of Mercury on a radar signal transmitted at a wavelength of 10 cm. The equivalent frequency of this radar wave is given by

$$f = c/\lambda = 3 \times 10^9 \text{ Hz} = 3 \text{ GHz}$$

where c is the velocity of light (3×10^8 m/s).

Mercury has a diameter of 4878 km. Its circumference is 1.5×10^4 km $= 1.5 \times 10^7$ m. If it rotates in 59 days (5×10^6 s), the equatorial rotational speed is a little more than 3 m/s. Thus one edge will be moving away from us at 3 m/s, and the other edge will approach at 3 m/s, relative to the center of mass of the planet. The total range in apparent velocity is 6 m/s, corresponding to a frequency (or wavelength) spread in the reflected radar signal given by the Doppler formula:

$$\Delta\lambda/\lambda = \Delta f/f = v/c = 2 \times 10^{-8}$$

If the transmitted wavelength is exactly 10 cm and there is no motion of the center of mass of Mercury relative to the Earth, the wavelengths of the returned signals would lie between 9.9999990 and 10.0000010 cm. By measuring this bandwidth in the returned signal, the rotation rate of the planet can be determined.

Note that one would rarely measure the particular apparent rotation calculated in the above example. At any given time, the apparent rotation of Mercury has two components, one due to its actual rotation, and the other resulting from our changing perspective as Mercury passes by the Earth. Thus we must know the relative orbits of Mercury and the Earth before we can utilize such radar observations to derive the true rotation rate.

EXERCISE

The Arecibo radar is sometimes used to study Earth-approaching comets at a wavelength of 12 cm. Suppose the nucleus of such a comet has a diameter of 5 km and its period of rotation is 10 hours. What are the relative speeds of its leading and trailing edges as seen from Arecibo? What is the total broadening of the 12-cm transmitted signal?

A.9 Atmospheric Structure (Chapter 7)

The variation of temperature and pressure with altitude in an atmosphere is called the structure of the atmosphere. Under certain circumstances, the structure of an atmosphere can be calculated from the basic physics that describes the behavior of gases. We begin with the equation of state for a perfect gas, written in the form

$$P\alpha = RT$$

where P is the pressure, α is the specific volume (1/density), T is the temperature, and R equals R^*/μ (the universal gas constant divided by the molecular weight of the gas). In a planetary atmosphere, the pressure is always greatest at the base of the atmosphere and declines smoothly with height. The temperature, however, is often a complicated function of altitude.

The variation of pressure with altitude z is given by the *hydrostatic equation* for a planet with surface gravity g:

$$dP = -(g/\alpha)dz$$

(This is a differential equation. If you have not had calculus, go straight to the end of this discussion.) This equation is readily solved using calculus for the particular case of an isothermal atmosphere, that is, an atmosphere that does not change in temperature with height. The result is

$$P = P_0 e^{-h/H}$$

where H, called the *scale height*, is defined to be

$$H = RT/g = R^*T/\mu g$$

The scale height is thus the height over which atmospheric pressure decreases by a factor of $1/e$.

The scale height has further significance: it is the physical height of a fictitious homogeneous atmosphere, that is, of an atmosphere in which the density remains uniform with height. The amount of atmosphere above some unit area at the surface of the Earth (or any planet) is equivalent to a column with uniform density of atmosphere of height H. On Earth, $H = 8$ km.

Some parts of an atmosphere, like the stratosphere of the Earth, are nearly isothermal. The lower parts of most atmospheres, however, are ruled by convection. Here, in the region called the troposphere, warm air rises and cooler air sinks. These convective motions represent a process that is approximately *adiabatic*, that is, a process in which there is no exchange of energy between a rising parcel of warm air and the surrounding atmosphere. The first law of thermodynamics allows us in the case of adiabatic convection to calculate the rate at which the temperature drops with altitude. This rate of change, dT/dz, is given (in degrees per meter) by

$$dT/dz = -g/c_P$$

where c_P is the heat capacity of the gas measured at constant pressure. Meteorologists call this temperature gradient the *lapse rate* of the troposphere. As long as convection is taking place, the troposphere will maintain this lapse rate, independent of the rate of energy being carried upward by the convection. As we see, the lapse rate depends only on the acceleration of gravity and the heat capacity of the gas.

EXERCISES

1. Calculate the lapse rates for Earth, Jupiter, and Titan, assuming that the value of c_P for each is about 1000 J kg^{-1} deg^{-1}, a value typical for a diatomic gas such as N_2 or H_2.

2. On Earth, Mt. Everest rises 9 km above sea level. Using the equation for an isothermal atmosphere, find the atmospheric pressure at the summit. Now, assuming that convection determines the atmospheric structure, find the temperature difference between the summit and sea level.

3. If air temperature at the base of Olympus Mons on Mars is -10 C, what is the temperature at the summit?

A.10 Building a Giant Planet (Chapter 11)

Suppose we try to transform the Earth into a planet whose composition has the same relative abundances of the elements as found in the Sun. How massive would that planet be?

We begin by referring to Tables 2.1 and 2.2. These emphasize that the greatest deficiencies of our planet are its lack of hydrogen and helium. Let us adopt silicon (Si) as our index element and add enough hydrogen and helium to yield a solar ratio of these elements to silicon. The solar ratio of H to Si (by number of atoms) is about 2×10^4. However, we must convert this to a mass ratio by multiplying the numbers of H and Si atoms by their respective atomic masses. This gives us a mass ratio of H to Si in the Sun of about 850. A similar calculation for helium gives a mass ratio of He to Si of about 250.

We could continue this process for each element, but the fact is that the others add very little, relative to hydrogen and helium. Even though their individual masses are greater than those of atoms of H and He, their abundances are lower by an even larger factor.

To build our planet, we begin with the mass of silicon in the Earth, which is 7×10^{23} kg. We then add the appropriate ratios of H (850) and He (250) and multiply each by the mass of Si:

$$(1 + 850 + 250) \times 7 \times 10^{23} = 8 \times 10^{26} \text{ kg}$$

This is equivalent to 120 times the mass of the Earth. In other words, by restoring the missing light elements to the Earth, we have created a planet with a mass greater than that of Saturn. We have indeed formed a giant planet!

EXERCISES

1. Use Table 2.1 to show that the addition of missing carbon and nitrogen to our artificial planet has only a small effect on its calculated mass.

2. Referring to Section A.2, show that hydrogen and helium could not escape from this new planet once it formed, even though it is closer to the Sun and hence much warmer than Jupiter or Saturn. What does this result tell you about the origin of the differences in composition between the terrestrial and giant planets?

◆

A.11 The Tidal Stability Limit (Section 14.9)

The *tidal stability limit*, also called the *Roche limit*, is the distance from a planet at which an unconsolidated satellite (one made of liquid, for example) will be disrupted by tidal forces. We can calculate this distance by finding the balance between the self-gravitation of an object and the differential gravitational forces from the planet that tend to pull it apart.

For the sake of simplicity, consider the situation for two identical spherical particles, each with mass m and radius r, that are just in contact with each other. The gravitational force between them is

$$F_g = G\frac{m^2}{4r^2}$$

These two particles will be at the tidal stability limit when the differential gravitational force on them is equal to F_g. Suppose they are in orbit about a planet of mass M at a distance d from the center of the planet. If they remain aligned as they orbit so that one is at a distance from the planet of $d + r$ and the other is at $d - r$, the differential force on them is simply

$$F_d = G\frac{Mm}{(d - r)^2} - G\frac{Mm}{(d + r)^2}$$

Equating the differential force to the self-gravitational force we obtain

$$G\frac{m^2}{4r^2} = GMm\left(\frac{1}{(d - r)^2} - \frac{1}{(d + r)^2}\right)$$

Suppose that these are two particles of a planetary ring. Farther from the planet than this limit,

the gravitational attraction between them can allow them to merge into a larger particle that can in turn continue to grow. But inside this tidal stability limit the particles will not stick together because of their mutual gravitational attraction.

The precise distance at which these forces balance depends on the densities of the particles and, to some extent, on their exact shapes. The term Roche limit is usually applied to the case considered by Roche, in which the orbiting particles have the same density as the planet itself. Algebraically, we can derive it from the above equation by expanding the right side and simplifying, to obtain

$$G\frac{m^2}{4r^2} = GMm(4d_R^{-3}r)$$

Solving for d_R, the Roche limit, for the case where the densities of planet and particle are the same, we have

$$d_R = 2.5R$$

where R is the radius of the planet.

EXERCISES

1. Find the distance in kilometers above the surface (or cloud tops) corresponding to the tidal stability limit for Saturn, Mars, and Earth.

2. Compare the positions of the A, F, and E Rings of Saturn with the tidal stability limit for that planet. Similarly, compare the orbit of Phobos with the stability limit for Mars. Finally, consider the orbit of the Soviet space station Mir compared to our planet's limit. Was it wise to place Mir so far inside the stability limit for Earth?

◆

A.12 Angular Momentum in the Solar Nebula (Section 15.3)

The distribution of angular momentum in the solar system has always challenged astronomers investigating the origin of the planets. To see why, it is instructive to calculate a table of the angular momentum associated with the Sun, Jupiter, Saturn, and the Earth, relative to the center of the solar system. For the Sun, this momentum is rotational, and it is given approximately by $0.1\,mR^2/p$, where m is the mass of the Sun, R is its radius, and p is its rotation period. For the individual planets, almost all of the angular momentum is contained in their orbital motion, given approximately by mr^2/p, where r is the radius of the orbit and p is the period required for one orbital revolution. With these formulas and the data given in the endpaper of this book, we calculate the numbers shown in Table A.1. Examining these values, we see that there is much more angular momentum in the planets, especially Jupiter, than in the Sun. The disparity is even more dramatic if we consider the angular momentum per kilogram, obtained by dividing the numbers in the last column by the mass: for the Sun this is 2×10^{10} while for Jupiter it is 2×10^{15}, a hundred thousand times greater.

For comparison, we can calculate the angular momentum for a condensing solar nebula of about one solar mass (2×10^{30} kg). Let us assume that the original nebula has dimensions typical of interstellar distances, say 1 light year or about 10^{16} m. Its minimum rotation period corresponds to one orbit around the center of the galaxy in 200 million years, or 6×10^{15} s. Then its angular momentum is that given by the formula above for the Sun, or about 3×10^{45} kg m²

s^{-1}, which is a hundred thousand times greater than the angular momentum of the Sun today. The angular momentum of the solar nebula per kilogram is, however, about 10^{15}, similar to that of Jupiter. Therefore we can conclude that the planets have about the correct quantity of angular momentum, while the Sun is highly deficient. Hence some process in the past must have robbed the Sun of its angular momentum, but most of this momentum was not transferred to the planets; it was simply lost from the system.

EXERCISES

1. The outer planets with their regular satellite systems offer us many opportunities for comparison with the solar system as a whole. Calculate the rotational angular momentum of Jupiter and the orbital angular momentum of each of the four Galilean satellites. How do these values compare with the situation for the Sun and planets?

2. Jupiter and its Galilean satellites are thought to have formed from the collapse of a proto-jovian nebula. Suppose this nebula had an initial radius of 2 AU and shared the orbital motion of Jupiter around the Sun. Use the same methods as those described for the solar nebula to calculate the angular momentum of this proto-jovian nebula, and compare with the angular momenta found in Exercise 1. What can you conclude about the loss of angular momentum in the jovian system? Why might the situation here differ from that of the Sun and planets?

TABLE A.1 Angular momenta of the Sun, Jupiter, Saturn, and Earth

	m (kg)	r (m)	p (s)	Momentum (kg m² s⁻¹)
Sun	2×10^{30}	7.0×10^{8}	2×10^{6}	5×10^{40}
Jupiter	2×10^{27}	7.8×10^{11}	4×10^{8}	3×10^{42}
Saturn	6×10^{26}	1.4×10^{12}	9×10^{8}	1×10^{42}
Earth	6×10^{24}	1.5×10^{11}	3×10^{7}	5×10^{39}

Units and Dimensions

Throughout this book, we have tried to emphasize the metric system, since it is the system of measurements commonly used in science at the present time. It is also the most widely used system in the world, with only the stubborn Americans, Burmese, and South Africans adhering to the foot-pound system. That adherence makes metric values unfamiliar to many of us, however, so we present tables below with some equivalences.

Note that in the metric system, units of measurement increase by factors of 10, 100, 1000, and so on:

1 centimeter = 10 millimeters

1 meter = 100 centimeters = 1000 millimeters

1 kilometer = 1000 meters = 100,000 centimeters = 1,000,000 millimeters.

English Unit	Metric Equivalent
1 inch	25.4 millimeters
	2.54 centimeters
1 foot	30.48 centimeters
1 yard	91.44 centimeters
	0.9144 meter
1 mile	1609.3 meters
	1.6093 kilometers
1 ounce	28.4 grams
1 pound	453.6 grams
	0.454 kilogram
1 quart	1136.5 cubic centimeters
	1.137 liters
1 mile/hour	1.609 kilometers/hour
	1609 meters/hour
	0.45 meters/second

Metric Unit	English Equivalent
1 millimeter	0.04 inch
1 centimeter	0.39 inch
1 meter	39.37 inches
1 kilometer	0.62 mile
1 gram	0.04 ounce
1 kilogram	35.27 ounces
	2.2 pounds
1 liter	0.88 quart
1 kilometer/hour	0.62 mile/hour
1 meter/second	2.2 miles/hour

In discussing the dimensions and masses of atoms, planets, and satellites, as well as distances between objects in the solar system and between waves of electromagnetic radiation, we encounter some very large and very small numbers. Instead of coping with a lot of zeros, it is easier to resort to what is called exponential notation, or powers of 10. This way we can write one million as 10^6 instead of 1,000,000. If this is unfamiliar, you can think of the exponent as giving you the number of zeros following the 1. Thus $10^0 = 1$.

Similarly for small numbers: $\frac{1}{10}$ is simply 10^{-1}; $\frac{1}{10,000}$ is 10^{-4}. Now the negative exponent tells you the number of places to the right of the decimal point. The diameter of a hydrogen atom is $0.00000001 = 10^{-8}$ cm.

Still further simplifications are used. The distance from the Earth to the Sun is 93 million miles or 93×10^6 miles or 149.6 million kilometers or 149.6×10^6 kilometers or 1.5×10^{13} centimeters or 1 astronomical unit. This last convenient unit is often used to give other distances in the solar system. For example, Jupiter is 5.203 astronomical units from the Sun.

At the short end of the scale, 10^{-8} centimeters is defined to be 1 angstrom (abbreviated Å, or simply A), in honor of Anders Ångstrom, a Swedish physicist who was a pioneer in spectroscopy. (Hence the diameter of a hydrogen atom is 1 angstrom). This unit is convenient for describing wavelengths of visible and ultraviolet light. The longer wavelengths of infrared light are often given in terms of micrometers (sometimes microns) where 1 micrometer (abbreviated μm) equals 10^4 angstroms, or 10^{-4} centimeters or 10^{-6} meters.

Glossary

accretion. Gravitational accumulation of mass in a planet or proto-planet. (Ch. 3)

ALSEP. Apollo Lunar Surface Experiment Package — seismometers and other instruments deployed on the Moon by the Apollo astronauts. (Ch. 5)

albedo. Reflectivity of a surface. *Bond albedo* and *geometric albedo* are defined in the Quantitative Supplement.

angular momentum. A measure of the momentum associated with rotational motion about an axis, or the revolution of bodies around their common *center of mass*. (Ch. 3)

anorthosite. Primary *igneous* rocks of the lunar *highlands*, composed almost entirely of plagioclase feldspar. (Ch. 5)

anticyclonic. The type of rotation induced in a high-pressure mass of atmosphere by *Coriolis* forces; clockwise in the northern hemisphere of a directly rotating planet. (Ch. 7)

aperture. The diameter of the primary lens or mirror of a telescope; hence, the best single measure of the light-gathering power of a telescope. (Ch. 2)

aphelion. The point in an elliptical orbit of a planet, asteroid, or comet that is farthest from the Sun. (Ch. 2)

apogee. The point in the orbit of an Earth satellite that is farthest from the Earth. (Ch. 2)

asteroid family. A group of asteroids with similar orbital elements, indicating a probable common origin in a collision sometime in the past. (Ch. 4)

astrometric method. An approach to searching for external planets or other invisible companions by measuring the tiny periodic shifts in a star's position caused by the gravitational pull of objects in orbit about it. (Ch. 15)

astronomical unit (AU). The average distance of the Earth from the Sun (1 AU = 1.4959787×10^{11} m), frequently used as a unit of length for measuring distances within the solar system. (Ch. 1)

atmospheric window. The part of the *electromagnetic spectrum* within which a planetary atmosphere is more-or-less transparent; for Earth, the wavelength regions where astronomical observations can be carried out from the ground. (Ch. 2)

aurora. Light radiated by atoms and ions in a planetary *ionosphere*, mostly in the magnetic polar regions; also called polar lights. (Ch. 7)

bar. Unit of pressure equal to 10^6 dyne/cm^2, or approximately the atmospheric pressure at the surface of the Earth. (Ch. 7)

basalt. *Igneous* rock, composed primarily of silicon, oxygen, iron, aluminum, and magnesium, produced by the cooling of lava. Basalts make up most of Earth's oceanic crust and are also found on other planets that have experienced extensive volcanic activity. (Ch. 3)

bending wave. A spiral wave induced in a flat, spinning system of self-gravitating bodies (such as the rings of Saturn) by the out-of-plane gravitational pull of an exterior satellite in an inclined orbit. (Ch. 14)

biological contamination. The accidental carrying of terrestrial microorganisms to another planet. Viking avoided the biological contamination of Mars by sterilizing the spacecraft before launch. (Ch. 10)

breccia. Any rock made up of recemented fragments of material, usually the result of extensive impact cratering. (Ch. 3)

brown dwarf. A hypothetical object intermediate between a planet and a star; roughly, with mass less than 0.07 solar masses and greater than a few times the mass of Jupiter. Also called an *infrared dwarf*. (Ch. 15)

caldera. Volcanic crater, often resulting from the partial collapse of the summit of a *shield volcano*. (Ch. 7)

capture theory. The theory for the origin of the Moon in which the Moon is formed separately from the Earth and later captured into Earth orbit. (Ch. 6)

captured atmosphere. A *primitive* planetary atmosphere obtained from the gas in the *solar nebula*. Examples of captured atmospheres are the atmospheres acquired by the giant planets from collapse of nebular gas onto their solid cores. (Ch. 2)

carbonaceous meteorite. A *primitive meteorite* made primarily of *silicates* but often including chemically bound water, free carbon, and complex organic compounds. Also called carbonaceous chondrites. (Ch. 3)

carbonate. A chemical compound that contains CO_3, such as calcium carbonate ($CaCO_3$), the primary constituent of the shells of ma-

rine organisms. When decomposed by heating, carbonates release carbon dioxide. (Ch. 7)

catastrophism. The concept that the geology of the Earth and planets has been greatly influenced by rare events of large magnitude, such as impacts by asteroids. In the eighteenth and nineteenth centuries catastrophism was associated specifically with attempts to explain most geological features as results of the Biblical flood, but today the term is used in the much broader sense given above. (Ch. 5)

center of mass. The average position of the various mass elements of a body or system, weighted according to their distances from that center of mass; also called center of gravity. (Ch. 15)

central peak. The mountain or group of mountains (sometimes forming a ring) produced in the center of a large impact *crater* or *impact basin*. (Ch. 5)

chaotic rotation. A recently coined term for rotation that does not have a fixed or nearly fixed period, such as that of the Saturn satellite Hyperion. (Ch. 14)

chaotic terrain. The regions of the martian uplands consisting of jumbled depressions and isolated hills, thought to have been produced by collapse induced by the withdrawal of subsurface water or ice. (Ch. 9)

chondrule. A small silicate spherule (typically a few millimeters in diameter) commonly found in *primitive meteorites*, perhaps representing droplets of material that condensed in the *solar nebula* 4.5 billion years ago. (Ch. 3)

cinder cone volcano. A steep-sided volcanic mountain produced by fallback of fountains of lava. (Ch. 7)

coma. The diffuse gaseous component of the head of comet; often used interchangeably with *comet head*. (Ch. 4)

comet head. The main part of the atmosphere of a comet, forming a visible halo around the *nucleus*. (Ch. 4)

comet nucleus. The solid part of a comet, typically a few kilometers (up to tens of kilometers) in diameter, consisting of a mixture of ices and solid silicate and carbonaceous grains. (Ch. 4)

compound. A substance composed of two or more chemical elements, such as H_2O (formed from hydrogen and oxygen). (Ch. 2)

constellation. Originally a configuration of stars; now a specific area of the sky with internationally agreed-upon boundaries. (Ch. 1)

continental drift. A gradual motion of the continents over the surface of the Earth due to *plate tectonics*. (Ch. 7)

co-orbital satellite. Informal term for the two Saturn satellites Janus and Epimetheus, which share almost the same orbit, or for any other satellites that may be found in similar dynamical situations. (Ch. 14)

Coriolis effect. The deflection of material moving across the surface of a rotating planet, producing in the case of the Earth's atmosphere the familiar *cyclonic* and *anticyclonic* patterns that characterize our mid-latitude weather. (Ch. 7)

cosmic rays. Atomic nuclei (mostly protons) that are observed to strike the Earth's atmosphere with exceedingly high energies. (Ch. 5)

crater. A circular depression (from the Greek word for cup), generally of impact origin. (Ch. 5)

crater density. The degree to which impact *craters* are packed together on a planetary surface, measured in units of the number of craters of a given size per unit area (for example, the number of craters larger than 10 km in diameter per million square kilometers of surface). (Ch. 5)

crater ray. Ejecta from an impact that extends out to many crater diameters; specifically, the long bright rays on the Moon radiating from recent large craters such as Tycho or Copernicus. (Ch. 5)

crater retention age. The time over which a planetary surface has accumulated impact craters; hence, the time since the surface was mobile, or since extensive erosional or tectonic forces were acting to destroy craters. (Ch. 5)

crust. The outer solid layer of a planet; on Earth, the upper 10 to 60 kilometers. (Ch. 7)

C-type asteroid. One of the most populous group of *main belt asteroids*, characterized by dark, spectrally neutral or slightly red surfaces; thought to be *primitive*, similar in composition to the *carbonaceous meteorites*. (Ch. 4)

cyclonic. The type of rotation induced in a low-pressure mass of atmosphere by *Coriolis forces*; clockwise in the southern hemisphere of a directly rotating planet. (Ch. 7)

daughter theory. The class of theories for formation of the Moon in which the lunar material was once part of the Earth; hence the Moon is the "daughter" of the Earth. (Ch. 6)

deuterium. A "heavy" form of hydrogen, in which the nucleus of each atom consists of one proton and one neutron instead of one proton only. (Ch. 2)

differentiated meteorite. A meteorite from a *differentiated parent body*, as contrasted with a *primitive meteorite*. (Ch. 3)

differentiation. The gravitational separation or segregation of different kinds of material in different layers in the interior of a planet, as a result of heating. (Ch. 2)

Doppler effect. Apparent change in wavelength (and frequency) of the radiation from a source due to its relative motion toward or away from the observer. (Ch. 6)

dust tail. A *cometary* tail, usually broad, somewhat curved, and yellow-white in color, made up of dust grains released from the *nucleus* of the comet. (Ch. 4)

Earth-approaching asteroid. An asteroid with an orbit that crosses the Earth's orbit, or that will at some time cross the Earth's orbit as it evolves under the influence of the planets' gravity. (Ch. 4)

eccentric ring. A planetary ring that has the form of an ellipse, as a result of the alignment of the individual elliptical orbits of the particles that comprise it. (Ch. 14)

eccentricity (of an ellipse). The degree to which an orbit is non-circular; specifically, the ratio of the distance between the foci to the major axis of the ellipse. (Ch. 1)

eclipse of the Moon. The phenomenon visible when the Moon passes wholly or in part through the shadow of the Earth. (Ch. 1)

eclipse of the Sun. The blocking of all or part of the light of the Sun by the Moon. (Ch. 1)

ecliptic. The apparent annual path of the Sun on the celestial sphere. (Ch. 1)

ejecta blanket. Rough, hilly region surrounding an impact *crater* made up of ejecta that has fallen back to the surface — usually extending one to three crater radii from the rim. (Ch. 5)

electromagnetic radiation. Radiation consisting of electric and magnetic waves; these include radio, infrared, visible light, ultraviolet, X rays, and gamma rays. (Ch. 2)

electromagnetic spectrum. The whole array or family of electromagnetic waves, usually ordered by wavelength (or equivalently by frequency or energy). (Ch. 2)

embedded satellite. A small (perhaps invisible) satellite orbiting within a ring system and gravitationally influencing the structure of the ring. (Ch. 14)

epicycle. A theoretical component of the Ptolemaic model for the orbits of the planets, consisting of a small circle, the center of which revolves along the rim of another circle (the deferent). (Ch. 1)

equinox. One of the intersections of the *ecliptic* and the celestial equator; occupied by the Sun on about March 21 and September 21. (Ch. 1)

escape velocity. The velocity required to escape entirely from the gravitational field of an object; also the minimum impact velocity for any body arriving from a very great distance. Within the solar system, escape from a planet results in an orbit about the Sun; still higher energies are required to escape from the Sun as well. (Ch. 2)

eucrite. One of a class of *basaltic meteorites* believed to have originated on the asteroid Vesta. (Ch. 3)

fall. A *meteorite* or group of meteorites that are seen to fall, as contrasted with a *find*. (Ch. 3)

fault. In geology, a crack or break in the crust of a planet along which slippage or movement can take place, accompanied by seismic activity. (Ch. 7)

find. A *meteorite* that is found without its fall being witnessed, perhaps decades or centuries later, as contrasted with a *fall*. (Ch. 3)

first quarter moon. The phase of the Moon about eight days after new moon, when it is half illuminated as seen from Earth. (Ch. 1)

fluidized ejecta. Ejecta from an impact *crater* that flows along the surface like a liquid rather than arching freely through space, apparently a common phenomenon in the past on Mars. (Ch. 9)

fragmentation. Breaking apart by impact. (Ch. 3)

full moon. The phase of the Moon when it is at opposition (180 degrees from the Sun) and its full daylight hemisphere is visible from the Earth. (Ch. 1)

Gaia hypothesis. The suggestion that life on Earth responds to changing external circumstances (such as variations in the luminosity of the Sun) in ways that preserve conditions suitable for life. (Ch. 7)

gas exchange (GEX). One of the Viking life-detection experiments, in which martian soil was brought into contact with nutrients to determine if metabolism took place. (Ch. 10)

gas retention age. A measure of the age of a rock, defined in terms of its ability to retain radioactive argon. (Ch. 3)

GCMS instrument. One of the Viking lander instruments, the gas chromatograph-mass spectrometer was a powerful device for the chemical analysis of the martian atmosphere and soil. (Ch. 10)

geocentric. Earth-centered, as in the pre-Copernican idea that the Sun, Moon, and planets all circled the Earth, which was located at the center of the universe. (Ch. 1)

geologic time scale. The history of the Earth over the past 4.5 billion years, as determined from the rocks deposited in its crust. (Ch. 7)

geyser. A periodic eruption of steam and hot water, produced when groundwater is heated in certain volcanic areas such as Yellowstone National Park. (Ch. 13)

glass. An amorphous *igneous* rock, produced when a hot melt cools very rapidly. (Ch. 5)

granite. An *igneous* rock associated primarily with the Earth's continental crust, composed chiefly of quartz and alkali feldspar. (Ch. 7)

gravity anomaly. A feature in the gravity field of a planet that indicates a part of the crust is out of equilibrium — that is, the crust has not yet fully adjusted to a geologically young feature, either a mountain or a depression. (Ch. 8)

gravity-assisted trajectory. An interplanetary trajectory of a spacecraft that uses the close flyby of one planet to provide the acceleration to continue to another target. (Ch. 2)

greenhouse effect. The blanketing of infrared radiation near the surface of a planet by, for example, carbon dioxide in its atmosphere, producing an elevated surface temperature. (Ch. 8)

Hadley cell. A theoretical model for the circulation of a planetary atmosphere, in which air heated near the equator rises and moves toward the poles, where it descends and flows back toward the equator. The circulation of the atmosphere of Venus approximates this situation. (Ch. 8)

half-life. The time required for half of the *radioactive* atoms in a sample to disintegrate. (Ch. 3)

heliocentric. Sun-centered; specifically, the Copernican theory that the planets are in orbit around the Sun rather than the Earth. (Ch. 1)

highlands (lunar). The older, heavily cratered crust of the Moon, covering 83 percent of its surface and composed in large part of *anorthositic breccias.* (Ch. 5)

ice age. One of the periods in the Earth's climatic history when global cooling led to the formation of extensive ice sheets over polar and even temperate land masses. (Ch. 7)

igneous rock. Any rock produced by cooling from a molten state. (Ch. 2)

imaging radar. *Radar* carried on a moving platform (airplane or spacecraft) that can, with suitable data processing, produce images of the ground beneath; also called synthetic aperture radar. (Ch. 8)

impact basin. A large impact feature, usually 200 km or more in diameter. (Ch. 5)

inclination (of an orbit). The angle between the orbital plane of a revolving body and some fundamental plane—usually the plane of the celestial equator or of the ecliptic. (Ch. 1)

infrared dwarf. See *brown dwarf.* (Ch. 15)

Io plasma torus. The prominent feature of the jovian *magnetosphere* located near the orbit of Io that consists of relatively dense concentrations of sulfur and oxygen ions carried around the planet with its rotating magnetic field. (Ch. 11)

ionosphere. The upper region of the Earth's atmosphere in which many of the atoms are ionized, or any similar feature of the atmosphere of another planet. (Ch. 7)

iron meteorite. A *meteorite* composed primarily of metallic iron and nickel and thought to represent material from the core of a *differentiated parent body*. (Ch. 3)

irregular satellite. A planetary satellite with an orbit that is *retrograde*, or of high *inclination* or *eccentricity*. (Ch. 13)

isotope. Any of two or more forms of the same element, whose atoms all have the same number of protons but different numbers of neutrons. (Ch. 2)

K/T event. The major break in the history of life on Earth (a *mass extinction*) that occurred 65 million years ago, between the Cretaceous and Tertiary periods, apparently due to the impact of an asteroidal object. (Ch. 7)

labeled release (LR). One of the Viking life-detection experiments, which tested for metabolism by exposing the martian soil to nutrients labeled with radioactive carbon atoms. (Ch. 10)

last quarter moon. The phase of the Moon about eight days before new moon, when it appears half illuminated as seen from Earth. (Ch. 1)

lithosphere. The upper layer of the Earth, to a depth of 50 to 100 km, involved in *plate tectonics*. (Ch. 7)

long-period comet. A comet with a period of revolution of 200 years or more. (Ch. 4)

magnetic braking. A process proposed to account for the slow rotation of some stars, in which *angular momentum* is transferred from the star to the surrounding plasma through its magnetic field. (Ch. 15)

magnetic flux tube. A feature of the jovian *magnetosphere* consisting of a loop of electrical current connecting Io and the planet. (Ch. 11)

magnetosphere. The region around a planet that is occupied by that planet's magnetic field, and within which the planetary field dominates over the interplanetary field associated with the *solar wind*. (Ch. 7)

magnetotail. The region of a planetary or cometary *magnetosphere* in which the magnetic field lines stream away from the object as they are carried "downstream" by the *solar wind*. (Ch. 11)

main belt asteroids. Asteroids that occupy the main asteroid belt between Mars and Jupiter, sometimes limited specifically to the most populous parts of the belt, from 2.2 to 3.3 AU from the Sun. (Ch. 4)

mantle. The part of a planet between its crust and core; on Earth, the mantle is the most massive part of the planet, with about 65 percent of the Earth's material. (Ch. 7)

mare (plural: **maria**). Latin for "sea"; name applied to the dark, relatively smooth features that cover 17 percent of the Moon. (Ch. 5)

mass extinction. The sudden disappearance in the fossil record of a large number of species of life, to be replaced by new species in subsequent layers. Mass extinctions are indications of catastrophic changes in the environment, such as might be produced by a large impact on the Earth. (Ch. 7)

metamorphic rock. Any rock produced by the physical and chemical alteration (without melting) of another rock that has been subjected to high temperature and pressure. (Ch. 2)

meteor. The luminous phenomenon observed when a *meteoroid* enters the Earth's atmosphere and burns up; popularly called a "shooting star." (Ch. 4)

meteor shower. Many *meteors* appearing to radiate from a common point in the sky caused by the collision of the Earth with a swarm of meteoritic particles. (Ch. 4)

meteorite. A portion of a *meteoroid* that survives passage through the atmosphere and strikes the ground. (Ch. 3)

meteoroid. An object of debris in space, before it impacts a planet or is heated by friction in the planet's atmosphere; to be distinguished from a *meteor* or a *meteorite*. (Ch. 3)

mineral. A solid *compound* (often primarily silicon and oxygen) that forms rocks; the term could also be applied to condensed *volatiles* such as ice. (Ch. 2)

molecule. A combination of two or more atoms bound together; the smallest particle of a chemical compound or substance that exhibits the chemical properties of that substance. (Ch. 2)

new moon. The phase of the Moon when its longitude is the same as that of the Sun, rendering it invisible from Earth. (Ch. 1)

noble gas. One of the chemically inactive or inert elements: helium, neon, argon, krypton, xenon, radon. (Ch. 8)

non-thermal radiation. Any radiation from an astronomical body that is not thermal in origin, such as radio emission (*synchrotron radiation*) from electrons spiraling in the *magnetosphere* of Jupiter. (Ch. 11)

nuclear winter. A term coined in the 1980s for the period of global environmental disturbance, including blockage of sunlight and decline in surface temperature, that would follow a major nuclear war. (Ch. 7)

occultation. The passage of an object of large angular size in front of a smaller object, such as the Moon in front of a distant star, or the rings of Saturn in front of the Voyager spacecraft. (Ch. 11)

Oort comet cloud. The spherical region around the Sun from which most "new" comets come, representing objects with aphelia at about 50,000 AU, or extending about a third of the way to the nearest other stars. (Ch. 4)

orbital velocity. The speed required to maintain a satellite in a low-altitude circular orbit, approximately equal to the *escape velocity* divided by 1.41. (Ch. 2)

organic compounds. Any compounds containing carbon and hydrogen, especially the complex compounds often associated with life, either as chemical precursors or as the product of living things. (Ch. 3)

outflow channel. A martian channel, typically several kilometers wide and hundreds of kilometers long, that once drained large floods of water from the southern uplands to the northern lowlands. (Ch. 9)

outgassed atmosphere. A planetary atmosphere formed by release of gases from the interior, often in association with volcanic activity. Also called a degassed atmosphere. (Ch. 2)

ozone. The allotrope of oxygen, O_3. (Ch. 2)

parent body. In planetary science, any larger original object that is the source of other objects, usually through breakup or ejection by impact cratering — for example, the asteroid Vesta is thought to be the parent body of the *eucrite meteorites*. (Ch. 3)

parent/daughter isotopes. In referring to the process of *radioactive* decay, the parent is the radioactive *isotope* that is destroyed in the decay process to produce a new, or daughter, isotope. (Ch. 3)

perigee. The point in the orbit of an Earth-satellite that is closest to the Earth. (Ch. 2)

perihelion. The point in the orbit of a planet, asteroid, or comet that is closest to the Sun. (Ch. 2)

perturbation. The gravitational effect of one object on the orbit of another, if this effect is very small. (Ch. 12).

Phanerozoic period. The most recent eon in the Earth's history, covering approximately the past 600 million years. It is divided into the Paleozoic, Mesozoic, and Cenozoic eras. (Ch. 7)

photochemistry. Chemical reactions that are caused or promoted by the action of light — usually ultraviolet light that excites or dissociates some compounds and leads to the formation of new compounds. (Ch. 8)

photon. A discrete unit of electromagnetic energy. (Ch. 2)

planetesimals. The hypothetical objects, from tens to hundreds of kilometers in diameter, that formed in the *solar nebula* as an intermediate step between tiny grains and the larger planetary objects we see today. The comets and some asteroids may be leftover planetesimals. (Ch. 15)

plasma. A mixture of ionized and neutral gas. (Ch. 2) .

plasma tail. A cometary tail, usually narrow and bluish in color, extending straight away from the Sun and consisting of *plasma* streaming away from the head under the influence of the *solar wind*. Also called an ion tail. (Ch. 4)

plastic deformation (of ice). The tendency of ice to flow or deform slowly under pressure; even at low temperatures, topographic relief will gradually be reduced and smoothed by this process on an icy object such as Callisto or Ganymede. (Ch. 13)

plate tectonics. The motion of segments or plates of the outer layer of the Earth (the *lithosphere*) driven by slow convection in the underlying *mantle*. (Ch. 7)

plume eruption. On Io, a large *geyser*-like volcanic eruption apparently produced by the release of hot sulfur and sulfur dioxide vapor. (Ch. 13)

Precambrian period. The period of Earth history from the formation of the planet until the beginning of the *Phanerozoic*, about 600 million years ago. The Precambrian is divided into the Priscoan, Archean, and Proterozoic eons. (Ch. 7)

primitive. In planetary science and meteoritics, a type of object or rock that is little changed chemically since its formation — hence representative of the conditions in the *solar nebula* at the time of formation of the solar system. Also used to refer to the chemical composition of an atmosphere that has not undergone extensive chemical evolution. (Ch. 2)

primitive meteorite. A meteorite that has not been greatly altered chemically since its condensation from the *solar nebula* — called in meteoritics a chondrite (either ordinary chondrite or carbonaceous chondrite). (Ch. 3)

primitive rock. Any rock that has not experienced great heat or pressure and thus remains representative of the original condensates from the *solar nebula* — never found on any object large enough to have undergone melting and *differentiation*. (Ch. 2)

pyrolytic release (PR). One of the Viking life-detection experiments, in which martian soil is brought into contact with radioactive CO_2 to determine if any of the gas is absorbed into microorganisms. (Ch. 10)

quantitative. Numerical or exact, as in the case of a quantitative theory, which makes mathematical predictions about nature; generally speaking, well-developed scientific theories are quantitative. (Ch. 1)

radar. Transmitted and received radio-wave signals used to measure the properties of a distant target, especially its distance and motion. (Ch. 6)

radioactive age dating. The technique of determining the ages of rocks or other specimens by the amount of decay of certain *radioactive* elements contained therein. (Ch. 3)

radioactivity. The process by which certain kinds of atomic nuclei naturally decompose (with a specific *half-life*) with the spontaneous emission of subatomic particles and gamma rays. (Ch. 3)

regolith. The broken or pulverized upper layers of a planetary surface, fragmented by impacts; specifically the lunar soil. (Ch. 5)

regular satellite. A planetary satellite that has an orbit of low or moderate *eccentricity* in approximately the plane of the planet's equator. (Ch. 13)

remote sensing. Any technique for measuring an object at a distance; used particularly to refer to measurements (imaging, spectrometry, radar, etc.) of a planet carried out from an orbiting spacecraft. (Ch. 2)

resolution. The degree to which fine details in an image are separated or resolved. Resolution can be specified in either angular or linear units, but is used here usually in the sense of the linear dimensions (in km) of the smallest features that can be studied on a planet. (Ch. 5)

resonance. An orbital condition in which one object is subject to periodic gravitational *perturbations* by another — most commonly arising when two objects orbiting a third have periods of revolution that are simple multiples or fractions of each other. (Ch. 4)

resonance gap. A location in an ensemble of orbiting particles that is empty because it corresponds to periods that are simple fractions of the period of a perturbing external body. Examples of resonance gaps are the Kirkwood gaps in the main asteroid belt and several of the gaps in the rings of Saturn. (Ch. 4)

retrograde motion. An apparent westward motion of a planet with respect to the stars, caused by the motion of the Earth. (Ch. 1)

retrograde rotation (or **retrograde revolution**). Movement that is backwards with respect to the common direction of motion in the solar system; counterclockwise as viewed from the north; going from east to west rather than from west to east. (Ch. 1)

rift. In geology, a place where the crust is being torn apart by internal forces, generally associated with injection of new material from the *mantle* and with the slow separation of *tectonic plates*. (Ch. 7)

runaway greenhouse effect. A process whereby the heating of a planet leads to an increase in its atmospheric *greenhouse effect* and thus to further heating, quickly altering the composition of its atmosphere and the temperature of its surface. (Ch. 8)

runoff channel. A branching river channel with many tributaries, found in the old martian uplands and presumably formed at a time when higher temperatures and a more massive atmosphere permitted rain on Mars. (Ch. 9)

scientific model. A description of processes in nature, usually expressed mathematically, that should be self-consistent, provide a framework for understanding observations and experiments already carried out, and predict additional observations or experiments with which the model can be tested against alternatives. (Ch. 6)

secondary crater. An impact *crater* produced by ejecta from a primary impact. (Ch. 5)

sedimentary rock. Any rock formed by the deposition and cementing of fine grains of material. On Earth, sedimentary rocks are usually the result of erosion and weathering, followed by deposition in lakes or oceans, but *breccias* formed on the Moon by impact processes should also be considered sedimentary rocks. (Ch. 2)

sedimentation. The process on Earth in which eroded material is transported (by water or wind) to lower areas, often the seafloors, where it slowly accumulates to form new *sedimentary rock*. (Ch. 7)

seismic waves. Vibrations induced by earthquakes or impacts that travel through the Earth or other planet and can be used to probe its interior. (Ch. 7)

SETI. The search for extraterrestrial intelligence, carried out so far by looking for radio signals from other civilizations in the galaxy. (Ch. 15)

shepherd satellite. An informal term for a satellite that is thought to maintain the structure of a planetary ring through its close gravitational influence — specifically, the two Saturn satellites, Prometheus and Pandora, that orbit just inside and outside of the F Ring, and the two small satellites of Uranus that orbit on either side of its Epsilon Ring. (Ch. 14)

shield volcano. A broad volcano built up through the repeated non-explosive eruption of fluid *basalts* to form a low dome of shield shape, typically with slopes of only 4 to 6 degrees, often with a large *caldera* at the summit. Examples include the Hawaiian volcanoes on Earth and the Tharsis volcanoes on Mars. (Ch. 7)

short-period comet. A comet with a period of revolution of less than 200 years. (Ch. 4)

silicate. Any of a wide range of rocks and minerals composed in part of silica (silicon and oxygen). (Ch. 2)

sister theory. A theory for the origin of the Moon in which the Earth and Moon formed together out of the *solar nebula.* (Ch. 6).

SNC meteorite. One of a class of *basaltic meteorites* thought by many planetary scientists to be impact-ejected fragments from Mars. (Ch. 3)

solar day. The "day" as we usually think of it: the average interval between the times when the Sun crosses the meridian (noon). A solar day consists of 8.64×10^4 s. (Ch. 6)

solar nebula. The cloud of gas and dust from which the solar system formed. (Ch. 3)

solar wind. A radial flow of *plasma* leaving the Sun at an average speed of about 400 km/s and spreading through the solar system. (Ch. 2)

solidification age. The most common age determined by *radioactive* dating techniques — the time since the rock or mineral grain being tested solidified from the molten state, thus isolating itself from further chemical changes. (Ch. 3)

spectrograph (or **spectrometer**). An instrument for obtaining a *spectrum* (especially a spectrum recorded on a photographic plate); in astronomy, usually attached to a telescope to record the spectrum of a star or planet. (Ch. 2)

spectroscopy. The study of *spectra.* (Ch. 2)

spectrum. See *electromagnetic spectrum.* (Ch. 2)

spiral density wave. A wave phenomenon produced in a self-gravitating rotating sheet of material. In the rings of Saturn such waves, generated by resonant gravitational effects of the satellites, produce spiral patterns that wind around the rings like grooves in a phonograph record. (Ch. 14)

sputtering. The process by which energetic atomic particles striking a solid alter its chemistry and eject atoms or molecular fragments from the surface. (Ch. 11)

stony meteorite. A *meteorite* composed mostly of stony material. The term can be applied to either *primitive* or *differentiated* meteorites if they are made of *silicates.* (Ch. 3)

stony-iron meteorite. A fairly rare kind of *differentiated meteorite,* composed of a mixture of *silicates* with metallic iron-nickel, thought to have originated near the core-mantle boundary of a *differentiated parent body.* (Ch. 3)

stratigraphy. The study of rock strata, particularly the sequence of layers and the information this provides on the geologic history of a region. (Ch. 5)

stratosphere. The layer of the Earth's atmosphere above the *troposphere* (where most weather takes place) and below the *ionosphere*; also, the similar layer above the troposphere on any planet, in which temperatures are roughly constant. (Ch. 7)

S-type asteroid. The second most common class of asteroids, located primarily in the inner part of the main belt, and characterized by moderate reflectivities (20 percent) and spectra that indicate the presence of silicate minerals similar to those in many meteorites. It is not known whether the S-type asteroids are *primitive* or *differentiated*, however. (Ch. 4)

subduction. In terrestrial geology, the process whereby one crustal plate is forced under another, generally associated with earthquakes, volcanic activity, and the formation of deep ocean trenches. (Ch. 7)

summer solstice. The point on the celestial sphere where the Sun reaches its greatest distance north of the celestial equator (on about June 21). (Ch. 1)

synchronous rotation. Rotation of a body so that it always keeps the same face toward another object; the situation where the periods of rotation and revolution of an orbiting body are equal. (Ch. 6)

tectonic. Associated with the forces acting in a planet's *crust*; also, features such as *rifts* and *faults* produced by such forces. (Ch. 8)

thermal inversion. A situation in a planetary atmosphere where local heating causes the temperature to increase with altitude, rather than decreasing as it normally does. (Ch. 11)

tidal force. A differential gravitational force that tends to deform a body or to pull it apart, due to the tidal effect of its neighbor. (Ch. 6)

tidal heating. Heating of one body by tidal friction or repeated stressing resulting from its motion within the strong gravitational field of its neighbor, as in the tidal heating of Io. (Ch. 6)

tidal stability limit. The distance from a planet — approximately 2.5 planetary radii from the center — within which differential gravitational forces (or *tides*) are stronger than the mutual gravitational attraction between two adjacent orbiting objects. Within this limit, fragments are not likely to accrete or assemble themselves into a larger object. Also called the Roche limit. (Ch. 14)

tide. Deformation of a body by the differential gravitational force exerted on it by another body; in the Earth, the deformation of the ocean surface by the gravitational forces exerted by the Moon and Sun. (Ch. 6)

Titius-Bode rule. A numerical scheme by which a sequence of numbers can be obtained that give the approximate distances of the planets from the Sun in astronomical units. Also called Bode's law. (Ch. 1)

troposphere. Lowest level of the Earth's atmosphere, where most weather takes place; any region in a planetary atmosphere where convection normally takes place. (Ch. 7)

uncompressed density. The density that a planetary object would have if it were not subject to self-compression from its own gravity — hence the density that is characteristic of its bulk material independent of the size of the object. (Ch. 6)

uniformitarianism. The idea in geology that the landforms we see were formed through the action of the same processes that are acting today, continuing over very long spans of time. In the early nineteenth century this approach, as opposed to the Biblical *catastrophism* of previous generations, laid the foundations of modern geology. (Ch. 5)

vernal equinox. The point on the celestial sphere where the Sun crosses the celestial equator passing from south to north (on about March 21). (Ch. 1)

volatile. A relative term that indicates substances with low melting temperatures and high vapor pressures. Volatiles are usually used in this text to indicate the common ices (H_2O, CO_2, etc.), but the term can also be applied to such elements as cadmium, zinc, lead, and rubidium, which are gaseous at temperatures up to 1000 K; these are called volatile elements, as opposed to refractory elements. (Ch. 2)

volcanic hot spot. A location where there is a particularly large flow of heat from the interior of a planet; on Io, the name given to the areas covering about 1 percent of the surface from which most of the internal heat escapes to space. (Ch. 13)

water cycle. On Earth, the circulation of water from the oceans to the atmosphere, where part of it is deposited as precipitation on the land, to flow again into the ocean, thus maintaining the cycle. (Ch. 7)

winter solstice. The point on the celestial sphere where the Sun reaches its greatest distance south of the celestial equator; also the date (on about December 21) when the Sun reaches this point. (Ch. 1)

zodiac. A belt around the sky that is 18° wide and centered on the *ecliptic*, and within which are found the Moon and planets. (Ch. 1)

zodiacal dust cloud. A tenuous, flat cloud of small *silicate* dust particles, derived from comets and collisions in the *asteroid belt*, that pervades the inner solar system. (Ch. 4)

Figure Credits

Part One *Opener*: Palomar Observatory
photograph. *Chapter 1*: 1.1, Harvard College
Observatory photograph; 1.2, 1.3, Dale Smith;
1.4, Goldsmith, D. *The Evolving Universe*, 2nd ed.
© 1985 The Benjamin/Cummings Publishing Co.
Reprinted with permission. 1.7, NASA; 1.8,
Goldsmith, D. *The Evolving Universe*, 2nd ed.
© 1985 The Benjamin/Cummings Publishing Co.
Reprinted with permission. 1.9–1.11, Yerkes
Observatory; 1.13–1.16, Goldsmith, D. *The Evolving
Universe*, 2nd ed. © 1985 The Benjamin/Cummings
Publishing Co. Reprinted with permission. 1.17,
Yerkes Observatory; 1.18, *Le Baiser* (1886), marble
sculpture by A. Rodin, S.1002 in Musée Rodin,
Paris, photographed by Bruno Jarret (183.6 cm ×
110.5 cm × 118.3 cm). Copyright ARS NY/ADAGP
1987. Courtesy ARS and Musée Rodin. 1.19, 1.20,
Donald Goldsmith. *Chapter 2*: 2.2, David Morrison;
2.3a, University of Hawaii photo by M.W. Buie,
D.P. Cruikshank, and A. Storrs; 2.3b, Ann Ronan
Picture Library; 2.5, 2.7, Goldsmith, D. *The
Evolving Universe*, 2nd ed. © 1985 The Benjamin/
Cummings Publishing Co. Reprinted with
permission. 2.10, National Optical Astronomy
Observatories; 2.11, Courtesy NRAO/AUI; 2.12,
The Space Telescope Science Institute, operated by
AURA Inc., for NASA; 2.14, Goldsmith, D. *The
Evolving Universe*. © 1981 The Benjamin/Cummings
Publishing Co. Reprinted with permission. 2.15–
2.18, NASA.

Part Two *Opener*: Courtesy of H. Reitsema and A.
Delamere, Ball Aerospace. *Chapter 3*: 3.1, Mt.
Wilson and Las Campanas Observatories, Carnegie
Institution of Washington; 3.2, Goldsmith, D. *The
Evolving Universe*, 2nd ed. © 1985 The Benjamin/
Cummings Publishing Co. Reprinted with
permission. 3.4, Smithsonian Institution photo; 3.5,
NASA; 3.7, Photo of G.J. Wasserburg by Robert
Paz, Institute Photographer; 3.8a, From Wood, *The
Solar System*, p. 99, published by Prentice-Hall.
3.8b, *New Solar System*, Sky Publishing Co./
Cambridge University Press. 3.9, 3.10, Smithsonian
Institution photos; 3.11, NASA. *Chapter 4*: 4.1,
Discovered/photographed by Eleanor F. Helin,
Palomar Observatory; 4.3, Adapted from data
assembled by Benjamin Zellner; 4.4, Courtesy
Robert Millis (published in the *Astronomical
Journal*); 4.7, Reprinted with permission from
Geochim. et Cosmochim. Acta 40, 701–719, C.R.
Chapman (1976) Pergamon Journals, Ltd. 4.8,
University of Arizona photo by Gill McLaughlin;
4.12–4.15, NASA; 4.16, Avec autorisation speciale
de la Ville de Bayeux; 4.17, Mt. Wilson and Las
Campanas Observatories, Carnegie Institution of
Washington; 4.18a, Valery Barsukov, Vernadsky
Institute, and Roald Sagdeev, Institute for Cosmic
Research; 4.18b, Arlene Ammar, CNES; 4.19a,
David Morrison; 4.19b, Arlene Ammar, CNES;
4.20, H. Reitsema and A. Delamere, Ball Aerospace;
4.22, Mt. Wilson and Las Campanas Observatories,

Carnegie Institution of Washington; 4.23, Copyright
© 1986 Royal Observatory, Edinburgh; 4.24,
Palomar Observatory photograph; 4.25, NASA,
Courtesy Donald Brownlee; 4.26, NASA, Courtesy
Mark V. Sykes; 4.28, Photos from the SOLWIND
satellite coronagraph, Naval Research Laboratory,
Washington, D.C., Courtesy D.J. Michels; 4.29,
New Mexico State University Observatory.

Part Three   *Opener*: NASA. *Chapter 5*: 5.1, Mt.
Wilson and Las Campanas Observatories, Carnegie
Institution of Washington; 5.2, Lick Observatory;
5.3, NASA; 5.5, NASA; 5.6, SOVFOTO; 5.7–5.13,
NASA; 5.14, U.S. Geological Survey; 5.16, 5.17,
NASA; 5.19, U.S. Geological Survey; 5.20–5.29,
NASA. *Chapter 6*: 6.2, Courtesy of the National
Astronomy and Ionosphere Center; 6.3, 6.6,
Goldsmith, D. *The Evolving Universe*, 2nd ed. ©
1985 The Benjamin/Cummings Publishing Co.
Reprinted with permission. 6.7–6.10, NASA; 6.11,
Yosio Nakamura, from Nakamura *et al.*, *Proc. Lunar
Sci. Conf.*, 5th, 2883–2890 (1974); 6.12, Foster,
Earth Science, © 1982 The Benjamin/Cummings
Publishing Co. Reprinted with permission of
Addison-Wesley, Reading, MA.

Part Four   *Opener*: Galleria degli Uffizi, Florence.
Chapter 7: 7.1, NASA; 7.2, 7.3, Adapted from W.K.
Hartmann; 7.4 (left), G.W. Stose; 7.4 (right), U.S.
Geological Survey; 7.5, Foster, *Earth Science*, ©
1982 The Benjamin/Cummings Publishing Co.
Reprinted with permission of Addison-Wesley,
Reading, MA. 7.6a, University of Hawaii photo by
Clark R. Chapman; 7.6b, Foster, *Earth Science*, ©
1982 The Benjamin/Cummings Publishing Co.
Reprinted with permission of Addison-Wesley,
Reading, MA. 7.7, U.S. Navy, Courtesy Dale P.
Cruikshank; 7.9, 7.10, Foster, *Earth Science*, © 1982
The Benjamin/Cummings Publishing Co. Reprinted
with permission of Addison-Wesley, Reading, MA.
7.11, Nafi Toksoz; 7.12–7.15, Robert J. Foster;
7.16a, Goldsmith, D. and T. Owen. *The Search for
Life in the Universe*. © 1980 The Benjamin/
Cummings Publishing Co. Reprinted with
permission. 7.16b, David A. Usher; 7.17, Foster,
Earth Science, © 1982 The Benjamin/Cummings
Publishing Co. Reprinted with permission of

Addison-Wesley, Reading, MA. 7.18, NASA; 7.20,
Foster, *Earth Science*, © 1982 The Benjamin/
Cummings Publishing Co. Reprinted with
permission of Addison-Wesley, Reading, MA. 7.21,
SOVFOTO; 7.22a, Meteor Crater Enterprises,
Northern Arizona, U.S.A.; 7.22b, NASA; 7.23,
From the collection of Ronald A. Oriti, Department
of Earth Sciences, Santa Rosa Junior College; 7.24,
C.J. Orth *et al.*, in *Science 214*, 1341–1343 (1981).
7.25, Adapted from data tabulated by David M.
Raup in *The Nemesis Affair* (Norton, 1986); 7.26a,
Vic Hessler, University of Alaska; 7.26b, 7.27,
NASA. *Chapter 8*: 8.1, Palomar Observatory
photograph; 8.3, Tobias Owen; 8.4, James Pollack,
NASA Kuiper Airborne Observatory; 8.5, Based on
Moons and Planets, Second Edition, by William K.
Hartmann, © 1983 by Wadsworth, Inc. Adapted by
permission. 8.6, NASA, Courtesy Larry Travis; 8.9,
S.J. Limaye, Ph.D. thesis, Department of
Meteorology, University of Wisconsin, 1977; 8.11,
Institute of Radioengineering and Electronics, USSR
Academy of Science, Courtesy Valery Barsukov;
8.12, NASA/U.S. Geological Survey, Courtesy
Harold Masursky; 8.13, Courtesy D.B. Campbell,
Arecibo Observatory; 8.14, NASA/U.S. Geological
Survey, Courtesy Harold Masursky; 8.15, Byron
Crader/Tom Stack and Associates; 8.16, Institute of
Radioengineering and Electronics, USSR Academy
of Science, Courtesy Valery Barsukov; 8.17a, Valery
Barsukov; 8.17b, David Morrison; 8.18, 8.19,
Valery Barsukov.

Part Five   *Opener*: NASA/U.S. Geological Survey,
Courtesy Harold Masursky. *Chapter 9*: 9.1, Lowell
Observatory photo; 9.2, 9.3, NASA; 9.5, U.S.
Geological Survey, Courtesy Michael H. Carr; 9.6,
NASA, Courtesy Michael H. Carr; 9.7, NASA; 9.8,
NASA, Courtesy Michael H. Carr; 9.10, Brown
University, Courtesy James W. Head; 9.11–9.13,
NASA; 9.14, Adapted from data given by Michael
H. Carr in *The Surface of Mars* (Yale University
Press, 1981); 9.15–9.26, NASA; 9.27, NASA/U.S.
Geological Survey; 9.28, 9.29, NASA. *Chapter 10*:
10.1, Based on *Moons and Planets*, Second Edition,
by William K. Hartmann, © 1983 by Wadsworth,
Inc. Adpated by permission. 10.2, 10.3, NASA;
10.4, Lowell Observatory photograph; 10.5, 10.6,

Goldsmith, D. and T. Owen. *The Search for Life in the Universe.* © 1980 The Benjamin/Cummings Publishing Co. Reprinted with permission. 10.7–10.10a, NASA; 10.10b, David Morrison.

Part Six *Opener:* Painting by Paul Hudson, Courtesy National Geographic Society. *Chapter 11:* 11.1, 11.2, NASA; 11.3, NASA; 11.4, 11.5, NASA; 11.6, Tobias Owen; 11.7, Based on *Moons and Planets,* Second Edition, by William K. Hartmann, © 1983 by Wadsworth, Inc. Adapted by permission. 11.8, NASA; 11.10–11.16, NASA; 11.17, 11.18, Paintings by Davis Meltzer © National Geographic Society; 11.19, 11.20, 11.22, NASA. *Chapter 12:* 12.1, Charles T. Kowal; 12.2, Lowell Observatory photograph; 12.6, Robert Danehy and Tobias Owen; 12.7, 12.9, 12.10, NASA; 12.11, B.A. Smith, University of Arizona, and R.J. Terrile, Jet Propulsion Laboratory; Mt. Wilson and Las Campanas Observatories; 12.12, NASA; 12.13a, U.S. Navy Observatory photograph; 12.15, University of Hawaii computer image by Marc W. Buie.

Part Seven *Opener:* NASA. *Chapter 13:* 13.1, University of Arizona; 13.3–13.5, NASA; 13.6, Roger Clark; 13.7–13.10, NASA; 13.12–13.15, NASA; 13.16, R. Howell, D.P. Cruikshank, and T. Geballe; 13.17–13.21, NASA; 13.22, Virgil Kunde,

from Kunde *et al., Nature 292,* 686 (1981); 13.23, NASA. *Chapter 14:* 14.1–14.4, NASA; 14.5a, From W.A. Baum *et al., Icarus 47,* 84–96 (1981); 14.5b, NASA; 14.6–14.11, NASA; 14.13–14.15, NASA; 14.16a, Reproduced with permission from the *Annual Review of Astronomy and Astrophysics,* Vol. 17, © 1979 by Annual Reviews, Inc. 14.16b, From the *Astronomical Journal,* Vol. 83, No. 8 (1978), p. 984. 14.17–14.21, NASA; 14.22a, David Morrison; 14.22b, NASA.

Epilogue *Opener:* Rijksmuseum Amsterdam. *Chapter 15:* 15.1, National Optical Astronomy Observatories; 15.2, Goldsmith, D. *The Evolving Universe,* 2nd ed. © 1985 The Benjamin/Cummings Publishing Co. Reprinted with permission. 15.5, R.J. Terrile, Jet Propulsion Laboratory, and B.A. Smith, University of Arizona; Las Campanas Observatory; 15.7, NASA.

Color Plates 1, David Morrison; 2, Thomas R. McGetchin; 3, Dale P. Cruikshank; 4, NASA; 5a, NASA; 5b, Cyril Ponnamperuma; 6, NASA; 7, NASA/U.S. Geological Survey, Courtesy Harold Masursky; 8, 9, NASA/U.S. Geological Survey, Courtesy Alfred McEwen; 10a, 10b, NASA, Courtesy Ray Arvidson; 11–16, NASA; 17–19, NASA/U.S. Geological Survey, Courtesy Alfred McEwen.

Index

◆ D ◆

◆ E ◆

◆ F ◆

◆ G ◆

◆ H ◆

◆ N ◆

♦ T ♦

♦ U ♦

♦ V ♦

♦ W ♦

♦ Y ♦

♦ Z ♦

◆ SPACE EXPLORATION BY *SELECTED* UNMANNED SATELLITES AND PROBES ◆

Spacecraft	Country of Origin	Launch Date
Sputnik 1	USSR	4 October 1957
Sputnik 2	USSR	3 November 1957
Explorer 1	U.S.	31 January 1958
Luna 2	USSR	12 September 1959
Luna 3	USSR	4 October 1959
Mariner 2	U.S.	27 August 1962
Mars 1	USSR	1 November 1962
Ranger 7	U.S.	18 July 1964
Mariner 4	U.S.	28 November 1964
Mariner 5	U.S.	14 June 1965
Venera 3	USSR	16 November 1965
Luna 9	USSR	31 January 1966
Luna 10	USSR	31 March 1966
Surveyor 1	U.S.	30 May 1966
Lunar Orbiter 1	U.S.	10 August 1966
Venera 4	USSR	12 June 1967
Mariner 6	U.S.	24 February 1969
Venera 7	USSR	17 August 1970
Luna 16	USSR	12 September 1970
Mars 2	USSR	19 May 1971
Mariner 9	U.S.	30 May 1971
Pioneer 10	U.S.	2 March 1972
Venera 8	USSR	26 March 1972
Pioneer 11	U.S.	5 April 1973
Mariner 10	U.S.	3 November 1973
Venera 9	USSR	8 June 1975
Venera 10	USSR	14 June 1975
Viking 1	U.S.	20 August 1975
Viking 2	U.S.	9 September 1975
Voyager 1	U.S.	5 September 1977
Voyager 2	U.S.	20 August 1977
Pioneer Venus Orbiter	U.S.	20 May 1978
Pioneer Venus Probe Carrier	U.S.	8 August 1978
Venera 11	USSR	9 September 1978
Venera 12	USSR	12 September 1978
Venera 13	USSR	30 October 1981
Venera 14	USSR	4 November 1981
Venera 15	USSR	2 June 1983
Venera 16	USSR	7 June 1983
VEGA 1	USSR	15 December 1984
VEGA 2	USSR	21 December 1984
Sakigake	Japan	8 January 1985
Giotto	ESA★	2 July 1985
Suisei	Japan	18 August 1985
Phobos (2 spacecraft)	USSR	1988 (?)
Magellan	U.S.	1989 (?)
Galileo	U.S.	1989 (?)

★ESA = European Space Agency, a consortium of European countries.